Advanced Materials and Sustainable Energy Technologies AMSET2025

2nd International Conference on Advanced Materials & Sustainable Energy Technologies (AMSET2025), Research Centre for Nano-Materials and Energy Technology (RCNMET), Sunway University, Selangor Darul Ehsan, Malaysia, November 13-14, 2025

Editors

Adarsh Kumar Pandey[1], Saidur Rahman[2], Kalidasan Balasubramanian[3], Norulsamani binti Abdullah[2]

[1]Mechanical and Aerospace Engineering Department, College of Engineering, United Arab Emirates University, Al Ain 15551, United Arab Emirates

[2]Research Centre for Nano-Materials and Energy Technology (RCNMET), Faculty of Engineering and Technology, Sunway University, Bandar Sunway, Petaling Jaya, 47500, Selangor Darul Ehsan, Malaysia

[3]Sunway Centre for Electrochemical Energy and Sustainable Technology (SCEEST), Faculty of Engineering and Technology, Sunway University, No. 5 Jalan Universiti, Bandar Sunway, Selangor Darul Ehsan 47500, Malaysia

Peer review statement

All papers published in this volume of “Materials Research Proceedings” have been peer reviewed. The process of peer review was initiated and overseen by the above proceedings editors. All reviews were conducted by expert referees in accordance to Materials Research Forum LLC high standards.

Published under License by **Materials Research Forum LLC**
Millersville, PA 17551, USA

Published as part of the proceedings series
Materials Research Proceedings
Volume 66 (2026)

ISSN 2474-3941 (Print)
ISSN 2474-395X (Online)

ISBN 978-1-64490-414-5 (Print)
ISBN 978-1-64490-415-2 (eBook)

Distributed worldwide by

Materials Research Forum LLC
105 Springdale Lane
Millersville, PA 17551
USA
https://mrforum.com

Manufactured in the United States of America
10 9 8 7 6 5 4 3 2 1

Table of Contents

Preface

This special issue, titled " Advanced Energy Materials for Sustainable Technologies " delves into the cutting-edge research and developments in the fields of sustainable energy technologies and the applications of advanced materials to enhance their performance and accessibility. The issue aims to showcase the synergistic potential of integrating advanced sustainable energy systems with techniques to optimize energy efficiency, predict energy consumption patterns, facilitate better decision-making, and accelerate the global transition towards a more sustainable future.

Contributions to this issue cover a broad range of topics, including renewable energy generation, energy storage, smart grids, carbon capture, green buildings, electric vehicles, and the role of machine learning in achieving sustainable development goals. The purpose of this special issue is to promote sustainable development via sustainable materials and provide relevant opportunities for scholars, energy economists and scientists to share their knowledge on the perspective of sustainable development across the world with focus of Sustainable Energy Technologies. We also encourage integrative studies regarding the widespread functions of sustainable material, design methodologies and numerical simulations of phase change materials, thermal energy storage and disposal policies relevant to industry, techno economic analysis of phase change materials with solar thermal systems.

Committees

Organizing Committee

Adjunct Professor Adarsh Kumar Pandey	Organising Chair
Distinguished Professor Saidur Rahman	Co-Organising Chair
Assoc Prof Dr Zafar Said	Co-Organising Chair
Dr Tan Kim Han	Publicity & Promotion
Dr Norulsamani binti Abdullah	Secretary
Ts. Dr Kalidasan Balasubramanian	Paper & Publication
Ms Aisha Ridzuan	Registration & Finance
Dr Hazim Khir	Event Management & Publicity
Dr. Dharshini Mohanadas	Event Management & Publicity
Ms Atikah Aini Abdullah	Event Management & Publicity
Mr Md Abu Zaed	Event Management & Publicity
Dr Anas Islam	Event Management & Publicity
Mr Yasir Ali Bhutto	Event Management & Publicity
Mr Geoffrey Katiedzo Aligula	Event Management & Publicity
Mr Shahriar Raian	Event Management & Publicity
Mrs Basra Sultana	Event Management & Publicity
Dr Aman Yadav	Event Management & Publicity
Mr Muppana Veera Nagendra	Event Management & Publicity
Mr. Noraz Khairi	Event Management & Publicity
Dr John Paul	Event Management & Publicity
Ms Munirah Zulkifli	Event Management & Publicity
Ms Ainin Sufiya binti Sofian	Event Management & Publicity
Ms Sahar Ghafarian	Event Management & Publicity
Ms Siti Nabila binti Abdul Rahim	Event Management & Publicity

International Technical Committee Members

Prof. Yogendra Kumar Mishra, Professor, University of Southern Denmark, Denmark.
Prof. Zhenjun Ma, Professor, University of Wollongong, Australia
Dr. S.K. Tyagi, Professor, Indian Institute of Technology Delhi, India.
Dr. Vineet V. Tyagi, Associate Professor, Shri Mata Vaishno Devi University, India.
Dr. Richa Kothari, Professor, Central University Jammu, India
Dr. Tangellapalli Srinivas, Associate Professor, Dr B R Ambedkar National Institute of Technology Jalandhar, India.
Prof. P Abdul Salam, Professor, Asian Institute of Technology (AIT), Thailand.
Dr. Iftikhar Hussain, Postdoctoral Researcher, City University of Hong Kong, Hong Kong.
Prof. Dr. Wenye Lin, Professor, Guangzhou Institute of Energy Conversion, Chinese Academy of Sciences, China.
Prof. Dr. R Jeyalakshmi, Professor, SRM Institute of Science and Technology, India
Dr. Hafiz Muhammad Ali, Associate Professor, King Fahd University of Petroleum and Minerals, Saudi Arabia
Dr. Therese May G. Alejandrino, Director - Technopreneurship and Collaborative Applied Research (TechnoCoRe), Technological Institute of the Philippines, Philippines
Dr. DJ Donn Matienzo, DOST PCIEERD Balik Scientist / Business Development Head - TechnoCoRe Catalyst, Technological Institute of the Philippines, Philippines
Dr. C. Subramaniyan, Assistant Professor, Bannari Amman Institute of Technology, India
Assoc. Prof. Ruben Sudhakar, Associate Professor and former HoD, Department of Energy and Environment, NIT Tiruchirappalli, India
Prof. Dr. R. Senthil, Professor, SRM Institute of Science and Technology, Chennai, India.
Dr. M. Kathiresan, Senior Scientist, CSIR-Central Electrochemical Research Institute, Karaikudi 630003, Tamil Nadu, India.
Assoc. Prof. Dr. S. Senthil Raja, Associate Professor, Department of Mechatronics Engineering, SRM Institute of Science and Technology, Tamil Nadu, India. P

National Technical Committee Members

Arshad Adam Salema, Associate Professor and Course Director of Master of Applied Engineering, Monash University Malaysia
Assoc. Prof. Ts. Dr Farihahusnah Hussin, Associate Professor, Sunway University, Malaysia
Dr. Muhammad Shakeel Ahmad, Senior Lecturer, UMPEDAC, University of Malaya, Malaysia
Prof. Ir. Ts. Dr. Kumaran Kadirgama, Professor, Universiti Malaysia Pahang Al Sultan Abdullah, Malaysia
Ir. Dr. Amares Singh Gill, Senior Lecturer II, Taylor's University, Malaysia
Dr. Nur Izzah Nabilah Haris, Senior Lecturer, Universiti Putra Malaysia, Malaysia
Assoc. Prof. Dr. Md. Hasanuzzaman, Associate Professor, UMPEDAC, University of Malaya, Malaysia
Dr. Nabila A Karim, Senior Research Fellow, Universiti Kebangsaan Malaysia, Malaysia
Dr. Nor Asrina Binti Ramlee, Senior Lecturer, School of Engineering and Technology, University of Technology Sarawak
Ms. Foo Jinny, Teaching Fellow, School of Computing and Artificial Intelligence, Faculty of Engineering and Technology, Sunway University

Development of Polylactic Acid (PLA) Reinforced with Water Hyacinth Fiber (WHF) as Filament for 3D Printing

Ruem Garrote ARRIBAS[1,a *], April Marcelo JANIEL[1,b]

[1]Technological Institute of the Philippines – Quezon City, 938 Aurora Boulevard, Cubao, Quezon City, 1109 Metro Manila, Philippines

[a]arribasruem@gmail.com, [b]amarcelo.me@tip.edu.ph

Keywords: Water Hyacinth, Polylactic Acid, Twin-Screw Extrusion, Composite Filament

Abstract. Polylactic acid (PLA) is a widely used filament for 3D Printing because of its cost-effectiveness and wide availability, but its brittleness limits its mechanical performance. This study explored water hyacinth fiber (WHF) as an alternative additive material to PLA to create filament for Fused Filament Fabrication (FFF) 3D printing method that can improve composite mechanical and physical properties while addressing environmental concerns. In this study, PLA was mixed with WHF in different proportions (10:90, 15:85, 20:80, 25:75 and 30:70) through a twin-screw extruder. Out of the applied combinations, the 10WHF90PLA (10% WHF and 90% PLA) was determined as the best combination according to the tensile test results. This blend performed best with a tensile strength of 47.99 N/mm^2, which is an 8.54% improvement over pure PLA. The strength of impact also increased by 16.75% while the flexural strength reduced slightly. One-Way ANOVA results statistically verified that fiber content was a significant influence of tensile strength ($p < 0.05$), where the 10% WHF blend (10WHF90PLA) had the highest F-value of samples. SEM-EDS analysis indicated that the fibers were evenly distributed, well bonded to the PLA matrix and had low porosity. The composite was also found to be highly printable with a good surface finish and no signs of nozzle clogging during the 3D printing process.

Introduction

One of the most significant discoveries in the production and prototyping industry is 3-Dimensional (3D) Printing. It has gained increasing interest in research and innovations towards sustainable and green manufacturing. It is used in different sectors including goods, automobile manufacturing and biomedical. Fused Filament Fabrication (FFF) is known to be the economical and most widespread method in 3D printing and is mostly used by individuals that started to explore the additive manufacturing world, where objects are produced by printing thermoplastic polymer layer-by-layer [1]. Polylactic acid (PLA) is the most used polymer available in the market owing to its biodegradability, printability, and world production. It is produced by different methods using cornstarch, sugarcane or tapioca different processes, such as saccharification, fermentation, purification, and synthesis [2]. It is also extensively applied to natural fiber fillers [3]. Although it has its drawbacks, such as brittle nature, low toughness, and flexibility, these can be addressed by incorporating natural waste fibers such as kenaf, bamboo, hemp, or flax [4].

Natural fibers are abundant agricultural by-products in many developing countries, offering excellent physical and mechanical properties that improve composites for various applications [5]. Classified as seed, leaf, bast, grass, straw, wood, mineral, or animal fibers, they contain cellulose, hemicellulose, and lignin, which contribute to flexibility and strength [6]. Their biodegradability and recyclability provide an advantage over synthetic fillers. However, their hydrophilic nature poses compatibility challenges with hydrophobic polymers, often addressed through surface treatments such as NaOH [7].

In additive manufacturing, natural fibers have gained attention as reinforcements in PLA matrices to develop sustainable bio-composites [6], [8]. Studies reveal that fillers such as rice husk, coffee grounds, wood, coconut shell, and hemp chips can be pretreated, ground, and compounded with PLA for filament production [8], [9]. This process often involves twin-screw extrusion and subsequent 3D printing via FFF to evaluate mechanical properties [10]. Results consistently show improvements in biodegradability and selected mechanical properties compared to neat PLA.

Water hyacinth (Eichhornia crassipes), an invasive aquatic weed common in tropical regions such as the Philippines, represents a promising natural fiber source [11], [12]. Despite causing blockages in waterways and environmental degradation, its high cellulose content (60–70%) and low lignin content (~4%) make it suitable for fiber extraction and composite reinforcement [13], [14]. Traditional uses include fertilizer, animal feed, paper, handicrafts, and household products, but recent studies highlight its potential in producing lightweight, strong, and thermally stable composites [13]–[16]. Pretreatment and drying methods are crucial in this process to minimize moisture content (ideally <13%), ensuring compatibility with the polymer matrix and avoiding extrusion defects [17].

Applications of WHF in composites remain limited to injection molding and resin-based composites [15], [16], although some studies have tested WHF-PLA blends with additional reinforcements like nano-silica, achieving tensile strengths of up to 36.93 MPa [15]. Recent work even explored 3D-printed PLA composites with WHF at low loadings, showing improvements in tensile strength, impact resistance, and sound absorption properties [18]. This emphasizes the untapped potential of WHF as a reinforcement for 3D printing filaments.

Producing filaments for additive manufacturing typically involves extrusion through single- or twin-screw extruders, with twin-screw extrusion favored for its superior mixing capability [19], [20]. Manufacturing filaments in the additive production process is commonly performed with the aid of an extrusion process via single- and twin-screw extruders, where twin-screw extrusion is more popular owing to its enhanced mixing capacity [19], [20]. After processing, filaments are printed and tested according to the ASTM D638, ASTM D790, and ASTM D5420 standards of tensile, flexural, and impact test respectively, and microstructural analyses (e.g., SEM-EDS) [21]. This method of WHF-PLA composites has been confirmed by previous studies that indicate that natural fiber-polymer compositions enhance the performance and properties of the composite [16], [18], [22].

Methodology

Concept of the Study. The study's concept was developed by identifying the problem: the challenges associated with the fast growth of water hyacinth plants and the brittleness and low toughness of PLA. The research study was established by reviewing existing literature and studies regarding these two materials and exploring the reinforcement of natural fibers to strengthen polymers, specifically PLA.

As shown in Fig. 1, the realization of this study started from material preparation, which involved the water hyacinth fiber extraction and preparation of polylactic acid filament. The mixtures of these two materials in different ratios were processed under the twin-screw extruder to create the thin-wire filament. The experiment design consisted of 5 trials with five samples each, considering one factor only, the combination ratio of the composites. The output filament was then printed through the FFF method to fabricate the shape required for standard testing (ASTM D638 Type V) throughout the project's evaluation phase. Samples from each trial were printed using the same printing parameters and tested according to the tensile strength (ASTM D638).

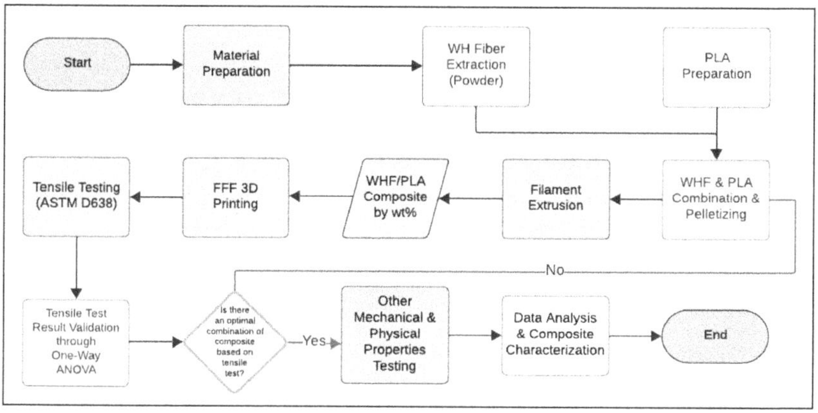

Fig 1. Project concept and procedure for developing and optimizing WHF/PLA composite filaments.

The optimal WHF and PLA composite combination was determined based on the testing results and analyzed through One-Way ANOVA. to determine the significance of fiber content on tensile strength. A decision point was introduced: Is there an optimal composite. If no optimal combination was identified, the process allowed for refinement of ratios and re-extrusion of filaments. If an optimal combination was confirmed, which this study achieved, samples were printed again according to the optimal combination of PLA and WHF for another set of standard testing and characterization of the other properties of the optimal composite. At the same time, pure PLA underwent the same process as the composite and served as the basis of comparison and evaluation to see if there was an effect on the addition of WHF to PLA to enhance its properties.

Materials & Equipment. WHF and PLA were the main materials used to conduct this study. The PLA employed is Grade BC2003RE to 3D print, with a melting temperature of 185 °C. PLA materials with a weight of 5 kilograms were purchased in D&L industries and the Taguig City Water Lily Livelihood Center was where the bundles of dried water hyacinth plants were purchased. The composite was first extruded and pelletized in a twin-screw extruder (LABTECH type LTE20-40) held by DOST Taguig. Water hyacinth was processed into an electric grinder of 200-mesh powder. The final filament was made in a machine known as ReDeTec PhotoCycler. The end samples to test were fabricated using Creality Ender 3 3D printer and the testing and evaluation process was done by using DOST-ITDI equipment.

Material Preparation. The water hyacinth plants have undergone several processes, such as drying, grinding, and sieving, to ensure compatibility with PLA for filament production. The plants were sun-dried for three days, cut into 1-inch pieces, and ground using an electric grain miller for 10 minutes. The resulting powder was sieved through a 200-mesh strainer and stored until the required quantity was obtained. Five composite formulations of WHF/PLA (10:90, 15:85, 20:80, 25:75, 30:70) were prepared, each trial containing 500 g of material (see Table 1).

Table 1. *Amount of WHF and PLA per trial.*

TRIAL	*WHF (%)*	*PLA (%)*	*WHF [g]*	*PLA [g]*	*Mixture [g]*
10WHF90PLA	0.10	0.90	50	450	500
15WHF85PLA	0.15	0.85	75	425	500
20WHF80PLA	0.20	0.80	100	400	500
25WHF75PLA	0.25	0.75	125	375	500
30WHF70PLA	0.30	0.70	150	350	500
Total	-	-	**500**	**2000**	**2500**

To prevent moisture-related defects during extrusion, WHF powder and PLA pellets were oven-dried at 90 °C overnight before compounding. This step followed the procedures recommended in [17], [20], which emphasize the importance of proper drying to maintain material quality. Moreover, as noted in [16], reducing natural fiber moisture to below 13% improves polymer compatibility and prevents processing flaws. The dried materials were then stored in sealed containers with desiccants until extrusion.

Extrusion and 3D Printing. The WHF and PLA mixtures were extruded in a twin-screw extruder at 185 °C, producing pelletized composites. The pellets were re-extruded into filaments using the ReDeTec PhotoCycler V3 under controlled conditions: melt temperature 187 °C, auger speed 13 rpm, puller speed 40 ft/min, and fan level 70%. The produced filaments maintained a uniform 1.75 ± 0.05 mm diameter. The Creality Ender 3 printer was then used to create standard tensile test samples (ASTM D638 Type V) using the composites at a nozzle temperature of 215 °C, bed temperature of 60 °C, speed of 100 mm/s, and infill ratio of 100%.

Testing and Evaluation. The composite specimens underwent standardized testing and characterization. Tensile strength and strain were evaluated using ASTM D638, while flexural performance was assessed following ASTM D790. Impact resistance was measured according to ASTM D5420 to determine the material's energy absorption capacity. Furthermore, SEM-EDS analysis was carried out to investigate the microstructural features and fiber–matrix interaction of the WHF/PLA composites.

Results and Discussion

PLA–WHF composite filaments were successfully fabricated, showing distinct differences in surface appearance and extrusion behavior depending on fiber content. Lower WHF content (10WHF90PLA) resulted in shinier, smoother strands like pure PLA, while higher WHF content (30WHF70PLA) produced matte, rough-textured filaments (see Fig. 2). During extrusion, mixtures above 30% WHF became brittle and fragmented, making them unsuitable for pelletizing. Stable and continuous filaments were achieved with 30% WHF or less, leading to the selection of 10–30% WHF blends for testing. In terms of printability, 10WHF90PLA exhibited the best results with smooth printing and no clogging, whereas higher WHF blends (25–30%) frequently clogged and fractured.

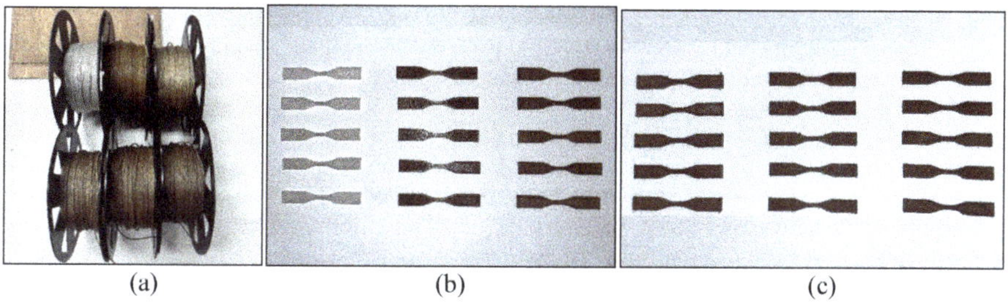

Fig. 2. Spools of PLA–WHF composite filaments with varying ratios, and 3D-printed specimens for tensile test (a), left to right – 10WHF90PLA to 15WHF85PLA (b), and left to right – 20WHF80PLA to 30WHF70PLA (c).

Tensile Test. Tensile testing following ASTM D638 revealed that pure PLA had an average tensile strength of 44.21 N/mm^2 and strain of 15.81% (Fig. 3). The 10WHF90PLA composite achieved the highest tensile strength of 47.99 N/mm^2, surpassing pure PLA by 8.54%, though strain decreased slightly to a value of 14.96%. Increasing WHF content beyond 10% reduced tensile strength significantly, with 15WHF85PLA dropping to 33.33 N/mm^2 and 30WHF70PLA to 30.69 N/mm^2, confirming weak fiber–matrix adhesion at higher loading.

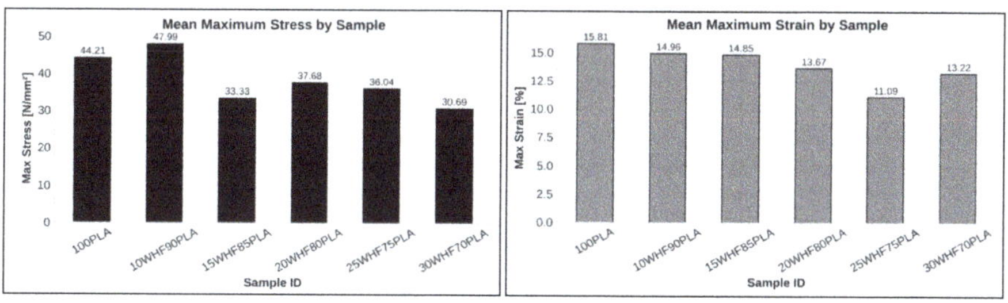

Fig. 3. Comparison of Mean Maximum Stress and Strain for PLA-WHF Composites

A one-way ANOVA using Design Expert Software confirmed that the variations in tensile strength among the composite groups were statistically significant shown in Table 2, with a p-value < 0.0001 and an F-value of 16.90, indicating that fiber content had a decisive effect on tensile performance, with 10% WHF identified as the optimal reinforcement level.

Table 2. *One-Way ANOVA results summary.*

Source	*Sum of Squares*	*df*	*Mean Square*	*F-value*	*p-value*	*Significance*
Model	1084.26	5	216.85	16.90	< 0.0001	Significant
A-A (Treatment)	1084.26	5	216.85	16.90	< 0.0001	Significant
Pure Error	307.99	24	12.83	—	—	—
Total	1392.25	29	—	—	—	—

Flexural and Impact Test. Flexural testing (ASTM D790) compared 100PLA and 10WHF90PLA, yielding average flexural strengths of 44.91 N/mm^2 and 42.15 N/mm^2, respectively. While the average strength decreased slightly, one composite sample achieved 51.12 N/mm^2, indicating potential performance gains when fiber bonding is optimized. Impact testing (ASTM D5420) showed improved failure energy in 10WHF90PLA (46 in-lbf) compared to pure PLA (39.4 in-lbf), reflecting an approximately 16% increase in impact resistance.

Surface Morphology of 10WHF90PLA Composite. Physical characterization supported these mechanical findings. SEM analysis (Fig. 4a) revealed a rough surface with cracks and voids, indicating limited fiber–matrix bonding. EDS spectrum confirmed the composite's organic composition with carbon (61.76%) and oxygen (38.24%) as primary elements (Fig. 4b). The carbon map (Fig. 4c) shows a uniform red coloration, indicating a generally even dispersion of carbon-rich phases, which include both the PLA and WHF constituents. These results emphasize the need for compatibility enhancers to improve interfacial adhesion.

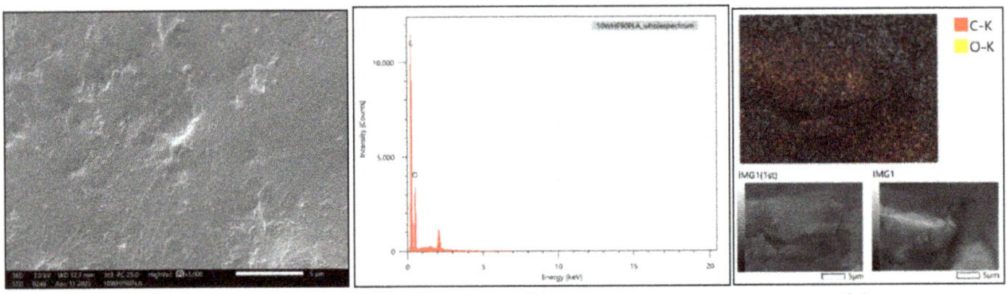

(a) (b) (c)

Fig. 4. Surface Morphology (a) and Elemental Composition of 10WHF90PLA Composite (b, c)

Conclusion

The study successfully developed PLA-based filaments reinforced with water hyacinth fiber (WHF) for FFF 3D printing. Filaments were consistently produced across varying WHF/PLA ratios, with the optimal 10WHF90PLA formulation achieving a tensile strength of 47.99 N/mm^2, representing an 8.54% improvement over pure PLA. Although higher fiber loadings reduced tensile performance, the optimal composite showed enhanced impact resistance (16.75% increase) and acceptable flexural properties. SEM-EDS confirmed uniform fiber distribution, good matrix adhesion, and minimal porosity, validating the material's compatibility with extrusion and printing processes.

The findings highlight WHF as a sustainable reinforcement for PLA, offering both mechanical benefits and ecological value by utilizing an invasive species for additive manufacturing. Future research should optimize 3D printing parameters, investigate hybrid fiber systems to enhance interfacial bonding, and conduct long-term environmental durability studies to support practical applications in sustainable product design.

References

[1] N. Shahrubudin, T. C. Lee, and R. Ramlan, "An Overview on 3D Printing Technology: Technological, Materials, and Applications," Procedia Manufacturing, vol. 35, pp. 1286–1296, 2019. https://doi.org/10.1016/j.promfg.2019.06.089

[2] M. Żenkiewicz, P. Rytlewski, and R. Malinowski, "Compositional, physical and chemical modification of polylactide," Polymers for Advanced Technologies, vol. 43, pp. 192–199, 2016.

[3] B. Baghaei, M. Skrifvars, M. Rissanen, and S. K. Ramamoorthy, "Mechanical and thermal characterization of compression moulded polylactic acid natural fiber composites reinforced with

hemp and lyocell fibers," J. Appl. Polym. Sci., vol. 131, no. 15, 2014. https://doi.org/10.1002/app.40534

[4] R. Gunti, A. V. Ratna Prasad, and A. V. S. S. K. S. Gupta, "Mechanical and degradation properties of natural fiber-reinforced PLA composites: Jute, sisal, and elephant grass," Polym. Compos., vol. 39, no. 4, pp. 1125–1136, 2018. https://doi.org/10.1002/pc.24041

[5] I. Elfaleh et al., "A comprehensive review of natural fibers and their composites: An eco-friendly alternative to conventional materials," Results in Engineering, vol. 19, p. 101271, 2023. https://doi.org/10.1016/j.rineng.2023.101271

[6] R. Ilyas et al., "Natural Fiber-Reinforced Polylactic Acid, Polylactic Acid Blends and Their Composites for Advanced Applications," Polymers, vol. 14, no. 1, p. 202, 2022. https://doi.org/10.3390/polym14010202

[7] M. H. M. Nasir et al., "Effect of Chemical Treatment of Sugar Palm Fibre on Rheological and Thermal Properties of the PLA Composites Filament for FDM 3D Printing," Materials, vol. 15, no. 22, p. 8082, 2022. https://doi.org/10.3390/ma15228082

[8] M. N. Ahmad, M. R. Ishak, M. Mohammad Taha, F. Mustapha, and Z. Leman, "A Review of Natural Fiber-Based Filaments for 3D Printing: Filament Fabrication and Characterization," Materials, vol. 16, no. 11, 2023. https://doi.org/10.3390/ma16114052

[9] D. Deb and J. M. Jafferson, "Natural fibers reinforced FDM 3D printing filaments," Mater. Today Proc., vol. 46, pp. 1308–1318, 2021. https://doi.org/10.1016/j.matpr.2021.02.397

[10] V. Mishra, S. Negi, S. Kar, and A. Sharma, "Recent advances in fused deposition modeling-based additive manufacturing of thermoplastic composite structures: A review," Advances in Polymer Technology, 2022. https://doi.org/10.1177/08927057221102857

[11] A. Ajithram, J. T. W. Jappes, and N. C. Brintha, "Water hyacinth (Eichhornia crassipes) natural composite extraction methods and properties – A review," Mater. Today Proc., vol. 45, pp. 1626–1632, 2021. https://doi.org/10.1016/j.matpr.2020.08.472

[12] A. Arivendan, W. J. Jebas Thangiah, S. Irulappasamy, and B. N. Chrish, "Study on characterization of water hyacinth (Eichhornia crassipes) novel natural fiber as reinforcement with epoxy polymer matrix material for lightweight applications," J. Ind. Text., vol. 51, no. 5_suppl, pp. 8157S–8174S, 2022. https://doi.org/10.1177/15280837211067281

[13] S. George, S. Thomas, N. N. Nedumpillil, and S. Jose, "Extraction and Characterization of Fibers from Water Hyacinth Stem Using a Custom-Made Decorticator," J. Nat. Fibers, vol. 20, no. 2, 2023. https://doi.org/10.1080/15440478.2023.2212927

[14] S. Tanpichai, S. K. Biswas, S. Witayakran, and H. Yano, "Water Hyacinth: A Sustainable Lignin-Poor Cellulose Source for the Production of Cellulose Nanofibers," ACS Sustainable Chem. Eng., vol. 7, no. 23, pp. 18884–18893, 2019. https://doi.org/10.1021/acssuschemeng.9b04095.

[15] P. Kumar Panda et al., "Exploration on Mechanical Behaviours of Hyacinth Fibre Particles Reinforced Polymer Matrix-Based Hybrid Composites for Electronic Applications," Adv. Mater. Sci. Eng., vol. 2021, p. 4933450, 2021. https://doi.org/10.1155/2021/4933450

[16] P. Kongkaew, P. Praneekrit, T. Rudchapo, and K. Khampui, "Mechanical and physical properties of water hyacinth and cogon grass fiber reinforced epoxy resin composites," J. Phys.: Conf. Ser., vol. 2145, p. 012036, 2022. https://doi.org/10.1088/1742-6596/2145/1/012036

[17] C. Yamoum and R. Magaraphan, "Effect of peanut shell content on mechanical, thermal, and biodegradable properties of peanut shell/polylactic acid biocomposites," Polym. Compos., vol. 38, no. 4, pp. 682–690, 2017. https://doi.org/10.1002/pc.23627

[18] N. K. Wijekoon et al., "Facile fabrication of 3D-printed cellulosic fiber/polylactic acid composites as low-cost and sustainable acoustic panels," Sustain. Chem. Environ., vol. 8, p. 100168, 2024. https://doi.org/10.1016/j.scenv.2024.100168

[19] A. Dey, I. N. Roan Eagle, and N. Yodo, "A Review on Filament Materials for Fused Filament Fabrication," J. Manuf. Mater. Process., vol. 5, no. 3, p. 69, 2021. https://doi.org/10.3390/jmmp5030069

[20] A. J. Professor Kaleeswaran, N. S. Natraj, and N. R. Kumar, "Analyzing the Quasi-Static Behavior of Fiber Composite Filament Used in 3D Printing Applications," Int. J. Sci. Res. Eng. Trends, vol. 10, no. 3, 2024.

[21] A. M. Forster, "Materials Testing Standards for Additive Manufacturing of Polymer Materials: State of the Art and Standards Applicability," NIST Interagency/Internal Report (NISTIR) 8059, Gaithersburg, MD, 2015. https://doi.org/10.6028/NIST.IR.8059

[22] W. Ahmed et al., "Implementing FDM 3D Printing Strategies Using Natural Fibers to Produce Biomass Composite," Materials, vol. 13, no. 18, p. 4065, 2020. https://doi.org/10.3390/ma13184065

Extended Graph Energies in Structure–Property Modeling

Sourav MONDAL[1,a], Zahid RAZA[2,b*]

[1]MASEP Research Group, RISE, University of Sharjah, Sharjah 27272, UAE

[2]Department of Mathematics, College of Sciences, University of Sharjah, Sharjah 27272, UAE

[a]souravmath94@gmail.com, [b]zraza@sharjah.ac.ae

Keywords: Molecular Graph, Graph Energy, Molecular Descriptor, QSPR Analysis

Abstract. Graph-theoretic descriptors are well-known to display molecular topology. The degree-based and graph spectrum-based invariants have attracted considerable attention in chemical graph theory. Here, we investigate numerous graph energies generated from weighted adjacency matrices. We compute these extended energies for a family of benzenoid hydrocarbons and examined in relation to their experimental boiling points. We find from a comparative regression analysis that certain degree-based energies track the variation of boiling point across the series with notable consistency. We observe that incorporating local degree information into the spectral framework enhances sensitivity to changes in ring fusion and molecular growth. Our findings imply that such extended graph energies may serve as practical descriptors in QSPR analysis.

Introduction

Throughout this work, we consider $\Gamma = (V(\Gamma), E(\Gamma))$ as a simple connected graph. For a vertex $u \in V(\Gamma)$, we denote its degree by d_u. A topological index is a graph invariant that reflects aspects of molecular connectivity and remains unchanged under graph isomorphism. These indices form a fundamental component of QSPR investigations. We found the systematic use of topological descriptors through the Wiener index in 1947 [1], which demonstrated that boiling points of paraffins could be correlated with graph-based structural measures. Since that time, a wide variety of indices have been developed and studied within mathematical chemistry.

Degree-based topological indices [2-7] form an important subclass of these descriptors. Introduced originally through the Zagreb indices, they have proven effective due to their simplicity and strong chemical relevance. Most degree-based indices can be expressed in the form

$$I(\Gamma) = \sum_{uv \in E(\Gamma)} \phi(d_u, d_v)$$

where ϕ is a symmetric function of the degrees of adjacent vertices. Many such indices, including the Randić, harmonic, atom–bond connectivity, augmented Zagreb, and forgotten indices, have been shown to perform well in modeling molecular properties.

Table 1. Closed-Form Expressions of ϕ for Selected Degree-Based Topological Indices.

Descriptors	$\phi(x,y)$	Descriptors	$\phi(x,y)$
First Zagreb index (M_1)	$x+y$	Second Zagreb index (M_2)	xy
Forgotten index (F)	$x^2 + y^2$	Harmonic index (H)	$\frac{2}{x+y}$
Randić index (R)	$\frac{1}{\sqrt{xy}}$	Reciprocal Randić index (RR)	$\sqrt{xy}$

Symmetric division deg index (SDD)	$\frac{x}{y}+\frac{y}{x}$	Atom bond sum connectivity index (ABS)	$\sqrt{\frac{x+y-2}{x+y}}$
Augmented Zagreb index (AZI)	$\left(\frac{xy}{x+y-2}\right)^3$	Modified second Zagreb index (mM_2)	$\frac{1}{xy}$
Atom bond connectivity index (ABC)	$\sqrt{\frac{x+y-2}{xy}}$	Inverse sum indeg index (ISI)	$\frac{xy}{x+y}$

Spectral graph theory examines how graph structure is reflected in the eigenvalues of associated matrices. The adjacency matrix $A(\Gamma)$ is one of the most fundamental such matrices. Its eigenvalues are real, and their absolute sum defines the graph energy, introduced by Gutman in 1978 [13]. Initially motivated by applications in quantum chemistry, graph energy has since become an important invariant in both algebraic graph theory and chemical graph theory.

Motivated by the success of degree-based indices and classical graph energy, several modified adjacency matrices have been introduced by incorporating degree-based weights [8-12]. Each such matrix gives rise to a corresponding extended graph energy. In this work, we study graph energies associated with several well-known degree-based indices and investigate their usefulness in structure–property modelling through regression analysis.

Methodology

The molecular structures of the selected benzenoid hydrocarbons were drawn and verified using ChemDraw Professional, after which each compound was represented as a hydrogen-suppressed molecular graph where vertices correspond to carbon atoms and edges represent covalent bonds. From these graphs, weighted adjacency matrices were generated. To define the extended graph energies, modified adjacency matrices were constructed by incorporating degree-based weight functions into the classical adjacency matrix framework. For a given molecular graph Γ, the standard adjacency matrix $A(\Gamma)$ was first generated. Subsequently, degree-dependent weight functions corresponding to the selected topological indices were assigned to each edge uv as symmetric functions of the degrees d_u and d_v. We use these resulting weighted adjacency matrices to compute eigenvalues. We calculate corresponding extended graph energies as the sum of absolute eigenvalues. All descriptor computations were implemented through custom scripts in MATLAB R2023b to ensure accuracy and reproducibility. The linear fitting between descriptors and properties are made by MATLAB plotting library.

Structure-Property Relationship Analysis

Although a large number of topological indices have been proposed, not all of them are equally effective in chemical applications. In practice, a descriptor is useful only if it can reproduce observed trends in experimental data. For this reason, QSPR analysis is employed here to examine how degree-based graph energies relate to boiling point variations within a homologous series of benzenoid hydrocarbons.

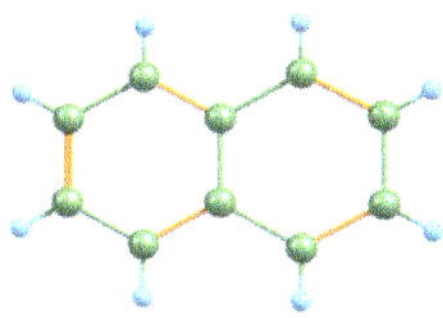

BHC1

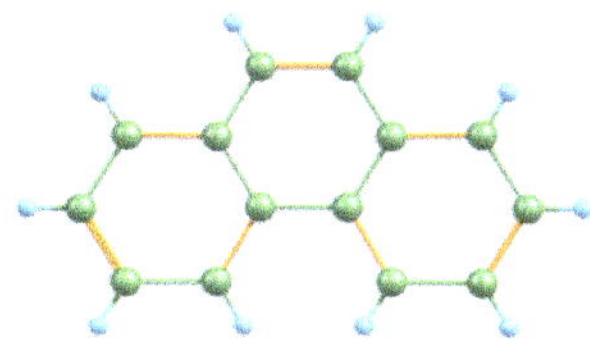

BHC2

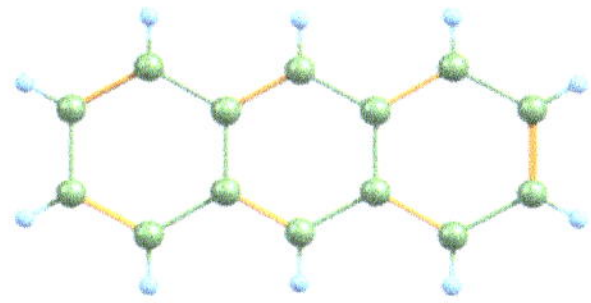

BHC3

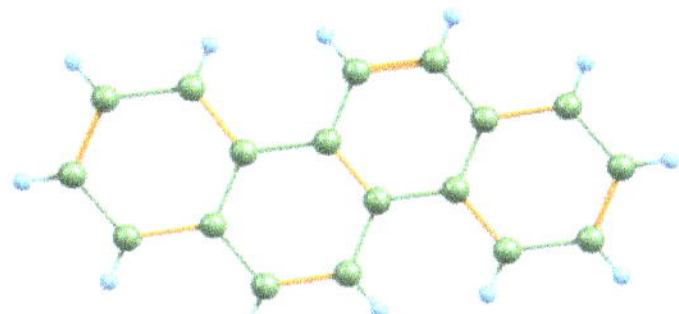

BHC4

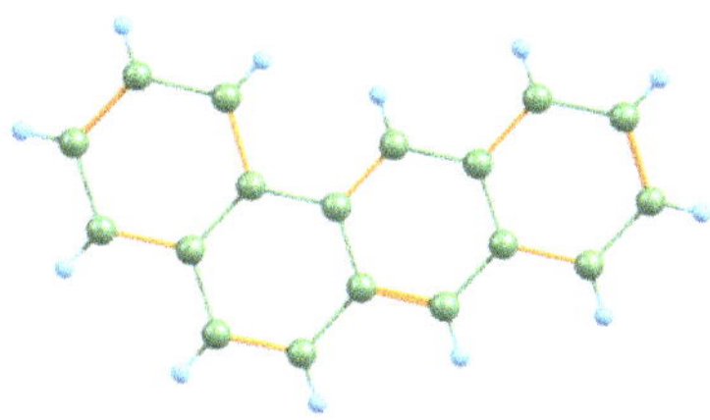

BHC5

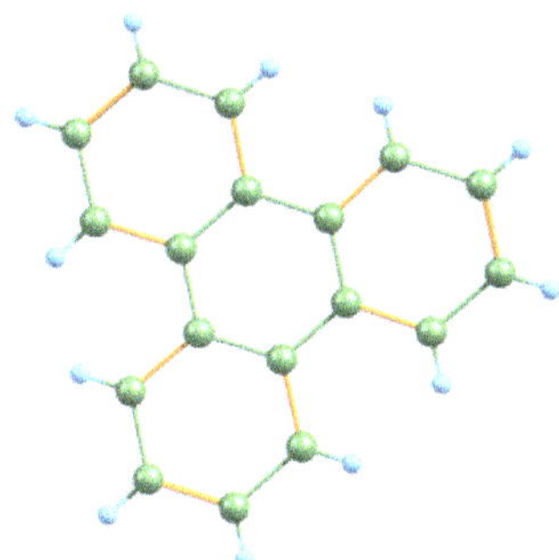

BHC6

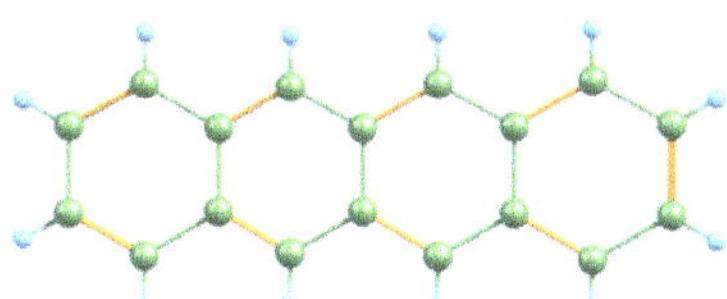

BHC7

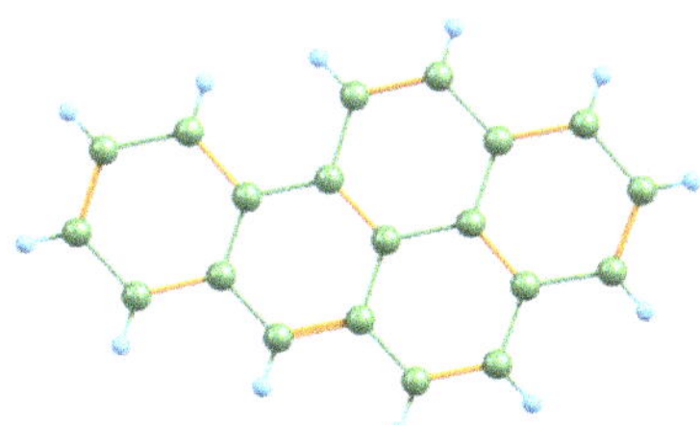

BHC8

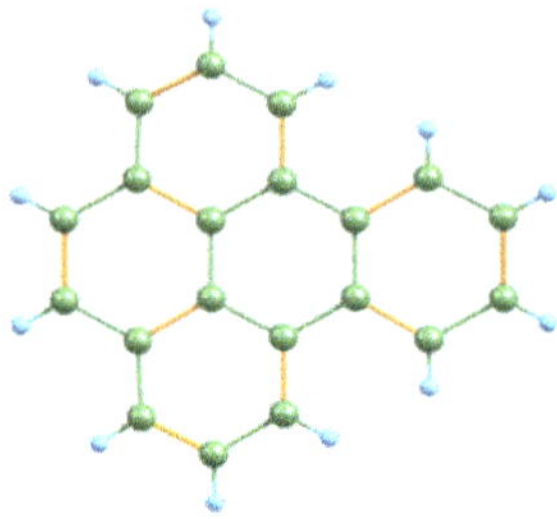

BHC9

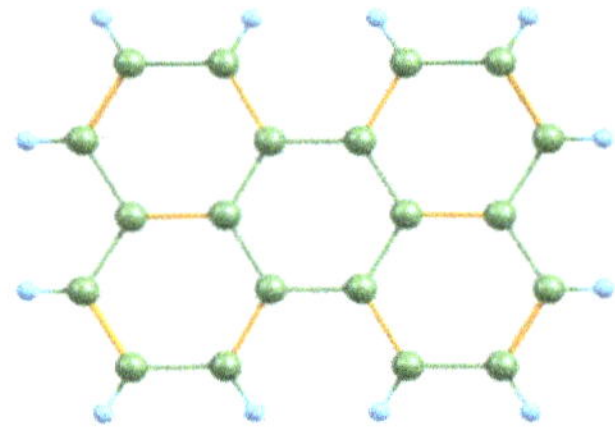

BHC10

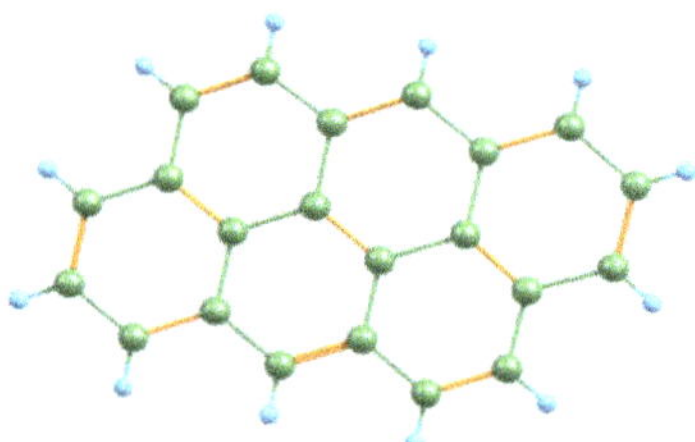

BHC11

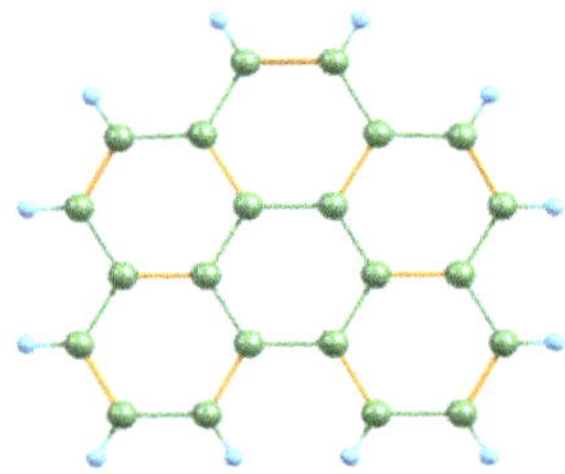

BHC12

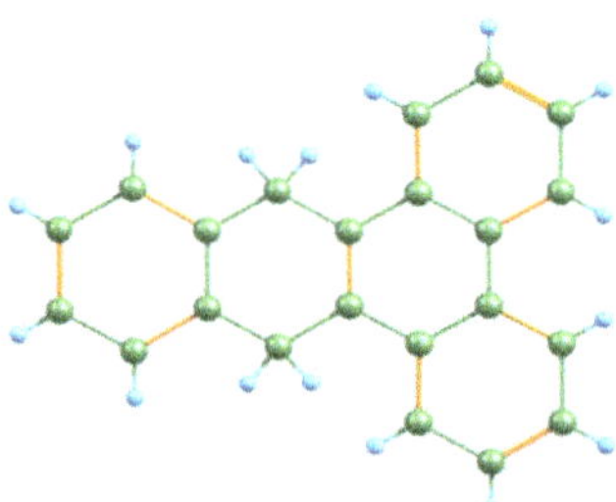

BHC13

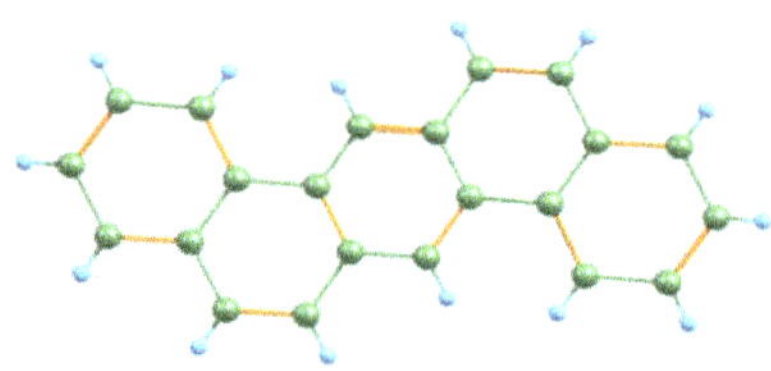

BHC14

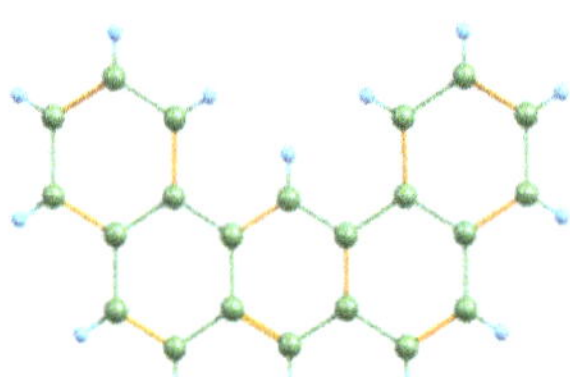

BHC15

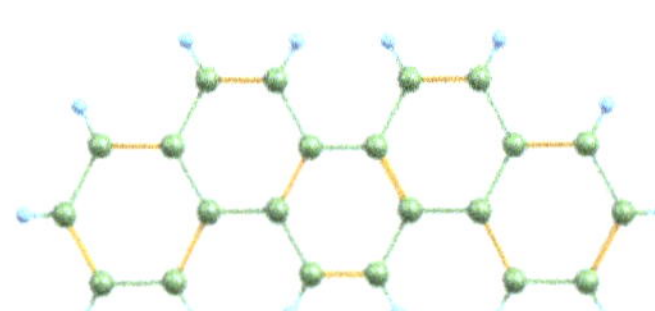

BHC16

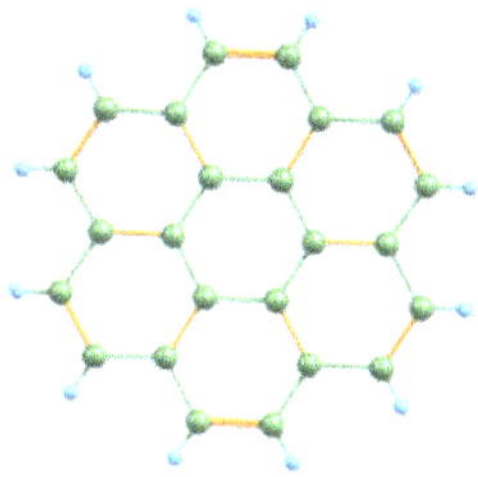

BHC17

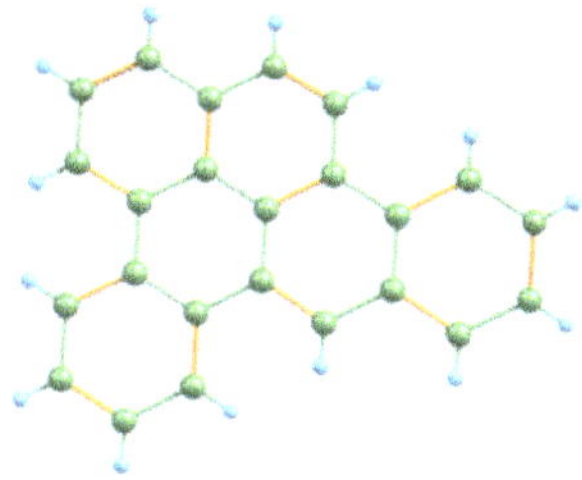

BHC18

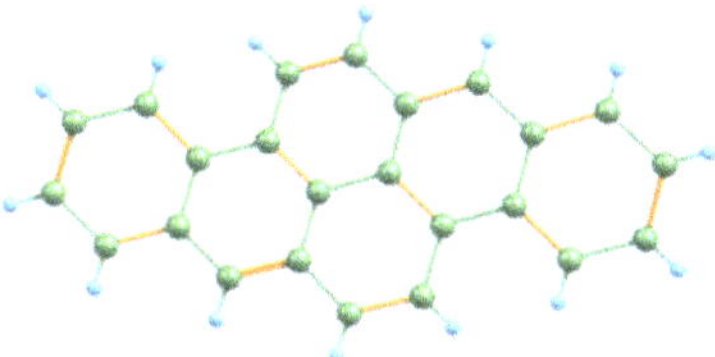

BHC19

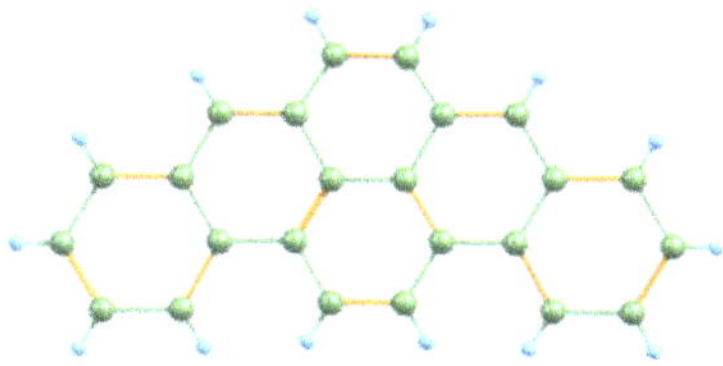

BHC20

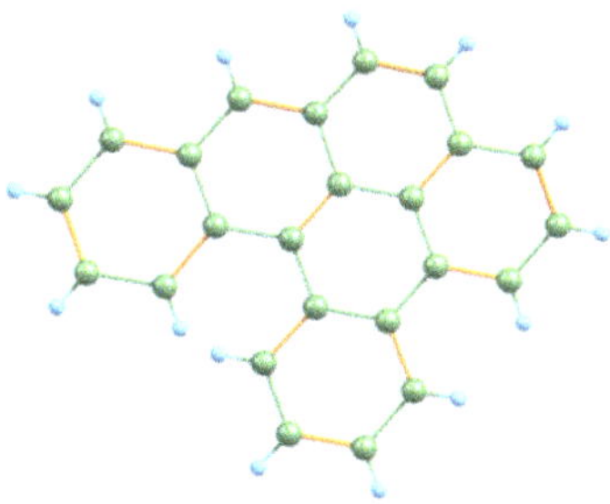

BHC21

The calculated energy values of these compounds are listed in Tableau 2, 3.

Table 2: Various Graph Energies of Certain Benzenoid Hydrocarbons.

BHC	M_1E	M_2E	FE	mM_2E	RE	$ISIE$
BHC1	61.2722	68.9769	142.3375	2.9103	6.2637	15.0979
BHC2	90.7148	106.9667	220.5065	3.8831	8.6001	22.3483
BHC3	90.3388	105.7300	220.2930	3.8060	8.5010	22.1358
BHC4	120.0696	144.8832	298.4665	4.8512	10.9261	29.5803
BHC5	119.8665	143.7999	298.6406	4.7861	10.8502	29.4064
BHC6	120.3072	145.9754	298.7469	4.9014	10.9910	29.7281
BHC7	119.3466	142.4163	298.1457	4.6994	10.7329	29.1573
BHC8	138.6265	171.9377	354.8946	5.1313	11.8862	34.1039
BHC9	138.9804	173.1511	355.3120	5.1930	11.9684	34.2890
BHC10	138.6323	172.7361	354.6812	5.1672	11.9214	34.1893
BHC11	157.1999	199.0184	411.3685	5.4111	12.8460	38.6314
BHC12	157.7418	200.4576	412.0359	5.4969	12.9636	38.8792

BHC13	149.5166	182.8639	377.0130	5.8083	13.2487	36.7991
BHC14	149.3608	181.8429	376.9077	5.7641	13.1951	36.6702
BHC15	149.3550	181.8379	376.8814	5.7639	13.1947	36.6695
BHC16	149.4561	182.8281	376.5172	5.8209	13.2555	36.8186
BHC17	176.7264	228.0341	469.2339	5.8193	13.9904	43.5306
BHC18	168.2811	211.0087	433.2492	6.1545	14.2860	41.4992
BHC19	158.7122	189.1822	395.9916	6.2742	14.3035	38.7935
BHC20	167.9442	209.8121	432.8758	6.0933	14.2052	41.3179
BHC21	168.1275	210.8264	432.7552	6.1458	14.2682	41.4699

Table 3: Various Graph Energies of Certain Benzenoid Hydrocarbons.

BHC	SDDE	HE	ABSE	AZIE	RRE	ABCE
BHC1	28.1173	6.2274	10.0891	113.3178	30.4128	9.6342
BHC2	40.0322	8.5454	14.5425	166.6867	45.0219	13.6282
BHC3	40.1643	8.4254	14.4865	161.7538	44.7147	13.5796
BHC4	51.8894	10.8539	18.9799	220.0020	59.5909	17.6054
BHC5	52.1270	10.7560	18.9556	215.2472	59.3654	17.5902
BHC6	51.7883	10.9308	19.0111	223.5279	59.7987	17.6339
BHC7	52.1905	10.6176	18.8744	210.0158	58.9854	17.5166
BHC8	58.3252	11.7952	21.5036	251.1477	68.7529	19.6636
BHC9	58.2434	11.8920	21.5542	255.1274	69.0275	19.7092
BHC10	58.1036	11.8420	21.4939	253.8668	68.8411	19.6509
BHC11	64.7625	12.7358	24.0287	282.2574	77.9231	21.7226
BHC12	64.7343	12.8729	24.1136	287.3259	78.3071	21.8015
BHC13	63.9147	13.1486	23.4346	272.1114	74.1696	21.6067
BHC14	64.0657	13.0826	23.4184	268.7187	74.0010	21.5941
BHC15	64.0596	13.0823	23.4174	268.7170	73.9989	21.5930
BHC16	63.7687	13.1656	23.4230	273.3388	74.1738	21.5884
BHC17	71.3303	13.8867	26.7119	320.3425	87.7030	23.9329
BHC18	70.1118	14.1894	25.9835	308.0422	83.5609	23.6811
BHC19	69.3748	14.1573	25.1112	280.0332	78.4602	23.3080
BHC20	70.2042	14.0938	25.9357	304.1100	83.2942	23.6382
BHC21	69.9946	14.1740	25.9565	307.9565	83.4938	23.6511

We discuss the linear regression model of the form $A = cB + D$, where A, B, D, c donote property, index, intercept and slope, respectively. We consider coefficient of determination (r^2), standard error (SE), F-statistic (F), and significance value (SF) to examine in details. The energies are found to have strong correlation with boiling point (BP). We obtain the following regression relations.

$\mathrm{BP} = 3.338\, M_1E + 30.318, r^2 = 0.984,\ SE = 13.109,\ F = 1142.979,\ SF = 1.945 \times 10^{-18}$.
$\mathrm{BP} = 2.449\, M_2E + 75.193, r^2 = 0.962,\ SE = 20.021,\ F = 479.206,\ SF = 6.13 \times 10^{-15}$.
$\mathrm{BP} = 1.197\, FE + 71.489, r^2 = 0.968,\ SE = 18.420,\ F = 569.574,\ SF = 1.25 \times 10^{-15}$.
$\mathrm{BP} = 112.458\, \mathrm{m}M_2E - 98.716, r^2 = 0.963,\ SE = 19.793,\ F = 490.762,\ SF = 4.93 \times 10^{-15}$.
$\mathrm{BP} = 45.792\, RE - 60.663, r^2 = 0.986,\ SE = 11.884,\ F = 1395.123,\ SF = 3.01 \times 10^{-19}$.
$\mathrm{BP} = 13.530\, ISIE + 31.538, r^2 = 0.981,\ SE = 13.977,\ F = 1003.196,\ SF = 6.58 \times 10^{-18}$.
$\mathrm{BP} = 8.649\, SDDE - 15.775, r^2 = 0.994,\ SE = 7.911,\ F = 3171.805,\ SF = 1.32 \times 10^{-22}$.
$\mathrm{BP} = 46.176\, HE - 61.048, r^2 = 0.986,\ SE = 12.073,\ F = 1351.165,\ SF = 4.06 \times 10^{-19}$.
$\mathrm{BP} = 22.776\, ABSE + 0.544, r^2 = 0.991,\ SE = 9.631,\ F = 2133.810,\ SF = 5.54 \times 10^{-21}$.
$\mathrm{BP} = 1.837\, AZIE + 30.965, r^2 = 0.972,\ SE = 17.032,\ F = 669.460,\ SF = 2.82 \times 10^{-16}$.

$$\text{BP} = 6.721\ RRE + 30.911, r^2 = 0.982,\ SE = 13.534,\ F = 1071.245,\ SF = 3.56 \times 10^{-18}.$$
$$\text{BP} = 25.963\ ABCE - 22.528, r^2 = 0.993,\ SE = 8.383,\ F = 2822.618,\ SF = 3.96 \times 10^{-22}.$$

The linear fitting of the regression relations are depicted in Fig 1, Fig 2, Fig 3, Fig 4.

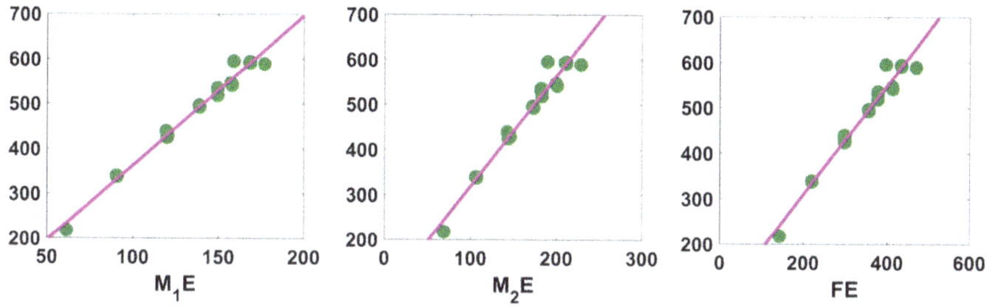

Figure 1: Linear fitting of M_1E, M_2E and FE with BP.

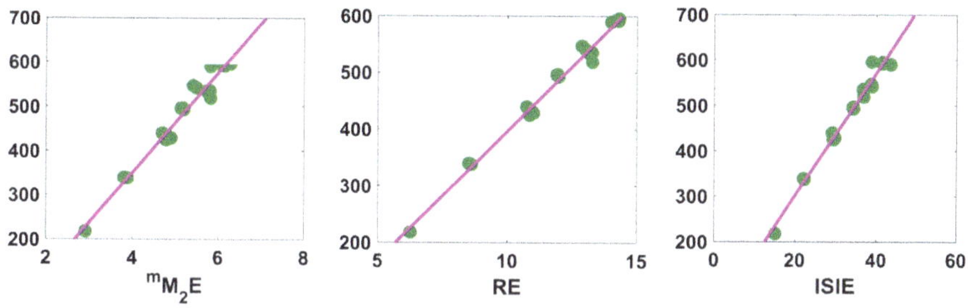

Figure 2: Linear fitting of mM_2E, RE and ISIE with BP.

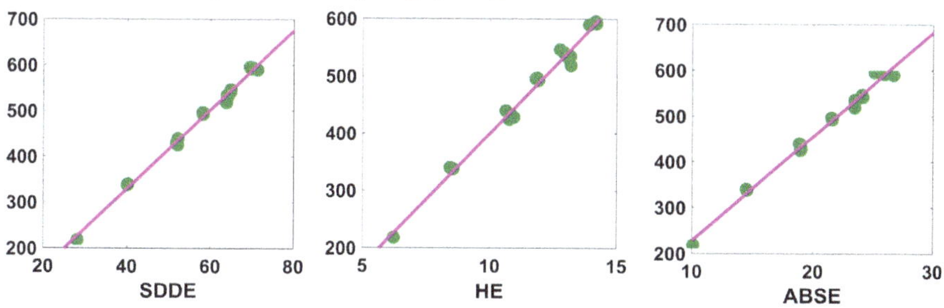

Figure 3: Linear fitting of SDDE, HE and ABSE with BP.

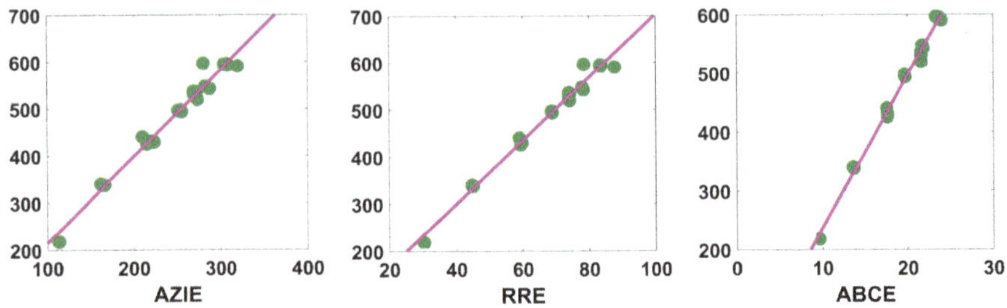

Figure 4: Linear fitting of AZIE, RRE and ABCE with BP.

We observe a strong linear dependence between boiling point and several degree-based graph energies. In particular, we find that SDDE, ABCE, and ABSE display minimal dispersion and consistently low standard errors, which implies that these energies are especially sensitive to structural growth in fused-ring systems.

Concluding Remarks

In this paper, we examined several graph energies derived from degree-based modifications of the adjacency matrix and studied their relationship with the boiling points of benzenoid hydrocarbons. The statistical analysis shows that a number of these energies vary in close agreement with the experimental data.

From these observations, we can claim that introducing degree-based weights into the adjacency framework provides a meaningful extension of the classical notion of graph energy. We found that the resulting quantities are easy to compute and retain a clear structural interpretation, which makes them suitable for structure–property studies. As a future direction of research, we may include nonlinear modelling or the examination of other families of hydrocarbons and heteroatomic compounds in order to better understand the range of applicability of these degree-weighted energies.

Acknowledgements

Z.R. is supported by the University of Sharjah Research Grant No. 25021440174 and MASEP Research Group.

References

[1] H. Wiener, Structural determination of paraffin boiling points, J. Am. Chem. Soc. 69 (1947) 17-20. https://doi.org/10.1021/ja01193a005

[2] I. Gutman, Degree-Based Topological Indices, Croat. Chem. Acta 86 (2013) 351-361. https://doi.org/10.5562/cca2294

[3] S. Mondal, Z. Raza, Complete solution to open problem on exponential arithmetic-geometric index, Appl. Math. Comput. 513 (2026) 129809. https://doi.org/10.1016/j.amc.2025.129809

[4] Z. Raza, L. Saha, K.C. Das, On inverse symmetric division deg index of graphs, RAIRO Oper. Res. 57 (2023) 3223-3236. https://doi.org/10.1051/ro/2023181

[5] Z. Raza, M. Imran, Expected values of some molecular descriptors in random cyclooctane chains, Symmetry 13 (2021) 2197. https://doi.org/10.3390/sym13112197

[6] Z. Raza, The expected values of some indices in random phenylene chains, The Eur. Phys. J. Plus 136 (2021) 1-15. https://doi.org/10.1140/epjp/s13360-021-01082-y

[7] Z. Raza A. Ali, Bounds on the Zagreb Indices for Molecular (n,m)-Graphs, Int. J. Quantum Chem. 120 (2020) e26333. https://doi.org/10.1002/qua.26333

[8] P. Das, S. Mondal, B. Some, A. Pal, Extension of adjacency matrix in QSPR analysis, Chemom. Intell. Lab. Syst. 243 (2023) 105024. https://doi.org/10.1016/j.chemolab.2023.105024

[9] P. Das, S. Mondal, A. Pal, On Second Zagreb Energy of Graphs, MATCH Commun. Math. Comput. Chem. 92 (2024) 105-131. https://doi.org/10.46793/match.92-1.105D

[10] I. Gutman, B. Furtula, S.B. Bozkurt, On Randic energy, Linear Algebra Appl. 442 (2014) 50-57. https://doi.org/10.1016/j.laa.2013.06.010

[11] S. Mondal, S. Barik, N. De, A. Pal, A note on neighborhood first Zagreb energy and its significance as a molecular descriptor, Chemom. Intell. Lab. Syst. 222 (2022) 104494. https://doi.org/10.1016/j.chemolab.2022.104494

[12] S. Mondal, P. Das, Z. Raza, A. Pal, M. Ghorbani, (2026). Graph Spectrum of Neighbourhood Sombor Matrix and Structure-Property Modelling, Theor. Comput. Sci. 1068 (2026) 115758. https://doi.org/10.1016/j.tcs.2026.115758

[13] I. Gutman, The energy of a graph, Ber. Math. Statist. Sekt. Forsch. Graz 103 (1978) 1-22.

Advanced Materials and Sustainable Energy Technologies – AMSET2025 Materials Research Forum LLC
Materials Research Proceedings 66 (2026) 18-29 https://doi.org/10.21741/9781644904152-3

Design and Simulation of Plastic Solid Waste Shredder with Continuous Variable Transmission

Angelo G. FERRER[1,a *], John Gerald B. MAZO[1,b] and Luvisminda D. MARCELO[1,c]

[1]Mechanical Engineering Department, National University, Manila, Philippines

[a]ferrerangelog@gmail.com, [b]jgbmazo@national-u.edu.ph, [c]ldmarcelo@national-u.edu.ph

Keywords: Continuous Variable Transmission, Ansys, Simulink, Von-Mises, Plastic Solid Waste, Shredder

Abstract. Problems with plastic solid waste have been a major increasing problem not only in local towns but in the whole Philippines and traditional PSW shredders have been adequate but there is still a huge room for improvements. This study presents an innovative approach to increasing the thermal efficiency of small scale PSW shredders. Continuously Variable Transmissions (CVTs) have only been exclusively adapted to the automotive industry and in doing so, hinders its advantages to other fields such as solid waste management. As a modern approach to designing, a proof of concept in the form of stress and performance simulation when a CVT is integrated into a PSW shredder is therefore presented in this paper. In addition to the calculations in designing the parts, the CVT PSW shredder assembly is also tested under von-Mises, Soderberg and Goodman criteria for failure in Ansys Mechanical to ensure safe use while under operation and Matlab Simulink is used for performance simulation. After the design process regarding the parts' dimensions, materials, and other considerations within the scope, the result from performance simulation showed promising results in that the major KPIs fuel consumption and theoretical throughput are better in the model with CVT when compared to the same model using a fixed gear ratio transmission.

Introduction

Plastic solid waste or PSW is a common reason for many types of pollution. Most common sources of PSW are from urbanized neighborhoods which are considered as Municipal Solid Waste or MSW. Philippines daily waste generation per capita ranges from 300 to 700 grams which also increases with population, income, family expenditure, and human development index [1]. Inefficiency in the use of plastic also adds to the pollution. Single use plastics found in food packaging, water bottles, and other PSW contribute significantly in the stream of PSW being generated. To shred the PSW of materials efficiently, the energy loss on the delivery of energy from prime mover up to the cutter blades must be minimal. Because of this, continuous variable transmission or CVT is studied and integrated into the assembly to increase its efficiency in shredding.

Currently, plastic shredders in market supply mostly use 3-phase AC motors as their prime movers. Variable speed drives or VSDs have been implemented on shredders using AC motors which is a complex system of torque, current, and voltage to feed into a program that decides what action on the transmission to take. It may reverse the transmission or shift into a larger gear ratio to reduce the angular speed in return of a higher torque [2].

The solution proposed aims to design and simulate a plastic solid waste shredder for the 6 types of recyclable plastic, from polyethylene terephthalate or PET to polystyrene or PS, utilizing a CVT to take advantage of the increased efficiency both thermally and mechanically while taking into account all of the factors aforementioned for consideration of actual fabrication's feasibility.

The design of an effective plastic solid waste shredder improved upon by the integration of continuous variable transmission technology starts as a testament that CVTs are not only exclusive for the application in the automotive industry but to the waste management sector as well.

Materials and Methods

To design and simulate the plastic solid waste shredder with cvt, herein referred as CVT PSW shredder, the three main methods of investigation are usage of standard formulas, von-Mises yield criterion through FEA in Ansys, and Matlab Simulink performance simulation. The methods are discussed below briefly.

The types of material the CVT PSW Shredder is designed for are recyclable thermoplastics with recycling number 1 to 6 and the ultimate tensile strengths vary from the average tensile strength of the 6 types of recyclable plastic ranging from 8.3 MPa to 72 MPa [3]. To design and simulate the parts of the CVT PSW shredder, the engineering software Solidworks, Ansys Mechanical, and Matlab/Simulink are used. The design process is iterative in nature as reflected in the operational framework below.

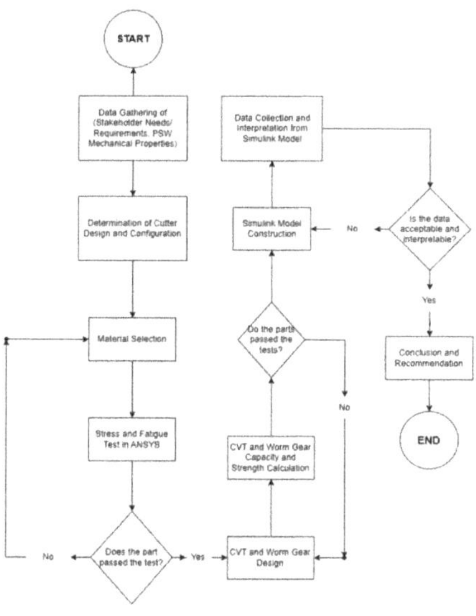

Fig. 1. Operational framework for design and simulation phases

Design Phase. In the designing of the CVT PSW shredder, the researchers opted for concave cutters variant of blades as they are more cost effective and easier to service in that when the blades are worn out, the blade can simply be rotated to expose an unused side of the blade. Data gathered from the stakeholder suggests that their area generates about 8 MT of MSW per day wherein approximately 60% of it is considered as PSW according to the stakeholder barangay captain. This serves as a base for the capacity of the shredder and the benchmark that the researchers will use to determine the soundness of the design.

To start the design phase, a baseline design is done and the assembly part that will be mostly iterated is the transmission assembly because it is the point where changes impact the overall performance and characteristics of the CVT PSW shredder.

Fig. 2. CVT PSW shredder assembly without housing (left) and with housing (right)

The essential baseline dimensions and characteristics of the parts of the assembly are also discussed with the exception of the CVT and other transmission elements which will be discussed separately.

Concave Cutter and Stator Blades. The blades are made of M2 tempered tool steel for durability and longevity of the part. It has an edge length of 2 in. with a thickness of 1.18 in. The stator blade is a matching blade that matches the concave cutters shape to simulate cutting action.

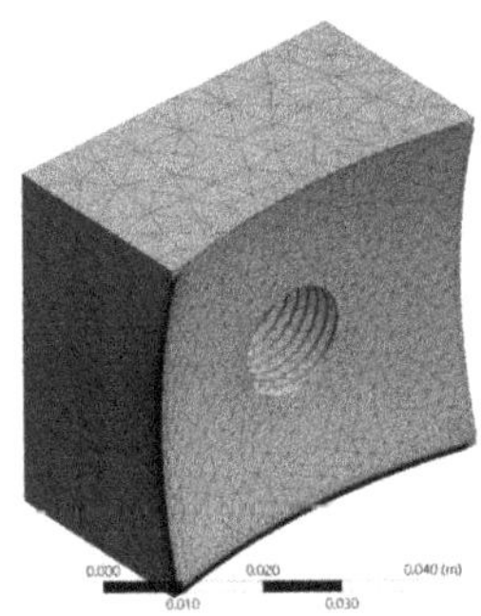

Fig. 3. Isometric view of meshed 3D model of concave cutter

Drum holder. The dimensions of the drum holder are 1 ft. in diameter and 470 mm. It is made with common structural steel to and is purposed for holding the concave cutters in place. It is designed to be as hollow as it can be without compromising its mechanical integrity under operation.

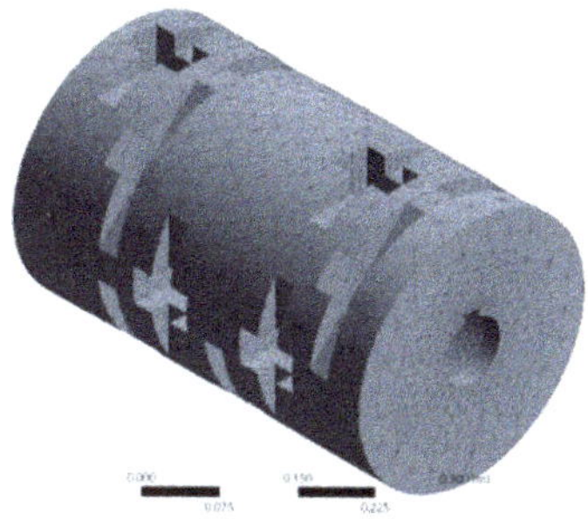

Fig. 4. Isometric view of meshed 3D model of drum holder

Prime Mover. The prime mover used in the shredder is a GX270 8.4 HP 4-Stroke OHV Petrol Engine. The engine will deliver a max net torque of 19.14 N-m at 2500 rpm. During the calculations and designing process, the engine will be aimed to run constantly at 2500 rpm since this is the angular speed at which the engine produces the maximum torque at the lowest rpm. The net torque of 19.14 N-m indicated in the power curve is the torque produced at the net brake power at that specified rpm of the engine. The definition provided is from the standardized testing code SAE J1349 which is now the standard testing code of Honda applied to all their commercially available engines [4].

CVT Design. To appropriately design the necessary dimensions of the CVT given the baseline designs and aforementioned data about the capacity. The following flow of formula is used to build up the necessary speed ratio, dimensions of the CVT, capacity of belt, and other essential characteristics of the CVT. The starting point is from the edge of the blades in contact with the PSW feed moving towards the transmission and ending up in the energy source prime mover. The first data needed is the torque and shear stress induced on the shaft of the drum holder, the safe practice is to consider the maximum torque which is calculated from by using the maximum shear strength, the active edge length of blade, considered thickness of feed, and radius of drum holder. Moreover, t_p is equivalent to 1 mm which is twice the thickness a Nasr and Yehia used in their study [5].

$$T_{drum} = \boldsymbol{\tau}_{PSW} \times t_p \times L_{blade} \times R_{drum}. \tag{1}$$

With this, the shear stress in the shaft for shaft design can be calculated as follows.

$$\boldsymbol{\tau} = [16T \times km]/\pi D^3. \tag{2}$$

Where km is the numerical combined shack and fatigue factor for torsional moment equating to 2.0 for sudden loading and heavy shocks. From these values, the speed ratio or SR can be calculated through the formula below.

$$SR = T_{drum}/T_{engine}. \tag{3}$$

For CVT design, the researchers based the calculations from Faires' Design of Machine Elements [6]. The formulas revolve around the design of the belt for safe and long-life span and the design of the sheave for accommodating the speed ratio determined from the presented formulas above. For belt design, the design hp is determined first by using a multiplying service factor N_{sf}. After the design hp is determined, the general dimensions of the CVT consisting of the sheave diameters, belt length, center distances are determined.

$$D_{max} = SR \times D_{min}. \tag{4}$$

$$C = [D_{max} + D_{min}]/2 \text{ or } C = D_{max}. \tag{5}$$

$$L = 2C + \pi[D_{max} + D_{min}]/2 + [D_{max} + D_{min}]^2/4C. \tag{6}$$

After these values are determined, the general belt formula applied to v-blets can be used to calculate the rated hp per belt of the chosen type of belt that will be used in the CVT PSW shredder.

$$\text{Adjusted rated hp per belt} = [\, a(10^3/v_m)^{0.09} - c/K_d\, D_{min} - e(v_m^2/10^6)\,] \times (v_m/10^3) \times K_t K_L. \tag{7}$$

$$\text{Number of Belts} = \text{Design hp}/\ \text{Rated hp per belt}. \tag{8}$$

The number of belts is only limited to two as it may complicate manufacturing and designing. Although it is preferable to have only 1 belt for a CVT setup, there is an indirect relationship between the maximum speed attainable and the characteristics of the belt, including the quantity of belt used.

Simulation Phase. To simulate the equivalent stresses or von-mises stresses and the fatigue stresses, Ansys Mechanical is used under the structural analysis workflow. For performance simulation, Matlab/Simulink is used to generate the results. Under stress simulation, the parts are simulated between two theories. The first theory of failure is based on Soderberg's criterion of failure for ductile materials that are used for the parts such as the drum holder. The other theory is based on Goodman's theory of failure for brittle materials that are applied to the concave cutters. The setup of the simulation resembles the forces that are acting on the parts. However, important notice must be given to the setup since forces cannot be applied directly on edges as it may simulate stress singularities and can invalidate the whole simulation. To overcome this, the researchers opted for an edge represented by a 0.5 mm thin surface which is the thinnest simulation setup the researchers' equipment can handle.

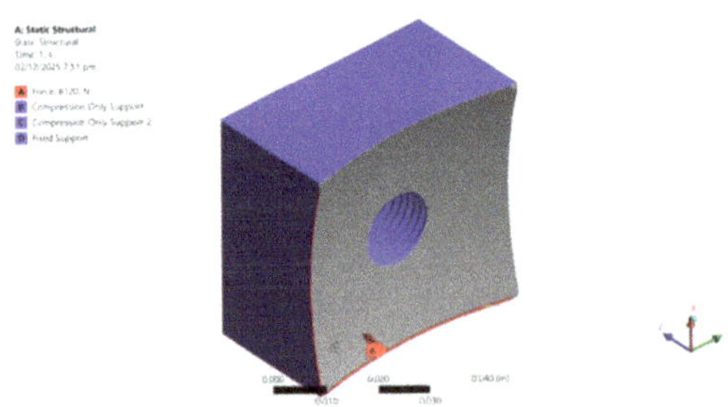

Fig. 5. Concave cutter stress simulation setup

The 0.5 mm thin surface depicted in the figure is where the force from the cutting action of the blade is directed to. This minimizes the chance of a simulation test resulting in a stress singularity. For performance simulation, Matlab through its Simulink add-ons application is set up to simulate normal operating conditions as closely as possible. Randomness of the feed is done through a fluctuating signal that generates a torsional resistance on a band brake.

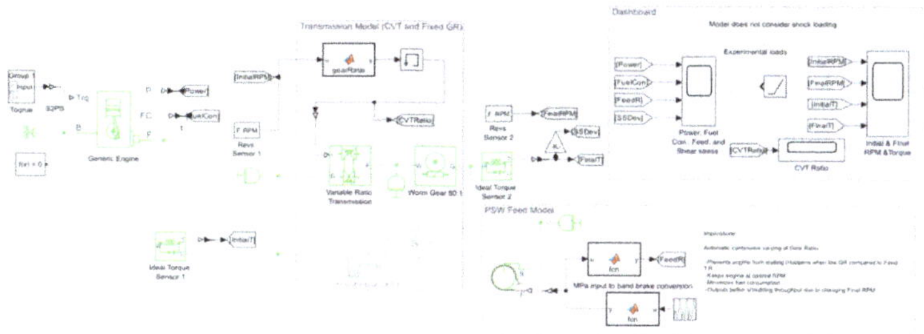

Fig. 6. Matlab/Simulink simulation setup

The simulation monitors the angular velocities of the drum and the internal combustion engine or ICE. More importantly, the ICE block on the left side also generates the fuel consumption and the power generated which is useful in determining the model's efficiency in shredding.

Results

Stress Simulation Results. In this section, the resulting data from the methods presented in the previous section is presented. To ensure safe and sound design, the stress simulations must have a factor of safety when the stress under normal operating conditions is simulated. Moreover, the result must converge to a certain value or limit in relation to varying resolution of the model simulation mesh. The mesh is important as it affects how good the simulation resembles actual results in practice. Therefore, the convergence of value is a must in order to validate the simulation.

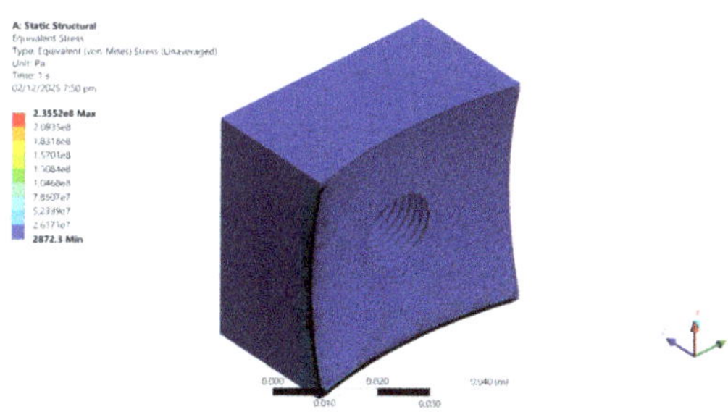

Fig. 7. Concave cutter blades von-mises stress simulation results

The resulting values from the simulation of von-mises stress is also fed through the fatigue stress calculator inside Ansys wherein it is analyzed under Goodman's theory. The parameters of the fatigue stress analysis are that the load is cyclic varying from 0 to 100% of the load.

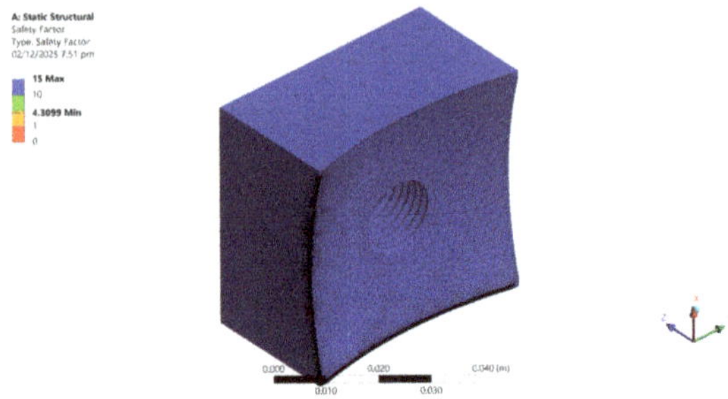

Fig. 8. Concave cutter blade fatigue stress safety factor results

Safety factor results proved that the design is safe and will have a long lifespan. The summary of results with regards to stress simulation and fatigue stress analysis is summarized in the table below.

Table 1. Summary of stress and fatigue simulation results.

Part	Minimum equivalent stress safety factor	Part life [cycles]	Convergence test [% of change]
Concave Cutter	4.31	10^{12}	5.42
Stator Blade	1.70	10^{12}	1.773
Blade Holder	15	10^{6}	0.88
Drum	3.04	10^{6}	0.87

Performance Simulation Results. To arrive at the performance simulation, complete characteristics of the parts of the CVT PSW shredder must be provided. Since the overall capacity of a v-belt depends on its type, length, sheave diameter, and speed, iterative process of design is a must and the researchers took advantage of the theoretical formulas to come up with a safe design. The CVT design presented utilizes 2 belts with a flyball shifting mechanism in which the weight of the flyballs can be obtained by calculating the centrifugal force required to shift the sheaves. Although this can be calculated, it is still best to manually increment the flyballs to obtain the right weight.

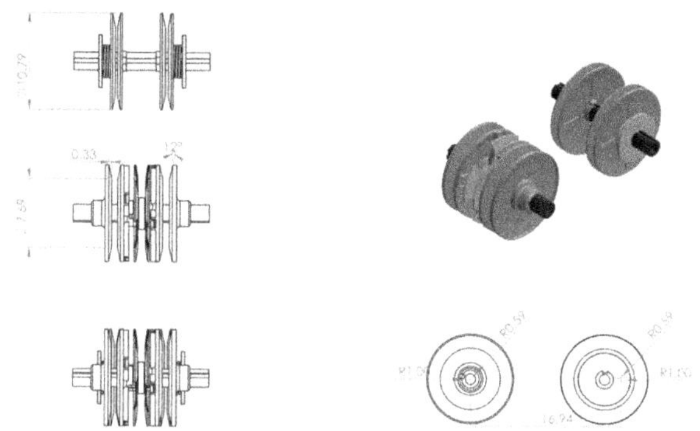

Fig. 9. CVT design

With these data, the parameters for the Simulink simulation can now be set and run. For the CVT control, a user defined formula coded by the researchers is made so as to closely follow the behavior of an automatically shifting CVT due to the flyballs centrifugal force. The grayed out path or blocks from Figure 6 represents the model without a CVT. This path can be activated while disabling the path with CVT to simulate a model that runs in a fixed gear transmission. This is done for comparison of efficiency between a model with CVT integrated and a model without CVT. Moreover, a summary of the parts constructed and their corresponding materials or characteristics is hereby presented.

Table 2. Bill of materials

Part	Material/ Description
Concave Cutter Blade (16)	M2 Tool Steel Quenched at 2200 F
Blade Holder (16)	ASTM A36
Drum	ASTM A36
Stator Blade	M2 Tool Steel Quenched at 2200 F
CVT Sheaves (8)	ASTM A36
Belt (2)	B60 V-belt or equivalent CVT belt
Shafts (2)	C1030 steel
Pillow Block Bearing (6)	Chrome Steel GCr15
Fly ball (8)	10.5 g mass
KAS97	Helical Bevel Gear

Materials Research Proceedings 66 (2026) 18-29 https://doi.org/10.21741/9781644904152-3

The summary of parts is obtained through rigorous consideration of alternatives, especially on the reducer KAS97. Additionally, the helical bevel reducer will be paired with a simple fixed 3-belt pulley to reduce the initial RPM to a magnitude that can be handled by the reducer.

Simulation Results Data. The scope blocks seen on Figure 6 reads the data with respect to time.

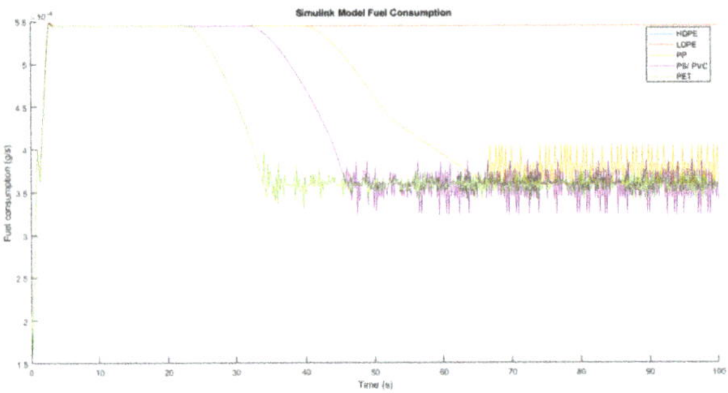

Fig. 10. CVT integrated model fuel consumption simulation results comparison

The figure above showcases the realistic limitation of the model in that it best performs under a certain range of PSW strengths. Shredding PSW that goes below the 40 MPa observed threshold only results in maximized fuel consumption since the engine attempts to run at its maximum capacity.

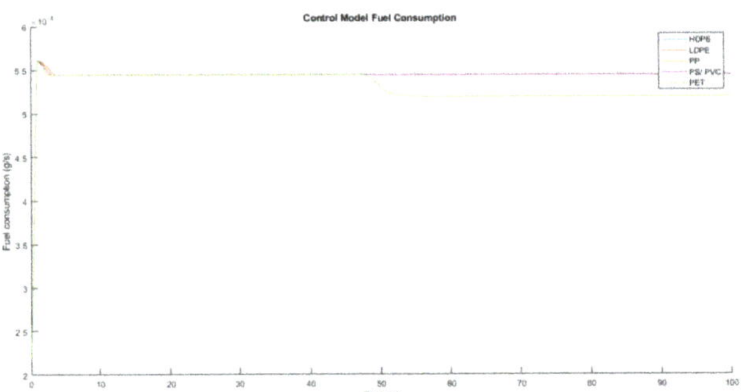

Fig. 11. Control model fuel consumption simulation results comparison

As seen from the figure above, the control model outputs constant maximum fuel consumption which makes it inferior to the model with CVT. The parameters for both the results such as the feed strength, its values, and other parameters remain unchanged to ensure unbiased results.

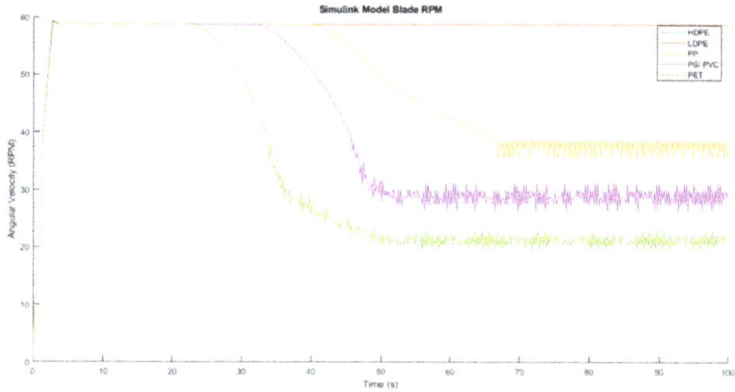

Fig. 12. CVT integrated model drum angular velocity simulation results

In the figure depicted above, it is evident that the model being tested engages the CVT when the plastics are equal or greater in strength than polypropylene recyclable plastics (PP). In essence, the model with CVT automatically adjusts its torque to match the required torque for shredding.

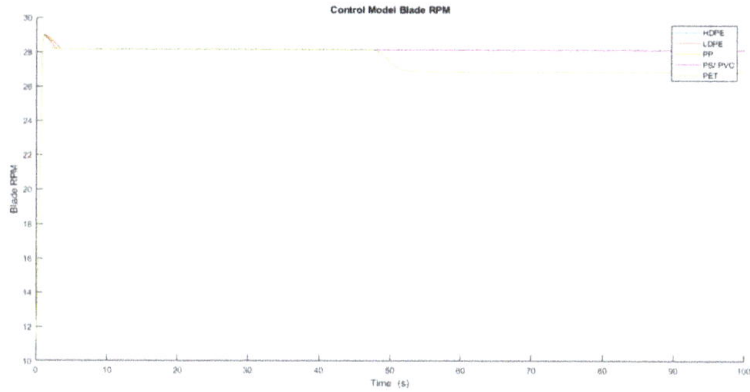

Fig. 13. Control model drum angular velocity simulation results

The former model outperforms the control in terms of thermal efficiency with the shredding rate in consideration. Given that all parameters of the setup for simulation are the same for the CVT integrated model and control, it can be seen from Fig 12-13 that not only the designed model has better fuel consumption but it also outputs higher rate of shredding under the same circumstances.

Discussion

Design and simulation of the plastic solid waste shredder with CVT using Solidworks, Ansys, and Matlab Simulink was successfully accomplished and shown in the previous sections. The design of the parts is done in coordination with the stress simulation analysis to make sure the parts are safe for normal operating conditions. After the design is finalized, the Matlab Simulink model is then constructed based on the 3D modelling and stress simulation stage. The performance data gathered from this simulation is then reported and interpreted.

To discuss in depth the design, the aspects of the designed CVT must be studied. Due to the dimension of the B section v-belts, the speed reduction ratio was limited to 1.4:1. The CVT ratio when paired with KAS97 bevel helical reducer and a triple-belt pulley assembly, produces a total speed ratio of 134.204:1 which is an acceptable reducer ratio given that the recommended ratio to shred PET is 129:1 when using an 8.4 hp GX270 Honda engine.

The mechanical characteristics of the CVT is then utilized inside MATLAB Simulink to use for simulation. An automatic controller composed of a CVT block from the driveline block set and the user defined function block allows the researchers to simulate the performance the model can achieve.

With regards to the base reducer, the researchers opted to use a pulley and a commercially available bevel helical reducer. The pulley utilizes three B section v-belts with a driven and driving pulley diameters of 10.8 in and 5.4 in respectively. This setup yields a reduction ratio of 2:1 which is then connected to KAS97 with a gear ratio of 47.93:1. The reasoning for the usage of pulley instead of a direct one-unit reducer is to save cost. With these changes in the assembly, the researchers were able to bring down the overall cost by 20 000 pesos.

The concave cutting blades used M2 tool steel based on the fact that tool steels are designed for cutting purposes. Since heat treatment is a variable that the research was not able to tackle due to limitations, approximations of the strength of materials based on ASTM 370-17 literature were used. In line with this, stress simulations were done by calculating von-mises stresses. For fatigue simulations, they were done based on the characteristics of the material. Brittle materials such as the concave blade and stator blades were analyzed under Goodman's theory of failure while ductile materials such as the holders and drum are analyzed under Soderberg's criteria of failure. Mesh independence study is also conducted to ensure realistic results and efficient simulation.

Conclusion

During the performance simulation inside Matlab Simulink, the integration of the Continuous Variable Transmission to the Plastic Solid Waste Shredder showed an increase in the overall efficiency of the shredder when compared with a control setup wherein the transmission system is of the fixed nature. This increase in overall efficiency can be observed through the key performance indicators of the PSW CVT shredder plotted from figures 10 through 13. The fuel savings was observed to be 36.36% when the PSW strength falls between 80 MPa to 50 MPa. or the plastics PP, PS, PVC, and PET. With regards to feed shredded, the throughput efficiency or the mass of recycled PSW per unit time [3] is also increased with the model wherein the CVT is integrated. As seen from Figure 12, the final RPM varies depending on the feed torsional resistance or the recyclable plastic type. As the torsional resistance or the strength of feed decreases, the blade is allowed to rotate faster since it requires less torque to shred the soft feed. When compared to the control design, the RPM is at a constant 28 RPM regardless of the feed parameters which makes it inferior when compared to the performance of the model with CVT.

Recommendations. As seen from the resulting graphs for the performance simulation, higher ranges of speed ratio from CVT can result in a wider range of recyclable plastic solid waste that can be shredded at the optimal RPM of the engine. Due to this relationship, the researchers recommended further study of CVT types that can be implemented practically. Toroidal CVT's, metal chain belts, and other advancements that were not tackled in this paper due to computing power, resources, and other limitations may be studied to see if any of them are a viable and more effective replacement for the elastomer belt CVT used in this study.

With regards to stress simulation, a more in-depth mesh independence study could be done to increase the elements' quality, skewness, and aspect ratio. Materials used in the blade are assumed to be quenched as per the manufacturers data. In depth study of the effect of different heat treatments such as precipitation hardening, cold working, nitriding, case hardening, and other approaches should be studied to better improve the accuracy of the stress and fatigue simulations done in Ansys.

With regards to the PSW feed being processed, bulk movement of the PSW cannot be identified and thus may affect the operation of the shredder if multiple stacks of PSW clogs in one area of the cutting blades. To eliminate this, prototyping should consider devising a mechanism that could prevent the said clogging of the feed. Although the

Exploration of other multibody simulation software can also be considered such as Altair MotionSolve. This can be used not only with the cutting actions but with the CVT assembly as well.

Integration of a clutch, Arduino throttle controller, and a flywheel can also be done to increase ease of maintenance. The flywheel can also smoothen the operation of the shredder by dampening shocks and storing kinetic energy for which it can dump whenever the strength of the plastic spikes up.

Further study on the types of cutters and different types of cutting profile should also be given notice as it may affect the cutting effectiveness of the shredder. Additionally, the material optimization of the parts was not highlighted and can be improved upon next researchers to improve the cost-effectiveness of the shredder. Significant volume of parts that were tested for stress simulations show that it does not contribute to the bearing of stress meaning that strategic material removal could retain the parts integrity while retaining its strength relative to its function.

References

[1] Lagman-Bautista, J. (2020). Solid waste generation in the Philippines: A correlational analysis on economic and social metrics. International Journal of GEOMATE, 19(72).

[2] Using VSDs for shredder control. Hub4. (n.d.). https://hub-4.com/news/using-vsds-for-shredder-control

[3] Wong, J. H., Gan, M. J. H., Chua, B. L., Gakim, M., & Siambun, N. J. (2022). Shredder machine for plastic recycling: A review paper. IOP Conference Series: Materials Science and Engineering, 1217(1), 012007. https://doi.org/10.1088/1757-899x/1217/1/012007

[4] Rating Engine Power. Honda engines. (n.d.). https://www.honda-engines-eu.com/en/products/rating-engine-power

[5] Nasr, M. F., & Yehia, K. A. (2019). Stress Analysis of a Shredder Blade for Cutting Waste Plastics. Journal of International Society for Science and Engineering, 1(1), 9–12. https://doi.org/10.21608/jisse.2019.20292.1017

[6] Faires, V.M. (1941). Design of Machine Elements. 4th ed. New York: The Macmillan Company.

Advanced Materials and Sustainable Energy Technologies – AMSET2025 Materials Research Forum LLC
Materials Research Proceedings 66 (2026) 30-33 https://doi.org/10.21741/9781644904152-4

Study of Bioelectrode Performance in Microbial Electrolysis Cell for Clean Energy and Carbon Reduction

Yashawini Phriya Rauichandran[1], Swee Su Lim[1*], Wai Yin Wong[1], Mimi Hani Abu Bakar[1], Ebrahim Mahmoudi[2], Peer Mohamed[2], Eileen Yu[4], Jean Marrie[5], Hassan Mohamed[3]

[1]Fuel Cell Institute, Universiti Kebangsaan Malaysia, 43600 Bangi, Malaysia

[2]Department of Chemical and Process Engineering, Faculty of Engineering and Built Environment, Universiti Kebangsaan Malaysia, 43600 Bangi, Malaysia

[3]Department of Sustainable Energy (ISE), Universiti Tenaga Nasional, 43000 Kajang, Selangor

[4]Department of Chemical Engineering, University of Southampton, Southampton, United Kingdom

[5]Department of Chemical Engineering, Chemical Engineering Research Center of Toulouse, France

*limss@ukm.edu.my

Keywords: Microbial Electrosynthesis Cell, Bioanode, Biocathode, Three Chamber, Gas Chamber

Abstract: Microbial electrolysis cells (MECs) are emerging technologies that are capable of simultaneous wastewater treatment and hydrogen generation. This research involves studying bioelectrodes enrichment and interactions to enhance MEC performance and feasibility. Bioanodes and biocathodes were enriched separately at +0.2 V and -0.9 V. The bioanode achieved a maximum current density of 0.30 ± 0.05 A/m^2 within potential range of -0.2 and +1.0 V, while the biocathode began producing hydrogen below -0.8 V vs. SHE; thus, it is optimal to operate both bioelectrodes at these potentials in MEC system. In the two-chamber (2cMEC) system, different ranges of applied potentials had affected performance: hydrogen production started at below 0.7 V, peaked between 0.7 and 1.2 V and was limited by the anode above 1.2 V. Highest hydrogen production rate was achieved within 0.9 V and 1.2 V, with the contribution of bioanode supply of only 25% and external power supply of 75%. A three-chamber (3cMEC) system with a gas chamber included next to the cathode was developed to overcome pH imbalance issue and support CO_2 capture. This setup enabled dissolution of CO_2 gas directly into the catholyte resulting in a higher carbonate concentration (550 ± 200 mg/L) and improved pH stability. The internal resistance was 66.55 Ω due to the formation of biofilm layer on the electrodes in the system. According to the results, 3cMEC showed good potential for CO_2 dissolution to facilitate carbon capture and allow for the simultaneous hydrogen production in the same system.

Introduction

Microbial electrolysis cells (MECs) present a promising approach for wastewater treatment, renewable energy production and CO_2 reduction [1]. MECs, in contrast to traditional electrolysis, oxidize organic matter using bioanodes enriched with electrochemically active microorganisms, producing electrons that are then transferred to the cathode to produce hydrogen [2]. Nevertheless, there is very limited study on systems that combine bioanode and biocathode and understanding their interaction is significant for achieving balanced redox reaction, stable current generation and efficient product recovery [3]. Challenges in systems include weak anodes, suboptimal growth conditions, and unknown electron transfer pathways which lead to high energy consumption and lower performance [4,5]. Thus, this study aims to optimize the bioanode and biocathode, study their synergistic interactions and develop a three-chamber MEC for direct CO_2 dissolution and hydrogen production.

Materials and Methods

Microbial electrolysis cells (MECs) were constructed with nylon rods and polyacrylic blocks with Ag/AgCl reference electrodes and a cation exchange membrane (CMI-7000) to separate the chambers. Carbon felt, and carbon cloth served as the electrodes with titanium current collectors. The 2cMEC setup included a gas collector for hydrogen gas measurement, while the 3cMEC added an extra gas chamber as CO_2 gas reservoir next to the cathode chamber.

The anodic and cathodic media were prepared based on the modified Atlas and Rozendal formulations with vitamins, minerals, and phosphate buffers [6,7]. For the 3cMEC, CO_2 was directly diffused as the carbon source in the catholyte. Enrichment was done at +0.2 V vs SHE at bioanodes and at -0.7 V vs. SHE is using a Quad potentiostat. Inoculum dominated with *Geobacter* and *Desulfovibro* species were being enriched in both anode and cathode simultaneously. All cells were operated at 26.5 ± 2.5 °C with controlled feeding and voltage application to stabilize bioelectrode activity.

Electrochemical characterization done included electrochemical impedance spectroscopy (EIS), chronoamperometry (CA), and cyclic voltammetry (CV) to investigate cell resistance, redox activity and power performance. Gas and liquid analyses were conducted via gas chromatography, ion chromatography and spectrophotometry, while the calculation on the total energy recovery using Coulombic efficiency, Faraday's law and energy balance equations.

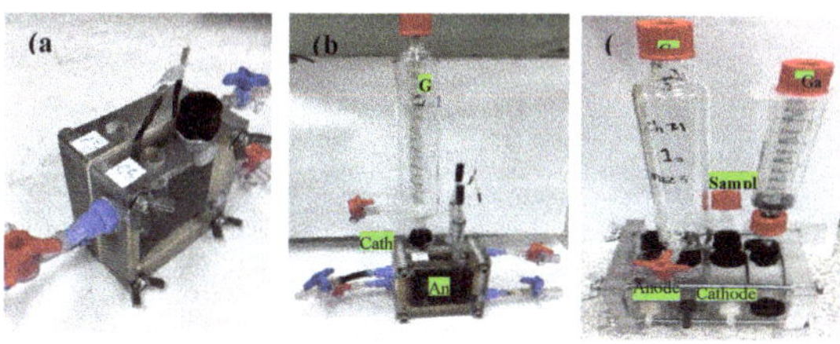

Figure 1. Lab-scale Bioelectrochemical system (BES): (a) microbial fuel cell (MFC), (b) microbial electrolysis cell (MEC) and(c) three-chamber microbial electrolysis cell (3cMEC)

Results and Discussion

The bioanode's oxidation efficiency was strongly influenced by internal resistance and enrichment composition. The optimal enrichment medium ratio of phosphate buffer, acetate, and ammonium was identified as 50:10:10 mM, achieving the lowest internal resistance and highest current output. At an applied potential of +0.2 V vs. SHE, the bioanode exhibited dual catalytic redox peaks at −0.2 and +0.2 V and reached a maximum current density of 0.30 ± 0.05 A/m². Prolonged enrichment had further increased current density to 1.68 A/m² after six weeks, demonstrating robust microbial electrocatalytic activity. However, applied potentials beyond 1.2 V led to abiotic reactions that dominate over bioanode activity, reducing selectivity and leading to energy losses. This highlights the need to apply optimal potential to maintain the bioelectrodes' performance.

On the other hand, the biocathode, enriched at −0.9 V vs. SHE, achieved a stable hydrogen-producing biofilm with activity initiating below −0.8 V and optimal hydrogen evolution near 1.0 V. Hydrogen production peaked at 15.0 L H_2/m²/day in fed-batch operation. Sulfate concentrations above 3 mM inhibited hydrogen evolution unless cathode potentials were lowered below −1.0 V, indicating a threshold for effective simultaneous sulfate and proton reduction. Addition of bicarbonate (10 mM) enhanced carbon buffering capacity and proton availability, improving hydrogen yield. Despite promising results, challenges such as mass transfer limitations and catalyst deactivation at low potentials (≤0.3 V) and over long-term operation remained, emphasizing the importance of operational control and medium concentration.

Chronoamperometric testing with the two-chamber (2cMEC) identified three distinct potential operational regions that affects the system performance. At voltages below 0.7 V, both bioanode and biocathode begin to generate electrochemical activity but hydrogen production was still minimal. Withing 0.7 V and 1.2 V, the MEC system operated within its optimal hydrogen range, achieving the highest hydrogen with a peak generation rate of 20.0 ± 5.0 dm³ H_2/m²/day observed near 0.9 to 1.0 V. At voltages above 1.2 V, the system entered the bioanode limitation phase. At low applied voltages (0.3V), bioanode supplied almost 100% of the energy required for hydrogen production, relying fully on microbial catalysis. However, as the voltage increases (> 1.2 V), the bioanode's contribution dropped to 25%, with the remaining 75% supplied by external electrical supply. This shift significantly proves that at higher potentials, the system depends majorly on external power than on bioelectrochemical energy, reflecting a transition in how the MEC operates and balances biological and electrical energy sources. Additionally, the electrolyte properties varied with applied voltages, causing catholyte pH to rise 11.0 which adversely affected hydrogen production.

The three-chamber MEC (3cMEC) addressed rising catholyte pH by integrating a gas chamber for in-situ CO_2 dissolution, stabilizing pH at 10.0 and increasing carbonate to 550 ± 200 mg/L, enhancing cathode performance without lowering hydrogen output. Fed-batch feeding restored bioanode activity within 1–2 hours and maintained current densities around 2.3 A/m², compared with other feeding modes. Energy recovery (RCE, Rcat) peaked between 0.7–1.2 V. Although external power exceeded 65% at higher voltages, enhanced bioanode function reduced this reliance and stabilized hydrogen production. The 3cMEC matched 2cMEC performance while enabling effective CO_2 capture, making it a promising technology for combined energy and production and carbon management.

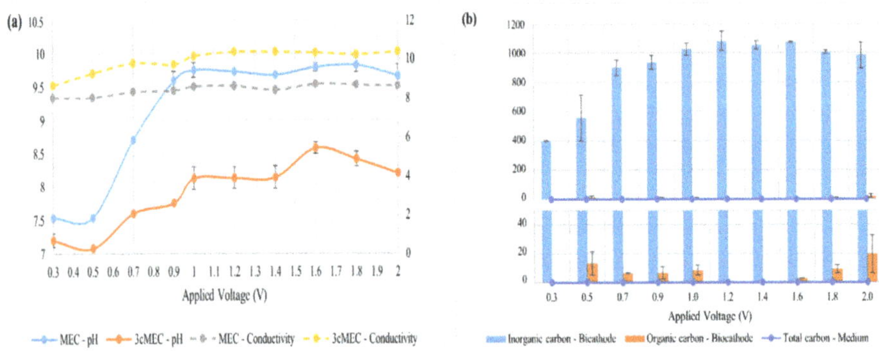

Figure 2 (a) The pH and conductivity profile compared to 3cMEC and (b) total carbon content in the cathode of 3cMEC (small figure is inserted below the original figure for showing the readable organic carbon values).

Conclusion

This study highlights key limitations of bioanodes in MECs, notably internal resistance and reduced biocatalytic activity at high voltages, which constrain overall system efficiency. The enriched bioanode achieved a maximum current density of 0.30 ± 0.05 A/m² within the potential range of −0.2 to +1.0 V, while the biocathode initiated hydrogen production below −0.8 V vs. SHE. Biocathode activity is influenced by sulfate levels and applied potential, requiring precise control for optimal hydrogen production. In the 2cMEC configuration, hydrogen production began below 0.7 V and peaked between 0.9 and 1.2 V, where external power contributed approximately 75% of the total energy input.

The innovative 3cMEC design enables effective in-situ CO_2 capture alongside hydrogen generation by stabilizing catholyte pH and enhancing carbonate accumulation. The 3cMEC system achieved a carbonate concentration of 550 ± 200 mg/L and exhibited a total internal resistance of 66.55 Ω, thereby showing better performance compared to 2cMEC. Future research should focus on minimizing internal resistance, advancing scale-up with real wastewater, and developing selective membranes for optimized integrated energy and carbon management.

Acknowledgments
This project is funded by GUP-2023-004 and FRGS grant.

References

[1] Rahimi, N., & Eicker, U. (2023). Mathematical Modeling of microbial electrolysis cells for enhanced urban wastewater treatment and hydrogen generation. Processes, 11(4), 1157. https://doi.org/10.3390/pr11041157

[2] Santoro, C., Arbizzani, C., Erable, B. and Ieropoulos, I. (2017). Microbial fuel cells: From fundamentals to applications. A review', Journal of Power Sources, 356, 225-244. Santoro, C., Arbizzani, C., Erable, B. and Ieropoulos, I. (2017). Microbial fuel cells: From fundamentals to applications. A review', Journal of Power Sources, 356, 225-244. https://doi.org/10.1016/j.jpowsour.2017.03.109

[3] Lim, S.S., Yu, E.H., Daud, W.R.W., Kim, B.H. and Scott, K. (2017). Bioanode as a limiting factor to biocathode performance in microbial electrolysis cells, Bioresource Technology, 238, 313-324. https://doi.org/10.1016/j.biortech.2017.03.127

[4] Kumar, G., Bakonyi, P., Zhen, G., Sivagurunathan, P., Koók, L., Kim, S.-H., Tóth, G., Nemestóthy, N. and Bélafi-Bakó, K. (2017). Microbial electrochemical systems for sustainable biohydrogen production: Surveying the experiences from a start-up viewpoint, Renewable and Sustainable Energy Reviews, 70, 589-597. https://doi.org/10.1016/j.rser.2016.11.107

[5] Rago, L., Monpart, N., Cortes, P., Baeza, J.A. and Guisasola, A. (2016). Performance of microbial electrolysis cells with bioanodes grown at different external resistances, Water Sci Technol, 73(5), 1129-1135. https://doi.org/10.2166/wst.2015.418

[6] Atlas, R.M. (2010). Handbook of Microbiological Media, Fourth Edition. CRC Press. https://doi.org/10.1201/9781420039726

[7] Rozendal, R.A., Jeremiasse, A.W., Hamelers, H.V.M. and Buisman, C.J.N. (2008). Hydrogen Production with a Microbial Biocathode, Environmental Science & Technology, 42(2), 629-634. https://doi.org/10.1021/es071720

Experimental Study on the Compatibility and Miscibility of a Compressor Nano-Lubricant with Sustainable Refrigerant Mixtures for Efficient Cooling Systems

Nayra ELHELW[1,a], Naser ALI[2,b], Ammar M. BAHMAN[1,c *]

[1]Mechanical Engineering Department, College of Engineering and Petroleum, Kuwait University, P.O. Box 5969, Safat 13060, Kuwait

[2]Kuwait Institute for Scientific Research, Energy and Building Research Center, Nanotechnology Applications Program, Safat 13109, Kuwait

[a]n.elhelw@ku.edu.kw, [b]nmali@kisr.edu.kw, [c]a.bahman@ku.edu.kw

Keywords: AC, Compressor, Refrigerants, Nano-lubricant, Compatibility, Miscibility

Abstract. According to the International Energy Agency, by 2027 global building energy consumption is expected to increase by nearly 8% compared to 2025. Since air conditioners (ACs) are among the primary contributors to energy use, research has increasingly focused on improving energy efficiency, particularly through enhancing compressor lubrication, thereby supporting environmental sustainability. In real-world refrigeration systems, refrigerants and compressor lubrication oil often mix, which can compromise compressor reliability; therefore, AC systems should be designed to account for such scenarios. This study examines a nano-enhanced lubricant (0.025 vol.% short SWCNTs in POE32 lubricating oil) in combination with environmentally sustainable refrigerants: propane (R-290), and R-410A to assess compatibility and miscibility. Compatibility tests revealed that the addition of nanoparticles had a positive impact on the tested materials by repairing some surface microcracks and generally polishing the surfaces. In miscibility testing, vapor pressure measurements confirmed that the refrigerants were miscible within the typical operating ranges of refrigeration systems when nanoparticles were present. Their presence even had beneficial effects on the vapor pressure, as evidenced by the local immiscibility that R-290 experienced with the pure lubricant, but not with the nanolubricant. This, in turn, reduces the likelihood of oil buildup in heat exchangers, leading to improved heat transfer efficiency, while also aligning with environmental regulations that promote the adoption of environmentally friendly refrigerants. Accordingly, the promising results from using nano-enhanced lubricants combined with sustainable refrigerant mixtures suggest that building energy demand could be reduced, thereby contributing to a greener future. As a next step, a mathematical model will be developed to predict miscibility through vapor pressure measurements rather than relying solely on experimental testing. In addition, deeper investigation of compatibility will be conducted using high-resolution material characterization techniques such as X-ray diffraction (XRD), X-ray fluorescence (XRF), and scanning electron microscopy (SEM).

Introduction

In a move towards sustainability, governments have been adopting policies to protect the environment, such as the Montreal Protocol, which aims to protect the ozone layer by phasing out ozone-depleting substances (ODS). The ODP substances include chemical compounds such as chlorofluorocarbons (CFCs) and hydrochlorofluorocarbons (HCFCs). The 2016 Kigali Amendment extends this protocol to include hydrofluorocarbons (HFCs), potent greenhouse gases mainly used in refrigeration and air conditioning industries [1-3] .In tropical countries such as Malaysia, the climate is hot and humid almost year-round, with relative humidity levels reaching very high values (up to 90%). However, when compared with Hot Ambient Temperature (HAT) countries, Malaysia's average temperature is relatively lower, at 27.6 °C [4]. Despite this fact, the high moisture content in the air makes installing an AC unit essential to ensure thermal comfort [5]. According to the International Energy Agency (IEA), 80% of households in Malaysia have at least one installed air conditioning unit [6]. This relates to the energy consumption analysis conducted in Malaysia, which predicts an annual increase of nearly 4.9% in residential building energy demand [7].

On the other hand, the necessity of installing AC units in HAT countries arises from the elevated temperatures and the prolonged duration of peak hours. Nonetheless, this also reflects the pattern of energy consumption in these countries, which underscores the importance of improving the efficiency of AC systems to save energy. For instance, this aligns with the Sustainable Development Scenario (SDS) developed by the IEA, which indicates that improving the efficiency of AC systems will contribute to energy savings greater than the total energy produced in Malaysia [6]. In return, researchers have been investigating technologies to reduce energy usage while contributing to sustainability as well. Some studies shed light on enhancing compressor performance by adding nano-additives to compressor oil [8]. Althalab et al. [9] investigated carbon-based nano-lubricants, dispersing short-SWCNTs, short-MWCNTs, and graphene in polyester oil (POE32). They recommended 0.025 vol% short-SWCNTs, as it exhibited promising tribological properties, reducing mechanical losses by 2.85%, and improved thermophysical properties, increasing thermal conductivity by 1.46% and reducing heat capacity by 1.6% compared to pure oil. The nano-lubricant mixture can thus reduce compressor power consumption and effectively dissipate heat [9].

In real-world setups, oil leakage in RAC systems creates a refrigerant-lubricant mixture, altering fluid properties and affecting AC materials, hindering reliability (Han et al., 2017; Zeng et al., 2022). Accordingly, researchers have conducted miscibility tests for refrigerant-lubricant mixtures and compatibility tests on AC materials [10, 11]. In this research, different nano-lubricant-refrigerant combinations were experimentally investigated using compatibility and miscibility tests. The nano-lubricant was 0.025 vol.% SWCNTs dispersed in POE32 oil with refrigerants R-290, and R-410A. Compatibility tests exposed AC materials to lubricant-refrigerant mixtures, while miscibility tests used vapor pressure measurements to evaluate refrigerant miscibility and identify miscibility gaps.

Materials and Methods

Material Compatibility Test. The material compatibility test was conducted using sealed tube experiments. The materials tested included Copper (Cu), Polytetrafluoroethylene (PTFE), Aluminum, and Stainless Steel. The 30 specimens were encapsulated in cylindrical containers filled with a lubricant-refrigerant mixture. These containers were then heated to a temperature of 175 °C under atmospheric pressure and maintained under these conditions for two consecutive weeks. At the end of the exposure period, optical evaluation was used to assess the effects of the different lubricant-refrigerant mixtures on the materials, along with measurements of the length, diameter, and mass of each specimen before and after the experiments.

Miscibility Test. Miscibility was assessed by experimentally measuring the vapor pressure of lubricant-refrigerant mixtures across a wide range of conditions: temperatures from -20 °C to 80 °C and refrigerant mass fractions of 5%, 20%, 50%, and 80%. The pressure was measured using pressure sensors. Miscibility gaps were identified whenever a distinct interface in the liquid solution known as phase separation was observed in the glass testing cell, which was placed inside a transparent temperature-controlled chamber.

Results and Discussion

Material Compatibility Test Results. Based on the dimensional and weight measurements, the changes remained within the range of uncertainty; the relative deviation did not exceed 5%. This was further confirmed by visual observation. In addition, among the 30 tested specimens, two samples exposed to the R-410A/SWCNT nanolubricant mixture showed surface discoloration, likely due to the presence of moisture inside the capsules and possibly due to the polishing effect of the nanoparticles. As shown in Fig. 1, the nanoparticles healed some microcracks in the test sample, most likely due to their mending effect on worn surfaces. Moreover, the addition of nanoparticles increased the surface area available for lubrication, which reduced contact stresses.

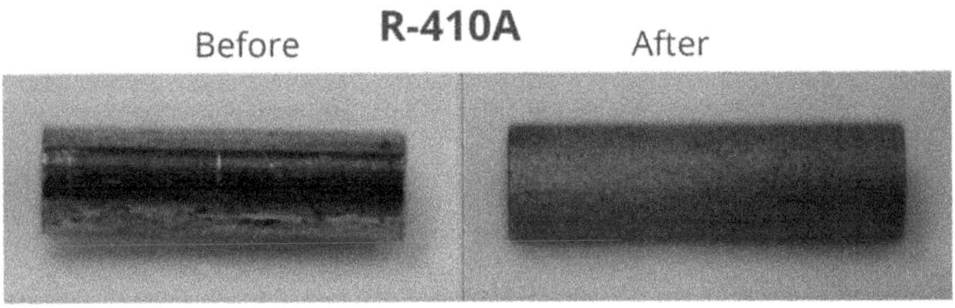

Figure 1: Compatibility test results of a steel material subjected to a nano lubricant-refrigerant mixture.

Miscibility Test Results. The vapor pressure values were measured at each set temperature and refrigerant mass percentage. As expected, the vapor pressure increased with increasing refrigerant content. For the case of refrigerant R-290 with pure lubricant, one anomalous point was recorded at 81.1 °C and 15 wt.%: the pressure was higher than that measured at approximately the same temperature with a higher refrigerant concentration (27 wt.%). This deviation at 15 wt.% of refrigerant stands out as a possible indication of partial phase separation, although it was not visually observed. The visually observed miscibility gaps are represented by cross markers in Fig. 2. As illustrated, visual phase separation was recorded at concentrations below 3 wt.% and at temperatures below 0 °C which far from typical operating ranges of this specific refrigerant. However, these phase separation cases were not observed when nanoparticles were present. From Fig. 3, it can be seen that the presence of nanoparticles had a beneficial effect on the pressure. Unlike the case of the refrigerant with pure lubricant, no unexpected pressure behavior was observed around 80°C between 15 and 27 wt.%. The pressure also rose more monotonically as the refrigerant percentage increased. Nonetheless, whether nanoparticles were present or not, all the empirical data followed a reasonable pattern: as the temperature increased, the data points shifted to the left. This behavior was expected because as the liquid mixture was heated, the refrigerant molecules absorbed the heat and underwent phase change into the vapor phase, resulting in a leaner mixture. There are exactly four sets of data that follow the leftward shift with increasing temperature, as each corresponds to one of the four experimental cells (5 wt.%, 20 wt.%, 50 wt.%, and 80 wt.%). However, the data set at 100 wt.% was calculated using REFPROP [12]. It was also observed that there were minor temperature deviations between the cells, even though they were placed together in the chamber. This is likely due to small differences in heat transfer depending on the location of the cells.

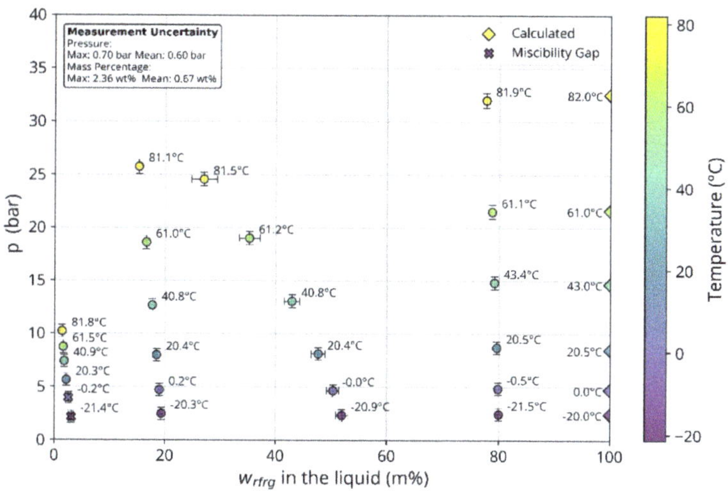

Figure 2: Vapor-pressure of lubricant SL32 + R-290 (baseline).

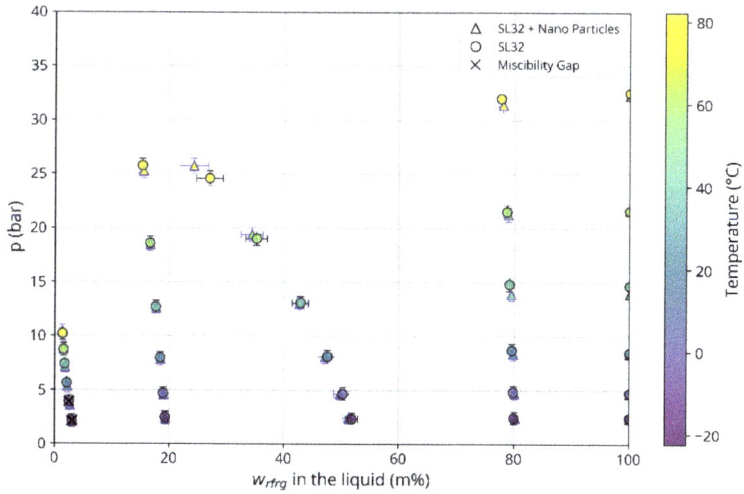

Figure 3: Comparison of lubricant SL32 + R-290 with and without nanoparticles.

For refrigerant R-410A, miscibility gaps were recorded at 75-78 wt.% for the nanoparticle-containing mixtures at 41 and 61°C. However, the mixtures without nanoparticles remained fully miscible in all cases, except at 81.8°C, 56.47 bar, and 55 wt.%. This exception should not be a concern, since the critical temperature and pressure of R-410A are 72.8°C and 48.6 bar, respectively, and such conditions would not be encountered in real-life applications.

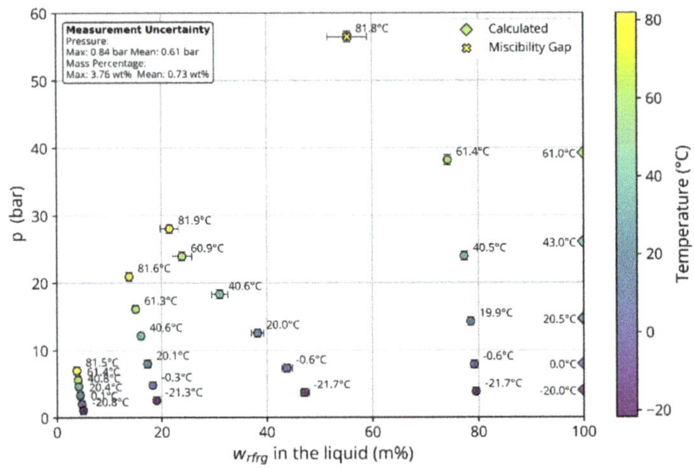

Figure 4: Vapor-pressure of lubricant SL32 + R-410A (baseline).

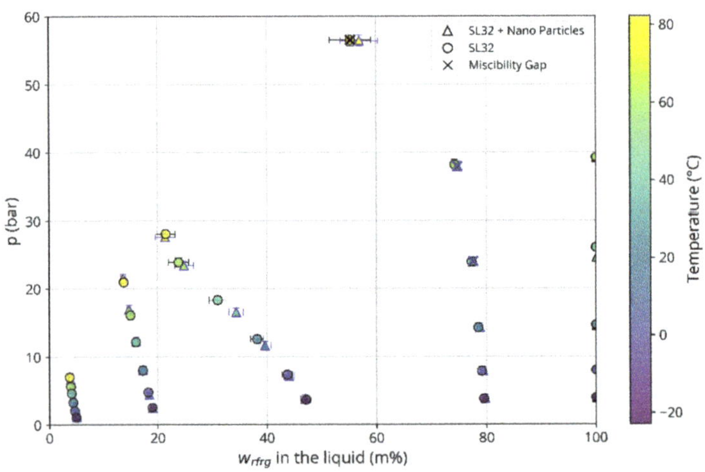

Figure 5: Comparison of lubricant SL32 + R-410A with and without nanoparticles.

Conclusions

In line with environmental regulations and efforts to enhance compressor lubricant properties, this study conducted compatibility and miscibility tests on mixtures of R-290, R-410A, and a nano-lubricant. The goal was to assess their effects on common compressor materials and refrigerant-oil miscibility. Results showed that the nano-lubricant may enhance surface structure through polishing and mending effects, and all refrigerants demonstrated miscibility consistent with Raoult's law behavior. The addition of nanoparticles helped resolve the local instability observed for refrigerant R-290, thereby reducing the probability of oil film buildup. In the end, the promising results from both the compatibility and miscibility tests were reassuring in supporting the use of a combination of eco-friendly refrigerants and enhanced lubrication, all serving the goal of sustainability. Future work should include mathematical modelling of vapor pressure to get more insights at critical points where experimental readings are unstable and to also predict the vapor pressure data outside the testing range. Besides carrying out some material characterization tests like SEM, XRF, and XRD analyses for deeper material insight.

Acknowledgement

The authors would like to thank the Kuwait Foundation for the Advancement of Sciences (KFAS) for their financial support under project no. CN22-35EM-1547. Additional support was provided by Kuwait University General Facility projects (Grant No. GE01/07 and GS03/01) for sample preparation and SEM/EDS measurements, as well as the Kuwait Institute for Scientific Research (KISR) for assistance with characterization of thermophysical properties and physical dispersion stability. Special thank you to the Schaufler Chair of Refrigeration, Cryogenics and Compressor Technology at the Technical University of Dresden (TUD) for their assistance in conducting the compatibility and miscibility experimental tests.

References

[1] "The Montreal Protocol on Substances That Deplete the Ozone Layer," United States Department of State. Accessed: May 03, 2025. [Online]. Available: https://www.state.gov/the-montreal-protocol-on-substances-that-deplete-the-ozone-layer/

[2] W. Y. Park *et al.*, "Ensuring the climate benefits of the Montreal Protocol: Global governance architecture for cooling efficiency and alternative refrigerants," *Energy Res. Soc. Sci.*, vol. 76, p. 102068, Jun. 2021. https://doi.org/10.1016/j.erss.2021.102068

[3] A. Woodcock, "Hydrofluorocarbons, Climate, and Health — Moving the Montreal Protocol beyond Ozone-Layer Recovery," *N. Engl. J. Med.*, vol. 388, no. 26, pp. 2404–2406, Jun. 2023. https://doi.org/10.1056/NEJMp2302197

[4] V. Phung, K. Oka, Y. Honda, and Hijioka, "Daily temperature effects on under-five mortality in a tropical climate country and the role of local characteristics," *Environ. Res.*, vol. 218, p. 114988, Feb. 2023. https://doi.org/https://doi.org/10.1016/j.envres.2022.114988

[5] Y. H. Yau, B. T. Chew, and A. Shaifullah, "(PDF) Thermal comfort temperature range study for workers in a factory in Malaysia," *ResearchGate*, 2012, Accessed: Sep. 26, 2025. [Online]. Available: https://www.researchgate.net/publication/288641683_Thermal_comfort_temperature_range_study_for_workers_in_a_factory_in_Malaysia

[6] "The Future of cooling in Southeast Asia – Analysis," IEA. Accessed: Sep. 26, 2025. [Online]. Available: https://www.iea.org/reports/the-future-of-cooling-in-southeast-asia

[7] A. Khamis, A. Alamshah, A. Ahmad, and A. Rahman, "(PDF) Energy & electricity consumption analysis of Malaysian power demand." Accessed: Sep. 26, 2025. [Online].

Available:
https://www.researchgate.net/publication/224170349_Energy_electricity_consumption_analysis_of_Malaysian_power_demand

[8] Z. Said, S. Rahman, and M. Sohail, "Nano-refrigerants and nano-lubricants in refrigeration: Synthesis, mechanisms, applications, and challenges," *Appl. Therm. Eng.*, vol. 233, p. 121211, Oct. 2023. https://doi.org/10.1016/j.applthermaleng.2023.121211

[9] H. Althalab, N. Ali, S. A. Ebrahim, and A. M. Bahman, "Air conditioning compressor lubricant enhancement using carbon-based nanomaterials for hot climates," *Renew. Sustain. Energy Rev.*, vol. 217, p. 115672, Jul. 2025. https://doi.org/10.1016/j.rser.2025.115672

[10] J. A. Majurin, W. Gilles, and S. J. Staats, "Materials Compatibility of HVACR System Materials with Low GWP Refrigerants," 2014.

[11] A. D. Sumner and A. J. Barthel, "Clearing the Cloudiness on Lubricant-Refrigerant Miscibility," 2021.

[12] E. W. Lemmon, I. H. Bell, and M. L. Huber, "NIST Standard Reference Database 23: Reference Fluid Thermodynamic and Transport Properties-REFPROP, Version 10.0, National Institute of Standards and Technology, Standard Reference Data Program, Gaithersburg, 2018," *NIST*, 2018, Accessed: Oct. 10, 2025. [Online]. Available: https://www.nist.gov/srd/refprop

Nano-enhanced Phase Change Materials (NePCM) for Thermal Energy Storage in Buildings

Ayah EISA[1,a], Aman YADAV[2,b], Zafar SAID[3,c], A. K. PANDEY[2,3,d], Naser ALI[4,e], Ammar M. BAHMAN[1,f*]

[1]Mechanical Engineering Department, College of Engineering and Petroleum, Kuwait University, P.O. Box 5969, Safat 13060, Kuwait

[2]Sunway Centre for Electrochemical Energy and Sustainable Technology (SCEEST), Faculty of Engineering and Technology, Sunway University, No. 5 Jalan Universiti, Bandar Sunway, 47500, Selangor Darul Ehsan, Malaysia

[3]Mechanical and Aerospace Engineering Department, College of Engineering, United Arab Emirates University, Al Ain, 15551, United Arab Emirates

[4]Nanotechnology Applications Program, Energy and Building Research Center, Kuwait Institute for Scientific Research, Safat 13109, Kuwait

[a]ayah.eisa@ku.edu.kw, [b]aman49936@gmail.com, [c]zafar@uaeu.ac.ae, [d]adarshp@uaeu.ac.ae, [e]nmali@kisr.edu.kw, [f]a.bahman@ku.edu.kw

Keywords: Phase Change Material, Thermal Conductivity, Fluctuations

Abstract. With the increasing demand for energy due to the rise in temperature, passive cooling strategies are implemented to help achieve thermal comfort with an optimal energy reduction. One of the effective approaches can be the implementation of Nano-enhanced Phase Change Materials (NePCM) into a building envelope and active HVAC systems to help in temperature control by thermal energy storage. With pure PCM, low conductivity has been a major drawback that caused it to have a slow discharge rate. The inclusion of nanomaterials has dramatically improved its thermal conductivity which led to a shorter discharge time and better performance. Carbon-based nano-additives has shown the greatest enhancement in thermal conductivity while being one of the most expensive option. This review article delves into the characteristics that makes the integration of NePCM into buildings a promising thermal energy storage strategy that can lead to energy savings and reduce temperature fluctuations, especially for arid and hot weather climates without the use of an energy source.

Introduction

Buildings thermal regulation has become a necessity due to the increasing rate of global warming and temperature rise which makes it consumes most of the generated energy; around 40% of the energy used worldwide is consumed by buildings [1]. The need to cutdown on energy consumption and achieve thermal regulation has led to the focus on passive cooling strategies. One of the promising strategies to be incorporated into buildings is Phase Change Materials (PCMs). PCMs can absorb heat at either constant temperature while transitioning from one phase to another (*i.e.*, latent heat) or as sensible heat [2]. Such a mechanism of absorbing and releasing thermal energy without significant temperature changes can help in controlling temperature fluctuations in buildings [3]. By absorbing latent heat during the day and releasing it at night, PCMs can minimize thermal fluctuations which corresponds to less cooling load due to the less demand of HVAC system regulation [4].

There are three main categories of PCMs that are used in thermal applications: organic, inorganic and eutectic compounds [4]. Organic PCMs include paraffin and non-paraffin and are the most versatile in building envelope applications [5]. Inorganic PCMs such as salt hydrates and metallic PCMs have higher energy storage capacity and a higher thermal conductivity that in organic PCMs [6]. Eutectic PCMs can handle the drawbacks of a single-phase change material such as phase separation, unadjustable temperature, and supercooling

One of the drawbacks of pure PCM is its low thermal conductivity which can be as low as 0.2 W/(m-K) for PEG-1000 [7]. Therefore, pure PCM should be altered to minimize the response time which can be achieved by implementing nanotechnology to develop nano-enhanced phase change materials (NePCM) that can speed up its freezing and melting rates [7]. NePCMs has advanced thermal energy storage (TES) especially during its charging and discharging modes [7] which not only enhanced its thermal conductivity but also diminished supercooling. The combination of passive cooling and nanotechnology has a high potential in overcoming the increasing demand in energy due to the accelerating global warming. In this work, its integration into building envelopes is investigated as a viable solution to this issue. Future challenges, along with techno-economic and lifecycle analyses of current literature are studied herein as well to provide a comprehensive overview of the proposed strategy.

Nano-enhanced Phase Change Materials: Fundamentals

Phase change materials (PCM) have emerged in the buildings development research sector as one of the most promising strategies for thermal regulation [8]. Pure PCMs are constrained by their relatively low thermal conductivity which can limit their ability of an optimal rate of heat absorption and release [9]. Therefore, highly conductive nano-additives were introduced, and with the help of the characteristics of the PCMs, highly conductive NePCMs were developed [10].

NePCM can be integrated to a building for the goal of thermal regulation through macro-micro or nano-capsulation and then be used in walls, windows, roofs or the construction materials in between [11], [12], [13], [14]. By the release and absorption of heat, the building's temperature fluctuations can be decreased which could lower the load on the HVAC systems. Innovative integration and material selection such as bio-PCM and hybrid composites has led to better thermal properties and effective performance [10].

During operation, NePCM absorbs heat from the surroundings as the temperature rises during the day and reaches a temperature above its melting point. With the help of the nano-additives, the process of heat absorption can be accelerated in comparison with pure PCMs. On the other hand, when the temperature cools down at night to a temperature below its freezing point, the NePCM solidifies and releases the stored heat into the atmosphere [15]. The mechanism of a NePCM panel can be simply illustrated in Fig. 1. By undergoing such a cycle during the day, the integration of NePCM can help in substantially reducing the cooling load and peak energy demands as it acts as a thermal energy storage system that operates with the use of zero energy [16].

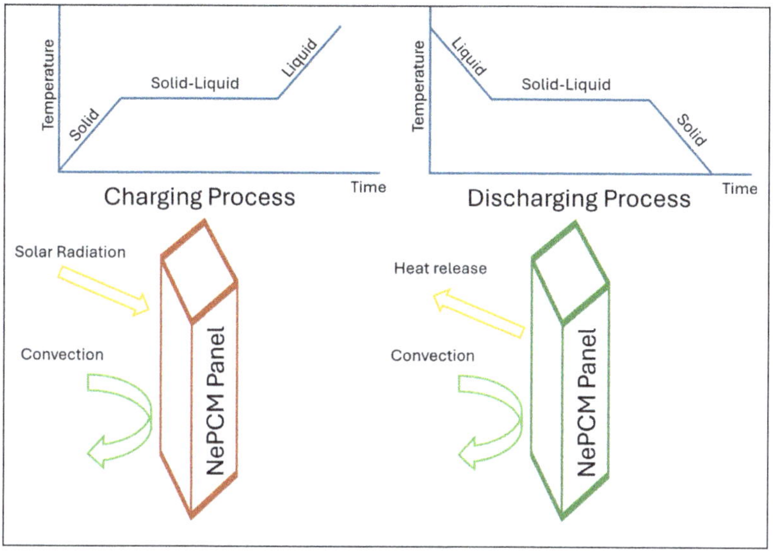

Fig. 1. Mechanism of NePCM thermal storage cycle.

Classification of Nano-enhanced Phase Change Materials

The incorporation of nanotechnology to traditional PCM applications in thermal regulation resulted in an advanced technology that can overcome the limitations of pure PCMs by improving their thermophysical properties [17]. Different materials and different incorporation techniques can lead to different performance results, especially in arid climate conditions.

Base PCMs and Nano-additives. The selection of the base PCM is the first step in the development of NePCMs. PCMs can range from organic, inorganic and eutectic PCM. Organic PCMs such as paraffin and fatty acids are known to be mostly non-toxic, non-corrosive and easily accessible. Their temperature range can go from 15-45°C which makes it applicable for most situations [2]. Although organic PCM has a high latent heat storage capacity [18], its thermal conductivity is very low (~0.2 W/(m-K)) [7], which results in high installation [17]. On the other hand, inorganic PCMs such as salt hydrates can offer a higher latent heat capacity but are prone to sub-cooling and phase separation [19]. Eutectic PCMs, on the other hand, are a mixture of two or more types of PCMs. The development objective of such PCMs is to alter the range of operation temperature to be tailored for a specific application [4]. When selecting a base PCM for a NePCM application for thermal regulation, one key criterion is the melting range of the PCM and the building's thermal fluctuations. Based on the location of the integrated PCM into the building envelope, the target range can vary.

The nano-additives in a NePCM depend on the desired enhancement to meet that a pure PCM lack in a certain application. In general, high-conductivity nanoparticles are incorporated to enhance the thermophysical properties [10]. The most commonly used nanoparticles (NPs) in NePCMs can be categorized into metallic, metal oxide and carbon-based NPs.

Metallic and metal oxide nanoparticles (Cu, CuO Ag, Al_2O_3, TiO_2, SiO_2, ZnO) can have very high thermal conductivity that can decrease super-cooling in PCM. Metallic NPs can increase the overall suspension's thermal conductivity by 46% with minor changes in phase transition temperatures [15]. Carbon-based nanomaterials such as graphene nanoplatelets, carbon nanotubes

(CNTs), graphite can be used as thermophysical properties enhancement to PCMs. Stonehouse and Abeykoon [20] concluded that even at values of less than 5 wt.%, 3D carbon-based NPs can have a thermal conductivity enhancement of around 300% when compared to that of lower dimensional carbon-based materials where the enhancement is around 50%[20].

Encapsulation strategies. In order to integrate the NePCMs into a building envelope, the suspension is encapsulated using an appropriate method that prevents leakage based on the desired application. Macroencapsulation, microencapsulation or shape-stabilization are all methods of encapsulation strategies that can be applied [21]. Macroencapsulation is when large quantities of NePCM is being contained in a shell such as panels, tubes or boxes. This strategy can be incorporated in concrete bricks in building cavities and building envelopes and act as a thermal regulation module [22], [23]. On the other hand, microencapsulation is the ability to contain the NePCM in a protective spherical micron size shell [24]. It was found that microencapsulation can have improved mechanical and chemical stability, extended thermal cycling and can prevent chemical reactions due to the core-shell characteristics [25]. Shape-stabilization of PCMs (SSPCMs) is another form of encapsulation where the PCM is absorbed or embedded into a matrix where it cannot flow when melted. The capillary forces within the porous material keep the liquid from leaking [26]. Shape-stabilization has shown excellent chemical and thermal cycling stability even under crucial thermal reliability tests [27], [28].

The placement of the NePCM panels can have an influence on its performance in thermal regulation. Achaku et al. [12] found that the optimal PCM placements were in an arrange of 341 mm and 356 mm away from the external surface of the wall through which an energy consumption reduction of 34–37% was achieved. Moreover, it was found that the closer the placement is to the inner wall, the more stabilization of indoor temperatures and achieved due to thermal storage of NePCM that causes a reduction in heat transfer. Fig. 2(a) illustrates the heat transfer rate versus the placement of the PCM. As a result, a significant energy consumption reduction corresponds to the ideal range of 341-356 mm as shown in Fig. 2(b). Although at 100 mm away from the external surface shows the most optimal results compared with the baseline (NO PCM), a temperature analysis showed that at the range of 341-356 mm the PCM reaches its phase change temperature which shows how the PCM can act as a thermal buffer by the storage and release of energy to the external wall [12].

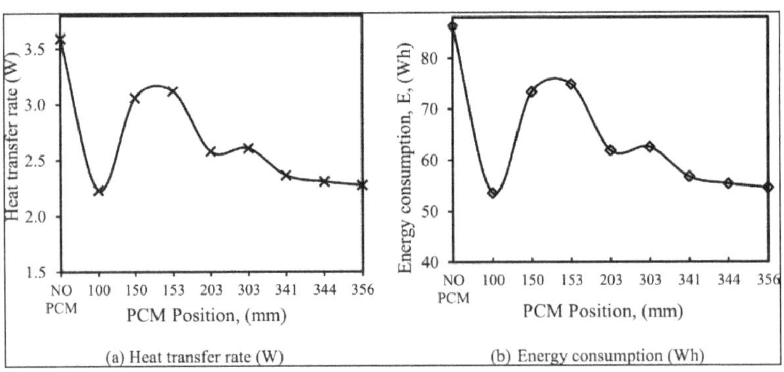

Fig. 2. Effects of PCM placement positions on: (a) Heat transfer rate (b) Energy consumption [12].

Performance Analysis

Thermal Properties. In literature, NePCMs have been characterized based on their thermophysical properties in comparison with pure PCM in a thermal regulation application. Studies have shown that thermal conductivity can be enhanced drastically through the incorporation of nanoparticles. Table 1 shows a comparison of a number of the studies where the level of enhancement of thermal conductivity varies based on the type of nanoparticle used in a NePCM. Carbon-based nanoparticles could achieve higher enhancement percentages due to their high thermal conductivity and low density, making them highly efficient additives [29]. One drawback of carbon-based nanomaterials is their cost. NPs' addition (CuO, CNT, Al_2O_3, TiO_2) to PCM has achieved significant enhancements in thermal conductivity and heat storage density which corresponds to the rapid charging and discharging rate, thereby facilitating a more responsive TES system that is applicable for dynamic thermal load regulation [30].

Table 1: Nanoparticle Influence on Thermal Conductivity for Different PCMs.

Nanoparticle	PCM	Thermal Conductivity Enhancement	Ref.
Copper and graphene oxide	Paraffin RT31	12.5%	[31]
Graphene and SiO_2	Paraffin 98	50%	[32]
GNT and MWCNT	Paraffin RT54HC	40%	[33]
CNT	Methyl stearate (MS)	2319%	[34]
Graphene nanoplatelets	Lauric acid	230%	[35]
Nano-sheet Graphene	Polyurethane–paraffin	97.4%	[36]

In addition to the change in thermal conductivity, the latent heat of NePCM was found to be lower than that of pure PCM. For instance, a reduction of 12% in latent heat capacity after the incorporation of titanium dioxide TiO_2 into paraffin PCM was exhibited due to a proportional relation between the amount of NPs added and the reduction percentage [37]. The overall efficacy of NePCM is also dependent on the phase transition temperature, spatial distribution within the building envelope, and local climatic conditions.

Mechanical properties. The addition of NPs to NePCM have shown that it can have an effect on the mechanical strength and properties when incorporated using a form stabilization encapsulation strategy [38]. The addition of NePCMs to building materials such as concrete or cement mortar can help improves their thermal and mechanical properties which can aid in the thermal regulation of the building [39]. For instance, the addition of Nano-Al_2O_3 to spherical bio-based encapsulated PCM into gypsum not only enhanced thermal properties and lowered subcooling degree, but it also enhanced compressive strength (1.28 kN) attributable to the structural reinforcement [40].

Application

The properties of NePCM can lead to a numerous range of applications in passive thermal energy system (TES). Researchers have investigated its integration into passive systems (building envelope elements) and active systems (HVAC components and energy systems) [23], [41] in order to thermally regulate the indoor temperature of a building. The integration of PCMs has been studied and analyzed in research where studies focusing on Paraffin wax micro-encapsulated into a building envelope have shown enormous results of energy reduction of cooling load up to 20% and excellent thermal insulation performance [42], [43].

An experimental study of a hybrid CPV/T system with a heat exchanger and a spectral splitting unit has incorporated glycerol/silver nanofluid (NF) and dotriacontane/gold (NePCM) which is an example of NePCM integration in active systems. The Dotriacontane/gold NePCM's role was to act as an absorptive filter. By optimizing the system, an annual maximum energy and exergy efficiencies of 123.9 % and 17.08 % were achieved, indicating overall thermal improvements [41].

For passive systems, the NePCM can be integrated to the building walls, roof or windows. Yang et al. [13] evaluated the integration of a hybrid NePCM containing Paraffin with nano Al_2O_3-CuO particles with a ratio of 4:6 found a temperature deviation of 9.31 °C [13]. Moreover, integration of NePCM into roofing systems can aid in temperature and thermal regulation. Abbas and Muthulingam [23] integrated macroencapsulated nano-enhanced PCM (NePCM) in a two-way hollow concrete roof (HCR) where the NePCM samples tested consists of different NPS wt.% of 2 % and 4 % mass fractions of multi-walled carbon nanotubes (MWCNT) and graphene nanoplatelets (GNP). The resulted thermal conductivity as shown in Fig. 3 is improved compared with the baseline PCM of 0.22 W/m K. MWCNT–NePCM exhibited an increase of 27.3 % and 36.4 % for the 2 % and 4 %, respectively. Morever, the GNP–NePCM exhobited higher thermal conductivity values with an increase of 36.4 % and 45.5 %. This increase can be correlated to the two-dimensional planar geometry of GNPs [23].

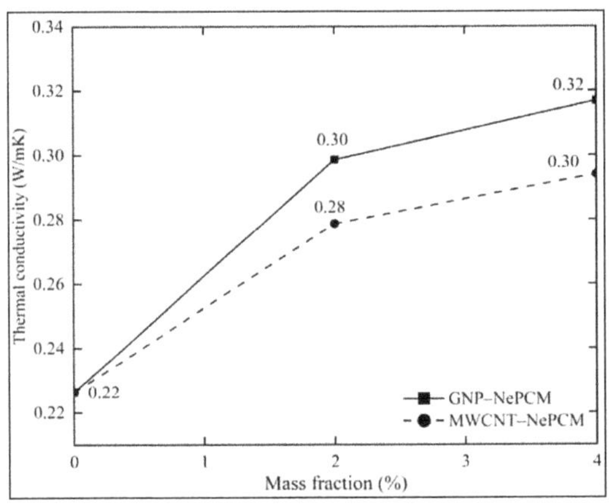

Fig. 3. Enhancement of thermal conductivity of MWCNT–NePCM and GNP–NePCM at different concentrations [23].

Limitations and Challenges

Dispersion and Stability. The incorporation of NePCM in a building makes it harder to maintain frequently, therefore, ensuring long-term stability of the NPs is one of the primary concerns in the development of NePCM. Agglomeration, sedimentation or even clustering can all lead to a non-uniform property throughout the fluid. This can degrade the enhanced thermal properties of the suspension and require researchers to do cycling stability tests in addition to properties characterization. AGuan and Chen [44] prepared a nano-encapsulated PCM with SiO_2 as the shell and conducted 2000 thermal cycles. The results highlighted that the melting enthalpy maintained its high value with only a 6.8 J/g reduction which corresponds to its remarkable cycling stability[44]. In NePCM fabrication, surfactants are used in order to assist stabilization and prevent agglomeration and/or phase separation [45].

Nanomaterials Concentrations. While the addition of nanomaterials into PCM has achieved an increase in thermal conductivity as discussed, higher concentrations can set some drawbacks that will affect its efficacy. A decrease in latent heat was observed with increasing graphene nanoplatelets (GNP) and multiwall carbon nanotubes (MWCNT) hybrid nanoparticles weight fraction in RT-54HC organic PCM [9]. Moreover, Ho and Gao [46]evaluated the incorporation of 10 wt.% of Al_2O_3 NPs to paraffin and resulted in a 6% enhancement of thermal conductivity while a reduction of 13% in latent heat capacity was a major side-effect[46]. Higher nanomaterial content can lead to more agglomeration and sedimentation during the repetitive cooling periods which can negatively affect the thermal properties of the NePCM [47].

Other challenges can be related to the compatibility of the NePCM and the shell materials in all types of encapsulations [23], [43], [44]. Taken together, the limitations of NePCM are being investigated rapidly in the research field. The selection of PCM, nano-additives, method of encapsulation can all help in minimizing the drawbacks of NePCM given their effective performance in thermal regulations.

Conclusion

The growing global energy demand, combined with rising ambient temperatures, has made passive cooling a critical component in building design. The use of nano-enhanced phase change materials (NePCM) in buildings can offer a promising pathway towards thermal regulation and managing temperature fluctuations while reducing energy consumption. By taking advantage of the latent heat capacity of phase change materials (PCMs) and elevating the thermal conductivity through the integration of highly conductive nano-additives, an advanced suspension is developed with enhanced thermal properties and enhanced phase stability. This enhancement can help effectively accelerate the heat transfer rate from and to the NePCM enabling rapid charging and discharging cycles which are a crucial requirement for maintaining uniform interior thermal comfort in buildings. Although the addition of NPs to PCM has led to enhancement in thermal conductivity, high concentrations can have a negative effect on latent heat capacity which leads to the need to investigate the proper concentration that will have the optimal enhancement without compromising any other necessary properties. Moreover, lab-scale studies have investigated NePCM stability in short periods of time compared to the expected lifespan of the NePCM [44]. Future studies should focus more on the dispersion methods of NPs in PCM that can guarantee long-term stability through life-cycle assessments. The lab-scale data in literature show promising thermal energy storage results but more depth into the techno-economic analysis and life-cycle environmental impacts are underexplored for the future of integrating it into buildings. Due to that, scalability and commercialization remain a significant barrier to the future of NePCM within the building sector and thermal regulation. With the prioritization of the thermal conductivity enhancement and performance of NePCM using experimental and numerical approaches [48], [49], [50], [51] , the next step in research should aim to highlight it's compatibility in a real-life scenario through large-scale validation and long-term performance analysis under extreme circumstances that can navigate the development of such strategy in the future of thermal regulation. By addressing these limitations and challenges discussed herein, NePCM technology can be transitioned from being a lab-scale research objective to a reliable commercial building solution for thermal regulation.

Acknowledgement

The authors would like to acknowledge the Kuwait Foundation for the Advancement of Sciences (KFAS) for the financial support under Project Grant No. CN23-35EM-2074.

References

[1] S. Christopher *et al.*, "Renewable energy potential towards attainment of net-zero energy buildings status – A critical review," *J Clean Prod*, vol. 405, p. 136942, Jun. 2023. https://doi.org/10.1016/J.JCLEPRO.2023.136942

[2] R. A. Lawag and H. M. Ali, "Phase change materials for thermal management and energy storage: A review," *J Energy Storage*, vol. 55, Nov. 2022. https://doi.org/10.1016/j.est.2022.105602

[3] X. Q. Zhou *et al.*, "Zinc Oxide Nanoparticles: Synthesis, Characterization, Modification, and Applications in Food and Agriculture," Apr. 01, 2023, *Multidisciplinary Digital Publishing Institute (MDPI)*. https://doi.org/10.3390/pr11041193

[4] K. Jiao, L. Lu, L. Zhao, and G. Wang, "Towards Passive Building Thermal Regulation: A State-of-the-Art Review on Recent Progress of PCM-Integrated Building Envelopes," *Sustainability 2024, Vol. 16, Page 6482*, vol. 16, no. 15, p. 6482, Jul. 2024. https://doi.org/10.3390/SU16156482

[5] D. G. Atinafu, B. Y. Yun, E. E. Kwon, S. J. Chang, and S. Kim, "Unveiling the effect of molecular chain length on the thermal energy storage capacity and transition temperature of alkane-based phase change composites," *Chemical Engineering Journal*, vol. 462, p. 142303, Apr. 2023. https://doi.org/10.1016/J.CEJ.2023.142303

[6] S. A. Mohamed *et al.*, "A review on current status and challenges of inorganic phase change materials for thermal energy storage systems," *Renewable and Sustainable Energy Reviews*, vol. 70, pp. 1072–1089, Apr. 2017. https://doi.org/10.1016/J.RSER.2016.12.012

[7] R. Singh, S. Sadeghi, and B. Shabani, "Thermal conductivity enhancement of phase change materials for low-temperature thermal energy storage applications," *Energies (Basel)*, vol. 12, no. 1, Jan. 2019. https://doi.org/10.3390/en12010075

[8] R. Baetens, B. P. Jelle, and A. Gustavsen, "Phase change materials for building applications: A state-of-the-art review," *Energy Build*, vol. 42, no. 9, pp. 1361–1368, Sep. 2010. https://doi.org/10.1016/J.ENBUILD.2010.03.026

[9] A. Islam, A. K. Pandey, K. Sharma, Y. A. Bhutto, R. Saidur, and D. Buddhi, "Enhancing thermo-physical properties of hybrid nanoparticle-infused medium temperature organic phase change materials using graphene nanoplatelets and multiwall carbon nanotubes," *Discov Mater*, vol. 4, no. 1, pp. 1–19, Dec. 2024. https://doi.org/10.1007/S43939-024-00130-5/FIGURES/9

[10] Z. Said *et al.*, "Nano-enhanced phase change materials: Fundamentals and applications," *Prog Energy Combust Sci*, vol. 104, p. 101162, 2024. https://doi.org/https://doi.org/10.1016/j.pecs.2024.101162

[11] L. Jiang *et al.*, "Experimental and numerical study on thermal performance of phase-change-material cool roofs in summer," *Sustain Cities Soc*, vol. 99, Dec. 2023. https://doi.org/10.1016/j.scs.2023.104936

[12] R. Achaku, L. Li, and Y. Chen, "An experimental investigation of phase change material (PCM)-enhanced cavity walls with integrated windows in office buildings: Optimising energy savings," *Sustainable Energy Technologies and Assessments*, vol. 80, p. 104381, Aug. 2025. https://doi.org/10.1016/J.SETA.2025.104381

[13] X. Yang *et al.*, "Comprehensive performance evaluation of double-glazed windows containing hybrid nanoparticle-enhanced phase change material," *Appl Therm Eng*, vol. 223, p. 119976, Mar. 2023. https://doi.org/10.1016/J.APPLTHERMALENG.2023.119976

[14] K. Biswas, P. Soroushian, and S. S. Shrestha, "Combined experimental and numerical evaluation of a prototype nano-PCM enhanced wallboard," 2013. https://doi.org/10.1016/j.apenergy.2014.02.047

[15] B. Adera, V. R. Ancha, T. Tadiwose, and E. Getahun, "Nano enhanced phase change materials for thermal energy storage system applications: A comprehensive review of recent advancements and future challenges," *International Journal of Thermofluids*, vol. 30, p. 101418, Nov. 2025. https://doi.org/10.1016/J.IJFT.2025.101418

[16] P. K. S. Rathore, N. K. Gupta, D. Yadav, S. K. Shukla, and S. Kaul, "Thermal performance of the building envelope integrated with phase change material for thermal energy storage: an updated review.," *Sustain Cities Soc*, vol. 79, p. 103690, Apr. 2022. https://doi.org/10.1016/J.SCS.2022.103690

[17] S. K. Jha, A. Sankar, Y. Zhou, and A. Ghosh, "Incorporation of Phase Change Materials in Buildings," *Construction Materials*, vol. 4, no. 4, pp. 676–703, 2024. https://doi.org/10.3390/constrmater4040037

[18] F. Lygerakis, C. Gioti, D. Gournis, I. V. Yentekakis, M. Karakassides, and D. Kolokotsa, "Enhancing Building Energy Efficiency with Innovative Paraffin-Based Phase Change Materials," *Energies 2024, Vol. 17, Page 4155*, vol. 17, no. 16, p. 4155, Aug. 2024. https://doi.org/10.3390/EN17164155

[19] Y. Lin, G. Alva, and G. Fang, "Review on thermal performances and applications of thermal energy storage systems with inorganic phase change materials," Dec. 15, 2018, *Elsevier Ltd.* https://doi.org/10.1016/j.energy.2018.09.128

[20] A. Stonehouse and C. Abeykoon, "Thermal properties of phase change materials reinforced with multi-dimensional carbon nanomaterials," *Int J Heat Mass Transf*, vol. 183, p. 122166, Feb. 2022. https://doi.org/10.1016/J.IJHEATMASSTRANSFER.2021.122166

[21] K. Jiao, L. Lu, L. Zhao, and G. Wang, "Towards Passive Building Thermal Regulation: A State-of-the-Art Review on Recent Progress of PCM-Integrated Building Envelopes," *Sustainability 2024, Vol. 16, Page 6482*, vol. 16, no. 15, p. 6482, Jul. 2024. https://doi.org/10.3390/SU16156482

[22] Z. Najam, M. El Alami, and M. Najam, "Experimental Study of Heat Transfer in a Real Scale Building Incorporating PCM in the Air Layer of the Vertical Walls," *Journal of Power and Energy Engineering*, vol. 07, no. 05, pp. 14–25, May 2019. https://doi.org/10.4236/JPEE.2019.75003

[23] P. J. Abass and S. Muthulingam, "Energy-efficient concrete roofs for buildings: Integrating macroencapsulated nano-enhanced PCMs for hot climate adaptation," *Case Studies in Thermal Engineering*, vol. 66, p. 105744, Feb. 2025. https://doi.org/10.1016/j.csite.2025.105744

[24] A. Ismail, J. Zhou, A. Aday, I. Davidoff, A. Odukomaiya, and J. Wang, "Microencapsulation of bio-based phase change materials with silica coated inorganic shell for thermal energy storage," *Journal of Building Engineering*, vol. 67, p. 105981, May 2023. https://doi.org/10.1016/J.JOBE.2023.105981

[25] J. Pereira, A. Moita, and A. Moreira, "An Overview of the Nano-Enhanced Phase Change Materials for Energy Harvesting and Conversion," *Molecules 2023, Vol. 28, Page 5763*, vol. 28, no. 15, p. 5763, Jul. 2023. https://doi.org/10.3390/MOLECULES28155763

[26] T. Liu *et al.*, "Thermoregulation, rheological properties and modification mechanism of asphalt modified with PUSSPCMs," *Constr Build Mater*, vol. 372, p. 130763, Apr. 2023. https://doi.org/10.1016/J.CONBUILDMAT.2023.130763

[27] H. Gao *et al.*, "Nanoconfinement effects on thermal properties of nanoporous shape-stabilized composite PCMs: A review," *Nano Energy*, vol. 53, pp. 769–797, Nov. 2018. https://doi.org/10.1016/J.NANOEN.2018.09.007

[28] C. Cárdenas-Ramírez, M. A. Gómez, F. Jaramillo, A. G. Fernández, and L. F. Cabeza, "Thermal reliability of organic-organic phase change materials and their shape-stabilized composites," *J Energy Storage*, vol. 40, p. 102661, Aug. 2021. https://doi.org/10.1016/J.EST.2021.102661

[29] G. Yang, Y. J. Yim, J. W. Lee, Y. J. Heo, and S. J. Park, "Carbon-Filled Organic Phase-Change Materials for Thermal Energy Storage: A Review," *Molecules 2019, Vol. 24, Page 2055*, vol. 24, no. 11, p. 2055, May 2019. https://doi.org/10.3390/MOLECULES24112055

[30] D. Christopher Selvam *et al.*, "Advances in nano-enhanced phase change materials and hybrid thermal energy storage systems: Paving the way for sustainable energy solutions," *Results in Engineering*, vol. 27, p. 105729, Sep. 2025. https://doi.org/10.1016/J.RINENG.2025.105729

[31] R. Calotă *et al.*, "A Novel Concept of Nano-Enhanced Phase Change Material," *Materials*, vol. 17, no. 17, Sep. 2024. https://doi.org/10.3390/ma17174268

[32] R. Bharathiraja, T. Ramkumar, M. Selvakumar, and N. Radhika, "Thermal characteristics enhancement of Paraffin Wax Phase Change Material (PCM) for thermal storage applications," *Renew Energy*, vol. 222, Feb. 2024. https://doi.org/10.1016/j.renene.2024.119986

[33] A. Islam, A. K. Pandey, K. Sharma, Y. A. Bhutto, R. Saidur, and D. Buddhi, "Enhancing thermo-physical properties of hybrid nanoparticle-infused medium temperature organic phase change materials using graphene nanoplatelets and multiwall carbon nanotubes," *Discov Mater*, vol. 4, no. 1, Dec. 2024. https://doi.org/10.1007/s43939-024-00130-5

[34] D. K. Döğüşcü, O. Güler, G. Hekimoğlu, A. Sarı, and O. Gencel, "High-performance CNT-integrated PolyHIPE networks enabling efficient PCM encapsulation via emulsion templating for advanced thermal energy storage," *J Energy Storage*, vol. 130, p. 117465, Sep. 2025. https://doi.org/10.1016/J.EST.2025.117465

[35] S. Harish, D. Orejon, Y. Takata, and M. Kohno, "Thermal conductivity enhancement of lauric acid phase change nanocomposite with graphene nanoplatelets," *Appl Therm Eng*, vol. 80, pp. 205–211, Apr. 2015. https://doi.org/10.1016/J.APPLTHERMALENG.2015.01.056

[36] B. Praveen, S. Suresh, and V. Pethurajan, "Heat transfer performance of graphene nano-platelets laden micro-encapsulated PCM with polymer shell for thermal energy storage based heat sink," *Appl Therm Eng*, vol. 156, pp. 237–249, Jun. 2019. https://doi.org/10.1016/J.APPLTHERMALENG.2019.04.072

[37] L. Ravasio, M. A. Hayat, R. K. Calay, R. Riise, and Y. Chen, "An experimental study on thermophysical properties of nano- TiO 2 -enhanced phase change materials for cold climate applications," *J Therm Anal Calorim*, vol. 149, no. 6, pp. 2549–2560, Mar. 2024. https://doi.org/10.1007/S10973-023-12859-X/TABLES/11

[38] Y. Huang, A. Stonehouse, and C. Abeykoon, "Encapsulation methods for phase change materials – A critical review," *Int J Heat Mass Transf*, vol. 200, p. 123458, Jan. 2023. https://doi.org/10.1016/J.IJHEATMASSTRANSFER.2022.123458

[39] S. Akhavan Mohseni, M. Akhavan Mohseni, A. Babapoor, and Z. Rahimi-Ahar, "Application of nano-phase change materials in thermal energy storage: An engineering approach to optimizing efficiency, sustainability, and industrial economics – A review," *J Energy Storage*, vol. 134, p. 118116, Oct. 2025. https://doi.org/10.1016/J.EST.2025.118116

[40] A. El Majd *et al.*, "Advanced gypsum/PCM composites incorporating biopolymer-encapsulated phase change materials for enhanced thermal management in buildings," *Constr Build Mater*, vol. 451, p. 138872, Nov. 2024. https://doi.org/10.1016/J.CONBUILDMAT.2024.138872

[41] J. Aboueian, A. Shahsavar, and H. R. Askarifard Jahromi, "Parametric assessment of a building-integrated PV/T system with a Nanofluid/NEPCM spectral splitter," *Journal of Building Engineering*, vol. 103, p. 112042, Jun. 2025. https://doi.org/10.1016/J.JOBE.2025.112042

[42] Z. Wu, Y. Xu, and B. Šavija, "Mechanical Properties of Lightweight Cementitious Cellular Composites Incorporating Micro-Encapsulated Phase Change Material," *Materials 2021, Vol. 14, Page 7586*, vol. 14, no. 24, p. 7586, Dec. 2021. https://doi.org/10.3390/MA14247586

[43] N. Lajimi, N. Ben Taher, and N. Boukadida, "Numerical simulation of heat and mass transfer of a wall containing micro-encapsulated phase change concrete (PCC)," *Front Environ Sci*, vol. 10, Jan. 2023. https://doi.org/10.3389/fenvs.2022.973725

[44] J. Guan and M. Chen, "Nanoencapsulation of binary fatty acids for high-stability phase change materials: Synergistic synthesis and thermophysical characterization," *Energy*, vol. 335, p. 138125, Oct. 2025. https://doi.org/10.1016/J.ENERGY.2025.138125

[45] W. Wei, F. Bai, and H. Fan, "Surfactant-Assisted Cooperative Self-Assembly of Nanoparticles into Active Nanostructures," *iScience*, vol. 11, pp. 272–293, Jan. 2019. https://doi.org/10.1016/J.ISCI.2018.12.025

[46] C. J. Ho and J. Y. Gao, "Preparation and thermophysical properties of nanoparticle-in-paraffin emulsion as phase change material," *International Communications in Heat and Mass Transfer*, vol. 36, no. 5, pp. 467–470, May 2009. https://doi.org/10.1016/J.ICHEATMASSTRANSFER.2009.01.015

[47] J. Paul, K. Kadirgama, M. Samykano, R. Saidur, A. K. Pandey, and R. V. Mohan, "An Updated Review on Low-Temperature Nanocomposites with a Special Focus on Thermal Management in Buildings," *Energy Engineering*, vol. 119, no. 4, pp. 1299–1325, May 2022. https://doi.org/10.32604/EE.2022.019172

Nano Cool Paint as a Passive Cooling Strategy for Net-Zero Energy Buildings (NZEBs)

Ayah EISA[1,a], Aman YADAV[2,b], Zafar SAID[3,c], A. K. PANDEY[2,3,d], Naser ALI[4,e], Ammar M. BAHMAN[1,f*]

[1]Mechanical Engineering Department, College of Engineering and Petroleum, Kuwait University, P.O. Box 5969, Safat 13060, Kuwait

[2]Sunway Centre for Electrochemical Energy and Sustainable Technology (SCEEST), Faculty of Engineering and Technology, Sunway University, No. 5 Jalan Universiti, Bandar Sunway, 47500, Selangor Darul Ehsan, Malaysia

[3]Mechanical and Aerospace Engineering Department, College of Engineering, United Arab Emirates University, Al Ain, 15551, United Arab Emirates

[4]Nanotechnology Applications Program, Energy and Building Research Center, Kuwait Institute for Scientific Research, Safat 13109, Kuwait

[a]ayah.eisa@ku.edu.kw, [b]aman49936@gmail.com, [c]zafar@uaeu.ac.ae, [d]adarshp@uaeu.ac.ae, [e]nmali@kisr.edu.kw, [f]a.bahman@ku.edu.kw

Keywords: Passive Cooling, Reflective Coating, Reflectance, Nano Cool Paint, Energy Reduction, Nanomaterial

Abstract. Energy consumption related to Air Conditioning systems in buildings has been rising rapidly due to increasing demand, climate change and the constant increase in temperature. An approach to address and minimize the consumption and its consequences is through the implementation of Net-Zero Energy Buildings (NZEBs). One technique to be used in a NZEB is radiative passive cooling and it has shown potential in overcoming such a challenge by aiding in thermal regulation which reduces energy consumption. The ability to formulate a radiative passive cooling coating with the help of nanotechnology results in a Nano Cool Paint (NCP) that can scatter incoming radiation and emit thermal radiation to the atmosphere. This work reviews the incorporation of nano-enhanced radiative passive cooling in exterior paint coatings on building facades and roofs based on available literature.

Introduction

In the recent years, studies have shown that buildings account for up to 40% of the global energy consumption [1] while relying on conventional energy sources that trigger global warming and causes overwhelming peak load demands in electricity. With the rise of population growth, urbanization and global temperatures, the cooling demand is expected to triple by 2050 if left unchecked [2]. This increasing demand required researchers to focus on energy reduction solutions rather than more efficient cooling units which has led to the implementation of Net-Zero Energy Buildings (NZEBs) due to its conceptual framework for balancing the energy being consumed and generated in a building [3]. Achieving NZEBs in regions with high demand can aggressively help in the reduction of energy consumption [4]. NZEBs concept revolves around two main designs integrated into the building, the active and passive designs. The active design corresponds to the ability of the building to provide as much power as it consumes while the passive design is the incorporation to building's envelope and structure to passively thermally regulate the building [5].

In order to implement it, passive cooling strategies – among other energy reduction strategies – are incorporated to reduce the cooling load especially in harsh arid climates. Passive cooling strategies can help in achieving high-efficiency thermal regulation with the need of zero energy

[6]. An example of a passive cooling technique is radiative cooling (PRC) such as reflective coatings. A simple exterior reflective coating consists of an acrylic emulsion binder, fillers and functional color pigments [7]. Recent studies have shown that incorporating nanotechnology to formulate the exterior coating can help in improving the properties that enhance radiative cooling which leads to the development of Nano Cool Paint (NCP) [8], [9]. NCP's enhanced properties are solar reflectance, thermal emissivity, UV resistance and durability [10], [11], [12]. Solar reflectance maximization is the main property that allowed painted surfaces to achieve sub-ambient temperatures under direct sunlight [13]. By carefully selecting the type of nanoparticles and binder, a reflective, durable exterior paint can be developed for building in hot weather conditions. Different nanomaterials and binders have been shown to give various results where some advanced thermo-optical properties while compromising mechanical properties. The goal to implement such a technique in NZEB leads to the need to investigate the development of NCPs with a high reliability factor.

Background and Mechanism

Nano Cool Paint (NCP) is a technique of radiative cooling applied on the exterior wall of a building that helps reflect incoming solar radiation and emit thermal infrared radiation to the atmosphere within the atmospheric transparency window (8 μm to 13 μm)[14]. To increase its effectiveness, it is formulated to have high solar reflectance and high emissivity. Solar reflectance refers to the percentage of sunlight radiation that is being reflected by the surface while thermal emissivity is a percentage of the capacity of thermal irradiation into the atmosphere [15]. The role of these two properties can help in maintaining the exterior surface temperature from gaining heat. Therefore, the heat flux going into the building is reduced and thermal regulation is accomplished with the need of zero energy [16]. Fig. 1 illustrates the mechanism of NCP under direct sunlight.

The incorporation of nanomaterials has helped to push solar reflectance and emissivity beyond what conventional coatings have achieved. Studies have shown that NCP has the potential to reflect visible light and near-infrared (NIR) wavelengths that can reduce the urban heat island phenomenon [17]. This can lead to a coating capable of sub-ambient temperature cooling under direct sunlight that contributes to the energy reduction of the building [18].

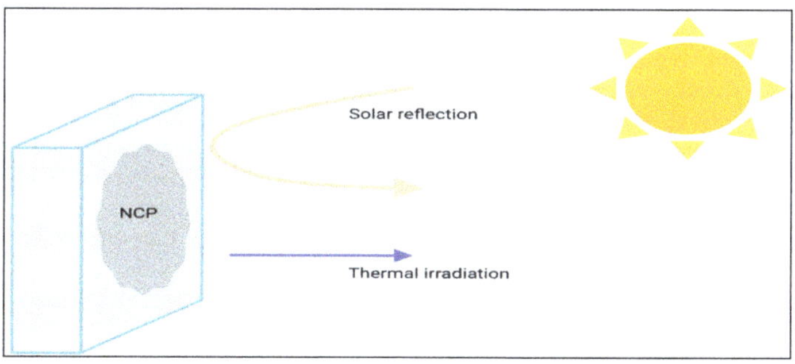

Fig. 1: Mechanism of Nano Cool Paint (NCP) under direct sunlight.

Materials and Methods

Just like any other nano fluid, Nano Cool Paint (NCP) consists of two primary materials which are a base fluid and nanomaterials. The base fluid is usually a binder or a paint emulsion where nanoparticles of any type can be dispersed in before application. In order to achieve optimum results, material selection is done based on the characteristics intended. The aim is to fabricate a paint that can scatter light due to its high reflectance and emit thermal irradiation. Moreover, the developed paint should not lose its durability and have reliable mechanical characteristics.

Nanomaterials. In the field of radiative coating, various nanoparticles have been studied and tested based on different criteria and climate conditions. Studies have shown that the developed NCPs can lead to a highly reflective surface which can also achieve sub-ambient temperatures under direct sunlight [19]. One of the most popular nanomaterials in NCP is metal oxides [20]. For instance, high-refractive-index metal oxides like titanium dioxide (TiO_2) have been used in the paint formulations and can provide strong scattering, but its moderate bandgap (~3.3 eV) leads to some UV absorption and potential photocatalytic degradation of organic binders [21]. TiO_2 is inexpensive and non-toxic, making it abundant in white paint. One of the downsides of nano-TiO_2 is that it can be photo catalytic which makes it need to be surface treated to minimize chemical reactivity [22]. Another common metal oxide is zinc oxide (ZnO). Although studies have shown that ZnO-based NCP have better mechanical properties and durability, its light scattering performance was not as good as TiO_2 which have a higher reflectance enhancement of up to 50% compared to 28% for ZnO [20], [23]. Researchers have also explored metal oxides with a higher bandgap such as aluminum oxide Al_2O_3 and zirconium oxide ZrO_2. A higher bandgap can result in better light scattering and reduce the solar absorption which can lead to the sub-ambient temperature reduction [24], [25]. In addition to the higher reflectance percentages, Al_2O_3 and ZrO_2 showed an enhancement to the mechanical properties in comparison with other NPs. Al_2O_3 addition can lead to an anti-corrosion and hydrophobic surface which can be useful in some harsh applications [26], [27]. Similarly, ZrO_2 can enhance the mechanical properties of the NCP, thermal stability, and anti-corrosion ability [28], [29].

Another category of nanomaterials used in NCP is metal sulfates. Metal sulfates such as barium sulfate ($BaSO_4$) and calcium carbonate ($CaCO_3$) are being used in commercial paint emulsions due to their ability to act a pigment in addition to their high reflectance performance of up to ~98% [19]. Lastly, silicon-based nanoparticles (SiO_2) are known for their broadband optical scattering, strong thermal emission, and improved coating stability against weathering [30], [31]. Silicon-coated oxides and hollow spheres nanoparticles have also been investigated and found that it can lead to enhanced dispersion and thermo-optical performance [32]. In addition to the low solar absorption, hollow spheres can act as thermal insulators due to the air core.

NCP Formulation. The development of NCPs follows a selection criterion based on the type of binder (base fluid) and nanoparticles (fillers) to be dispersed for enhancement. The most common type of paint to be used on exterior facades is water-based emulsions due to its physical properties [33]. Other binder options can range from expensive silicon resin to organic binders [21], [34]. Although the base fluid is important to show durability and weather resistance, NCPs light scattering characteristics are gained from the addition of the NPs. The material selection of both is based on the overall intended characteristics. The selection of the type of binder is crucially connected to the type of NPs chosen to ensure their compatibility with the intended application. The selected materials will define and conclude other main criteria to consider in formulation such as the particle volume concentration (PVC) and coating thickness. In addition, the dispersion of the NPs is highly dependent on the type of binder used as concluded by Qazi et al. [35] where different binders showed different agglomeration and clustering results in a SEM test. Well dispersed NPs ensures that a coating will have uniform optical and mechanical properties that can help in minimizing degradation in performance and inaccurate results in testing. In addition to optical performance, binder selection plays an important role in determining the durability of NCP under harsh environmental conditions. Water-based exterior coatings are known for their hydrophobic finish but can get affected by the prolonged UV exposure [36]. Therefore, researchers such as Song et al. used a surface modifier perfluorooctyl trichlorosilane (PFOTS) as their binder which aided in overcoming the drawback of polymers' reactivity to UV light [37]. Silicon-based binders can also provide improved UV-resistance and weather stability which can significantly enhance the long-term durability of NCP [38].

Discussion

The key properties that NCP acquire are related to its optical and thermal performance. High solar reflectance and high thermal emissivity are crucial to maximizing the coating's performance. Light-colored commercial exterior paints can have a solar reflectance of up to 80% [39]. Nano-enhanced coating has shown a great increase in solar reflectance in literature. For instance, when barium sulfate was added to a binder at a high concentration, solar reflectance of up to 97.6% and IR emissivity above 0.95 was achieved [40]. Different NPs have shown different results in enhancing optical and thermal performance. Another example is hollow Yttrium-Oxide Spheres (HYSs), which were able to enhance the reflectance and emissivity to reach superior cooling [41]. Hollow silicon-based spheres resulted in stronger reflectance in comparison with solid spheres which resulted in a temperature drop of 5.26 °C [42].

Low thermal conductivity can make NCPs induce a higher thermal insulation from the external heat [43]. Incorporating nanomaterials with low thermal conductivity has shown an increase in thermal performance. For instance, silica aerogel's significant low thermal conductivity (<0.05 W/mK) can outperform conventional insulators used in a building envelope [44].

Mechanical properties remain as important as the other properties of any exterior coating. With the focus being on the thermal and optical advancement of NCP, the neglect of focus on the adhesion and durability of the coating has led to NCPs that may not withstand harsh environmental conditions [45]. The addition of nanomaterials to a paint emulsion can negatively affect its adhesion and durability if not done at the appropriate ratio. A calcium carbonate-based NCP's adhesion strength can be maximized with the addition of ZnO [46].

To evaluate the performance of NCP in NZEBs, its passive cooling characteristics are taken into consideration. Solar reflectance and thermal emissivity that will contribute to exterior surface temperature reduction can thermally regulate the interior of the building. For instance, in Argentina, a reduction of 18°C on the exterior surface of a roof coated with NCP resulted in 5°C reduction to its internal side which corresponded to 3°C reduction in the indoor ambient temperature [47].

Application

Field experiments and implementation of NCP have shown that lower indoor temperatures, sub-ambient cooling and reduced power demand can be achieved due to its enhancement of solar reflectance and thermal emittance into the atmosphere resulting in a passive cooling approach that can be incorporated into NZEBs [24], [48], [49]. Although lab scale studies are available, especially in recent years, real-life applications are limited in hot arid climate conditions.

A simulation study of a building in Al Mizhar, UAE, was conducted using IES VE software and a reduction of energy of 47% was concluded [50]. Another simulation of different regions of the US showed that hot climate conditions resulted in more electricity savings of up to 9.31 kWh/m^2 [51]. A double-layer coatings composed of densely packed TiO_2 and SiO_2 or SiC nanoparticles, can cool an Al foil up to 8 °C at daytime and 5 °C at nighttime in Shanghai, China. It also helps to cool the black surface for more than 30°C at daytime [52]. An experimental performance evaluation of a porous polymer radiative coating (PPRC) reflectance was found to be 93% in the visible wave band (0.38–0.78 μm) which shows that optical capability of the doped nano/micro-porous structures where a drop of temperature and cooling load of up to 7.6°C and 85.9 W/m^2 was achieved at noon and up to 12.8 °C and 120.9 W/m^2 at night under moderate relative humidity and a clear sky as shown in Fig. 2 [53]. Although the performance of this developed NCP was highly affected by Shanghai humid weather, it still shows evidence of the role radiative cooling can contribute in NZEBs. A similar study consists of TiO_2 and SiO_2 as NPs was found to result in a solar reflectance of 92.68 % and surface temperature reduction of 9.5 °C [11].

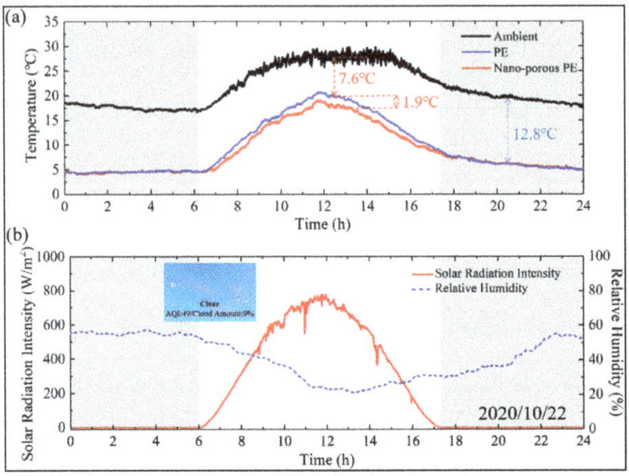

Fig. 2: (a) The temperature variation of the PPRC with PE film (blue) or nano-porous PE film (red) (b) The weather conditions during the test [53].

Different types of nanomaterials can result in different cooling results. Nanoporous Mg-doped SiO_2 nanoparticles were used to develop a coating that has resulted in tunable infrared emissivity and 86.2% solar reflectance. The coating exhibited high radiative cooling performance and a temperature reduction of 17.8 °C lower than the control [54]. A solar reflectance of 97.6% was achieved by the formulation of an NCP that consists of properly designed $BaSO_4$, $CaCO_3$ and SiO_2. The results showed a surface temperature of 5.5 C lower than commercial white paints and a theoretical RC power of 119.3 W/m^2 and 94.3 W/m^2 at nighttime and daytime, respectively [55]. Moreover, nanomaterials made of broken glass bubbles (BGB) were incorporated in nanoporous polyethylene (NanoPE) film resulted in solar reflectance of 97.3 % and 8.5 °C temperature reduction at noon and up to a 14.1 °C reduction at night. Compared with the glass bubbles (GB) and broken glass bubbles (BGB), the NanoPE film can boost the cooling performance of the coating during the whole day significantly as shown in Fig. 3 [56].

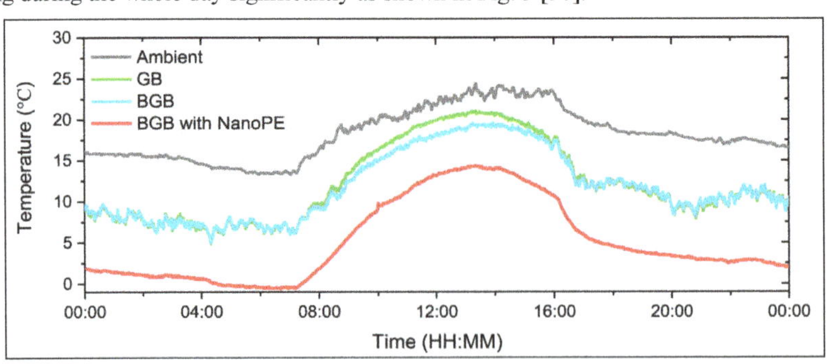

Fig 3: Temperature of ambient air/GB/BGB/BGB with NanoPE over a 24-h period in a field test in Hong Kong [56].

Limitations and Challenges

Aesthetic. High reflectance values have mostly been achieved with ultra-white coatings using white fillers which can correspond to more light scattering [13], [57]. In the buildings sector, architects and individuals often prefer colored exterior surfaces for aesthetic purposes. Recent research is investigating the possibility of incorporating color without compromising the properties of NCP. For instance, an optimized TiO_2 to reduce glare and brightness has led to negatively affecting the energy reduction by 7% with only a 30% difference in brightness [58]. Ultra-white NCP coatings can achieve a solar reflectance of up to 0.98 [40]; however, darker pigments can degrade such enhancements due to the increased absorption in the visible spectrum. A bilayer radiative coating technique with different pigments such as yellow, green and red was tested and a gradual decrease in solar reflectance is observed as the color darkens as shown in Table 1. A recent study found that fluorescent colored radiative cooling coatings present a scalable alternative to ultra-white coatings with maintaining a reflectance of 95%. The downside of such coatings is their high cost and degradation problems [59]. Click or tap here to enter text.

Table 1. Effect of coating color on solar reflectance of nano cool paint (NCP) coatings.

Coating color	Solar Reflectance	Ref.
Ultra-white	0.97	[40]
Yellow	0.91	
Green	0.89	[60]
Red	0.84	

Aging. Degradation of the mechanical, optical and thermal properties is a great limitation with existing NCPs. While continuous studies have found that various materials can reach high IR emissivity values in the longwave infrared spectrum [61]. However, reflectance deterioration might occur from aging. Since the coating is applied on an exterior wall, deterioration is prone to happen due to the exposure to the open atmosphere. Surface soiling due to the buildup of dust or other particles can cause the aged reflectance to decrease significantly [62]. Studies have concluded that most surfaces lose up to 25% of their initial solar reflectance index after 3 years of being applied if left unwashed [63]. To maintain long-term reliability and cooling performance in NZEBs applications, maintenance protocols should be considered. Scheduled or nature-assisted surface cleaning can help in maintaining the solar reflectance of the coating. Mesoporous silica aerogel (SA) infused coatings has shown great potential in enhancing the reflectance to the solar irradiance by up to 15% in the visible and near-infrared ranges (0.3-2.5 μm) while maintaining the surface hydrophobicity and self-cleaning characteristics which increases its durability against soiling and contaminations [61].Another factor that can lead to paint degradation is Ultraviolet (UV) radiation. Continuous exposure to UV radiation has caused a huge drop in energy savings of about 3.8 $kWhm^{-2}$ due to aging [64]. Although the addition of NPs such as ZnO has shown improved UV resistance, continuous exposure makes it a short-term solution in buildings sector [12]. Moreover, humid climate conditions can exhibit a form of degradation and yellowing in paints which makes it not applicable for tropical regions [65]. With the increase in relative humidity in tropical regions, the water vapor content in air increases and reduces the amount of heat radiated from the surface due to it being a major absorber of infrared radiation[66].For instance, radiative coolers that performed effectively in a dry region failed to achieve similar sub-ambient cooling performance in a hot humid region [49]. A study conducted in Malaysia's tropical climate has found that although the net cooling power of the paints had an inverse relationship with relative humidity, the sub-ambient cooling was achievable with coatings having a pigment PVC% at 60% and 70% for barium sulfate and calcium carbonate, respectively [19]. Moreover, surface moisture and condensation can degrade the solar reflectance by acting as a barrier between the NCP coating and the atmosphere which can be addressed by surface modification of NCP by making it more hydrophobic [67]. Researchers have examined and proposed different strategies such as high PVC%, surface modifications and structural strategies [19], [68], [69]. For instance, dual-layer coating has aided in sub-ambient cooling in Shanghai, China's humid climate [69].

Conclusion

The increase in demand toward Net-Zero Energy Buildings (NZEBs) due to the global climate conditions and electricity demands has led to recognizing the importance of incorporating innovative passive cooling strategies to its design. Nano-enhanced materials that can be integrated into the building's envelope can aid in thermal regulation and decrease the load put on the HVAC systems. Such approaches can be passive radiative cooling such as Nano Cool Paint (NCP) which is promised to lead toward a sustainable, Net-Zero energy future due to its proven efficacy in thermal regulation. NCP's high solar reflectance and emissivity contribute to radiative cooling that can help in energy and exterior surface temperature reduction. The application of NCP does not vary a lot from the standard painting processes which makes it a simple addition to either new or existing building that does not require any sort of structural modification. Existing studies and research have focused on the optical properties of NCPs as a fresh coat/sample. However, to be integrated into a NZEB and achieve its promised purpose, the long-term performance must be investigated as well. NCP outcome can vary considerably by climate and weather conditions which can have a negative effect on its mechanical and optical properties. Humid climates were found to face limitations due to existing water vapor that can interfere with the mechanism of NCP. In addition, accumulation of dust on exterior surfaces in dessert climates can decrease the solar reflectance significantly affecting its performance. Proper maintenance can help in securing the optimal efficacy of NCPs as a part of NZEB operational life. Further investigation in the material selection and application procedures will emphasize the need for research to start focusing more on the mechanical properties' enhancement of NCPs such as self-cleaning, UV-resistance or adhesion. Succeeding to meet an enhancement in optical properties while maintaining or enhancing mechanical properties can lead to securing a long-term reliability of NCPs to be used as a part of a NZEB strategy. In addition, the ability to include a wider color range of darker colors will increase its demand in the architectural perspective.

Acknowledgement

The authors would like to acknowledge the Kuwait Foundation for the Advancement of Sciences (KFAS) for the financial support under Project Grant No. [CN23-35EM-2074].

References

[1] A. Mirzazade Akbarpoor and A. Haghighi Poshtiri, Cooling urban buildings with a combined meander EAHE–DEC system: Energy and water consumption analysis, *Journal of Building Engineering*, vol. 108, p. 112850, Aug. 2025. https://doi.org/10.1016/J.JOBE.2025.112850

[2] "The Future of Cooling Opportunities for energy-efficient air conditioning Together Secure Sustainable," 2018. Accessed: Sep. 21, 2025. [Online]. Available: www.iea.org/t&c/

[3] Q. Wang, H. Wu, S. Bai, and L. Wang, Towards net zero energy buildings: Thermal and energy performance of CdTe PV glazing in tropical island regions, *Energy Build.*, vol. 347, p. 116271, Nov. 2025. https://doi.org/10.1016/J.ENBUILD.2025.116271

[4] M. Eksi, R. T. Akarsu, and M. Ozcan, Net-zero energy building transformation: Techno-economic and environmental evaluation in the Mediterranean Region, *Environ. Dev. Sustain.*, 2025. https://doi.org/10.1007/s10668-025-06389-9

[5] M. R. Maghami, A. Maghoul, K. F. Thang, and S. Sundaram, Net zero energy buildings: An overview of passive and active designs, *Energy Reports*, vol. 14, pp. 1327–1348, 2025. https://doi.org/https://doi.org/10.1016/j.egyr.2025.07.028

[6] A. A. Y. Freewan and A. A. Y. Freewan, Advances in Passive Cooling Design: An Integrated Design Approach, *Zero and Net Zero Energy*, Sep. 2019. https://doi.org/10.5772/INTECHOPEN.87123

[7] J. Lei, K. Kumarasamy, K. T. Zingre, J. Yang, M. P. Wan, and E. H. Yang, Cool colored coating and phase change materials as complementary cooling strategies for building cooling load reduction in tropics, *Appl. Energy*, vol. 190, pp. 57–63, 2017. https://doi.org/10.1016/j.apenergy.2016.12.114

[8] A. Felicelli *et al.*, Efficient radiative cooling of low-cost BaSO4 paint-paper dual-layer thin films, *Nanophotonics*, vol. 13, no. 5, pp. 639–648, Mar. 2024. https://doi.org/10.1515/NANOPH-2023-0642/DOWNLOADASSET/SUPPL/J_NANOPH-2023-0642_SUPPL_001.DOCX

[9] J. Wang, D. Yan, L. Ding, J. An, and M. Santamouris, On the development of super cool paints for cooling purposes, *Solar Energy*, vol. 296, p. 113592, Aug. 2025. https://doi.org/10.1016/J.SOLENER.2025.113592

[10] B. Lipovšek, J. Krč, O. Isabella, M. Zeman, and M. Topič, Modeling and optimization of white paint back reflectors for thin-film silicon solar cells, in *Journal of Applied Physics*, Nov. 2010. https://doi.org/10.1063/1.3512907

[11] T. V. Nguyen *et al.*, Effects of nano-TiO2 and nano-SiO2 particles on the reflectance and weathering durability of solar heat reflectance coating, *J. Photochem. Photobiol. A Chem.*, vol. 442, p. 114752, Aug. 2023. https://doi.org/10.1016/J.JPHOTOCHEM.2023.114752

[12] R. Bansal, S. Nair, and K. K. Pandey, UV resistant wood coating based on zinc oxide and cerium oxide dispersed linseed oil nano-emulsion, *Mater. Today Commun.*, vol. 30, p. 103177, Mar. 2022. https://doi.org/10.1016/J.MTCOMM.2022.103177

[13] W. R. Joseph, J. Y. Tan, A. K. Koyande, I. Khoiroh, J. Joynson, and S. Willis, Subambient passive radiative cooling effects of barium sulfate and calcium carbonate paints under Malaysia's tropical climate, *Environmental Science: Advances*, vol. 2, no. 12, pp. 1662–1679, Oct. 2023. https://doi.org/10.1039/d3va00161j

[14] G. Gangisetty and R. Zevenhoven, A Review of Nanoparticle Material Coatings in Passive Radiative Cooling Systems Including Skylights, *Energies (Basel).*, vol. 16, no. 4, 2023. https://doi.org/10.3390/en16041975

[15] R. C. Veloso *et al.*, Impact of SiO2, TiO2 and ZnO nanoparticles incorporation on the thermo-optical properties of dark-coloured façade coatings, *Journal of Building Engineering*, vol. 84, p. 108517, May 2024. https://doi.org/10.1016/J.JOBE.2024.108517

[16] C. Y. Wai, H. W. Chau, P. Paresi, and N. Muttil, Experimental Analysis of Cool Roof Coatings as an Urban Heat Mitigation Strategy to Enhance Thermal Performance, *Buildings 2025, Vol. 15, Page 685*, vol. 15, no. 5, p. 685, Feb. 2025. https://doi.org/10.3390/BUILDINGS15050685

[17] S. A. Mansour and A. H. Farha, A Review of Near-Infrared Reflective Nanopigments: Aesthetic and Cooling Properties, *Crystals 2025, Vol. 15, Page 271*, vol. 15, no. 3, p. 271, Mar. 2025. https://doi.org/10.3390/CRYST15030271

[18] B. Dopphoopha, K. Li, C. Lin, and B. Huang, Aqueous double-layer paint of low thickness for sub-ambient radiative cooling, *Nanophotonics*, vol. 13, no. 5, p. 659, Mar. 2024. https://doi.org/10.1515/NANOPH-2023-0664

[19] W. R. Joseph, J. Y. Tan, A. K. Koyande, I. Khoiroh, J. Joynson, and S. Willis, Subambient passive radiative cooling effects of barium sulfate and calcium carbonate paints under Malaysia's tropical climate, *Environmental Science: Advances*, vol. 2, no. 12, pp. 1662–1679, Oct. 2023. https://doi.org/10.1039/D3VA00161J

[20] R. C. Veloso *et al.*, Impact of SiO2, TiO2 and ZnO nanoparticles incorporation on the thermo-optical properties of dark-coloured façade coatings, *Journal of Building Engineering*, vol. 84, p. 108517, May 2024. https://doi.org/10.1016/J.JOBE.2024.108517

[21] S. Li *et al.*, An inorganic water-based paint for high-durability passive radiative cooling, *J. Mater. Chem. C Mater.*, vol. 13, no. 8, pp. 4137–4144, Feb. 2025. https://doi.org/10.1039/D4TC04108A

[22] S. Sindhu and A. Singh, Developing Mechanism with Photo Reflective and Photocatalytic Properties for Cool Roofs by Using Tio2 Nanoparticles, 2023. [Online]. Available: www.jchr.org

[23] A. H. Farha, A. F. Al Naim, and S. A. Mansour, Cost-Effective and Efficient Cool Nanopigments Based on Oleic-Acid-Surface-Modified ZnO Nanostructured, *Materials*, vol. 16, no. 6, Mar. 2023. https://doi.org/10.3390/ma16062159

[24] C. Lin *et al.*, Pushing Radiative Cooling Technology to Real Applications, *Advanced Materials*, vol. 37, no. 23, Jun. 2025. https://doi.org/10.1002/ADMA.202409738

[25] H. Gong *et al.*, Zirconia submicrosphere/potassium silicate metacoating with high irradiation stability for radiative cooling, *Adv. Compos. Hybrid Mater.*, vol. 8, no. 1, Feb. 2025. https://doi.org/10.1007/s42114-024-01130-y

[26] M. V. Krishna Mohan, T. V. K. Bhanuprakash, and A. Mukherjee, Al2O3 and CuO nano particulate-based paints for marine applications, *Engineering Research Express*, vol. 4, no. 3, p. 035056, Sep. 2022. https://doi.org/10.1088/2631-8695/AC8FCD

[27] S. Sharifi Golru, M. M. Attar, and B. Ramezanzadeh, Studying the influence of nano-Al2O3 particles on the corrosion performance and hydrolytic degradation resistance of an epoxy/polyamide coating on AA-1050, *Prog. Org. Coat.*, vol. 77, no. 9, pp. 1391–1399, Sep. 2014. https://doi.org/10.1016/J.PORGCOAT.2014.04.017

[28] H. Gong *et al.*, Zirconia submicrosphere/potassium silicate metacoating with high irradiation stability for radiative cooling, *Adv. Compos. Hybrid Mater.*, vol. 8, no. 1, Feb. 2025. https://doi.org/10.1007/s42114-024-01130-y

[29] D. P. Hung *et al.*, EFFECT OF ZIRCONIA NANOPARTICLES MODIFIED BY SILANE COUPLING AGENT ON SOME PROPERTIES OF EPOXY COATING, *Vietnam J. Sci. Technol.*, vol. 60, no. 4, pp. 664–674, 2022. https://doi.org/10.15625/2525-2518/16999

[30] Y. Liu *et al.*, Design and scalable fabrication of core-shell nanospheres embedded spectrally selective single-layer coatings for durable daytime radiative cooling, *Solar Energy Materials and Solar Cells*, vol. 260, p. 112493, Sep. 2023. https://doi.org/10.1016/J.SOLMAT.2023.112493

[31] C. Takai-Yamashita and M. Fuji, Hollow silica nanoparticles: A tiny pore with big dreams, *Advanced Powder Technology*, vol. 31, no. 2, pp. 804–807, Feb. 2020. https://doi.org/10.1016/j.apt.2019.11.034

[32] E. P. Ferreira-Neto *et al.*, Thermally stable SiO2@TiO2 core@shell nanoparticles for application in photocatalytic self-cleaning ceramic tiles, *Mater. Adv.*, vol. 2, no. 6, pp. 2085–2096, Mar. 2021. https://doi.org/10.1039/D0MA00785D

[33] M. Eren and H. K. Can, Preparation of zinc methacrylate-methylmethacrylate-butyl acrylate emulsions and their application in exterior paints, *Prog. Org. Coat.*, vol. 135, pp. 424–437, Oct. 2019. https://doi.org/10.1016/J.PORGCOAT.2019.06.031

[34] H. A. El-Wahab *et al.*, Efficacy of zinc and copper oxide nanoparticles as heat and corrosion-resistant pigments in paint formulations, *Sci. Rep.*, vol. 14, no. 1, Dec. 2024. https://doi.org/10.1038/s41598-024-74345-0

[35] U. Qazi, D. F. Anjusa Putra, Y. Z. Lai, and S. J. Shih, Selection of binder on SiO2-based thick film for radiative cooling applications, *Ceram. Int.*, vol. 51, no. 16, pp. 22942–22954, Jul. 2025. https://doi.org/10.1016/J.CERAMINT.2025.02.209

[36] M. Eren and H. K. Can, Preparation of zinc methacrylate-methylmethacrylate-butyl acrylate emulsions and their application in exterior paints, *Prog. Org. Coat.*, vol. 135, pp. 424–437, Oct. 2019. https://doi.org/10.1016/J.PORGCOAT.2019.06.031

[37] J. Song *et al.*, Durable radiative cooling against environmental aging, *Nature Communications 2022 13:1*, vol. 13, no. 1, pp. 1–12, Aug. 2022. https://doi.org/10.1038/s41467-022-32409-7

[38] U. Qazi, D. F. Anjusa Putra, Y. Z. Lai, and S. J. Shih, Selection of binder on SiO2-based thick film for radiative cooling applications, *Ceram. Int.*, vol. 51, no. 16, pp. 22942–22954, Jul. 2025. https://doi.org/10.1016/J.CERAMINT.2025.02.209

[39] X. Li, J. Peoples, P. Yao, and X. Ruan, Ultrawhite BaSO4Paints and Films for Remarkable Daytime Subambient Radiative Cooling, *ACS Appl. Mater. Interfaces*, vol. 13, no. 18, pp. 21733–21739, May 2021. https://doi.org/10.1021/acsami.1c02368

[40] X. Li, J. Peoples, P. Yao, and X. Ruan, Ultrawhite BaSO4Paints and Films for Remarkable Daytime Subambient Radiative Cooling, *ACS Appl. Mater. Interfaces*, vol. 13, no. 18, pp. 21733–21739, May 2021. https://doi.org/10.1021/acsami.1c02368

[41] J. Yu *et al.*, Enhancing Passive Radiative Cooling Films with Hollow Yttrium-Oxide Spheres Insights from FDTD Simulation, *Macromol. Rapid Commun.*, vol. 46, no. 3, Feb. 2025. https://doi.org/10.1002/MARC.202400770

[42] D. Hu, S. Sun, P. Du, X. Lu, H. Zhang, and Z. Zhang, Hollow Core-Shell Particle-Containing Coating for Passive Daytime Radiative Cooling, *Compos. Part A Appl. Sci. Manuf.*, vol. 158, p. 106949, Jul. 2022. https://doi.org/10.1016/J.COMPOSITESA.2022.106949

[43] C. Takai-Yamashita and M. Fuji, Hollow silica nanoparticles: A tiny pore with big dreams, *Advanced Powder Technology*, vol. 31, no. 2, pp. 804–807, Feb. 2020. https://doi.org/10.1016/j.apt.2019.11.034

[44] Z. A. Abdul Halim, N. Ahmad, and M. A. Mat Yajid, "Modifying commercial water-based acrylic paint with Rice husk-derived silica aerogel fillers for building thermal insulation," *Prog. Org. Coat.*, vol. 201, p. 109114, Apr. 2025. https://doi.org/10.1016/J.PORGCOAT.2025.109114

[45] S. Yang, F. Wang, J. Ou, and C. Li, Yellow self-cleaning radiative cooling paint for building exterior with high adhesion, *Prog. Org. Coat.*, vol. 204, p. 109200, Jul. 2025. https://doi.org/10.1016/J.PORGCOAT.2025.109200

[46] M. Wurood Asaad, L. M. Shaker, and A. Mahdi, An Experimental Study on the Effect of Nano Zinc Oxide on Reflective Coatings and Building Roof Temperatures, *Materials Science Forum*, vol. 1131, pp. 109–118, 2024. https://doi.org/10.4028/P-GDBLU6

[47] N. L. Alchapar and J. Balter, Nanopaint application in residential buildings. A study of the effect of cool paint on cooling energy demand, *Nanopaints: Characterization, Formulation, and Emerging Applications*, pp. 333–357, Jan. 2025. https://doi.org/10.1016/B978-0-443-26512-9.00013-6

[48] L. Zhou *et al.*, Radiative cooling for energy sustainability: Materials, systems, and applications, *Phys. Rev. Mater.*, vol. 6, p. 90201, 2022. https://doi.org/10.1103/PhysRevMaterials.6.090201

[49] C. Y. Tso, K. C. Chan, and C. Y. H. Chao, A field investigation of passive radiative cooling under Hong Kong's climate, *Renew. Energy*, vol. 106, pp. 52–61, Jun. 2017. https://doi.org/10.1016/J.RENENE.2017.01.018

[50] D. Tachouali and H. Taleb, "Achieving energy reduction by applying passive cooling strategies in a hot arid region: a case study of the UAE," 2015. https://doi.org/10.2495/SDP150941

[51] J. Peoples, Y. W. Hung, Z. Fang, J. Braun, W. T. Horton, and X. Ruan, Energy savings of radiative cooling paints applied to residential buildings, *Int. J. Heat Mass Transf.*, vol. 194, p. 123001, Sep. 2022. https://doi.org/10.1016/J.IJHEATMASSTRANSFER.2022.123001

[52] H. Bao, C. Yan, B. Wang, X. Fang, C. Y. Zhao, and X. Ruan, Double-layer nanoparticle-based coatings for efficient terrestrial radiative cooling, *Solar Energy Materials and Solar Cells*, vol. 168, pp. 78–84, 2017. https://doi.org/10.1016/j.solmat.2017.04.020

[53] H. Bao, C. Yan, B. Wang, X. Fang, C. Y. Zhao, and X. Ruan, Double-layer nanoparticle-based coatings for efficient terrestrial radiative cooling, *Solar Energy Materials and Solar Cells*, vol. 168, pp. 78–84, Aug. 2017. https://doi.org/10.1016/J.SOLMAT.2017.04.020

[54] X. Zhang *et al.*, Nanoporous Mg-doped SiO2 nanoparticles with tunable infrared emissivity toward effective radiative cooling coatings, *J. Alloys Compd.*, vol. 940, p. 168905, Apr. 2023. https://doi.org/10.1016/J.JALLCOM.2023.168905

[55] Y. Dong *et al.*, A low-cost sustainable coating: Improving passive daytime radiative cooling performance using the spectral band complementarity method, *Renew. Energy*, vol. 192, pp. 606–616, Jun. 2022. https://doi.org/10.1016/J.RENENE.2022.04.093

[56] X. Yu, F. Yao, W. Huang, D. Xu, and C. Chen, Enhanced radiative cooling paint with broken glass bubbles, *Renew. Energy*, vol. 194, pp. 129–136, Jul. 2022. https://doi.org/10.1016/J.RENENE.2022.05.094

[57] D. P. Hung *et al.*, EFFECT OF ZIRCONIA NANOPARTICLES MODIFIED BY SILANE COUPLING AGENT ON SOME PROPERTIES OF EPOXY COATING, *Vietnam J. Sci. Technol.*, vol. 60, no. 4, pp. 664–674, 2022. https://doi.org/10.15625/2525-2518/16999

[58] M. Baneshi and S. Maruyama, The impacts of applying typical and aesthetically–thermally optimized TiO2 pigmented coatings on cooling and heating load demands of a typical residential building in various climates of Iran, *Energy Build.*, vol. 113, pp. 99–111, Feb. 2016. https://doi.org/10.1016/J.ENBUILD.2015.12.028

[59] T. Wang, Q. Xuan, Y. Dong, D. Lei, and J. G. Dai, Bilayer fluorescent colored radiative cooling coatings for building energy saving, *Energy*, vol. 332, p. 137292, Sep. 2025. https://doi.org/10.1016/J.ENERGY.2025.137292

[60] J. Xu *et al.*, Colored radiative cooling coatings using phosphor dyes, *Mater. Today Nano*, vol. 19, no. 6555, p. 100239, Aug. 2022. https://doi.org/10.1016/j.mtnano.2022.100239

[61] A. Al-Mahdouri *et al.*, Multifunctional silica aerogel-infused paint for self-cleaning and radiative cooling, *Journal of Materiomics*, vol. 11, no. 6, p. 101070, Nov. 2025. https://doi.org/10.1016/J.JMAT.2025.101070

[62] R. Paolini *et al.*, Effects of soiling and weathering on the albedo of building envelope materials: Lessons learned from natural exposure in two European cities and tuning of a laboratory simulation practice, *Solar Energy Materials and Solar Cells*, vol. 205, Feb. 2020. https://doi.org/10.1016/J.SOLMAT.2019.110264

[63] T. U. K. Nutakki, W. U. Kazim, K. Alamara, T. Salameh, and M. A. Abdelkareem, Experimental Investigation on Aging and Energy Savings Evaluation of High Solar Reflective Index (SRI) Paints: A Case Study on Residential Households in the GCC Region, *Buildings 2023, Vol. 13, Page 419*, vol. 13, no. 2, p. 419, Feb. 2023. https://doi.org/10.3390/BUILDINGS13020419

[64] J. Song *et al.*, Durable radiative cooling against environmental aging, *Nature Communications 2022 13:1*, vol. 13, no. 1, pp. 1–12, Aug. 2022. https://doi.org/10.1038/s41467-022-32409-7

[65] Y. Çakmak, Determination of Aging Parameters and Following Structure Changes in Red Oil Paints: A Comprehensive Study, *Microchemical Journal*, vol. 210, p. 113012, Mar. 2025. https://doi.org/10.1016/J.MICROC.2025.113012

[66] C. Y. Tso, K. C. Chan, and C. Y. H. Chao, A field investigation of passive radiative cooling under Hong Kong's climate, *Renew. Energy*, vol. 106, pp. 52–61, Jun. 2017. https://doi.org/10.1016/j.renene.2017.01.018

[67] A. K. Aljwirah *et al.*, "Water-Based Hydrophobic Paints for Daytime Radiative Cooling," *ACS Appl. Mater. Interfaces*, vol. 17, no. 22, pp. 32914–32927, Jun. 2025. https://doi.org/10.1021/acsami.5c05687

[68] A. Felicelli *et al.*, Efficient radiative cooling of low-cost BaSO4 paint-paper dual-layer thin films,*Nanophotonics*, vol. 13, no. 5, pp. 639–648, Mar. 2024. https://doi.org/10.1515/NANOPH-2023-0642/DOWNLOADASSET/SUPPL/J_NANOPH-2023-0642_SUPPL_001.DOCX

[69] B. Dopphoopha, K. Li, C. Lin, and B. Huang, Aqueous double-layer paint of low thickness for sub-ambient radiative cooling, *Nanophotonics*, vol. 13, no. 5, p. 659, Mar. 2024. https://doi.org/10.1515/NANOPH-2023-0664

Design of an Eco-Friendly Lubricant Using COSMO-RS Screened Ionic Liquid

Yuvatharshini V. ARUMUGAM[1,a*], Asiah Nusaibah MASRI[1,2,3,b*], Izni Mariah IBRAHIM[1,2,3], Azad Anugerah ALI RASOL[1,2]

[1]Faculty of Chemical and Energy Engineering, Universiti Teknologi Malaysia, 81310 Johor Bahru, Malaysia

[2]UTM-MPRC Institute for Oil & Gas, Universiti Teknologi Malaysia, 81310 Johor Bahru, Malaysia

[3]Energy Management Group, Universiti Teknologi Malaysia, 81310 Johor Bahru, Malaysia

[a]yuvatharshini-1999@graduate.utm.my, [b]nusaibah@utm.my

Keywords: Biolubricant, Ionic Liquid, Lubricant Additive, COSMO-RS, Eco-Friendly

Abstract. Biolubricants derived from palm oil face challenges such as poor thermal and oxidative stability, limiting their industrial applications. Additives such as ionic liquids (ILs) can enhance the performance of these biolubricants. The structure, molecular weight, and functional groups of ILs influence key properties including solubility, viscosity and density. This study aims to develop a green thermally stable lubricant by screening suitable ILs as additive in palm oil. COSMO-RS computational model was used to predict their solubility in methyl palmitate ester based on activity coefficients. However, some predicted do not align with experimental results due to differences in intermolecular interaction which are acceptable within the expected limitations of the COSMO-RS model. Guanidinium-based ILs with fatty acid anions demonstrated the highest solubility. The length of alkyl chain and molecular weight significantly affected compatibility, highlighting the importance of molecular design. This work provides a theoretical basis and practical approach for formulating eco-friendly lubricants with enhanced performance, contributing to sustainable lubrication solutions in industry.

Introduction

The global consumption of lubricants averages approximately 35 million tons annually, with an anticipated growth rate of 1.6% per year [1, 2]. Polycyclic aromatic hydrocarbons (PAHs) are organic compounds consisting of two or more fused aromatic rings of carbon and hydrogen atoms which will can collapse the entire ecosystem. Moreover, waste lubricating oil contains high concentrations of heavy metals such as lead (Pb), cadmium (Cd), copper (Cu) and zinc (Zn) which are highly toxic to aquatic organisms [3, 4]. The growing environmental concerns and strict regulations on petroleum-based lubricants, increased the demand for eco-friendly alternatives. The presence of unwanted compound in lubricant can increase the toxicity of lubricant.

The key criteria of eco-friendly lubricants are biodegradable and non-toxicity. Bio-based lubricants are derived from renewable and environmentally friendly raw materials such as plant oils, animal fats, or chemically modified hydrocarbons [5]. Several studies have been conducted using vegetable oils as base oils, both with and without additives. Bio ester-based oil (BIOE) has demonstrated lower coefficient of friction (COF) and small wear scar diameter (WSD), indicating favorable lubricating performance [6]. F.Murilo et al [7] evaluated physicochemical properties such as specific mass, pour point, total acid content, flash point, and kinematic viscosity of vegetable oil (castor oil) and mineral oil. Castor oil is a viable option for lubricant applications because of its excellent biodegradability and stability throughout a broad temperature range.

Palm oil has potential to use as lubricant due to its abundant and easily available in Malaysia. Besides, it also exhibits excellent lubricant performance in some specific application such as cold extrusion processes and journal bearings [8]. Although vegetable oil possesses significant potential as non-toxic biolubricant, they exhibit certain limitations such as oxidative stability, thermal stability and corrosiveness arising, due to higher concentration of unsaturated fatty acid [9]. The viscosity of palm oil is restricted as a lubricant because inconsistent behaviour across a range of temperature [10]. These drawbacks can be solved by adding additives in palm oil. Lubricants are typically composed of base oils (80-90%) and additives (10-20%).

Various additives have been investigated for lubricant applications including nanoparticle [11, 12], combination of nanoparticle and ionic liquid (IL) [13] and traditional additives such as zinc dialkyldithiophosphate (ZDDP) [14, 15]. Among these, ILs have emerged as promising lubricant additives because their properties can be tuned through various combination of organic cation and organic or inorganic anion [16]. ILs have been explored both as additive and as base fluid. However, considering their relatively high cost and limited applicability, their use as additives is generally more practical [17]. ILs have both advantages and drawbacks, nevertheless, their drawbacks can be fixed by types of cations and anions, alkyl chain length, and functional groups [18]. Therefore, finding a suitable ILs for additive in base oil is crucial before proceed with lubricant formulation.

Regarding the huge number of ionic liquids (ILs) for lubricant applications, the Conductor-like Screening Model for Real Solvents (COSMO-RS) computational allow to screen and predict their suitability as lubricant additive. COSMO-RS is a quantum chemical technique employed to determine the thermodynamic properties of solvents through the calculation of molecule interactions and surface charge distributions [19, 20]. Turbomole or Tmolex software used to optimized the structure before proceed to COSMO-RS for the LUMO values [20]. The aim of this study is to screen the suitable ILs to use as additive in vegetable oil by evaluating parameters including solubility of methyl palmitate ester as representative of palm oil, viscosity and density by using the COSMO-RS computational framework.

Methodology

COSMO-RS

The COSMO-RS is developed by Klamt and Eckert [21–23]. There are two main software involve in this study which are Tmolex (TURBOMOLE GUI) and COSMOthermX software (18.0.2 version). In Tmolex, structures of cation and anion of ILs that not available in COSMOthermX were drawn and geometrically optimized with BP_TZVP parameterization. Density functional theory (DFT) with B3-LYP were conduct to get accurate quantum calculation. The energy calculation includes triple zeta valence polarised (TZVP) basis set and the Becke and Perdew functional were performed to generate COSMO file for further use in COSMOThermX [19, 24, 25]. Then the generated COSMO file import to COSMOThermX software by applying the BP-TZVP parameterisation to identify the thermodynamic parameters of the solvents [19]. Physicochemical properties such as solubility in methyl palmitate ester, density and viscosity were obtain from COSMOThermX to find out suitable IL that can used as additive in lubricant.

Eq. 1 is capacity equation describing amount of IL needed to soluble in base oil. Compounds with high capacity can achieve full solubility in base oils.

$$\left(C_{12}^{\infty} = \frac{1}{\gamma_1^{\infty}}\right)^{\text{IL phase}} \quad (1)$$

The density of ILs, ρ_i is computed from Eq. 2 which MW_i is molecular weight of IL, V_i corrected molar liquid from Quantitative-Structure-Property-Relationship (QSPR) and N_a is Avogadro's constant.

$$\rho_i = \frac{MW_i}{V_i N_A} \tag{2}$$

The σ-profile profile obtain from COSMO-RS were used to evaluate solubility based on the intermolecular force, and viscosity and density of ILs. Viscosity and density were predicted in 25°C because easy to prepare the lubricant in room temperature.

List of Cations and Anion

In the COSMO-RS analysis, a total of 506 ionic liquids (ILs) were generated, comprising 22 distinct cations and 23 anions as listed in Table 1. Variations in both the type and alkyl chain length of the cations and anions were systematically examined. Several ILs were employed to validate the computational predictions of density, viscosity, and solubility against experimental data reported in previous studies. Furthermore, specific anions, particularly amino acid and fatty acid categories, were selected for further investigation based on the COSMO-RS results, due to their natural origin and low toxicity.

Table 1 list of cation and anion analysed from COSMO-RS with abbreviation and structure

Cation							
No.	Name	Abbr	Structure	No.	Name	Abbr	Structure
1	Tetramethyl guanidine	TMG	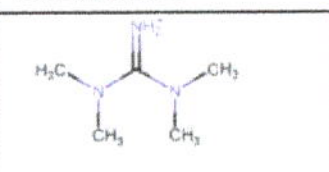	12	Tetrabutyl phosphonium	[P_{4444}]	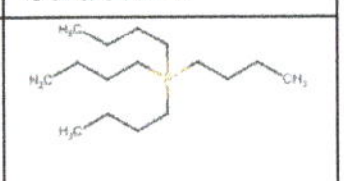
2	N,N,N,N-tetramethyl-N-methyl guanidine	TMG-MG	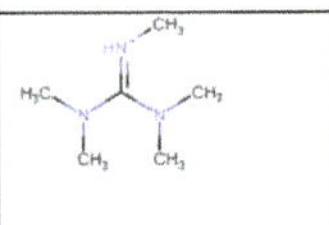	13	Trihexyl tetradecyl phosphonium	[P_{66614}]	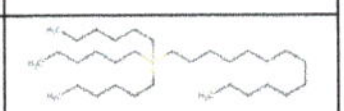
3	N,N,N,N-tetramethyl-N-ethyl guanidine	TMG-EG	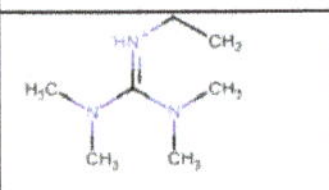	14	bis(2-hydroxyethyl) ammonium	BHEA	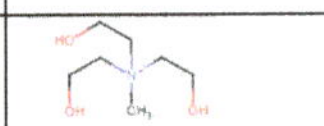
4	N,N,N,N-tetramethyl-N-propyl guanidine	TMG-PG	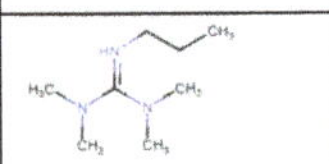	15	Diethyl methyl amine	DEMA	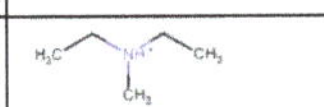
5	N,N,N,N-tetramethyl-N-butyl guanidine	TMG-BG	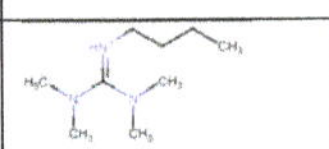	16	Tetrabutyl ammonium	TBA	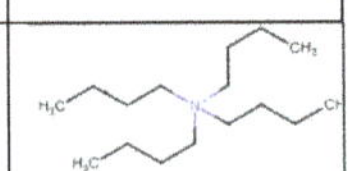
6	N,N,N,N-tetramethy	TMG-DBG	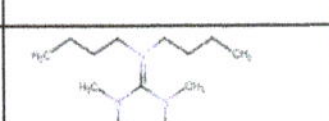	17	Trioctyl methyl ammonium	TOMAC	

	l-N-dibutyl guanidine						
7	N,N,N,N-tetramethy l-N-dihexyl guanidine	TMG -DHG	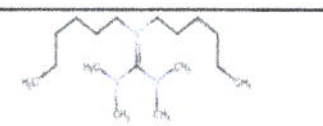	18	Tetraoctyl ammonium	TOA	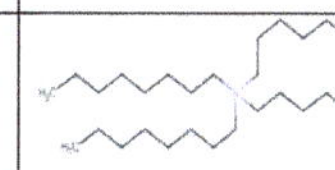
8	Choline	Cho	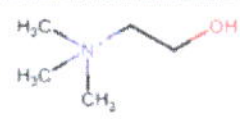	19	1-butyl-3-methyl imidazolium	C_4mim	
9	Tributyl methyl phosphonium	[P_{4441}]	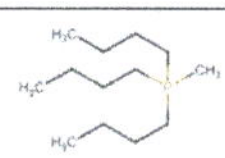	20	1-hexyl-3-methyl imidazolium	C_6mim	
10	Tributyl octyl phosphonium	[P_{4448}]	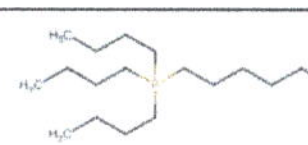	21	1-octyl-3-methyl imidazolium	C_8mim	
11	Tributyl tetradecyl phosphonium	[P_{44414}]	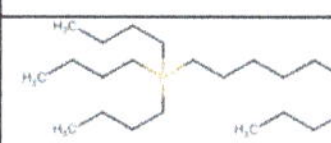	22	1-tetradecyl-3-methyl imidazolium	C_{14}mim	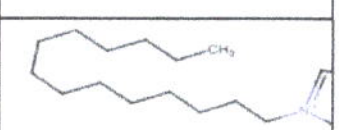
Anion							
No.	Name	Abbr	Structure	No.	Name	Abbr	Structure
1	Glycinate	Gly	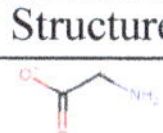	13	Histidinate	His	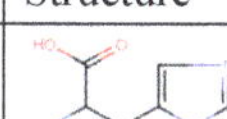
2	Alaninate	Ala	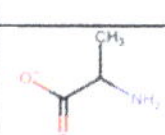	14	Caprate	C_8	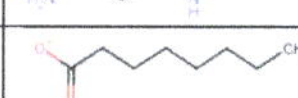
3	Lysinate	Lys	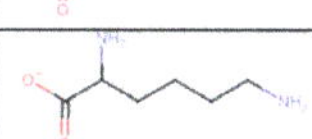	15	Decanoate	C_{10}	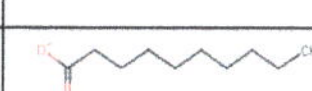
4	Serinate	Ser	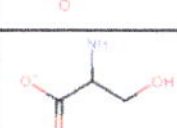	16	Laurate	C_{12}	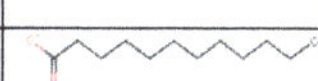
5	Isoleucinate	Iso-L	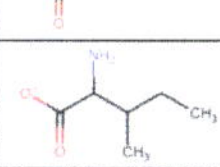	17	Myristate	C_{14}	
6	Dimethyl phosphate	DMPhos	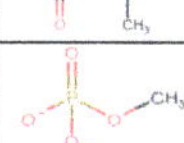	18	Palmitate	C_{16}	
7	Diethyl phosphate	DEPhos	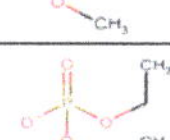	19	Oleate	C_{18}	
8	Dipropyl phosphate	DPPhos	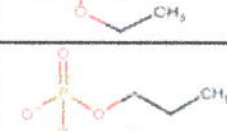	20	Acetate	OAC	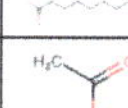

9	Dibutyl phosphate	DBPhos	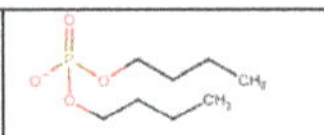	21	Trifluoro acetate	TFA	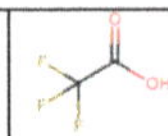
10	Diphenyl phosphate	DPP	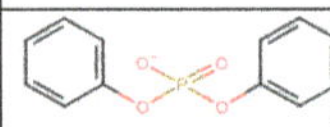	22	Hexafluorophosphate	PF_6	
11	Bis(2-ethyl hexyl) phosphate	DEHP	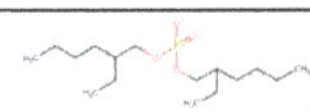	23	Chlorin	Cl	Cl^-
12	Bis(trifluoromethyl sulfonyl) amide	NTf_2	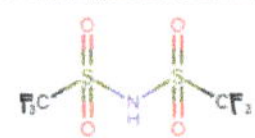				

Result and Discussion

Validation of COSMO-RS

Several ILs obtained from the COSMO-RS results were used to validate the computational predictions against experimental data reported in previous studies. The common physicochemical properties predicted by COSMO-RS include molecular weight, density, viscosity, conductivity and capacity (solubility). For lubricant applications, the key physicochemical parameters of interest are density, viscosity, thermal stability, corrosiveness, and solubility. In the present study, the predicted density, viscosity, and solubility of IL in ester oil were verified. Ravi T et al [19] reported that prediction of COSMO-RS exhibit a high correlation coefficient (R^2> 0.8) with experimental data, indicating strong agreement between computational and experimental results. Therefore, R^2 value greater than 0.8 is considered a valid indicator of COSMO-RS predictive accuracy.

Density

The density values of the analyzed ILs obtain from COSMO-RS calculations were compared with experimental data, as illustrated in Fig. 1. The dotted trendline represents the linear correlation between the two datasets, with a correlation coefficient (R^2) of 0.8648. The computational results showed good aligned to experimental findings, exhibiting a correlation coefficient (R^2) more than 0.8, which indicates a reasonably strong agreement. The prediction values were slightly lower than experimental result. This observation is consistent with other studies that have reported similar trends between predicted and experimental data [26, 27]. Alcantara M et al [26] reported an average relative deviation (ARD) ranging from 2.44% to 0.27% for density prediction using COSMO-RS, demonstrating a significant level of correlation. This finding suggest that COSMO-RS can be considered a reliable tool for estimating the density of ionic liquid. Consequently, density screening of lubricant additives via COSMO-RS constitutes an efficient computational method. However, experimental confirmation is crucial, and COSMO-RS predictions have to be regarded as guiding estimates rather than definitive values.

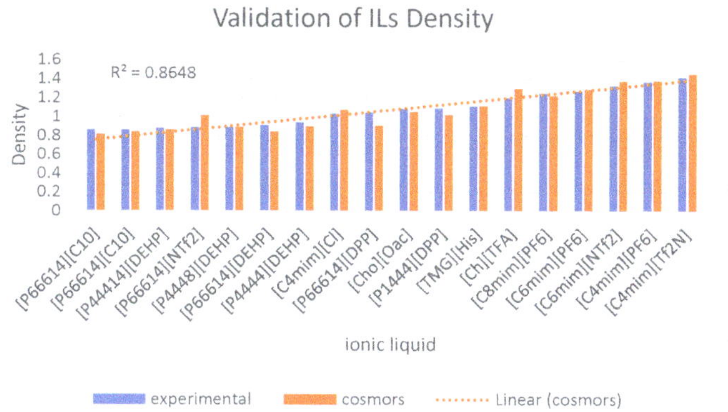

Fig. 1 The COSMO-RS estimated densities show a strong linear correlation with the experimental values ($R^2 = 0.8648$).

Viscosity

Fig. 2 illustrates the comparison between experimental viscosity and the predicted by the COSMO-RS model. The left y-axis (0-1200 cP) represents the experimental data (blue bars) obtained from various studies and the right y-axis (0-14000cP) represent the computational predictions (yellow bars) generated by COSMO-RS. The dotted trendline is the linear correlation of COSMO-RS prediction, with R^2 of 0.3034. Overall, the COSMO-RS model tends to overestimate the viscosity of most ionic liquids, particularly at higher viscosity values. Several studies have reported similar systematic error in COSMO-RS predictions, attributing them to limitations in accounting for specific intermolecular interactions especially van der Waals forces (E_{vdw}) and temperature which are dominant in viscosity of IL, followed by electrostatic/misfit (E_{MF}) and hydrogen-bonding (E_{HB}) [21, 22, 26]. Alcantara et al [26], observed comparable overestimations in viscosity predictions using COSMO-RS but improved agreement by corrected the room temperature to closely synchronized with experimental. Kurnia et al [22] developed quantitative structure-property relationship (QSPR) model based on COSMO-RS output to improve viscosity estimation accuracy. Similarly, Koi et al [21] constructed QSPR model to predict viscosity of IL based on three intermolecular interaction energies which are E_{MF}, E_{HB} and E_{vdw} derived from COSMO calculation.

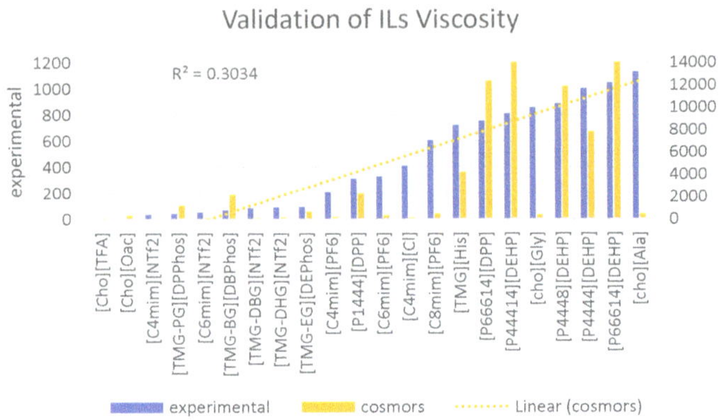

Fig. 2 The COSMO-RS estimated viscosity show a strong linear correlation with the experimental values ($R^2 = 0.3034$).

Solubility

According to Table 2, most COSMO-RS result demonstrate solubility values above 0.5, suggesting a high possibility of ionic liquids (ILs) dissolving in methyl palmitate ester while some of them are different from experimental result. These variations are primarily attributed to different types of base oil used in experimental studies such as mineral, synthetic and vegetable oil. Barnhill et al [28] investigated phosphonium-organophosphate ILs as additive in GTL 4 cSt (gas-to-liquid with kinematic viscosity of 4) derived from natural gas which not significant with COSMO-RS prediction due to distinct base oil characteristics. Nevertheless, COSMO-RS data suggest that these ILs have capability to soluble in methyl palmitate ester because the value is more than 0.5. The molecular structure of base oil also plays a critical role in solubility behaviour. For instance, [P66614][DS] is fully soluble in methyl palmitate ester based on COSMO-RS prediction result while less soluble in hexadecane and fully soluble in diethyl succinate based on experimental [29]. This variation is attributed to difference in structure of base oils which are hexadecane consist of a long alkane chain and diethyl succinate is contain ester functional group. Similarly, [TMG][His] exhibited inconsistent results between experimental and computation data because differences in the base oil used [30]. The experimental base oil, PEG200, is soluble in both organic solvents and water, [TMG][His] is classified as water-soluble but not oil-soluble due to the anion characteristic.

In several experimental studies, base oil were modified through epoxidation and emulsion processes to improve the solubility of ionic liquids (ILs) within the oil matrix [31, 32]. The epoxidation process enhances the polarity and stability of vegetable oil to improve their compatibility with other polar compound such as polymer and ILs [33]. Aviles et al [32] demonstrated that the epoxidation of avocado oil increased the interaction between IL and the base oil. Similarly, the microemulsion technique has been utilized to mix IL into castor oil formulation [31]. The COSMO-RS result is consistent with these findings, showing that [DEMA][Oms] and imidazolium-based ILs are insoluble in methyl palmitate ester but can dissolve effectively if the base oil is chemically modified. Amiril et al [34] verified that $[P_{66614}][(^iC_8)_2PO_2]$ and $[DEMA][NTf_2]$ are fully soluble in MJO which is produced through transesterification process to yield a base oil rich in methyl esters. Consequently, the COSMO-RS predictions have been verified by experimental results, with the observed inconsistencies being primarily attributed to variations in base oil type, molecular structure, and modification processes that increase the solubility of IL.

Table 2 Base oil used in ionic liquid from previous article.

no	Ionic liquid	Base oil	Solubility (experimental)	COSMO-RS (computational)	Ref
1	$[P_{66614}][DEHP]$	PAO 4 cSt, SAE%W-30	Fully soluble	1.136864	[35]
2	$[P_{1444}][DPP]$	Safflower oil, TMP ester, Hatcol 5033	Fully soluble	0.172333	[36]
3	$[P_{66614}][DPP]$		Fully soluble	0.963392	
4	$[P_{66614}][(^iC_8)_2PO_2]$	Modified jatropha oil (MJO)	Fully soluble	1.083228	[34]
5	[DEMA][NTf2]		Fully soluble	1.051711	
6	$[P_{66614}][DEHP]$	GTL 4cSt	Fully soluble	1.136864	[28]
7	$[P_{44414}][DEHP]$		Less soluble	1.029951	
8	$[P_{4448}][DEHP]$		Less soluble	0.608293	
9	$[P_{4444}][DEHP]$		Less soluble	0.632314	
10	$[P_{66614}][DS]$	Hexadecane, diethyl sucinnate	Less soluble, fully soluble	1.192477	[29]
11	[TMG][His]	PEG200	Fully soluble	0.015033	[30]
12	[DEMA][Oms]	Avocado oil	Fully soluble	0.034126	[32]
13	TBA-C10	SN-150	Fully soluble	0.431677	[37]
14	TBA-C12		Fully soluble	0.481395	
15	TBA-C14		Fully soluble	0.536231	
16	TBA-C16		Fully soluble	0.573462	
17	TBA-C18		Fully soluble	0.635651	
18	$[C_4mim][BF_4]$	TX-100, n-butanol, castor oil	Fully soluble	0.000637	[31]
19	$[C_8mim][Cl]$		Fully soluble	0.038478	

COSMO-RS Result

Solubility/Capacity Towards Methyl Palmitate Ester

Methyl palmitate ester is one of the structure consist in Trimethylolpropane (TMP) ester Table 3 and 4 present the solubility of fatty acid and amino acid- based ILs in methyl palmitate ester, analysed according to their respective cations and anions combinations. The colour scale ranging from green to red represents high to low solubility, respectively. Fatty acid- based ILs exhibit higher solubility than amino acid-based ILs due to the differences in the functional groups of their anionic structures. The greater compatibility of fatty acid–based ionic liquids can be related to the length of alkyl chains in anions [37]. Fatty acids anion consists of long hydrophobic alkyl chain with terminal carboxyl functional group, which tends to form strong hydrogen-bonded dimers further stabilized by van der Waals forces and hydrophobic interactions [38]. Amino acid anion contain non-polar side chain along with amino and carboxyl functional group as polar, which exhibit lower compatible with non-polar solvents due to the lack of favourable intermolecular interaction [39].

Although fatty acid-based ILs exhibit higher solubility than amino acid-based ILs, variations in the cation significantly influence the overall solubility of ionic liquids in ester-based oil. As shown in Table 3, cations such as $[TMG\text{-}DBG]^+$, $[TMG\text{-}DHG]^+$, $[P44414]^+$, $[P66614]^+$ and $[TOA]^+$ display high solubility across all fatty acid anions. If cations with long and highly branched combined with straight-chain fatty acid anions enhance compatibility with base oil due to increased hydrophobic surface area provided by the branched structure. The solubility trend increases with the elongation of the alkyl chain in the anion because the more extended molecular structure facilitates stronger van der Waals interactions with the hydrocarbon chains of the base oil [28]. Oleate anion demonstrates high solubility with most cations, due to presence of C=C double bond (unsaturated hydrocarbon) and a long alkyl chain in its structure. These features enhance compatibility with hydrophobic ester-based oils and reduce electrostatic repulsion with the cation [37]. Furthermore, cations contain branched alkyl chains exhibit higher capacity to soluble in base oils because steric hindrance reduces molecular packing efficiency and simultaneously increasing the hydrophobic surface area available for interaction.

Table 3 Heatmap analysis based on various cation and fatty acid anion in methyl palmitate ester.

Cation \ Anion	Caprate, $[C_8]^-$	Decanoate, $[C_{10}]^-$	Laurate, $[C_{12}]^-$	Myristate, $[C_{14}]^-$	Palmitate $[C_{16}]^-$	Oleate, $[C_{18}]^-$
$[TMG]^+$	0.096398	0.095583	0.119124	0.146947	0.168673	0.209823
$[TMG\text{-}MG]^+$	0.262058	0.248185	0.28259	0.316152	0.343747	0.396038
$[TMG\text{-}EG]^+$	0.333649	0.319651	0.358195	0.396125	0.426372	0.48215
$[TMG\text{-}PG]^+$	0.358145	0.347491	0.389163	0.431179	0.463711	0.52193
$[TMG\text{-}BG]^+$	0.288315	0.284829	0.324443	0.367881	0.400361	0.458363
$[TMG\text{-}DBG]^+$	0.856357	0.862255	0.912326	0.970216	1.013181	1.081398
$[TMG\text{-}DHG]^+$	0.975724	0.974535	1.037457	1.105085	1.152998	1.224277
$[Cho]^+$	0.01613	0.015691	0.026408	0.039601	0.051898	0.077525
$[P_{4441}]^+$	0.295818	0.30548	0.339854	0.387319	0.421381	0.48149
$[P_{4448}]^+$	0.390848	0.413727	0.439023	0.49078	0.52514	0.582757
$[P_{44414}]^+$	0.75488	0.78798	0.811265	0.872239	0.911782	0.97349
$[P_{4444}]^+$	0.420041	0.437352	0.470893	0.526347	0.564002	0.627284
$[P_{66614}]^+$	0.849361	0.888746	0.901208	0.958643	0.995081	1.051308
$BHEA^+$	0.003794	0.004087	0.008467	0.015229	0.022024	0.037677
$DEMA^+$	0.038397	0.038473	0.055177	0.075296	0.092296	0.126575
TBA^+	0.431677	0.447911	0.481395	0.536231	0.573462	0.635651
$TOMAC^+$	0.540122	0.567917	0.591057	0.644157	0.680241	0.740838
TOA^+	0.858894	0.898123	0.910212	0.967731	1.004127	1.058939
C_4mim^+	0.071743	0.073217	0.095503	0.120402	0.140305	0.181358
C_6mim^+	0.095931	0.099634	0.124278	0.152606	0.174685	0.219435
C_8mim^+	0.121617	0.12715	0.153533	0.184561	0.20844	0.256113
$C_{14}mim^+$	0.223755	0.23525	0.263365	0.300281	0.328013	0.38183

Table 4 Heatmap analysis based on various cation and amino acid anion in methyl palmitate ester.

Cation \ Anion	Glycinate $[Gly]^-$	Alaninate $[Ala]^-$	Isoleucinate $[iso\text{-}L]^-$	Lysinate $[Lys]^-$	Serinate $[Ser]^-$	Histidinate $[his]^-$
$[TMG]^+$	0.381506	0.122849	0.064119	0.08360	0.03512	0.015033
$[TMG\text{-}MG]^+$	2.0346	0.642816	0.257776	0.31380	0.21243	0.071568
$[TMG\text{-}EG]^+$	1.328587	0.565711	0.306041	0.33510	0.21429	0.076743
$[TMG\text{-}PG]^+$	0.815626	0.434449	0.297961	0.30156	0.18166	0.069821
$[TMG\text{-}BG]^+$	0.448147	0.267274	0.213968	0.21293	0.11536	0.048016
$[TMG\text{-}DBG]^+$	0.87195	0.720891	0.705503	0.57802	0.43669	0.174182
$[TMG\text{-}DHG]^+$	0.700891	0.680443	0.762342	0.58891	0.49652	0.216131
$[Cho]^+$	0.257992	0.016302	0.004247	0.00766	0.00146	0.001199

$[P_{4441}]^+$	0.170908	0.162702	0.196633	0.16462	0.10656	0.052787
$[P_{4448}]^+$	0.213346	0.229843	0.29259	0.23541	0.20720	0.10145
$[P_{44414}]^+$	0.498699	0.535615	0.636171	0.50926	0.53327	0.277892
$[P_{4444}]^+$	0.248359	0.250963	0.306785	0.24867	0.19419	0.091923
$[P_{66614}]^+$	0.611161	0.654375	0.757601	0.62402	0.69763	0.388512
$[BHEA]^+$	0.000337	0.000087	0.000234	0.00032	0.00001	0.000014
$[DEMA]^+$	0.113748	0.025239	0.014088	0.0193	0.00421	0.002827
$[TBA]^+$	0.268339	0.266406	0.319606	0.26064	0.20302	0.095046
$[TOMAC]^+$	0.319534	0.351886	0.431926	0.34366	0.35403	0.189278
$[TOA]^+$	0.621303	0.663154	0.767566	0.63187	0.70444	0.389459
$[C_4mim]^+$	0.044033	0.027153	0.029358	0.03055	0.00832	0.005015
$[C_6mim]^+$	0.030244	0.02858	0.040844	0.03561	0.01286	0.007511
$[C_8mim]^+$	0.033905	0.037075	0.056061	0.04508	0.02108	0.011847
$[C_{14}mim]^+$	0.08019	0.095016	0.133712	0.10112	0.08025	0.043214

However, some amino acids-based ILs are hydrophobic and polar characteristics [39]. As shown in Table 4, glycinate, alaninate and isoleucinate are classified as hydrophobic due to the presence of a single amino and carboxyl group, whereas lysinate, serinate and histidinate contain multiple amino and carboxyl group that increase overall polarity. Glycinate and alaninate anions paired with guanidine-based cations except $[TMG]^+$ and $[TMG\text{-}BG]^+$ demonstrate high solubility due to strong ionic interaction between the hydrogen bond donor and acceptor sites of the ionic liquid and the solvent molecules [40]. Phosphonium and ammonium-based cation paired with isoleucinate anions also show high solubility in methyl palmitate ester owing to the bulky and branched hydrophobic side chain present in both the anion and cation, which enhance the surface area available for interaction with the base oil. To further improve the solubility of amino acid-based ILs, the strength of cation and anion interactions is governed by electrostatic forces and orbital overlap, as cations that can significantly induce charge transfer from the anionic carbonyl group to their hydrogen-bond donor sites [41].

Sigma Profile

The σ-profile, derived from the probability distribution of molecular surface area and shielding density charge relative to charge density distribution in COSMO-RS, serves as a crucial indicator of the polarity distribution and interaction dynamics of molecular systems [25, 42]. This profile is essential for predicting and improving the solubility behaviour of ionic liquids in base oils. Fig. 3 (a), (b), and (c) illustrate the sigma profiles of various cations, guanidine, choline, phosphonium, ammonium and imidazolium-based, as well as those amino acid and fatty acid anions, respectively. The x-axis represents the screening charge density (σ), which is divided into three distinct regions, hydrogen bond donor (HBD) region ($\sigma < -1$), non-polar region ($-1 < \sigma < 1$) and hydrogen bond acceptor (HBA) region ($\sigma > 1$). The y-axis is sigma σ/sigma potential which reflects the capability of each cation or anion to interact and dissolve in methyl palmitate ester. The sigma profile shown Fig. 3 is correlated with Table 2 and 3, indicating that the intermolecular forces between cations and anions are controlling the solubility of ILs in base oil. Specifically, Van der Waals force is be in charge for non—polar region [19] and hydrogen bond force is predominant in the HBA and HBD region which increase the polarity of IL.

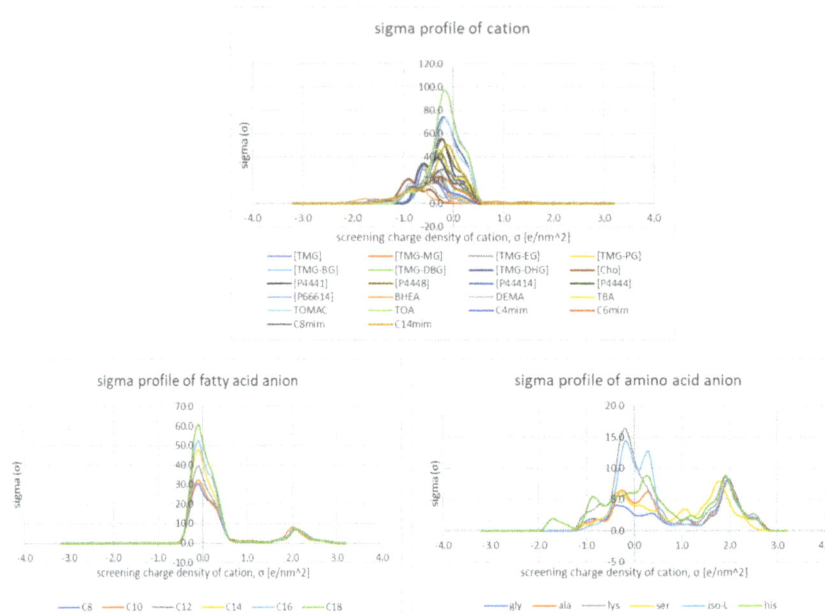

Fig. 3 Sigma profile generated from COSMO-RS (a) cation, (b) fatty acid anion and (c) amino acid anion.

Most of the cations illustrated in Fig. 3(a) exhibit high peak within the non-polar region. This observation is consistent with findings by Qi et al [43], who reported imidazolium-based cations primarily occupy the non-polar region, with peak intensities increasing as the alkyl chain lengthens. The sigma profiles of all cations indicate that elongation of the alkyl chain amplifies the central peak and shifts it toward $\sigma \approx 0$, indicating reduced polarity, except in the case of the choline cation. The choline cation display higher polarity with a peak is favorable to HBD region due to contain of trimethyl ammonium with end of tail is hydroxyl group. However, choline cations can act as hydrophobic depends on the functional group present in anion [44]. Guanidine cation display two notable peaks within non-polar region. cation containing one -NH group (TMG-MG, TMG-EG, TMG-PG and TMG-BG) are slightly shifted towards the HBD region, whereas cations with fully alkyl-substituted guanidine structures (TMG-DBG and TMG-DHG) exhibit their main peaks near to 0. The hydrogen atom in guanidine cation enable hydrogen bonding which slightly shift near to HBD region compared with fully alkylated analogues. As shown in Table 2, TMG-DBG and TMG-DHG cation exhibit higher solubility in ester base oil as these cations can participate in both van der Waals and hydrogen-bonding interactions. Phosphonium and ammonium cations are show high peak in non-polar region except BHEA cation because it has two hydroxyl groups at the end of structure.

Amino acid and fatty acid-derived anions are considered as environmentally sustainable alternatives due to their biodegradability and low toxicity. As shown in Fig. 3(b) and (c) fatty acid anion exhibit larger surface areas within the non-polar region compared to amino acid anions, indicating a greater tendency to dissolve in non-polar base oils. Moreover, peak of HBA in amino acid anion are higher than fatty acid anion, which is increasing polarity. This due to amino group and hydroxyl group (OH) contain lone pairs of electrons on nitrogen and oxygen atom [42] which tend to accept hydrogen atom to form hydrogen bond. Ser, Gly and Ala having high peak at HBA region but in less range of surface area which is 8. Glycinate anion with TMG-MG cation in Table 3 shows high capacity to soluble in methyl palmitate ester, might be due to intermolecular forces of van der Waals and hydrogen bonding between IL with ester base oil. Lys in hydrophilic amino acid, despite surface area of Lys is high at polar region due to strong ion–pair interactions, stabilized by hydrogen bonding [44].

Fatty acid anion are primarily non-polar because of their lengthy alkyl chains, with the terminal carboxyl group contributing slightly of polarity. Fig. 3(b) demonstrates high surface area within the non-polar region especially increasing length of alkyl chain. The alkyl chain length in fatty acid anion plays a major role because longer alkyl chain increase steric hindrance and alter the sigma profile by modifying the spatial distribution of charge densities [45]. These alkyl chains form van der Waals force with base oil, which strongly influence the characteristics of mixture, including adsorption, flowability, and rigidity. A small peak observed at the HBA region corresponds to the oxygen atom in the carboxyl group, which form hydrogen bond with ester base oil [42]. Consequently, most fatty acid anion have high solubility in ester base oil because their high peak in non-polar region compare to amino acid anion. Therefore, the solubility results presented in Table 3 and 4 result can be considered reliable. The charge distribution within ionic liquids significantly impacts their hydrogen-bonding characteristics, wherein the anion generally displaying a minor peak in the hydrogen-bond donor (HBA) region, while the cation more substantially contributes to the hydrogen-bond acceptor (HBD) region.

Conclusion

TMoleX and COSMOThermX were employed to predict the density, viscosity, and solubility of ionic liquids (ILs) in methyl palmitate ester, demonstrating strong correlation with experimental data ($R^2 > 0.8$) for density. Although COSMO-RS slightly overestimated viscosity, this can be improved by developing a QSPR model based on COSMO-RS energy values which are van der Waals forces (E_{vdw}), electrostatic/misfit (E_{MF}) and hydrogen-bonding (E_{HB}). Solubility predictions were generally consistent with experimental results, with variations attributed to differences in base oil types and modifications, such as epoxidation or emulsification, used to enhance IL miscibility. Solubility behaviour, analysed via σ-profiles, revealed that ILs with fatty acid anions exhibited higher solubility in methyl palmitate ester than those with amino acid anions, due to stronger nonpolar interactions from longer alkyl chains and bulky cations. Overall, COSMO-RS proves to be a reliable computational approach for predicting physicochemical properties and screening suitable ILs, enabling the rational design of eco-friendly lubricants with improved compatibility and performance.

Acknowledgement

The authors acknowledged UTM NEXUS POSTGRAD Grant (Q.J130000.5346.00L93), Universiti Teknologi Malaysia (UTM) under UTM Matching grant (Q.J130000.3046.05M07), Ministry of Higher Education (MOHE) under Fundamental Research Grant Scheme (FRGS/1/2023/TK08/UTM/02/36), Energy Management Group, UTM-MPRC Institute For Oil and Gas (IFOG) and Universiti Teknologi Malaysia for their support in conducting this study.

Declaration of Generative AI and AI-Assisted Technologies

During the preparation of this work used Scopus AI in order to ask question and retrieved factual information, which was later validated using Google Scholar. The authors also used Scispace in order to search for relevant content within academic papers to support the research. The authors used Quillbot in order to rephrase and refine text for improved clarity and readability. After using these tools, the authors reviewed and edited the content as needed and take full responsibility for the content of the publication.

References

[1] P. Nagendramma, S. Kaul, (2012). Development of ecofriendly/biodegradable lubricants: An overview. Renew. Sustain. Energy Rev., 16, 764–774. https://doi.org/10.1016/j.rser.2011.09.002

[2] J. Grace, S. Vysochanska, J. Lodge, P. Iglesias, (2015). Ionic liquids as additives of coffee bean oil in steel-steel contacts. Lubricants, 3, 637–649. https://doi.org/10.3390/lubricants3040637

[3] N. S. Battersby, (2000). The Biodegradability and Microbial Toxicity Testing of Lubricants - some recommendations. Chemosphere, 41, 1011–1027. https://doi.org/10.1016/s0045-6535(99)00517-2

[4] N. M. Abdullah, N. S. Mohd Aluwi, H. Park, N. Ahmad Kamal, (2018). Effect of Toxicity Concentration of Waste Lubricating Oil (Hazardous Noxious Substances) on Aquatic Life. Int. J. Eng. Technol., 7, 117–120. https://doi.org/https://doi.org/10.14419/ijet.v7i3.11.15942

[5] A.Z. Syahir, N.W.M. Zulkifli, H.H. Masjuki, M.A. Kalam, Abdullah Alabdulkarem, M.

Gulzar, L.S. Khuong, M.H. Harith, (2017). A review on bio-based lubricants and their applications. J. Clean Prod., 168, 997–1016. https://doi.org/10.1016/j.jclepro.2017.09.106

[6] N. A. A. M. Ariffin, S. Y. Foong, W. W. F. Chong, S. S. Lam, J.-H. Ng, H. Zhang, (2024) Ionic Liquid Lubricity Enhancement with Bio-oil Derived from Microwave Pyrolysis of Bamboo. Ind. Crops Prod., 214, 118543. https://doi.org/10.1016/j.indcrop.2024.118543

[7] F. M. T. Luna, J. B. Cavalcante, F. O. N. Silva, C. L. Cavalcante, (2015). Studies on biodegradability of bio-based lubricants. Tribol. Int., 92, 301–306. https://doi.org/10.1016/j.triboint.2015.07.007

[8] Z. Rasep, M. N. A. W. M. Yazid, S. Samion, N. A. C. Sidik, (2022). Potential of RBD Palm Oil as a Lubricant in Textured Journal Bearing using CFD with Consideration of Cavitation and Conjugate Heat Transfer. CFD Lett., 14, 98–110. https://doi.org/10.37934/cfdl.14.2.98110

[9] A. S. M. A. Haseeb, H. H. Masjuki, L. J. Ann, M. A. Fazal, (2010). Corrosion characteristics of copper and leaded bronze in palm biodiesel. Fuel Process. Technol., 91, 329–334. https://doi.org/10.1016/j.fuproc.2009.11.004

[10] T. Pathmasiri, G. Perera, (2020). Potential of using polyethylene as viscosity enhancer of palm oil to use as a lubricating oil. Adv. Mech. Eng., 12. https://doi.org/10.1177/1687814020970745

[11] A. Nugroho, M. Kozin, R. Mamat, Z. Bo, M. F. Ghazali, M. P. Kamil, P. Puranto, D. A. Fitriani, S. A. Azahra, K. P. Suwondo, P. S. Ashfiya, S. Daud, (2024) Enhancing tribological performance of electric vehicle lubricants: Nanoparticle-enriched palm oil biolubricants for wear resistance. Heliyon, 10. https://doi.org/10.1016/j.heliyon.2024.e39742

[12] A. Yadav, Y. Singh, P. Negi, (2021). A review on the characterization of bio based lubricants from vegetable oils and role of nanoparticles as additives. Mater. Today Proc., 46, 10513–10517. https://doi.org/10.1016/j.matpr.2021.01.046

[13] J. M. Liñeira del Río, E. R. López, J. Fernández, (2020). Synergy between boron nitride or graphene nanoplatelets and tri(butyl)ethylphosphonium diethylphosphate ionic liquid as lubricant additives of triisotridecyltrimellitate oil. J. Mol. Liq., 301, 112442 https://doi.org/10.1016/j.molliq.2020.112442

[14] A. Nassif, F. Georgi, P. Montmitonnet, F. Meyer, I. Lahouij, (2025). Synergistic anti-wear performance of TiO2 nanoparticles and ZDDP: Influence of dispersion methods. Tribol. Int., 210, 110791. https://doi.org/10.1016/j.triboint.2025.110791

[15] G. Huang, Q. Yu, Z. Ma, M. Cai, F. Zhou, W. Liu, (2019). Oil-soluble ionic liquids as antiwear and extreme pressure additives in poly-α-olefin for steel/steel contacts. Friction, 7, 18–31. https://doi.org/10.1007/s40544-017-0180-8

[16] A. N. Masri, M. I. Abdul Mutalib, W. Z. N. Yahya, N. F. Aminuddin, J. M. Leveque, (2018). Thermokinetics of SO_3H-functionalized dicationic ionic liquids: Effect of anions. IOP Conf. Ser.: Mater. Sci. Eng., 458, 012072. https://doi.org/10.1088/1757-899X/458/1/012072

[17] M. Sernaglia, D. Blanco, A. Hernández Battez, R. González, A. Fernández-González, M. Bartolomé, (2020). Two fatty acid anion-based ionic liquids - part II: Effectiveness as an additive to a polyol ester. J Mol Liq., 310, 113158. https://doi.org/10.1016/j.molliq.2020.113158

[18] A. N. Masri, A. A. Sulaimon, M. Z. Zakaria, A. S. Abidemi (2022). Imidazolium-based ionic liquids as demulsifier for water–crude oil emulsion. Mater. Today Proc. 64, 1597–1603. https://doi.org/10.1016/j.matpr.2022.03.098

[19] T. Ravi, A. N. Masri, H. Hasbullah, W. Z. N. Yahya, I. S. Azmi, I. M. Ibrahim, R. Mohsin, (2025). Systematic computational prediction and experimental confirmation of amino acid-based natural deep eutectic solvents for removal of sterically hindered trisulfur. Chem. Eng. Res. Des., 216, 270–281. https://doi.org/10.1016/j.cherd.2025.03.002

[20] S. A. S. Mohammed, W. Z. N. Yahya, M. A. Bustam, M. G. Kibria, A. N. Masri, N. D. M. Kamonwel, (2023). Study of the ionic liquids' electrochemical reduction using experimental and computational methods. J. Mol. Liq., 359, 119219. https://doi.org/10.1016/j.molliq.2022.119219

[21] Z. K. Koi, W. Z. N. Yahya, R. A. Abu Talip, K. A. Kurnia, (2019). Prediction of the viscosity of imidazolium-based ionic liquids at different temperatures using the quantitative structure property relationship approach. New J. Chem., 43, 16207–16217. https://doi.org/10.1039/c9nj03436f

[22] K. Adi Kurnia, M. Zunita, J. A. P. Coutinho, I. Gede Wenten, D. Santoso, (2022). Development of quantitative structure-property relationship to predict the viscosity of deep eutectic solvent for CO2 capture using molecular descriptor. J. Mol. Liq., 347, 118239. https://doi.org/10.1016/j.molliq.2021.118239

[23] J. Fan, Z. Dai, J. Cao, L. Mu, X. Ji, X. Lu, (2024). Hybrid data-driven and physics-based modeling for viscosity prediction of ionic liquids. Green Energy Environ., 9, 1878–1890. https://doi.org/10.1016/j.gee.2024.01.007

[24] H. W. Khan, A. V. B. Reddy, M. M. E. Nasef, M. A. Bustam, M. Goto, M. Moniruzzaman, (2020). Screening of ionic liquids for the extraction of biologically active compounds using emulsion liquid membrane: COSMO-RS prediction and experiments. J. Mol. Liq., 309, 113122. https://doi.org/10.1016/j.molliq.2020.113122

[25] J. Wang, Y. Guo, F. Liu, X. Zhang, W. Wang, Q. Peng, (2022). COSMO-RS prediction and experimental verification of deep eutectic solvents for water insoluble pesticides with high solubility. J. Mol. Liq., 349, 118139. https://doi.org/10.1016/j.molliq.2021.118139

[26] M. L. Alcantara, G. L. Bressan, P. V. A. Santos, M. F. V. Nobre, J. A. P. Coutinho, C. A. O. Nascimento, L. A. Follegatti-Romero, (2025). Density and viscosity of alkylammonium ionic liquids: Experimental and COSMO-RS. J. Mol. Liq., 417, 126616. https://doi.org/10.1016/j.molliq.2024.126616

[27] J. Palomar, V. R. Ferro, J. S. Torrecilla, F. Rodríguez, (2007). Density and molar volume predictions using COSMO-RS for ionic liquids. An approach to solvent design. Ind. Eng. Chem. Res., 46, 6041–6048. https://doi.org/10.1021/ie070445x

[28] W. C. Barnhill, J. Qu, H. Luo, H. M. Meyer, C. Ma, M. Chi, B. L. Papke, (2014). Phosphonium-organophosphate ionic liquids as lubricant additives: Effects of cation structure on physicochemical and tribological characteristics. ACS Appl. Mater. Interfaces, 6, 22585–22593. https://doi.org/10.1021/am506702u

[29]J. J. Buzolic, H. Li, Z. M. Aman, D. S. Silvester, R. Atkin, (2024). Surface-active ionic liquids as lubricant additives to hexadecane and diethyl succinate. Colloids Surf. A Physicochem. Eng. Asp., 699. https://doi.org/10.1016/j.colsurfa.2024.134669

[30]A. M. Sadanandan, P. K. Khatri, R. C. Saxena, S. L. Jain, (2020). Guanidine based amino acid derived task specific ionic liquids as noncorrosive lubricant additives for tribological performance. J. Mol. Liq., 313, 113527. https://doi.org/10.1016/j.molliq.2020.113527

[31]A. Wang, L. Chen, D. Jiang, H. Zeng, Z. Yan, (2014). Vegetable oil-based ionic liquid microemulsion biolubricants: Effect of integrated surfactants. Ind. Crops Prod., 62, 515–521. https://doi.org/10.1016/j.indcrop.2014.09.031

[32]M. D. Avilés, T. Caparros, P. Mostaza, M. D. Bermúdez, F. J. Carrión-Vilches, (2025). Novel vegetable biolubricants containing ionic liquid. Tribol. Int., 202. https://doi.org/10.1016/j.triboint.2024.110255

[33]A. N. Zakwan, R. S. F. Nazrah Hirzin, (2022). Comparative study on optimization of factors affecting epoxidation-hydroxylation reaction for the production of waste cooking oil based-polyol. Sci. Res. J., 19, 1–22. https://doi.org/10.24191/srj.v19i2.13748

[34]S. A. S. Amiril, E. A. Rahim, Z. Embong, S. Syahrullail, (2018). Tribological investigations on the application of oil-miscible ionic liquids additives in modified Jatropha-based metalworking fluid. Tribol. Int., 120, 520–534. https://doi.org/10.1016/j.triboint.2018.01.030

[35]J. Qu, D. G. Bansal, B. Yu, J. Y. Howe, H. Luo, S. Dai, H. Li, P. J. Blau, B. G. Bunting, G. Mordukhovich, D. J. Smolenski, (2012). Antiwear performance and mechanism of an oil-miscible ionic liquid as a lubricant additive. ACS Appl. Mater. Interfaces, 4, 997–1002. https://doi.org/10.1021/am201646k

[36]B. Khemchandani, A. Somers, P. Howlett, A. K. Jaiswal, E. Sayanna, M. Forsyth, (2014). A biocompatible ionic liquid as an antiwear additive for biodegradable lubricants. Tribol. Int., 77, 171–177. https://doi.org/10.1016/j.triboint.2014.04.016

[37]R. Gusain, A. Khan, O. P. Khatri, (2020). Fatty acid-derived ionic liquids as renewable lubricant additives: Effect of chain length and unsaturation. J. Mol. Liq., 301. https://doi.org/10.1016/j.molliq.2019.112322

[38]S. Yamamoto, H. Matsuda, Y. Kasahara, M. Iwahashi, T. Takagi, T. Baba, T. Kanamori, (2012). Dynamic molecular behavior of semi-fluorinated oleic, elaidic and stearic acids in the liquid state. J. Oleo Sci., 61, 649–657.

[39]C. J. Dongmo, M. Carrer, M. Houvet, T. Skrbic, G. Graziano, A. Giacometti, (2020). Can the roles of polar and non-polar moieties be reversed in non-polar solvents? Phys. Chem. Chem. Phys., 22, 25848–25858. https://doi.org/10.1039/d0cp02948c

[40]B. Yu, D. G. Bansal, J. Qu, X. Sun, H. Luo, S. Dai, P. J. Blau, B. G. Bunting, G. Mordukhovich, D. J. Smolenski, (2012). Oil-miscible and non-corrosive phosphonium-based

ionic liquids as candidate lubricant additives. Wear, 289, 58–64. https://doi.org/10.1016/j.wear.2012.04.015

[41]M. Shyama, S. Lakshmipathi, (2019). C–H···O interaction between cation and anion in amino acid-based ionic liquids—A DFT study in gas and solvent phase. Struct. Chem., 30, 185–194. https://doi.org/10.1007/s11224-018-1192-3

[42]M. A. Al-Maari, H. F. Hizaddin, A. Hayyan, (2023). COSMO-RS-based screening of organic solvents for efficient extraction of rubber seed oil: Computational prediction and experimental verification. Chem. Eng. Trans., 106, 1147–1152. https://doi.org/10.3303/CET23106192

[43]C. Qi, Z. Song, H. Cheng, L. Chen, Z. Qi, (2024). A systematic COSMO-RS study on mutual solubility of ionic liquids and C6-hydrocarbons. Green Chem. Eng., 5, 97–107. https://doi.org/10.1016/j.gce.2022.11.002

[44]L. Del Olmo, I. Lage-Estebanez, R. López, J. M. García De La Vega, (2016). Understanding the structure and properties of cholinium amino acid based ionic liquids. J. Phys. Chem. B, 120, 10327–10335. https://doi.org/10.1021/acs.jpcb.6b06969

[45]S. Xu, W. Ma, Q. Wei, X. Ren, (2025). Physicochemical properties and molecular interactions of long-chain 1-alkanols and N-dodecyl-2-pyrrolidone: Density, viscosity, thermodynamics, spectroscopy, and theoretical calculations. J. Chem. Eng. Data, 70, 1239–1249. https://doi.org/10.1021/acs.jced.4c00635

Carboxyl Versus Amide Groups Influence Towards Physicochemical and Thermal Properties of Choline Based Deep Eutectic Solvents

Theaveraj RAVI[1,a], Asiah Nusaibah Binte MASRI[1,2,3,b] *, Erny Izzaty MOHD NOR[1], Nur Ain Nabila HARMEN[1], Hasrinah HASBULLAH[1,4], Wan Zaireen NISA YAHYA[5], Izni Mariah IBRAHIM[1,2,3], Rahmat MOHSIN[1,2], Mohammad Afiq ZUBIR[1], Intan Suhada AZMI[6] and Jean-Marc LEVEQUE[7]

[1]Faculty of Chemical and Energy Engineering, Universiti Teknologi Malaysia, 81310 Skudai, Johor Bahru, Malaysia

[2]UTM-MPRC Institute for Oil & Gas, Universiti Teknologi Malaysia, 81310 Skudai, Johor Bahru, Malaysia

[3]Energy Management Group, Universiti Teknologi Malaysia, 81310 Skudai, Johor Bahru, Malaysia

[4]Advanced Membrane Technology Research Centre, Universiti Teknologi Malaysia, 8131 Skudai, Johor Bahru, Malaysia

[5]Universiti Teknologi PETRONAS, 32610 Seri Iskandar, Perak, Malaysia

[6]Universiti Teknologi MARA, 81750 Masai, Johor Bahru, Malaysia

[7]LRP, Universite Savoie Mont-Blanc, Le Bourget du Lac FR, 73376, France

[a]rktheaveraj@gmail.com, [b]nusaibah@utm.my*

Keywords: Thermogravimetric analysis, Deep Eutectic Solvents, Onset Temperature (T_o), Decomposition Temperature (T_{max}), Physicochemical Properties, Carboxyl, Amide

Abstract. This paper provides a thorough comparative analysis of the physicochemical and thermal properties of new deep eutectic solvent (DES); choline chloride – glutaric acid and the common DES; choline chloride – urea. The significance of choline chloride based deep eutectic solvents (DES) lies in their environmentally benign characteristics and their potential utilization in diverse industries owing to their low toxicity and ability to degrade naturally. Integrating ChCl with hydrogen bond donors having carboxyl group (GA) or amide group (urea) would open green application opportunities in bioprocessing, polymerization and extraction processes. In this study, the characterization encompasses of structural characterization through Fourier-transform infrared spectroscopy (FTIR), physicochemical characterization involving measurements of viscosity and refractive index and thermal characterization using techniques such as thermogravimetric (TGA) analysis and differential scanning calorimetry (DSC). Based on the viscosity and refractive index measurements, it can be observed that ChCl – GA has a higher value, followed by ChCl – urea. When it comes to melting temperatures, ChCl – GA has the lower melting point, which is 91.47 °C, whereas ChCl – urea has a slightly higher melting point of 98.02 °C. Thermal analysis was performed on both types of DES using TGA, with heating rates of 5, 10 and 15 °C/min, and a temperature range of 50 °C to 700 °C. The thermogravimetric study results showed that the ChCl – urea had greater thermal stability than ChCl – GA. By conducting a comprehensive examination of kinetic parameters, including activation energy (E_a) and pre-exponential factor (log A), using the Kissinger-Akahira-Sunose (KAS), Flynn-Wall-Ozawa (FWO) and Starink methods, it was concluded that the ChCl – urea compound exhibited lower average E_a and log A values compared to ChCl – GA. This difference can be attributed to the higher thermal stability of ChCl – urea. Our work established a robust correlation between the onset temperature (T_o), decomposition temperature (T_{max}), log A, E_a, and the values of DESs, indicating their thermal stability. In conclusion, ChCl – urea shows lower viscosity, lower density value, higher melting point and high thermal stability. Thus, it is suitable for any application that requires efficient flow properties and stability under high temperatures. ChCl – GA shows higher viscosity, higher density, higher melting point and low thermal stability. Thus, it would be suitable for applications requiring more robust, denser solvents that operate at lower temperatures.

Introduction

Solvents are essential in multiple fields of chemistry, such as synthetic, analytical, pharmaceutical, food and flavour chemistry, as well as materials and coating industries. By employing a solvent as a medium for a reaction, it becomes feasible to manipulate different reaction conditions. The circumstances encompass pressure, pH, temperature, concentration and kinetics [1]. Ionic liquids (ILs) have garnered significant interest during the last twenty years as a unique category of solvents [2]. These ionic liquids are compounds with low melting points, and some of them can even stay in a liquid form under normal conditions. They have a wide range of characteristics that make them beneficial for use as reaction solvents [3]. In addition, ILs generally have a low tendency to evaporate, which makes it possible to separate volatile reaction products through distillation. Moreover, their capacity to withstand high temperatures allows them to be used throughout a wide range of temperatures. Ionic liquids (ILs) have considerable benefits compared to traditional solvents in various applications. Nevertheless, ILs do have significant disadvantages. The primary drawbacks of ILs include their expensive starting materials and synthesis, toxicity, non-biodegradability and susceptibility to contaminants, especially water [4].

Deep eutectic solvents (DESs) might be seen as an alternative to ILs, as they are an expansion of the IL model and possess numerous similar features and physical qualities. Deep eutectic solvents (DESs) have attracted considerable research interest as a new type of solvent in recent years [3]. Extensive research has been conducted to examine the physical characteristics of different DESs, including as their melting temperature, density, viscosity, thermal stability and ionic conductivity. The purpose of this research is to investigate the potential commercial uses of these substances [5]. In previous investigations, a number of DESs were employed as eco-friendly solvents and catalysts [6]. Deep eutectic solvents (DESs) are formed through the combination of an organic salt, usually a quaternary ammonium salt (QAS), with other components such as a metal salt, a metal salt hydrate, or a hydrogen bond donor (HBD) [3]. The categorisation of DESs is as follows: Type I - combination of organic salt and metal salt, Type II - combination of organic salt and metal salt hydrate, Type III - combination of organic salt and hydrogen bond donor and Type IV - combination of metal salt and hydrogen bond donor [7].

The research in this work mostly focused on Type IV DESs. The predominant organic salts utilised in the preparation of deep eutectic solvents (DESs) are quaternary ammonium salts (QASs), with choline chloride (ChCl) being the most extensively researched QAS. When combined with the majority of hydrogen bond donors, it leads to the formation of eutectic mixtures that exhibit significantly enhanced physical properties compared to the use of other QASs [8]. The depression in freezing point observed was found to be significant, since it had one of the lowest viscosities. This can be attributed to the asymmetry and polar hydroxyl group of the choline cation. ChCl, sometimes referred to as provitamin B4, is a non-toxic substance that has been utilised in chicken feed for an extensive period of time [9]. This paper introduces ChCl – GA as a novel DES, marking the first thermal kinetics study for this solvent. GA, being a dicarboxylic acid and urea enhances the hydrogen bonding ability of the DES, which improves its capacity to dissolve complex molecules and maintain stability in harsh conditions. This makes the both DESs particularly useful in many applications. For ChCl – GA, application in biomass processing and polymerization reactions would be significant due to its carboxylic functional group, which promotes esterification and transesterification processes [4]. Besides, ChCl – urea is suitable for extraction of metals from ores or waste streams, as well as pharmaceutical synthesis due to its ability to form strong hydrogen bonds with various compounds [3].

Assessing the thermal stability of deep eutectic solvents (DESs) is essential prior to their implementation in certain applications. This parameter is important not just in academic contexts but also in high-temperature applications such as transesterification and dissolution. Research on the thermal stability of solvents commonly use methods such as UV-Vis spectroscopy, flame ionisation detection (FID) and mass spectrometry (MS) [10]. Nevertheless, the prevailing method involves estimating the onset temperature (T_o) and decomposition temperature (T_{max}) through the utilisation of a thermogravimetric analyser (TGA). The use of TGA allows for an isoconversional kinetic analysis, which provides accurate data for a full evaluation of kinetic parameters [11]. There are three isoconversional kinetics methods: maximum rate, differential and integral isoconversional. Although differential isoconversional methods based on Friedman can offer activation energy (E_a) values for each conversion (α), the calculation of E_a results in poor fitting and linearity. Moreover, the Friedman approach for computing demonstrates a wider spectrum of variability. However, the integral isoconversional technique, which includes the Flynn-Wall-Ozawa (FWO) and Kissinger-Akahira-Sunose (KAS) methods used by Starink and Coats-Redfern (CR), provides greater accuracy than the differential method [12].

Two forms of DES, namely choline chloride – glutaric acid (ChCl – GA) and choline chloride – urea (ChCl – urea), were prepared in this study. The three forms of DESs were structurally characterised using Fourier Transform Infrared (FTIR) spectroscopy [8]. Fourier Transform Infrared (FTIR) analysis is a useful technique for identifying the specific chemical groups present and verifying the creation of DESs by studying the interaction between ChCl and the hydrogen bond donors, namely glutaric acid and urea [3]. The distinctive peaks observed in the FTIR spectra offer valuable information on the molecular interactions and bonding, which are essential for comprehending the structural integrity and characteristics of the prepared DESs. The DESs were subjected to physicochemical characterisation in order to determine their density, viscosity and refractive index. The density of the DESs was determined to assess the impact of molecular interactions and structure on the compactness and mass per unit volume of the liquids [1]. Viscosity measurements, namely kinematic viscosity, were conducted to assess the impact on the flow behaviour and resistance of the DESs. The refractive index was determined with an automated refractometer, which yielded data regarding the propagation of light through the DESs [13]. The physicochemical properties play a crucial role in assessing the appropriateness of DESs for different industrial applications.

In addition, thermogravimetric analysis (TGA) was used to examine and explain the thermal stability of the DESs for thermal characterisation purposes. The TGA results yield information regarding the reduction in weight of the DESs as they undergo heating at different rates. In addition, the integral isoconversional approach was utilised to ascertain the kinetic characteristics, including the activation energy (E_a) and pre-exponential factors (log A), which are crucial for understanding the decomposition mechanisms and heat resistance of the DESs [12]. These criteria are crucial for accurately forecasting the thermal characteristics of DESs under varying circumstances. In addition, Differential Scanning Calorimetry (DSC) was employed to examine the temperature transitions of the DESs, including their melting and crystallisation characteristics. The DSC results yield useful data regarding the melting temperatures and eutectic properties of the DESs [3]. Differential scanning calorimetry (DSC) analysis provides valuable insights into the thermal stability and phase behaviour of deep eutectic solvents (DESs), which is crucial for their utilisation in processes that demand precise thermal profiles.

Methodology

Choline chloride ($C_5H_{14}ClNO$, 99%, Acros Organics), glutaric acid ($C_5H_8O_4$, 99%, QReC) and urea (CH_4N_2O, 99%, QReC) were used for the preparation of Deep Eutectic Solvents (DESs). The DESs were prepared by combining a hydrogen bond acceptor (HBA) and a hydrogen bond donor (HBD) at a suitable temperature to form a homogeneous liquid phase. In this study, choline chloride (ChCl) served as the HBA, while glutaric acid (GA) and urea acted as the HBDs in a 1:2 molar ratio. The preparation process involved initially melting the low-melting-point component, either GA (melting point 98 °C) or urea (melting point 133 °C), followed by the gradual addition of ChCl (melting point 302 °C). The mixture was continuously stirred and heated at 80 °C for 4 hours until a uniform, transparent liquid was obtained, indicating the successful formation of the DESs. This method follows the general procedure reported by previous researchers [3].

Morever, the DESs were analysed using a Perkin Elmer FTIR spectrometer, which had a wavenumber range of 450–4000 and performed four scans. Both DESs were examined using identical conditions. The DESs that will be analysed in the FTIR analysis are ChCl – GA and ChCl – urea. The kinematic viscosities of the DESs were tested using a Brookfield DV3T rheometer equipped with a temperature-controlled bath. The tests were conducted across a temperature range of 20 °C to 80 °C. In order to assess the behaviour of each DES in response to temperature, measurements were conducted at various temperatures. In order to ensure uniformity and the ability to make meaningful comparisons, both DESs, specifically ChCl – GA and ChCl – urea, were evaluated under the same conditions. The refractive indices of the DESs were measured using the ATAGO RX 5000 alpha automated refractometer. Prior to each series of measurements, the device was calibrated by determining the refractive index of millipore water.

Furthermore, the TGA measurements were performed via a Perkin Elmer TGA 4000 in a nitrogen environment to inhibit oxidation and other chemical processes. The DES samples underwent heating at three distinct rates: 5, 10 and 15 °C/min, spanning a temperature range from 50 °C to 700 °C. To provide an environment free from chemical reactions and to get precise measurements of thermal degradation and weight loss, a constant flow rate of 50 mL/min of nitrogen gas was maintained throughout the investigation. The thermal transitions of the DESs, including melting point, were evaluated using Mettler Toledo's differential scanning calorimetry (DSC). The DSC measurements were conducted in a nitrogen atmosphere using a gas flow rate of 50 mL/min. The specimens underwent a thermal cycling process at a rate of 10 °C per minute, spanning a temperature range from -50 °C to 250 °C.

The kinetics of the decomposition reaction for these deep eutectic solvents (DESS) were assessed using three different methods: Starink, Kissinger-Akahira-Sunose (KAS) and Flynn-Wall-Ozawa (FWO). The selection of these three methods was based on their widespread usage as p(y)-isoconversional methods, which allows for the calculation of kinetic parameters, including activation energy (E_a) and pre-exponential factor (log A), under various heating rates. Each technique produced a thermokinetic model using the equations presented by KAS, FWO and Starink, as shown in Equations 1-3 accordingly [12].

$$\ln\left(\frac{\beta}{T^2}\right) = \ln\left(\frac{AR}{E_a g(\alpha)}\right) - \frac{E_a}{RT}. \quad (1)$$

$$\ln(\beta) = \ln\left(\frac{AR}{E_a g(\alpha)}\right) - 5.331 - \frac{1.052E_a}{RT}. \quad (2)$$

$$\ln\left(\frac{\beta}{T^{1.92}}\right) = \ln\left(\frac{AR^{0.92}}{E_a^{0.92} g(\alpha)}\right) - 0.312 - \frac{1.008E_a}{RT}. \quad (3)$$

The symbol β in the equations denotes the heating rates, T represents the absolute temperature, g(α) represents a conversion factor function, A is the pre-exponential factor, R corresponds to the ideal gas constant with a value of 8.314 J/mol and E_a represents the activation energy. The computation to determine the activation energy (E_a) and pre-exponential factor (log A) includes using the average temperature obtained for each α value, as given in the supplementary data.

Results and Discussion

An extensive analysis was conducted to determine the functional groups present in deep eutectic solvents (DESs) used as functionalizing agents, followed by an investigation into the surface structural alterations of carbon. Fig. 1 illustrates the composition of the components of DESs. FTIR spectra were analyzed for two different deep eutectic solvents (DESs), specifically ChCl – GA and ChCl – urea. The identifiable peaks associated with the constituents of each DES, comprising two or more chemicals, were employed to interpret the spectrum. Fig. 2 presents the FTIR spectra of ChCl – GA and ChCl – urea.

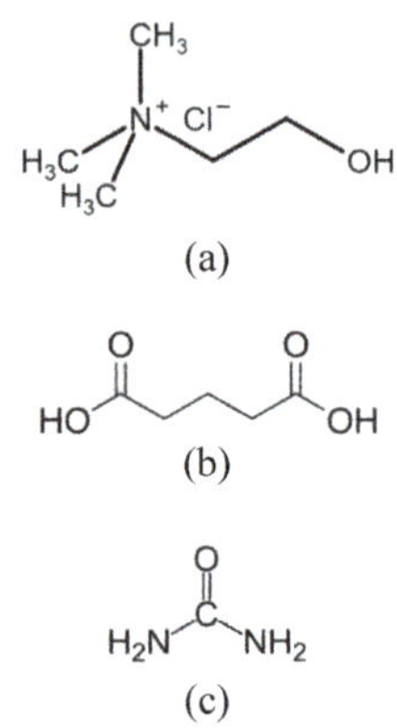

Fig. 1: Structure of component of DES (a) ChCl, (b) GA and (c) Urea

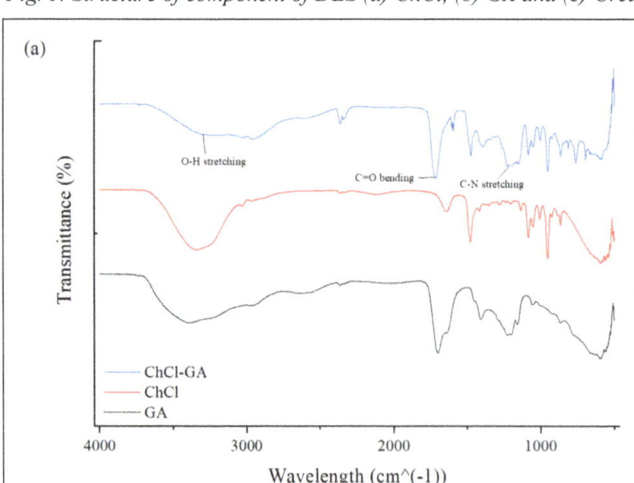

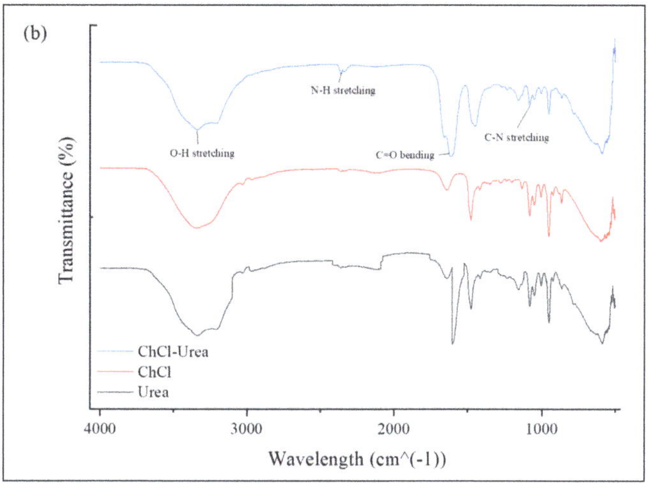

Fig. 2: FTIR spectra for (a) ChCl – GA and (b) ChCl – urea

According to Fig. 2(a), only three functional groups present in this DES comprised both ChCl and GA. Trace amounts of three peaks were detected in this DES, originating from O-H, C=O and C-N bonds derived from ChCl and GA sources. The primary functional groups found in ChCl – GA are quaternary ammonium, hydroxyl and carboxyl groups. Fig. 2(b) shows the presence of three functional groups in DES that contains ChCl and urea. The DES had trace levels of four peaks corresponding to O-H, N-H, C=O and C–N bonds. The ChCl – urea compound primarily contains functional groups such as quaternary ammonium, hydroxyl and amide groups.

In addition, the O-H stretching region of these DES demonstrates a reduced peak intensity (increased transmittance, T %) in comparison to the spectra of the individual components, namely ChCl, glutaric acid and urea. The observed drop in the highest level of intensity indicates the occurrence of hydrogen bond formation within the DES matrix. The greater peak intensity observed in the O-H stretching area of the individual components signifies more robust O-H stretching vibrations in the absence of hydrogen bonding [3]. Overall, the shift in the O-H stretching peak and the reduction in intensity in the FTIR spectra of the DES compared to the individual components conclusively demonstrate the formation of hydrogen bonds within the DES matrix, confirming its successful synthesis and unique structural characteristics.

The kinematic viscosities of both DESs were determined through measurements conducted at temperatures ranging from 20 °C to 80 °C. The temperatures employed were 20 °C, 25 °C, 30 °C, 35 °C, 40 °C, 45 °C, 50 °C, 55 °C, 60 °C, 65 °C, 70 °C, 75 °C and 80 °C. Fig. 3 displays the measured values of kinematic viscosities for both DESs (ChCl – GA and ChCl – urea) at the specified temperatures.

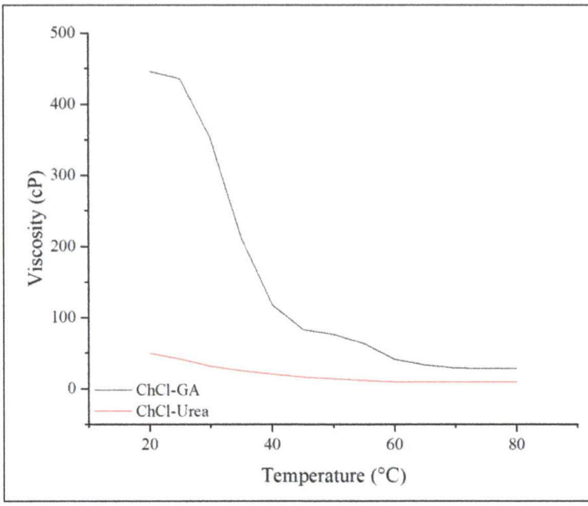

Fig. 3: Experimental kinematic viscosities of DESs at various temperatures

Fig. 3 demonstrates that the kinematic viscosity of both DESs reduces in a non-linear manner as the temperature increases within the examined temperature range. The data collected indicates that the DES containing urea has lower viscosity values in comparison to GA, possibly due to the smaller molecular size and reduced intermolecular tensions in the aliphatic structure of urea [14]. ChCl – GA has the highest viscosity compared to the other two DESs because of its larger molecular size and two carboxyl groups. These factors result in strong hydrogen bonding and increased internal friction [3]. These components lead to a more tightly bound liquid structure that opposes movement, resulting in a higher viscosity compared to ChCl – urea.

Furthermore, the refractive index serves as an indicator of the electronic polarisability of a molecule and offers valuable insights on the intermolecular forces and behaviour of molecules in a solution. Table 1 displays the refractive index values for both DESs. Understanding the refractive index is crucial for various applications including the utilisation of DESs in optical devices, sensor advancement and chromatography. It aids in guaranteeing precise identification and examination in procedures where the optical characteristics of the solvents are vital [15]. According to literature, the refractive index of DES does not have a direct correlation with the extraction process when it is used as an extracting agent. Nevertheless, in exceptional cases where the raffinate and extract phases possess identical refractive indices, it might be quite challenging to visually discern the existence of two distinct phases within the system [16].

Table 1: Refractive Index of DESs

DES	Refractive Index
ChCl – GA	1.47567
ChCl – Urea	1.44623

The objective of this work was to assess the thermal stability of ChCl – GA and ChCl – urea by analysing two key parameters: the onset temperature (T_o) and the decomposition temperature (T_{max}). T_o is the specific location where the baseline meets with the tangent of the sample's weight plotted against temperature. T_{max}, on the other hand, is the temperature at which the sample experiences the highest weight loss [17]. Fig. 4 presents the thermogravimetric (TG) and derivative thermogravimetry (DTG) curves of ChCl – GA and ChCl – urea. Table 2 presents the T_0 values for both DESs at different heating rates (5, 10 and 15°C/min). When comparing the T_0 values, it is observed that ChCl – urea has substantially higher values, ranging from 195 °C to 215 °C at different heating rates. This result indicates that ChCl – urea begins its thermal changes at elevated temperatures. Higher onset temperatures can serve as an indication of a material's stability during the early stages of heating. ChCl – GA demonstrates marginally lower initiation temperatures, varying between 190 °C and 210 °C. This result indicates that ChCl – GA experiences thermal transformations at lower temperatures in comparison to ChCl – urea. Lower onset temperatures may suggest a heightened reactivity to heat.

In addition, while comparing the T_{max} values, it is observed that the decomposition temperatures for ChCl – urea are higher, with a range of 270 °C to 310 °C. Higher decompostion temperature (T_{max}) results indicate that ChCl – urea can endure higher temperatures without experiencing major decomposition. This suggests an elevated level of thermal stability. ChCl – GA has lower decomposition temperature values, which range from 270 °C to 290 °C. Lower decomposition temperature values indicate that the ChCl – GA compound is more likely to decompose at lower temperatures, indicating relatively lower thermal stability. Overall, ChCl – urea has greater thermal stability, as evidenced by its higher T_o and T_{max} values. ChCl – GA demonstrates reduced thermal stability, leading to alterations at lower T_o and T_{max} values. Additionally, these DES exhibit a wider temperature range of stability, making them suitable for various applications across different temperature conditions.

The dissimilarities in thermal stability between the two deep eutectic solvents (ChCl – GA and ChCl – urea) can be ascribed to the distinct breakdown mechanisms, the characteristics of their functional groups and the strength of the bonds engaged [3]. The thermal stability of ChCl – urea is increased principally because of the presence of the amide group in urea. The amide group exhibits a higher affinity for the chloride ion, resulting in the formation of more robust hydrogen bonds that are less susceptible to breaking under high temperatures. The enhanced binding strength enhances the overall stability of the DES, rendering it more resilient to breakdown. On the other hand, ChCl – GA exhibits reduced thermal stability due to the comparatively weaker hydrogen bonds formed by its carboxyl groups, making it more susceptible to bond breakage when exposed to heat. The decomposition of these DESs entails the disruption of hydrogen bonds and the splitting of C-O or C=O bonds, contingent upon the particular DES [18].

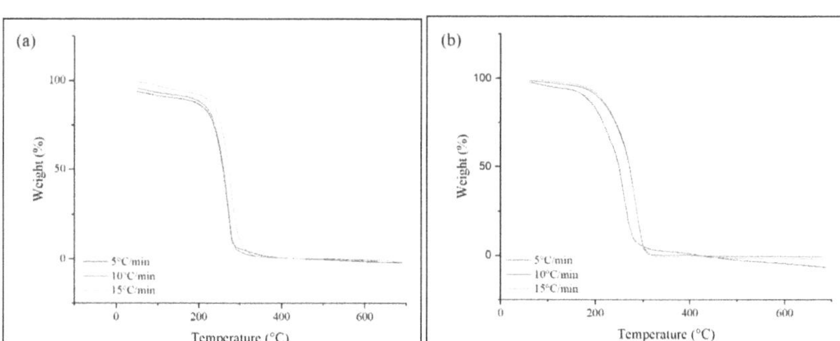

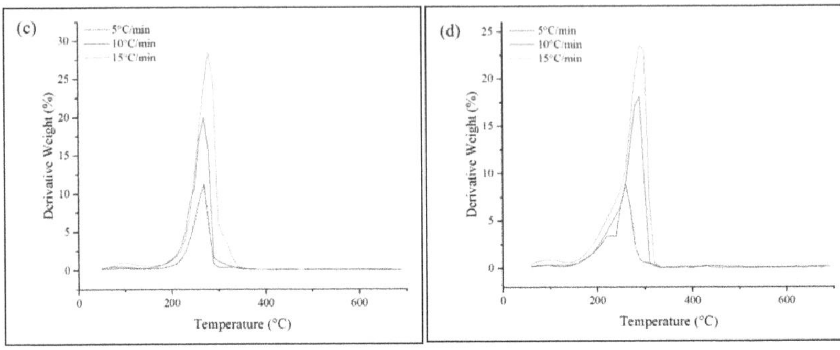

Fig. 4: The TG and DTG curves for (a,c) ChCl – GA and (b,d) ChCl – urea at heating rate of 5, 10 and 15 °C/min.

Despite variations in the DTG (derivative thermogravimetric) patterns between ChCl – GA and ChCl – urea, both DESs exhibited a comparable overall behavior in the TG and DTG curves, irrespective of the heating rates employed. This similarity can be attributed to a common decomposition process shared by these DESs. The thermal degradation of these DESs involves both breakdown and evaporation mechanisms [3]. During the breakdown step, nucleophilic substitution reactions, particularly SN_2 mechanisms, play a crucial role. For instance, in the case of ChCl – GA and ChCl – urea, nucleophilic attack by the chloride anion results in the formation of new compounds [19]. In general, the thermal degradation process of choline chloride based deep eutectic solvents (DESs) is mainly driven by reverse Menshutkin reactions and Hofmann eliminations [20].

Specifically, the reverse Menshutkin reaction in ChCl – GA and ChCl – urea leads to the formation of dimethylaminoethanol and methyl chloride, alongside glutaric acid and urea, respectively. This reaction occurs as the nucleophilic chloride anion attacks the quaternary ammonium cation, facilitating the release of methyl chloride [21]. In contrast, the Hofmann elimination mechanism yields trimethylamine and acetaldehyde, with glutaric acid and urea remaining as byproducts [22]. These are the proposed degradation mechanism inspired from other literature review as mentioned in Fig. 5. Both degradation pathways are accompanied by the conversion of the DES components into gaseous states, such as methyl chloride, trimethylamine and acetaldehyde. These processes are driven by the breakdown of the DES constituents through nucleophilic attack and subsequent thermal reactions, contributing to the overall thermal degradation profile observed.

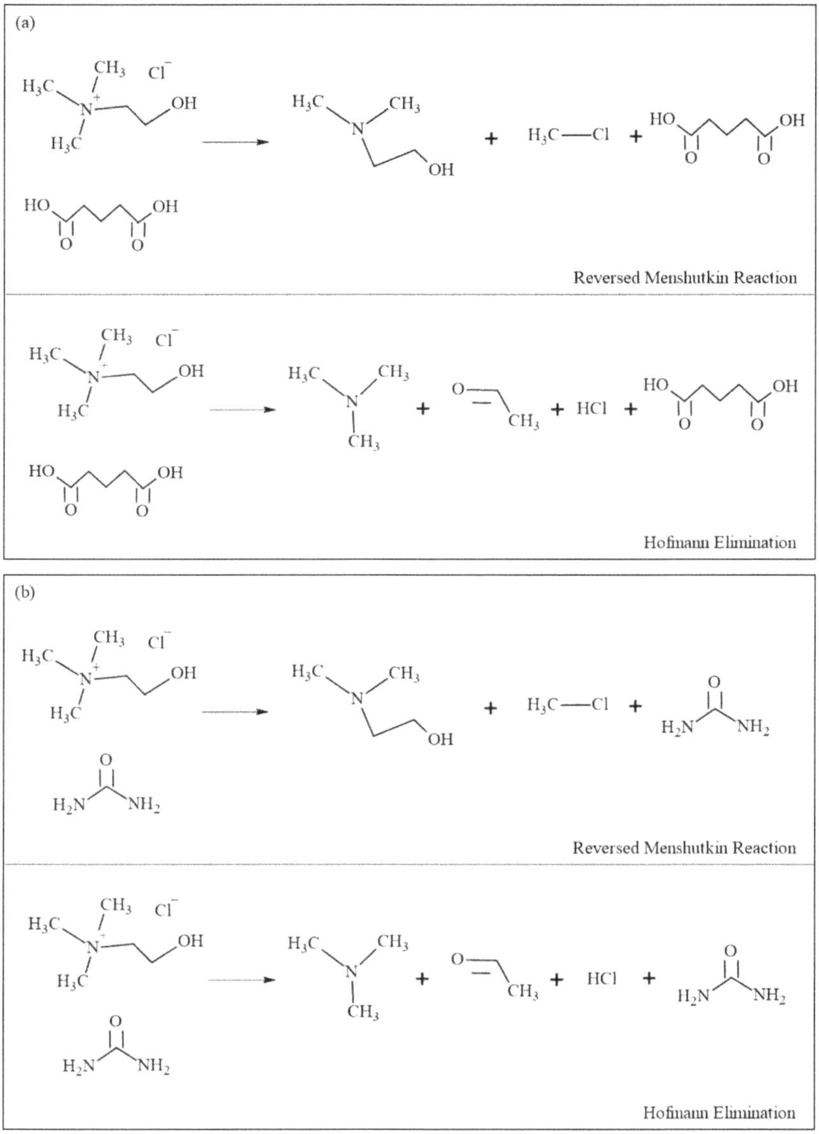

Fig. 5: Proposed degradation mechanism of (a) ChCl – GA and (b) ChCl – urea

Furthermore, the similarities found in the TG and DTG curves between ChCl – GA and ChCl – urea, despite their different structures, can be attributed to the shared breakdown mechanisms inherent to both deep eutectic solvents (DESs). The aliphatic nature and the presence of two carboxyl groups in ChCl – GA lead to a less stable structure. This is because the C-O and O-H bonds are more prone to breaking under heat stress due to the absence of significant resonance stabilisation, unlike the amide group in urea [12]. The decreased thermal stability of ChCl – GA can be ascribed to the presence of weaker bonds, which are more susceptible to attack by the nucleophilic chloride ion, leading to breakdown at lower temperatures [19]. On the other hand, ChCl – urea, which includes the amide group, has superior resistance to changes in temperature despite having aliphatic structure. The presence of the amide group in the DES increases its ability to form stronger hydrogen bonds, which in turn improves its resistance to nucleophilic attack and thermal degradation [20].

In addition, ChCl – urea, which contains a amide group, has greater thermal stability in comparison to ChCl – GA. The presence of the amide group in urea enhances hydrogen bonding and provides more resistance to nucleophilic attack compared to the carboxyl groups in glutaric acid. As a result, the DES structure is more resistant to thermal degradation [12]. Conclusively, the thermal stability trend can be described as follows: ChCl – GA (aliphatic compound with two carboxyl groups) has lower thermal stability compared to ChCl – urea (aliphatic compound with two amide group). This trend demonstrates the growing potency of hydrogen bonding and thermal resilience, which are directly related to the structural attributes of each DES.

Table 2: T_o and T_{max} at different heating rates

Heating Rate (°C/min)	ChCl – GA		ChCl – urea	
	T_o (°C)	T_{max} (°C)	T_o (°C)	T_{max} (°C)
5	190	270	195	270
10	200	280	205	300
15	210	290	215	310

Moreover, the preparation of DES resulted in the formation of a eutectic point, which is the temperature at which a mixture melts at a lower temperature than the individual components would melt on their own. The thermal analysis of DES was conducted using DSC to investigate the phase transitions at ChCl – GA and ChCl – urea, with the aim of determining the eutectic point. Fig. 6 illustrates the heating curve of both DESs.

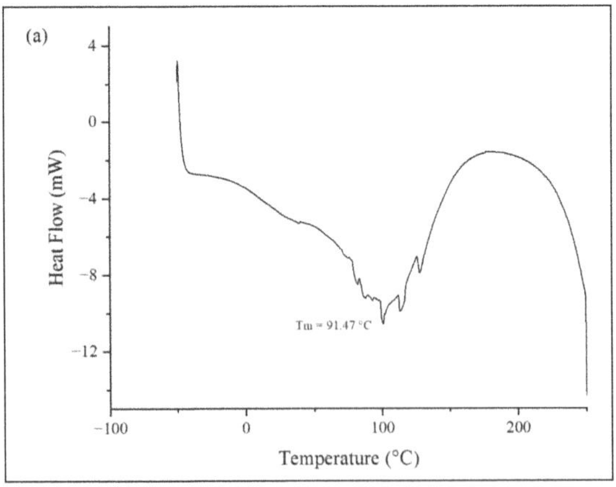

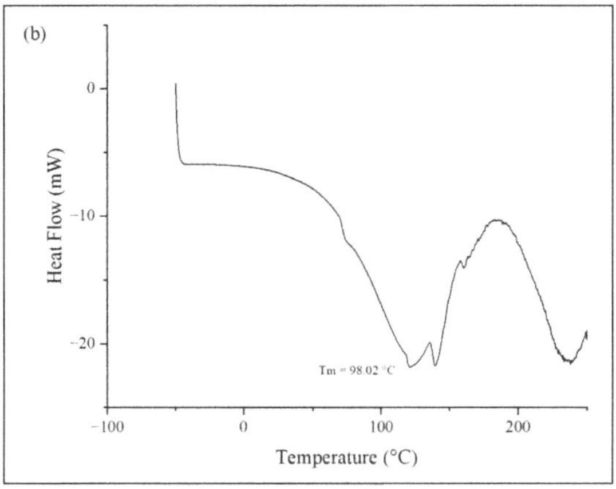

Fig. 6: Heating curve of DES (a) ChCl – GA and (b) ChCl – urea

Fig. 6(a) shows an observable endothermic effect for ChCl – GA, indicating the melting temperature (T_m). The phenomenon of melting point depression was a crucial aspect of our research on DES that we focused on extensively [3]. As anticipated, the melting temperature of ChCl – GA is lower than that of its corresponding hydrogen bond acceptor (HBA) and hydrogen bond donor (HBD). As shown in Table 3, the melting point of DES, which is 91.47 °C, is equivalent to its eutectic point. Glutaric acid has a melting temperature of 98 °C, but ChCl has a higher melting point of 302 °C. Fig. 6(b) shows that ChCl – urea exhibits a distinct endothermic peak, which indicates the melting temperature (T_m) of the deep eutectic solvent (DES). The differential scanning calorimetry (DSC) research indicates that the melting point of ChCl – urea is considerably lower in comparison to the melting points of its separate components, which is consistent with the eutectic properties of deep eutectic solvents (DESs). The recorded melting temperature of ChCl – urea is roughly 98.02 °C, indicating its eutectic point. The DES components further confirm this discovery, as urea has a melting temperature of around 133 °C, whereas ChCl has a significantly higher melting point of 302 °C. Thus, the huge disparity between the melting points of these substances and the observed eutectic point in all deep eutectic solvents (DES) indicates that the strong interactions between ChCl and all hydrogen bond donors (HBDs) greatly reduce the total melting point of the system [3].

Table 3: Melting temperature (T_m) of DESs

DES	T_m (°C)
ChCl – GA	91.47
ChCl – urea	98.02

The TG and DTG methods can be theoretically employed to determine the kinetic parameters for the thermal degradation of DESs. The TG technique was employed to compute the kinetic parameters, namely the activation energy (E_a) and the pre-exponential component (log A), in this study. The computations were conducted utilising fraction conversion (α) through the KAS, FWO and Starink methods. Table 4 displays the the value of E_a, ln(A) and R^2 for ChCl – GA and ChCl – urea at various fraction conversions using KAS, FWO and Starink methods.

Table 4: Values of E_a, ln (A) and R^2 of DESs at different fraction conversions

Method	α	E_a (kJ/mol)		ln (A)		R^2	
		ChCl – GA	ChCl – urea	ChCl – GA	ChCl – urea	ChCl – GA	ChCl – urea
KAS	0.1	170.479	129.715	26.903	17.608	0.8821	0.9784
	0.2	167.377	124.976	27.253	17.771	0.882	0.9783
	0.3	161.250	122.640	26.971	17.910	0.8819	0.9782
	0.4	158.232	120.329	26.916	17.933	0.8818	0.9782
	0.5	155.247	89.151	26.796	11.244	0.8818	0.9569
	0.6	152.279	85.684	26.634	11.011	0.8817	0.9567
	0.7	146.443	82.280	26.102	10.751	0.8816	0.9564
	0.8	82.570	66.369	11.513	7.614	0.9255	0.9963
	0.9	15.895	58.125	-5.237	6.633	0.9756	0.9961
FWO	0.1	170.919	132.194	40.849	32.090	0.8928	0.9811
	0.2	167.892	127.531	41.217	32.290	0.8929	0.981
	0.3	161.910	125.232	40.974	32.449	0.8929	0.981
	0.4	158.962	122.956	40.937	32.490	0.893	0.981
	0.5	156.038	93.193	40.836	26.350	0.893	0.964
	0.6	153.145	89.739	40.694	26.157	0.8931	0.9639
	0.7	147.439	86.349	40.200	25.937	0.8931	0.9639
	0.8	86.483	70.946	26.659	23.142	0.9381	0.997
	0.9	22.225	62.635	12.687	22.294	0.9902	0.996
Starink	0.1	169.497	129.057	27.554	18.282	0.8826	0.9785
	0.2	166.412	124.347	27.904	18.446	0.8825	0.9784
	0.3	160.333	122.029	27.626	18.587	0.8824	0.9784
	0.4	157.339	119.728	27.570	18.611	0.8823	0.9783
	0.5	154.362	88.798	27.451	11.945	0.8823	0.9572
	0.6	151.425	85.350	27.290	11.714	0.8822	0.957
	0.7	145.627	81.967	26.759	11.455	0.8821	0.9568
	0.8	82.249	66.170	12.215	8.334	0.9261	0.9964
	0.9	16.065	57.972	-4.398	7.359	0.9766	0.9962

The thermokinetic curve for KAS, FWO and Starink, along with E_a, for ChCl – GA, at various α, is displayed in Fig. 7. The average E_a of ChCl – GA is 134.42, 136.11 and 133.70 kJ/mol, according to KAS, FWO and Starink, respectively. Additionally, the pre-exponential factor was explored in terms of the log A, with KAS, FWO and Starink having respective values of 18.87, 36.12 and 19.22 min^{-1}. Guida and co-workers states that an imprecise approximation of temperature integration is the reason for the slight difference between the measured values of E_a and log A [23]. As can be seen in Fig. 7d, for α values ranging from 0.1 to 0.9, the three models show similar E_a patterns. E_a showed an decreasing tendency from α values of 0.1 to 0.9. The drop in E_a indicates the formation of intermediates from the nucleophilic substitution process [17]. Additionally, Fig. 8 displays the E_a profile and kinetic plot for ChCl – urea. The average E_a for the KAS, FWO and Starink models are 97.67, 101.20 and 97.27 kJ/mol, respectively. The average log A values for ChCl – urea are 13.16, 28.13 and 13.86 min^{-1}. Fig. 8d illustrates how the E_a trend of ChCl – urea decreases as the α value rises. E_a showed an decreasing tendency from α values of 0.1 to 0.9. The drop in E_a indicates the formation of intermediates from the nucleophilic substitution process [17].

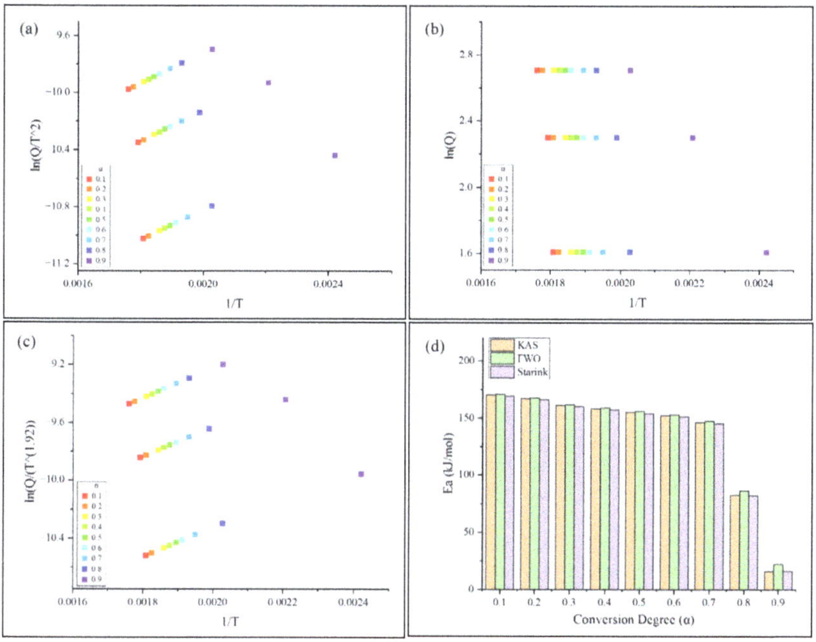

Fig. 7: Thermokinetic plot using (a) KAS, (b) FWO, (c) Starink method and (d) the E_a profile for ChCl – GA.

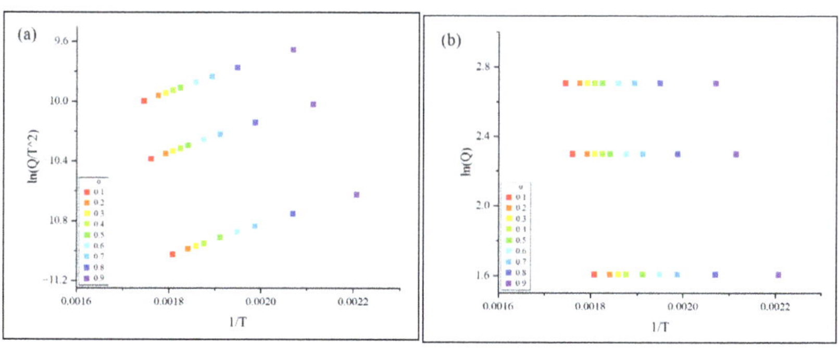

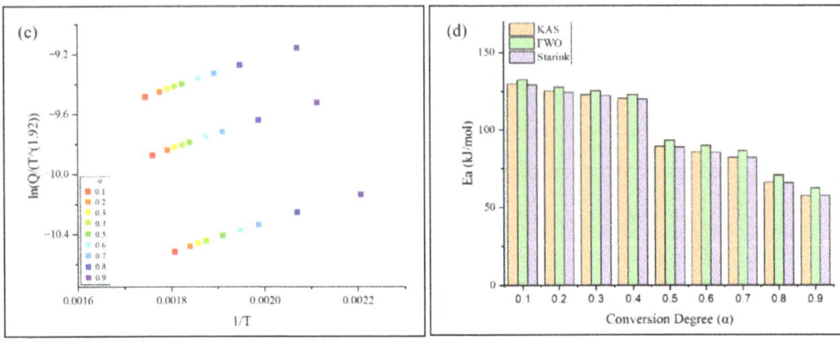

Fig. 8: Thermokinetic plot using (a) KAS, (b) FWO, (c) Starink method and (d) the E_a profile for ChCl – urea.

After extensive analysis of the kinetic properties of these DESs, it was shown that ChCl – GA had a larger average activation energy, indicating that it requires more energy to initiate a reaction. It tends to break down at lower temperatures since it has the lowest average T_{max}. The fact that ChCl – urea has the lowest average activation energy suggests that it reacts more readily. Moreover, its average T_{max} is the higher, indicating that it is more resistant to breakdown at lower temperatures. Because it reacts at lower temperatures (low activation energy) and can tolerate breakdown at higher temperatures (high T_{max}), ChCl – urea has a tendency to have higher thermal stability [10]. Additionally, ChCl – GA requires more energy to start reactions (high activation energy) and breaks down at lower temperatures (low T_{max}), in which it exhibits lesser thermal stability. For both DESs, the E_a values generated by the KAS, FWO and Starink models were almost equal, but the FWO model's log A value was found to be significantly higher than the values from the other two models, indicating that it was not very dependable. Compared to ChCl – GA, ChCl – urea has a lower average log A value. A material may have a higher susceptibility to thermal deterioration or reactions at a particular temperature if log A is noticeably higher for that substance. On the other hand, a lower log A could point to a substance that is more stable and reacts more slowly [12]. As a result, ChCl – urea has a higher thermal stability than ChCl – GA. Furthermore, the achieved R^2 value is higher than 0.8 which demonstrate that each α is the best fit with all of the chosen kinetic equations.

Furthermore, as a amide molecule, ChCl – urea exhibits greater thermal stability, as shown by its higher T_{max} values and starting temperatures. Under the studied conditions, ChCl – urea is more resistant to thermal breakdown, as indicated by its lower activation energy and T_{max} [12]. This indicates that it is thermally stable. Compared to ChCl – GA, which is classified as a carboxyl compound, exhibits low thermal stability. Thus, compared to amide compounds, carboxyl compounds often have reduced structural stability. Moreover, ChCl – GA has lower T_{max} and T_o values, which indicates that it has relatively weaker thermal stability and is more prone to thermal decomposition at lower temperatures.

Besides, there is still a lack of information on the maximum operating temperature (MOT) for DESs based on choline chloride. Recent studies have effectively used MOT to predict long-term thermal stability [19]. The MOT of DESs can be estimated using the kinetic parameter values found in this study. Equation (4) can be used to calculate the MOT of ChCl – GA and ChCl – urea as follows [12]:

$$\text{MOT} = \frac{E_a/R}{4.6+\ln(A.t_{max})}. \tag{4}$$

where the maximum exposure duration is denoted by t_{max}. The kinetic parameters derived from the KAS, FWO and Starink techniques were used to compute the MOT (maximum operating temperature) of ChCl – GA and ChCl – urea. Table 5 displays the computed MOT values for every DES.

Table 5: MOT of DESs

Methods	ChCl – GA	ChCl – urea
KAS	508.41 °C	501.21 °C
FWO	353.2 °C	316.86 °C
Starink	495.52 °C	484.64 °C

The highest temperature at which a DES can be used without experiencing appreciable thermal damage is determined by the maximum operating temperature (MOT), an essential characteristic. The trend in the MOT values, as shown in Table 5, indicates that ChCl – GA has a higher MOT than ChCl – urea. The proximity of these values indicates that each DES can sustain thermal stability throughout a range of temperatures, despite the modest changes in the MOT values [11]. Because of this, they are flexible choices for a range of high-temperature procedures where thermal stability is essential, like extractions and reactions. The comparable MOTs also suggest that there is little need to worry about temperature deterioration when using these DESs in applications that depend on other parameters like selectivity or solubility [12].

Conclusion
The objective of this study is to examine the structural, physicochemical, thermal stability of choline-based deep eutectic solvents (DESs). The empirical evidence regarding thermal stability indicates that ChCl – urea exhibits greater thermal stability compared to ChCl – GA. The DTG analysis also uncovered the single and multistage decompositions of ChCl – GA and ChCl – urea. In addition, the KAS, FWO and Starink methods were employed to do the kinetic evaluation. The obtained R^2 value is greater than 0.8, indicating that each α parameter is the most suitable fit for all of the selected kinetic equations. The thermal stability of the DESs has been found to have a linear correlation with the activation energy (E_a) and the pre-exponential factor (log A). By comparison, the mean activation energy (E_a) values for ChCl – GA using KAS, FWO and Starink methods are 134.42, 136.11 and 133.70 kJ/mol, respectively. Similarly, the average E_a values for ChCl – urea are 97.67, 101.20 and 97.27 kJ/mol. The log A derived from the FWO model greatly surpasses the values from the other two models, suggesting its probable lack of reliability. In terms of characterisation, the viscosity and refractive index results indicate that ChCl – GA has a greater value compared to ChCl – urea. ChCl – GA has a lower melting point of 91.47 °C, while ChCl – urea has a higher melting point of 98.02 °C. In conclusion, this study is essential for obtaining initial data about the thermal stability and kinetic properties of choline-based deep eutectic solvents (DESs) as an extractant.

Acknowledgements
The authors acknowledged the Universiti Teknologi Malaysia Encouragement Research (UTMER) Grant (QJ130000.3846.31J33), Ministry of Higher Education (MOHE) Fundamental of Research Grant Scheme (FRGS/1/2023/TK08/UTM/02/36), Energy Management Group, UTM-MPRC Institute For Oil and Gas (IFOG) and Universiti Teknologi Malaysia for their support in conducting this study.

References

[1] S. Khaidzir, et al., Ultrasonic-assisted technique as a novel method for removal of naphthenic acid from model oil using piperidinium-based ionic liquids, ACS Omega 6 (2021) 9629-9637. https://doi.org/10.1021/acsomega.1c00189

[2] A.A. Sulaimon, et al., Synthesis of dihydrogen phosphate-based ionic liquids: Experimental and COSMO-RS based investigation for methane hydrate inhibition, Journal of Molecular Liquids 319 (2020) 114092. https://doi.org/10.1016/j.molliq.2020.114092

[3] T. Ravi, A.N. Masri, I.M. Ibrahim, Choline-based Deep Eutectic Solvent for Extractive Oxidative Desulfurization of Model Oil, in: E3S Web of Conferences, EDP Sciences, 2024. https://doi.org/10.1051/e3sconf/202448803004

[4] S.K. Singh, A.W. Savoy, Ionic liquids synthesis and applications: An overview, Journal of Molecular Liquids 297 (2020) 112038. https://doi.org/10.1016/j.molliq.2019.112038

[5] K. Jafari, M.H. Fatemi, P. Estellé, Deep eutectic solvents (DESs): A short overview of the thermophysical properties and current use as base fluid for heat transfer nanofluids, Journal of Molecular Liquids 321 (2021) 114752. https://doi.org/10.1016/j.molliq.2020.114752

[6] F.M. Perna, P. Vitale, V. Capriati, Deep eutectic solvents and their applications as green solvents, Current Opinion in Green and Sustainable Chemistry 21 (2020) 27-33. https://doi.org/10.1016/j.cogsc.2019.09.004

[7] K.A. Omar, R. Sadeghi, Database of deep eutectic solvents and their physical properties: A review, Journal of Molecular Liquids 384 (2023) 121899. https://doi.org/10.1016/j.molliq.2023.121899

[8] R. Häkkinen, A.P. Abbott, Deep eutectic solvents-Teaching nature lessons that it knew already, in: Advances in Botanical Research, Elsevier, 2021, pp. 1-16. https://doi.org/10.1016/bs.abr.2020.09.013

[9] B. Janković, et al., Thermal decomposition kinetics of deep eutectic solvent (DES) based on choline chloride and magnesium chloride hexahydrate: New details on the reaction mechanism and enthalpy-entropy compensation (EEC), Journal of Molecular Liquids 374 (2023) 121274. https://doi.org/10.1016/j.molliq.2023.121274

[10] Y. Chen, et al., Evaluating thermal decomposition of ionic liquids and deep eutectic solvents for reliable and consistent thermal analysis: Issues, factors, classifications and suggestions, Thermochimica Acta 723 (2023) 179471. https://doi.org/10.1016/j.tca.2023.179471

[11] P. Wienchol, et al., Thermogravimetric and kinetic study of thermal degradation of various types of municipal solid waste (MSW) under N_2, CO_2 and oxy-fuel conditions, Energy 248 (2022) 123573. https://doi.org/10.1016/j.energy.2022.123573

[12] A.H. Ab Rahim, et al., Thermal kinetics of monocationic and dicationic pyrrolidinium-based ionic liquids, Materials 15 (2022) 1247. https://doi.org/10.3390/ma15031247

[13] M. Duerinck, Preparation of Ionic Liquids and Deep Eutectic Solvents for the Chemical Valorisation of Polylactide Waste, Universitat de Valéncia, 2020.

[14] G. Rajeswari, S. Jacob, Co-fermentation of lactic acid and acetone-butanol-ethanol (ABE) from the deep eutectic solvent-pretreated Aloe vera leaf rind through sequential valorization of holocellulose, Biomass Conversion and Biorefinery (2022) 1-19. https://doi.org/10.1007/s13399-022-03688-3

[15] D.S. Freitas, A. Cavaco-Paulo, C. Silva, Enhancing insights into the phenomena of deep eutectic solvents, Sustainable Materials and Technologies (2024) e01039. https://doi.org/10.1016/j.susmat.2024.e01039

[16] K.A. Omar, R. Sadeghi, Novel nonanol-based deep eutectic solvents: Thermophysical properties and their applications in liquid-liquid extraction and amino acid detection, Journal of Molecular Liquids 336 (2021) 116359. https://doi.org/10.1016/j.molliq.2021.116359

[17] A. Škulcová, et al., Physical properties and thermal behavior of novel ternary green solvents, Journal of Molecular Liquids 287 (2019) 110991. https://doi.org/10.1016/j.molliq.2019.110991

[18] A.M. da Costa Lopes, et al., Novel insights into biomass delignification with acidic deep eutectic solvents: A mechanistic study of β-O-4 ether bond cleavage and the role of the halide counterion in the catalytic performance, Green Chemistry 24 (2022) 2474-2487. https://doi.org/10.1039/C9GC02569C

[19] S. Dughera, et al., How do diazonium salts behave in deep eutectic solvents?, in: Diazo - Book of Abstracts, 2021, p. 43.

[20] B. Jóźwiak, et al., High operational functionality nanofluids from choline chloride, ethylene glycol, and multi-walled carbon nanotubes-Economical substitutes for ionanofluids, Journal of Molecular Liquids 409 (2024) 125551. https://doi.org/10.1016/j.molliq.2024.125551

[21] N. Peeters, et al., Choline chloride-ethylene glycol based deep-eutectic solvents as lixiviants for cobalt recovery from lithium-ion battery cathode materials: Are these solvents really green in high-temperature processes?, Green Chemistry 24 (2022) 6685-6695. https://doi.org/10.1039/D2GC02075K

[22] Y. Hariyanto, et al., Thermostable deep eutectic solvents for thermochemical reactions: Novel stability indicators, ACS Sustainable Chemistry & Engineering 12 (2024) 6774-6785. https://doi.org/10.1021/acssuschemeng.4c01620

[23] M. Guida, et al., Thermal degradation behaviors of sawdust wood waste: Pyrolysis kinetic and mechanism, Journal of Materials and Environmental Science 10 (2019) 742-755.

Evaluation of Phase Separation Models and Miscibility Behavior of Diol-Based Additives in Gasoline

Bakir ELOUARET[1,a], Asiah Nusaibah MASRI[1,2,3,b*], Hasrinah HASBULLAH[1,4,c], Izni Mariah IBRAHIM[1,2,3,d]

[1]Faculty of Chemical and Energy Engineering, Universiti Teknologi Malaysia, 81310 Johor Bahru, Johor, Malaysia

[2]UTM-MPRC Institute for Oil & Gas, Universiti Teknologi Malaysia, 81310 Johor Bahru, Johor, Malaysia

[3]Energy Management Group, Universiti Teknologi Malaysia, 81310 Johor Bahru, Johor, Malaysia

[4]Advanced Membrane Technology Research Centre, Universiti Teknologi Malaysia, 81310 Johor Bahru, Johor, Malaysia

[a]elouaret@graduate.utm.my, [b]nusaibah@utm.my, [c]hasrinah@utm.my, [d]iznimariah@utm.my

Keywords: Diol Additives, Gasoline Miscibility, Phase Separation, NRTL Model, Fuel Blending

Abstract. This study investigates the miscibility and phase separation behavior of diol-based oxygenated additives in gasoline through a combination of experimental and thermodynamic modeling approaches. Binary (methanol/hexane) and ternary (methanol/hexane/ethylene glycol) systems were experimentally tested to validate three thermodynamic models—Non-Random Two-Liquid (NRTL), UNIQUAC Functional-group Activity Coefficients (UNIFAC), and WILSON—using Aspen Plus simulations. Among them, the NRTL model demonstrated the highest accuracy, with percentage errors generally below 10%, confirming its suitability for predicting liquid–liquid equilibrium (LLE) in polar–nonpolar systems. The validated NRTL model was then applied to evaluate the miscibility of 23 diols with gasoline, including linear, branched, and unsaturated structures. Results demonstrated that miscibility is critically influenced by hydroxyl group positioning, carbon chain architecture, and the presence of unsaturation. Specifically, diols with vicinal or internal OH groups exhibited superior compatibility compared to those with terminal OH groups. Branching (e.g., 2-methyl-1,3-propanediol) and unsaturation (e.g., trans-2-butene-1,4-diol) further enhanced miscibility by reducing polarity and strengthening hydrocarbon interactions. In contrast, increasing carbon chain length alone provided only a modest improvement. Crucially, to provide direct design guidelines for fuel formulation, this study establishes a definitive structural ranking for diol–gasoline miscibility: Unsaturated diols > Vicinal diols > Branched diols > Longer-chain linear diols. These findings establish NRTL as a reliable predictive tool and provide design insights for developing diol-based additives that are fully compatible with hydrocarbon fuels.

Abbreviation
EG – Ethylene Glycol
LLE – Liquid-Liquid Equilibrium
NRTL – Non-Random Two-Liquid (thermodynamic model)
UNIFAC – UNIQUAC Functional-group Activity Coefficients (thermodynamic model)
WILSON – Wilson equation (thermodynamic model)

Introduction

Enhancing fuel performance is essential for improving engine efficiency, reducing emissions, and maximizing power output. One effective strategy is the incorporation of hydroxyl (–OH) groups into fuel molecules through oxygenated compounds, which contribute to lower emissions while maintaining high thermal efficiency [1]. The presence of –OH groups increases fuel polarity, enhancing air–fuel mixing and promoting more complete combustion. This results in improved thermal efficiency, power output, and reduced emissions. For instance, the addition of benzyl alcohol to gasoline has demonstrated higher fuel power and thermal efficiency compared to pure gasoline [2]. The reactivity of –OH groups supports more efficient combustion, thereby improving engine performance indicators such as power, torque, and brake thermal efficiency [3].

Increased oxygen content in fuels has also been shown to reduce nitrogen oxide (NOx) emissions [4]. However, the presence of –OH groups is beneficial, as higher concentrations can contribute to longer ignition delay; specifically, every 10% increase in –OH group concentration may extend the ignition delay by approximately 1.4%, depending on interactions with other fuel components.[5]. Given that alcohols and ethers improve combustion efficiency and reduce emissions by increasing fuel oxygen content [6], diols—containing two –OH groups—are similarly expected to support complete combustion and enhance engine performance. However, their high polarity can lead to phase separation when blended with non-polar hydrocarbon fuels [7].

However, this limitation, can be mitigated through strategic molecular design. The miscibility of such compounds is influenced not only by polarity but also by molecular structure. The position of –OH groups plays a key role: compact structures like 1,2-diols tend to aggregate with water, while more dispersed structures like 1,8-diols are more miscible with hydrocarbons [8]. To improve compatibility with gasoline, incorporating oxygenated compounds with longer carbon chains or cyclic structures can reduce polarity and enhance solubility. For example, higher alcohols such as n-butanol show better miscibility with gasoline than shorter-chain alcohols due to stronger interactions with non-polar hydrocarbon molecules [9].

This work aims to identify and classify diols based on their molecular structure to enhance miscibility with hydrocarbon components in fuel blends. A total of 23 structurally diverse diols including linear, branched, and cyclic variants were evaluated within a multicomponent gasoline matrix. By correlating key molecular features such as chain length, hydroxyl group positioning, and degree of branching with observed phase behavior, the study seeks to establish design principles for developing gasoline-compatible diol additives. To accurately predict the phase behavior between non-polar hydrocarbons and polar diols, thermodynamic models such as NRTL, UNIFAC, and WILSON were employed within Aspen Plus simulation tools. These models were used to simulate molecular interactions and to determine the most reliable framework for accurate phase behavior predictions.

Methodology of Work

Model Validation for Phase Behavior of Diol–Hydrocarbon Systems (Experimental Work). The aim was to experimentally observe phase behavior in both binary and ternary systems involving methanol (polar), hexane (non-polar), and ethylene glycol (polar) under ambient conditions. These systems were selected as simplified analogs of more complex mixtures anticipated in the upcoming simulation studies, particularly those involving gasoline-range hydrocarbons and diols.

Materials. Methanol (≥99.8%, analytical grade, GC assay) was purchased from Uchem; hexane (≥99%, industrial grade) from Zhong Ma Chemicals; and ethylene glycol (analytical grade) from Revlogi Materials. All chemicals were used without further purification.

Miscibility Testing of Alcohol-Hydrocarbon Blends. All chemicals were stored in sealed containers prior to use. Binary mixtures of hexane and methanol were prepared at volume ratios of 2:1, 2:2, and 2:3. Ternary mixtures containing methanol, hexane, and ethylene glycol were tested at 1:2:3, 2:1:3, and 3:2:1 volume ratio. Mixing was performed in 50 mL graduated glass cylinders. Each sample was stirred for 15 minutes using a magnetic stirrer, then allowed to rest for 15. Experiments were carried out at room temperature (~25°C), after which the phase behavior was visually inspected.

Model Validation for Phase Behavior of Diol–Hydrocarbon Systems (Simulation Work). The reliability of thermodynamic models for predicting phase behavior in polar and non-polar systems was validated by simulating the experimental mixtures using Aspen Plus (V11). The simulation aimed to replicate the binary and ternary systems consisting of methanol, hexane, and ethylene glycol under conditions identical to those used in the laboratory. All components were selected from the Aspen component library. Three thermodynamic models—Non-Random Two-Liquid (NRTL), Universal Quasi-Chemical Functional-group Activity Coefficients (UNIFAC), and Wilson Equation Model (WILSON)—were applied individually to evaluate their predictive capabilities. The liquid-liquid equilibrium (LLE) behavior of the systems was simulated using the decanter unit operation, which is specifically designed to separate immiscible liquid phases. Simulations were performed at 25°C and 1 atm to match the experimental temperature and pressure, and mole fractions were adjusted to reflect the exact compositions tested in the experimental mixtures. To assess the accuracy of each thermodynamic model, the simulated phase volumes were compared with the corresponding experimental phase volumes. The percentage error for each phase was calculated using the following equation:

$$Error\ (\%) = \left|\frac{Vsim - Vexp}{Vexp}\right| \times 100 \quad (1)$$

where Vsim is the simulated volume of the phase, and Vexp is the experimentally observed volume. This approach enabled quantitative evaluation of the deviation between predicted and observed results, thus providing a basis for selecting the most accurate model for subsequent simulations.

Application of a Validated Thermodynamic Model for Miscibility Analysis of Diol–Hydrocarbon Systems. Following the successful validation of thermodynamic models in the binary and ternary systems the model that demonstrated the best performance in previous experiments will be selected as the primary method for simulating the liquid–liquid equilibrium (LLE) behavior of fuel–additive systems. The methodology involved several sequential stages. First, a base gasoline composition was established using data obtained from the literature, as detailed in Table 1 [10] . This reference blend includes a range of hydrocarbons such as alkanes, cycloalkanes, and aromatics that are representative of conventional gasoline. This composition served as the hydrocarbon matrix for all miscibility tests and was treated as the baseline in every simulation case. Next, the selected diol additives were categorized and prepared for simulation. These included linear, branched, and cyclic diols, with their molecular formulas and structures listed in Table 2. The diversity of structural features allowed for a comprehensive evaluation of how molecular geometry influences miscibility in a hydrocarbon environment. The core of the simulation work was conducted using Aspen Plus V14, wherein the gasoline composition and each diol compound were defined as input components. The thermodynamic property method was set to the validated NRTL model. To simulate real mixing conditions, a decanter unit was used to perform two-phase flash separation. The gasoline and diol were fed into the decanter at various volumetric flow ratios and temperatures (see Fig. 1). Each simulation was performed by varying the diol volume fraction from 1% to 100% of the total flow. These flow conditions were intended to span a wide range of compositions, from low to high additive content. For each case, the software computed the final phase behavior and determined whether the mixture remained as a single homogeneous phase or separated into two distinct liquid layers. Importantly, the simulation calculated the volume of the separated phase, if present. These outputs were collected and tabulated for each diol and flow ratio combination. The results were then used to construct a phase diagram, which visually delineates the miscible region from the immiscible region based on the volumetric ratio of diol to gasoline. The diagram maps the transition between single-phase and two-phase regions as a function of temperature, providing a clear visualization of phase compatibility.

Table 1 Base Gasoline Composition Derived from Literature Source

Group	Component	Volume [%]
Paraffins	n-Butane	1.786
	n-Pentane	6.130
	n-Hexane	3.014
	n-Octane	1.909
	n-Nonane	0.206
Iso-Paraffins	Isopentane	11.665
	2,2-Dimethylbutane	5.148
	2,3-Dimethylbutane	1.585
	2-Methylpentane	5.193
	3-Methylpentane	2.412
	2,2-Dimethylpentane	0.242
	2,4-Dimethylpentane	0.397
	3,3-Dimethylpentane	0.212
	2-Methylhexane	2.550
	2,3-Dimethylpentane	0.768
	3-Methylhexane	2.846
	3-Ethylpentane	0.253
	2,2-Dimethylhexane	0.312
	2,5-Dimethylhexane	0.251
	2,2,3-Trimethylpentane	0.325
	2,3-Dimethylhexane	0.309
	2-Methylheptane	1.144
	3-Methylheptane	1.142
	3-Ethylhexane	0.176
Olefins	1-Pentene	0.120
Naphthenes	Cyclopentane	0.530
	Cyclohexane	2.251
	1,1-Dimethylcyclopentane	0.129
	Methylcyclopentane	2.996
	cis-1,3-Dimethylcyclopentane	0.833
	trans-1,3-Dimethylcyclopentane	0.703
	Methylcyclohexane	2.255
	Ethylcyclopentane	0.889
	trans-1,3-Dimethylcyclohexane	0.480
	cis-1,3-Dimethylcyclohexane	0.401
Aromatics	Benzene	1.525
	Toluene	10.606
	Ethylbenzene	2.335
	m-Xylene	4.995
	p-Xylene	2.268
	o-Xylene	3.070

	Isopropylbenzene	0.124
	Normal Propylbenzene	0.404
	1-Methyl-3-Ethylbenzene	1.267
	1-Methyl-4-Ethylbenzene	0.576
	1,3,5-Trimethylbenzene	0.460
	1,2,4-Trimethylbenzene	2.209
	1-Methyl-3-Isopropylbenzene	0.590
	1,2-Dimethyl-4-Ethylbenzene	0.136
	1,2,3,5-Tertmethylbenzene	0.137

Table 2 List of Diols Identified for Simulation, Categorized by Structure and Molecular Formula

Compound Name	**Molecular Formula**
1,2-Benzenediol	$C_6H_6O_2$
1,2-Butanediol	$C_4H_{10}O_2$
1,3-Benzenediol	$C_6H_6O_2$
1,3-Butanediol	$C_4H_{10}O_2$
1,3-Propanediol	$C_3H_8O_2$
1,4-Butanediol	$C_4H_{10}O_2$
1,4-Pentanediol	$C_5H_{12}O_2$
1,5-Pentanediol	$C_5H_{12}O_2$
1,6-Hexanediol	$C_6H_{14}O_2$
2,3-Butanediol	$C_4H_{10}O_2$
2,4-Pentanediol	$C_5H_{12}O_2$
2-Butyne-1,4-diol	$C_4H_6O_2$
2-Methyl-1,3-propanediol	$C_4H_{10}O_2$
cis-2-Butene-1,4-diol	$C_4H_8O_2$
Decane-1,10-diol	$C_{10}H_{22}O_2$
Ethylene Glycol	$C_2H_6O_2$
Hexylene Glycol	$C_6H_{14}O_2$
Isobutylene Glycol	$C_4H_{10}O_2$
Mono-tert-Butylhydroquinone	$C_{10}H_{14}O_2$
p-Hydroquinone	$C_6H_6O_2$
1,2-Propanediol	$C_3H_8O_2$
p-tert-Butylcatechol	$C_{10}H_{14}O_2$
trans-2-Butene-1,4-diol	$C_4H_8O_2$

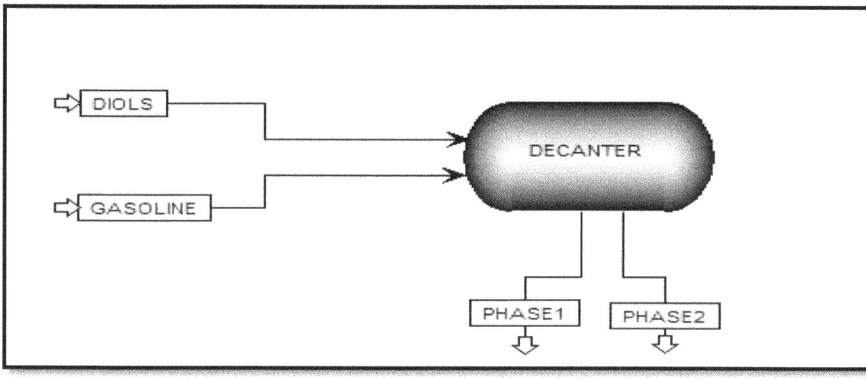

Fig. 1 Schematic representation of the decanter simulation

Plotting of Temperature–Composition (T–X) Miscibility Diagram. The miscibility line diagram (Temperature vs. Diol fraction) was developed to illustrate the variation in miscibility between the diol and gasoline as a function of temperature. The data used for constructing the diagram were obtained from Aspen Plus simulations using the NRTL thermodynamic model. In these simulations, mixtures of diol and gasoline were analyzed at various volumetric compositions ranging from 1% to 100% diol. For each composition, the temperature was varied between 0°C and 80°C to determine the phase behavior of the mixture. From the Aspen Plus outputs, the phase status of each mixture—whether single-phase or two-phase—was identified, and the corresponding temperatures were recorded. The temperature at which each composition first exhibited a single-phase condition was considered the miscibility boundary. formally representing the binodal curve of the system. These boundary data points were then exported from Aspen Plus into Microsoft Excel for further processing. In Excel, the temperature corresponding to the onset of miscibility was plotted against the diol composition in gasoline to obtain the (T vs. X) diagram. The x-axis represents the diol percentage in gasoline (1–100%), while the y-axis represents the temperature range from 0°C to 80°C. A smooth curve was drawn to connect the experimental boundary points, delineating the miscibility limit. In the resulting plot, the region below the curve represents the two-phase or immiscible region, whereas the region above the curve corresponds to the single-phase or fully miscible region. Because the maximum evaluated temperature was strictly capped at 80°C to prevent the volatilization of light hydrocarbon fractions within the gasoline matrix, these plotted binodal curves do not always capture the system's Upper Critical Solution Temperature (UCST). Consequently, for several diol blends, the absolute peak of the miscibility dome (the true UCST) exceeds this operational limit and naturally falls outside the plotted temperature range.

Results and Discussions

Model Validation for Phase Behavior of Diol–Hydrocarbon Systems. Phase volumes were recorded to provide a quantitative reference for miscibility. These experimental observations (summarized in Table 3 and Table 4, and illustrated in Fig. 2 and Fig. 3) serve as a baseline for evaluating the accuracy of thermodynamic models implemented in Aspen Plus. For the binary system (Hexane/Methanol), clear phase separation was observed, particularly at higher methanol content. The 2:1 ratio yielded two nearly equal phases, indicating partial miscibility. At the 2:2 ratio, a dominant hexane-rich phase was formed, while the methanol-rich phase decreased significantly. In the 2:3 ratio, only one phase was recorded, suggesting that the system may have approached a critical composition beyond which methanol becomes the continuous phase and hexane is fully dispersed. In the ternary system (Methanol/Hexane/Ethylene Glycol), phase separation was also evident in all tested ratios, although the volume differences between phases were less drastic compared to the binary system. The presence of ethylene glycol influenced the miscibility behavior by slightly reducing the extent of separation, particularly in the 2:1:3 mixture where the second phase volume dropped to 0.9 ml.

Table 3 Measured Phase Volumes for Binary Methanol/Hexane Mixtures at Room Temperature

Hexane/Methanol volume ratio [ml/ml]	**First Phase (Methanol-Rich) [ml]**	**Second Phase (Hexane-Rich) [ml]**
2 / 1	1.40	1.50
2 / 2	3.60	0.35
2 / 3	5.00	0.00

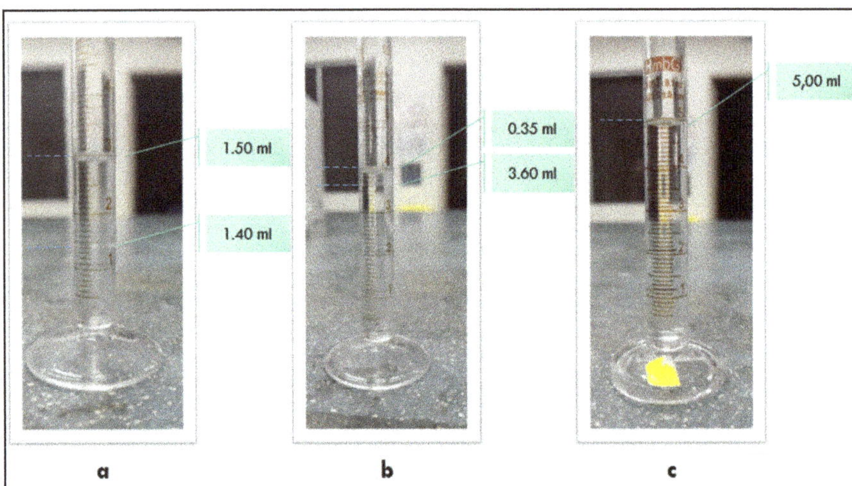

Fig. 2 Experimental images showing phase separation and measured volumes (in ml) for different Hexane/Methanol volume ratios: (a) 2:1 Hexane/Methanol, (b) 2:2 Hexane/Methanol, (c) 2:3 Hexane/Methanol

Table 4 Measured Phase Volumes for ternary Methano l / Hexane / Ethylene Glycol Mixtures at Room Temperature

Methanol / Hexane / EG volume ratio [ml/ml/ml]	**First Phase (Methanol/EG-Rich) [ml]**	**Second Phase (Hexane-Rich) [ml]**
1 / 2 / 3	4.15	1.85
2 / 1 / 3	5.10	0.90
3 / 2 / 1	4.35	1.65

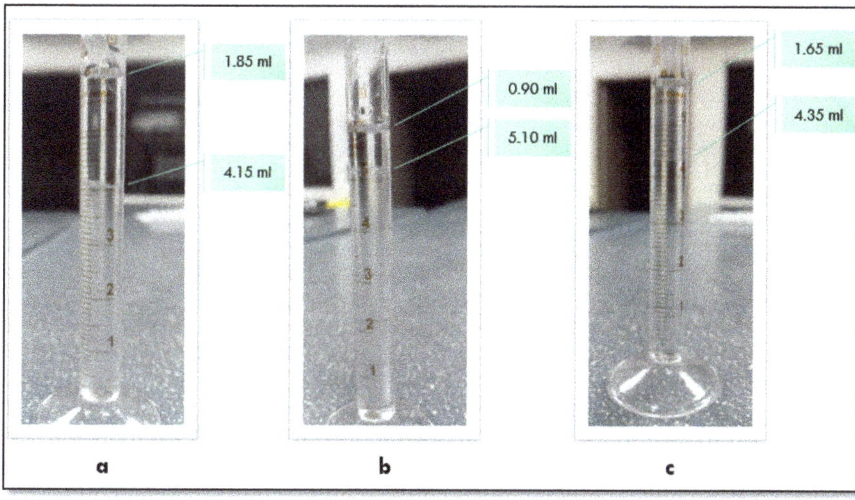

Fig. 3 Experimental images showing phase separation and measured volumes (in ml) for different Methanol/Hexane/Ethylene Glycol (EG) ratios: (a) 1:2:3, (b) 2:1:3, (c) 3:2:1

Model Validation Results. Simulated phase volumes were generated using the NRTL, UNIFAC, and WILSON thermodynamic models within Aspen Plus to evaluate their predictive accuracy against experimental data. These results (detailed in Tables 5 and 6) enable a comparative analysis of each model's ability to replicate the phase behavior observed experimentally. For the binary Methanol/Hexane system (shown in Tables 5), the NRTL and UNIFAC models closely approximated experimental results across most compositions, particularly at the 2:3 ratio, where both models predicted nearly complete miscibility. The WILSON model, however, consistently predicted single-phase behavior in conditions where phase separation was observed experimentally, indicating a limitation in its representation of partial miscibility for this system. In the ternary Methanol/Hexane/Ethylene Glycol system (shown in Tables 6), both NRTL and UNIFAC again showed reasonable agreement with experimental values, particularly in the 2:1:3 composition, where phase volumes differed by less than 6%. The WILSON model significantly overestimated the miscibility for several ternary mixtures, predicting no phase separation where a secondary phase was clearly present in experiments. Error analysis (Tables 5 and 6) confirms that the NRTL and UNIFAC models generally offer superior predictive accuracy, with percentage errors typically under 15% for both phases. In contrast, the WILSON model demonstrated large deviations, particularly in predicting the second phase, where errors often exceeded 100%, especially for compositions yielding near-equal phase volumes.

Based on a quantitative comparison of the experimental and simulated phase volumes, the NRTL model consistently demonstrated the highest accuracy, with percentage errors generally below 10% for most systems. For instance, in the binary 2:1 hexane/methanol mixture, the NRTL-predicted volumes (1.47 ml / 1.46 ml) closely matched the experimental values (1.4 ml / 1.5 ml), yielding percentage errors of 5% and 2.7% respectively. Similarly, for the ternary 2:1:3 methanol/hexane/EG system, the NRTL model predicted 5.1 ml and 0.9 ml for the two phases, almost identical to the measured volumes, resulting in negligible errors below 2%. However, despite this overall high accuracy, specific outliers were observed in the binary mixtures, notably a 120% error for the second phase of the 2:2 hexane/methanol ratio and a 100% error for the 2:3

ratio. These inflated percentage errors occur because the binary system is approaching its miscibility limit, where the experimental volume of the second phase is exceedingly small (0.35 ml) or entirely absent (0.00 ml). Consequently, even minute absolute volume deviations in the simulation (e.g., a difference of just 0.42 ml) translate into disproportionately high percentage errors. This mathematical sensitivity is a recognized characteristic in liquid-liquid equilibrium (LLE) modeling; as highlighted by Labarta et al., the high nonlinearity of the NRTL model near phase equilibrium boundaries makes calculations in these regions inherently challenging and highly sensitive to minute deviations [11]. In contrast, the ternary systems—stabilized by the addition of ethylene glycol—exhibit more distinct and measurable phase separations, allowing the NRTL model to predict phase volumes with much greater relative accuracy and without the mathematical distortion seen near the binary phase boundaries. In contrast, the UNIFAC model showed moderate deviations, while the WILSON model exhibited the highest discrepancies, particularly for binary mixtures where second-phase prediction was often absent or inaccurate. This observation aligns with the findings of Rakhatkyzy et al. (2023), who highlighted the NRTL model's superior predictive performance for mixtures involving polar solvents such as ethanol, propylene glycol, and 2-propanol [12]. They attributed this performance to the model's ability to incorporate local composition effects, which is critical for modeling highly non-ideal mixtures. Additional studies have also compared NRTL with models like UNIFAC and COSMO-SAC, revealing that, in some cases—particularly using modified versions like NRTL-SAC—the model exhibits even better performance in predicting solubility and phase behavior in polar mixed solvent systems [13] [14]. As shown in Table 5 and Table 6, NRTL yielded phase volume predictions that were closely aligned with experimental results for systems such as methanol (a polar protic solvent) and hexane (a non-polar hydrocarbon). This mirrors the accuracy reported by Hamamah & Grützner (2024) for other polar–nonpolar systems of water + n-propanol + toluene [15], further validating the model's reliability in complex scenarios. The UNIFAC model also produced acceptable predictions, although with slightly higher percentage errors, particularly in the second phase. This aligns with its known limitations in capturing specific hydrogen bonding and solvation effects, as noted in prior work by Fredenslund et al. (1977) and Rakhatkyzy et al. (2023), where they tested polar mixtures using Aspen Plus thermodynamic models. [12]. In contrast, the WILSON model exhibited poor predictive performance. It often overestimated miscibility, failed to predict phase splitting in some cases, and showed large percentage errors often reached up 100%. These limitations are consistent with observations by Rakhatkyzy et al. (2023), who reported similar failures when applying WILSON to solvents like acetone, where solvation effects play a dominant role [12]. Given the strong agreement between simulation and experimental results, supported by literature findings, the NRTL model is selected as the preferred thermodynamic approach for further simulations involving gasoline–diol systems. For example, NRTL successfully reproduced experimental tie-line and binodal curve data for the ternary system of water + n-propanol + toluene—a mixture involving both polar and non-polar components—while other models like UNIQUAC struggled to converge across temperatures (Hamamah & Grützner, 2024) [15].

Table 5 Simulated phase volumes and percentage errors for methanol/hexane binary mixtures using NRTL, UNIFAC, and WILSON thermodynamic models in Aspen Plus

Hexane/Methanol volume Ratio [ml/ml]	Phase	NRTL Volume [ml]	NRTL Error [%]	UNIFAC Volume [ml]	UNIFAC Error [%]	WILSON Volume [ml]	WILSON Error [%]
2 / 1	1[a]	1.47	5.00	1.55	10.71	2.88	105.71
	2[b]	1.46	2.67	1.37	8.67	0.00	100
2 / 2	1[a]	3.12	13.33	3.38	6.11	3.86	7.22
	2[b]	0.77	120.00	0.50	42.86	0.00	100
2 / 3	1[a]	4.78	4.40	4.85	3.00	4.85	3.00
	2[b]	0.07	100	0.00	100	0.00	100

a: Phase 1 refers to the Methanol-rich phase

b: Phase 2 refers to the Hexane-rich phase.

Table 6 Simulated phase volumes and percentage errors for methanol/hexane/ethylene glycol (EG) ternary mixtures using NRTL, UNIFAC, and WILSON thermodynamic models in Aspen Plus

Methanol/Hexane/ EG volume Ratio [ml/ml/ml]	Phase	NRTL Volume [ml]	NRTL Error [%]	UNIFAC Volume [ml]	UNIFAC Error [%]	WILSON Volume [ml]	WILSON Error [%]
1 / 2 / 3	1[a]	3.94	5.06	3.97	4.34	2.93	29.40
	2[b]	1.95	5.41	1.91	3.24	2.84	53.51
2 / 1 / 3	1[a]	4.99	2.16	5.04	1.18	5.78	13.33
	2[b]	0.85	5.56	0.80	11.11	0.00	100
3 / 2 / 1	1[a]	4.42	1.61	4.46	2.53	5.77	32.64
	2[b]	1.43	13.33	1.38	16.36	0.00	100

a: Phase 1 refers to the Methanol/EG-rich phase

b: Phase 2 refers to the Hexane-rich phase.

Influence of Chain Length towards Miscibility Testing of Diol-Hydrocarbon Systems. The miscibility of diols in gasoline was found to depend strongly on both the hydrocarbon backbone length and the positioning of hydroxyl (–OH) groups. Short-chain diols exhibited complete miscibility only at higher concentrations and elevated temperatures. For instance, ethylene glycol (C_2) showed good miscibility up to 98% (v/v) at 75 °C. As the carbon chain length increased from C_2 to C_6, miscibility generally improved, although the improvement was relatively modest. Specifically, 1,3-propanediol became miscible at 88% (v/v) at 90 °C, 1,4-butanediol at 87%, 1,5-pentanediol at 84%, and 1,6-hexanediol at 79% under the same temperature conditions. For 1,10-decanediol, miscibility with gasoline was limited to 1–4% (v/v); beyond this range, phase separation occurred, although partial miscibility reappeared above 55%, suggesting the formation of two distinct composition windows of miscibility. This behavior can be explained by the balance between hydrogen bonding and hydrocarbon–hydrocarbon interactions. Similar to primary alcohols such as methanol and ethanol, short-chain diols have high polarity and strong hydrogen-bonding tendencies, which favor association with polar solvents but hinder compatibility with hydrocarbons. As the carbon chain length increases, the hydrocarbon character of the molecule dominates, enhancing van der Waals interactions with gasoline components and thus improving miscibility [16, 17]. Overall, the results confirm that increasing the carbon chain length enhances hydrocarbon miscibility, while the position of hydroxyl groups and temperature further modulate the extent of phase compatibility. However, this enhancement becomes less pronounced beyond C_6, where the molecules begin to exhibit self-association and non-uniform phase behavior.

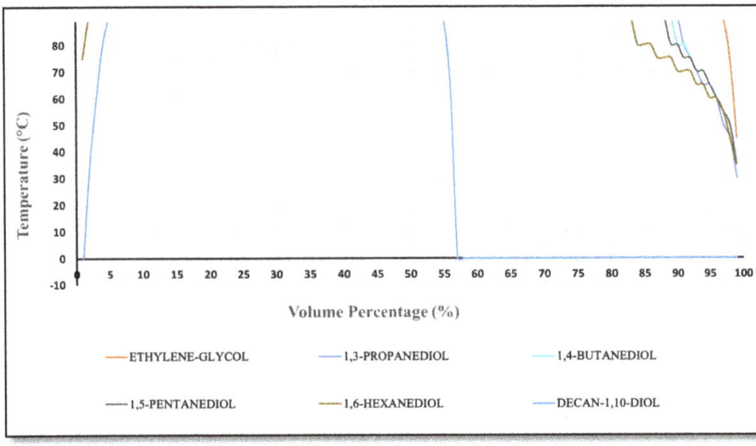

Fig. 4 Influence of carbon chain length of diols on their miscibility with gasoline at varying volume percentages and temperatures

Effect of Hydroxyl Group Position towards Miscibility Testing of Diol-Hydrocarbon Systems. The line graph illustrates the influence of hydroxyl (–OH) group positioning on the miscibility of diols with gasoline. Comparing 1,2-propanediol and 1,3-propanediol, it is evident that the vicinal hydroxyl groups (adjacent –OH groups) in 1,2-propanediol enhance miscibility. The mixture became miscible at 81% (v/v) at 90 °C, whereas 1,3-propanediol, with hydroxyl groups separated by one carbon atom, required 88% (v/v) at the same temperature to achieve miscibility. A similar trend was observed among butanediol isomers. 1,2-butanediol, in which the hydroxyl groups are attached to the first and second carbon atoms, exhibited the highest miscibility, becoming fully miscible at 73% (v/v) at 65 °C. 2,3-butanediol, containing hydroxyl groups on carbons 2 and 3, also showed good miscibility but required a slightly higher temperature of 70 °C. 1,3-butanediol displayed comparable miscibility (73%) at 70 °C, while 1,4-butanediol, where the hydroxyl groups are located at opposite ends of the molecule, demonstrated the poorest miscibility, requiring 87% (v/v) and 90 °C to achieve blending. Further comparison was made between p-tert-butylcatechol and mono-tert-butylhydroquinone, two aromatic diols containing bulky tert-butyl substituents. p-tert-butylcatechol exhibited miscibility at concentrations of 1–6% (v/v) at temperatures of 0, 25, 45, 65, 80, and 90 °C, respectively. Beyond these concentrations, phase separation occurred, but miscibility reappeared at a higher concentration of 50% (v/v) at 90 °C. Similarly, mono-tert-butylhydroquinone remained miscible at 1–6% (v/v) across 0–90 °C, with immiscibility occurring above this range, before miscibility returned at 53% (v/v) at 90 °C. These findings are consistent with previously reported trends for longer-chain diols in hydrocarbon systems. A study on octanediol isomers (C_8) in n-decane/water mixtures [8] found that all three isomers—1,2-diol, 4,5-

diol, and 1,8-diol—were insoluble in water and partitioned exclusively into the hydrocarbon phase. Both 1,2-diol and 4,5-diol, with internal or vicinal hydroxyl groups, dispersed effectively and formed small, disordered clusters, indicating good hydrocarbon compatibility. In contrast, 1,8-diol, which possesses terminal hydroxyl groups, showed strong self-aggregation into larger, structured assemblies, thereby reducing its effective miscibility [8]. These observations clearly indicate that the proximity of hydroxyl groups enhances molecular polarity and hydrogen-bonding capability, facilitating better interaction with gasoline molecules. As the hydroxyl groups move farther apart along the carbon chain, the molecule's overall polarity decreases, weakening intermolecular interactions and resulting in reduced miscibility.

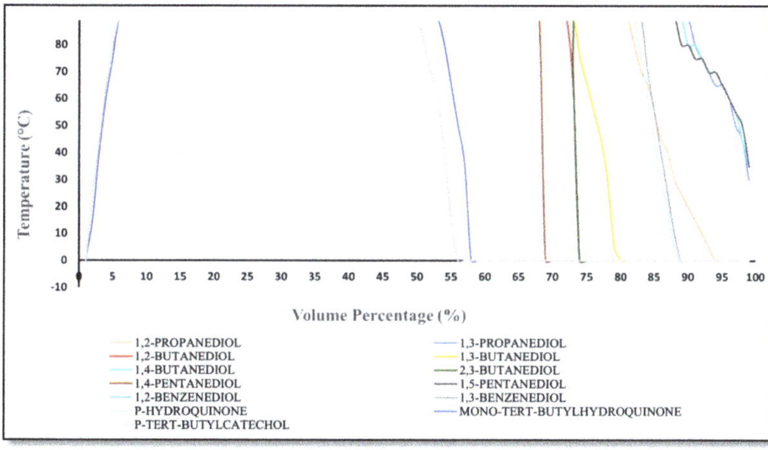

Fig. 5 Influence of OH group position in diols on their miscibility with gasoline at varying volume percentages and temperatures

Effect of Branching on Miscibility towards Miscibility Testing of Diol-Hydrocarbon Systems. As shown in Fig. 6 and Fig. 7, branched diols exhibited significantly higher miscibility with gasoline compared to their linear counterparts. For example, 2-methyl-1,3-propanediol achieved complete miscibility at 72% (v/v) at 90 °C, while its structural isomer, isobutylene glycol, showed similar miscibility at 73% (v/v) under the same conditions. In contrast, the linear analogue 1,4-butanediol remained biphasic even at high diol loadings—up to 87% (v/v) at 90 °C—clearly highlighting the positive influence of molecular branching on promoting miscibility with gasoline, especially at elevated oxygenate concentrations. A comparable trend was observed when comparing 1,6-hexanediol with its branched analogue, hexylene glycol (2-methyl-2,4-pentanediol). The linear 1,6-hexanediol exhibited poor miscibility, remaining biphasic up to 79% (v/v) at 90 °C, whereas hexylene glycol maintained a single-phase blend up to 65% (v/v) at the same temperature. This improvement can be attributed to the branched molecular structure, which reduces overall polarity and enhances van der Waals interactions with hydrocarbon molecules. This behavior mirrors that observed for branched monoalcohols such as isopropanol and tert-butanol, which preferentially partition into the hydrocarbon (gasoline) phase, unlike linear alcohols such as ethanol and methanol, whose stronger hydrogen-bonding tendency favors the polar phase [16].

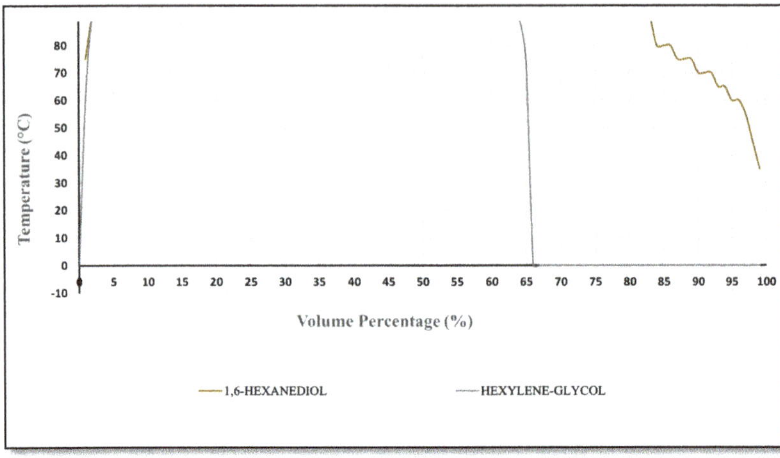

Fig. 6 Influence of molecular branching on the miscibility of diols with gasoline (1,2-Hexandiol and Hexylene Glycol)

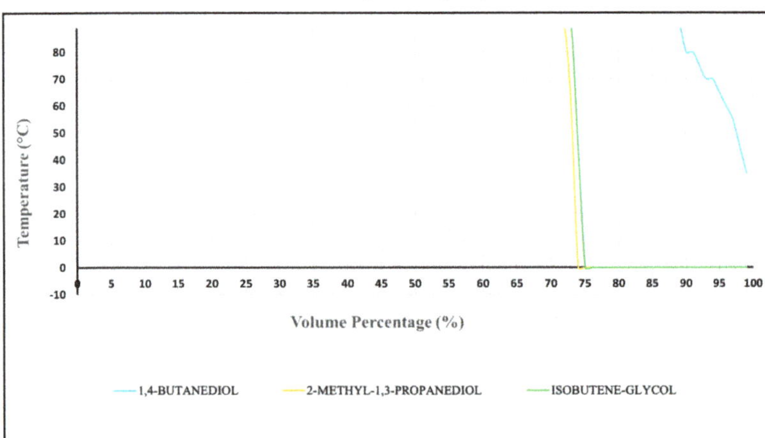

Fig. 7 Influence of molecular branching on the miscibility of diols with gasoline (1,4-Butanediol,2-methyl-1,3-Propanediol and Isobutene Glycol)

Effect of Unsaturation towards Miscibility Testing of Diol-Hydrocarbon Systems. As illustrated in the diagram line below (Fig. 8), 2-Butyne-1,4-diol becomes completely miscible with gasoline at about 64% v/v, followed by trans-2-Butene-1,4-diol at 66% v/v, while the saturated 1,4-Butanediol requires approximately 87% v/v to reach a single-phase region. This trend indicates that the presence of π-bonds enhances the compatibility of diols with the nonpolar hydrocarbon phase. The reduced polarity associated with unsaturated bonds weakens hydrogen-bonding interactions among diol molecules and increases dispersion interactions with hydrocarbons.

Consequently, as the degree of unsaturation increases (from single to double to triple bond), the diol exhibits improved solubility and lower phase-separation tendency

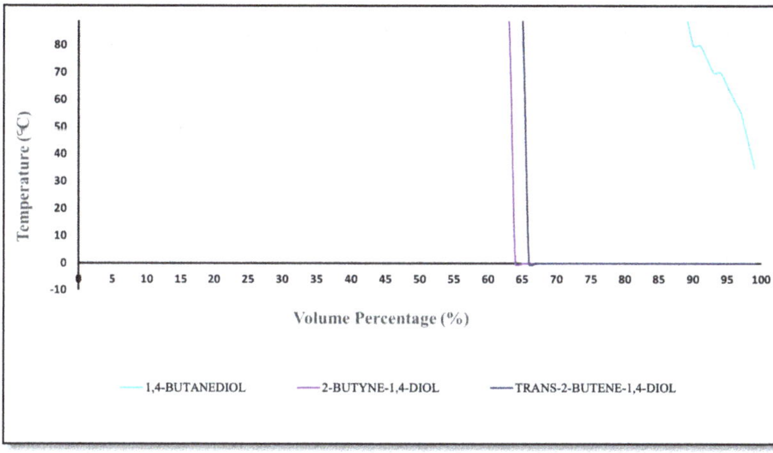

Fig. 8 Influence of unsaturation on the miscibility of diols with gasoline

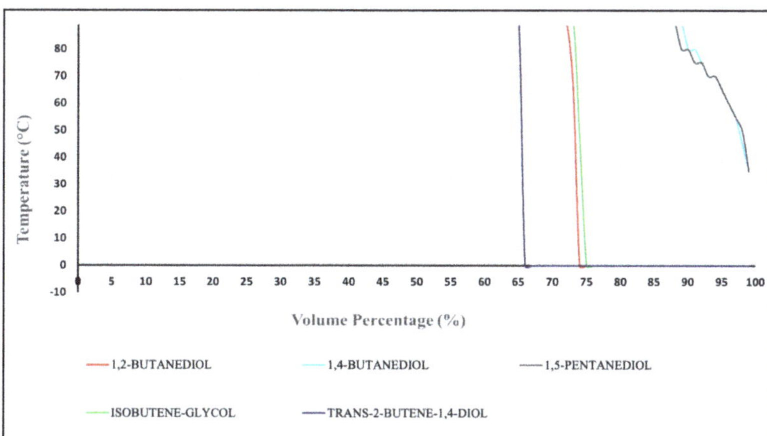

Fig. 9 Comparative influence of molecular structure on the miscibility of diols with gasoline

As illustrated in Fig. 9, the miscibility of butanediol derivatives with gasoline varies significantly depending on their molecular structure and the position of hydroxyl (–OH) groups. The unsaturated diol, trans-2-butene-1,4-diol, exhibits the highest miscibility, becoming completely soluble with gasoline at approximately 66 vol%, which suggests that the C=C double bond enhances molecular compatibility between the polar diol and non-polar hydrocarbon phases. The π-electrons in the double bond may strengthen dispersion interactions with gasoline molecules, thus minimizing phase separation. The vicinal diol, 1,2-butanediol, achieves complete miscibility at around 73 vol%, reflecting how the close proximity of hydroxyl groups promotes intramolecular hydrogen bonding. This internal interaction reduces overall polarity, improving solubility compared with longer-chain or terminal diols.

For the branched diol, isobutene glycol, miscibility begins near 73 vol%, indicating that branching can slightly enhance compatibility by limiting intermolecular hydrogen bonding within the diol phase. However, this improvement remains modest compared to the effect of unsaturation or hydroxyl positioning.

In contrast, the longer-chain linear diols—1,4-butanediol and 1,5-pentanediol—display the lowest miscibility, requiring diol concentrations above 84–87 vol% to achieve homogeneous blending. Their symmetrical structures and distant hydroxyl positioning contribute to higher polarity separation, resulting in weak interactions with non-polar gasoline molecules.

Overall, these findings confirm that unsaturation and hydroxyl positioning are dominant structural factors influencing miscibility. The observed trend—

Unsaturated diols > Vicinal diols > Branched diols > Longer-chain (linear) diols

Conclusion

The validation and application of thermodynamic models for simulating the phase behavior of diol–hydrocarbon systems have revealed critical insights for fuel formulation. Among the evaluated models, the NRTL model proved to be the most accurate and reliable, offering predictions that closely aligned with experimental phase separation data in both binary and ternary mixtures. Its superior performance justifies its selection as the preferred framework for simulating more complex gasoline–diol interactions. Subsequent miscibility testing using the validated NRTL model demonstrated that diol structure plays a pivotal role in blending behavior with gasoline. The validated thermodynamic simulations show that diol–gasoline miscibility is governed primarily by hydroxyl-group placement and molecular hydrocarbon character (branching and unsaturation), with carbon-chain length playing a secondary, moderating role. Vicinal or internally placed –OH groups reduce strong self-association and promote dispersion into the hydrocarbon phase, while branching and the presence of π-bonds increase hydrocarbon affinity and weaken extensive hydrogen bonding. Consequently, within the butanediol family the qualitative miscibility ordering is: Unsaturated diols > Vicinal diols > Branched diols > Longer-chain (linear) diols.

Acknowledgements

The authors acknowledged the Universiti Teknologi Malaysia (UTM) for UTM NEXUS POSTGRAD Grant (PY/2025/00519), Ministry of Higher Education (MOHE) Fundamental of Research Grant Scheme (FRGS/1/2023/TK08/UTM/02/36), UTM Encouragement Research Grant (QJ130000.3846.31J33), Energy Management Group and UTM-MPRC Institute For Oil and Gas (IFOG) for their support in conducting this study.

Declaration of Generative AI and AI-assisted Technologies

During the preparation of this work, the author utilized ChatGPT to ask questions and retrieve factual information, which was subsequently validated through reputable sources such as Google Scholar, ScienceDirect, Scopus, SciSpace, and various journal websites. Scopus AI played a crucial role in assisting the author in searching for specific information, ensuring that the retrieved content was published within the last five years and remained relevant to the research. Additionally, AI-powered tools like DeepSeek and ChatGPT were employed to refine and rephrase text for enhanced clarity and readability. Following the use of these tools, the author thoroughly reviewed and edited the content, assuming full responsibility for the final publication.

References

[1] Q. Wang, J. Ni, R. Huang, The potential of oxygenated fuels (n-octanol, methylal, and dimethyl carbonate) as an alternative fuel for compression ignition engines with different load conditions, Fuel 309 (2022) https://doi.org/10.1016/j.fuel.2021.122129

[2] D. Jesu Godwin, V. Edwin Geo, S. Thiyagarajan, M. Leenus Jesu Martin, T. Maiyalagan, C.G. Saravanan, F. Aloui, Effect of hydroxyl (OH) group position in alcohol on performance, emission and combustion characteristics of SI engine, Energy Conversion and Management 189 (2019) 195-201. https://doi.org/10.1016/j.enconman.2019.03.063

[3] Suyatno, H. Riupassa, S. Marianingsih, H.Y. Nanlohy, Characteristics of SI engine fueled with BE50-Isooctane blends with different ignition timings, Heliyon 9 (2023) e12922. https://doi.org/10.1016/j.heliyon.2023.e12922

[4] M.D. Bohon, M.J. Al Rashidi, S.M. Sarathy, W.L. Roberts, Experiments and simulations of NOx formation in the combustion of hydroxylated fuels, Combustion and Flame 162 (2015) 2322-2336. https://doi.org/10.1016/j.combustflame.2015.01.022

[5] J. Zhu, R. Li, Z. Wang, S. Liu, H. Lv, Decoupled analysis of the effect of hydroxyl functional groups on delay of ignition with fictitious hydroxyl, Process Safety and Environmental Protection 161 (2022) 285-294. https://doi.org/10.1016/j.psep.2022.03.028

[6] O.I. Awad, R. Mamat, T.K. Ibrahim, A.T. Hammid, I.M. Yusri, M.A. Hamidi, A.M. Humada, A.F. Yusop, Overview of the oxygenated fuels in spark ignition engine: Environmental and performance, Renewable and Sustainable Energy Reviews 91 (2018) 394-408. https://doi.org/https://doi.org/10.1016/j.rser.2018.03.107

[7] J.K.A. Bezerra, R.R. da Silva, E.L. de Barros Neto, G.G. de Medeiros, L.J.N. Duarte, H.N.M. de Oliveira, Evaluation of the Effect of Ethyl Alcohol Content In A Ternary Ethanol/Biodiesel/Diesel System, International Journal of Thermodynamics 27 (2024) 19-26. https://doi.org/10.5541/ijot.1372558

[8] M.V. Kirillov, A.A. Sizova, V.V. Sizov, Octanediols in confined hydrocarbon/water systems: A molecular simulation study of aggregation phenomena and liquid–liquid interface morphology, Journal of Molecular Liquids 428 (2025) 127503. https://doi.org/https://doi.org/10.1016/j.molliq.2025.127503

[9] C. Jin, X. Zhang, X. Wang, Y. Xiang, X. Cui, Z. Yin, X. Sun, J. Ji, G. Wang, H. Liu, Effects of polyoxymethylene dimethyl ethers on the solubility of ethanol/diesel and hydrous ethanol/diesel fuel blends, Energy Science & Engineering 7 (2019) 2855-2865. https://doi.org/https://doi.org/10.1002/ese3.466

[10] N. Ali, The Effect of Compositions (PIONA) on the Octane Numbers of Environmental Gasolines of Reformate, Isomerate and Hydrocracked Naphtha Blends by Using GC, International Journal of Oil, Gas and Coal Engineering 5 (2017) 167. https://doi.org/10.11648/j.ogce.20170506.17

[11] J.A. Labarta, M.M. Olaya, A.F. Marcilla, What does the NRTL model look like? Determination of boundaries for different fluid phase equilibrium regions, AIChE Journal 68 (2022) e17805. https://doi.org/https://doi.org/10.1002/aic.17805

[12] M. Rakhatkyzy, B. Suleimenova, M. Shaimardan, D. Shah, N. Nuraje, Modeling Solubility of Acetylsalicylic Acid in Aspen Plus, Engineered Science 26 (2023) 987. https://doi.org/10.30919/es987

[13] E. Sheikholeslamzadeh, S. Rohani, Solubility prediction of pharmaceutical and chemical compounds in pure and mixed solvents using predictive models, Industrial and Engineering Chemistry Research 51 (2012) 464-473. https://doi.org/10.1021/ie201344k

[14] H. Tun, Y. Hao, M. Haddix, C.C. Chen, Thermodynamic solubility modeling of 2, 2′, 4, 4′, 6, 6′ - hexanitrostilbene (HNS), Fluid Phase Equilibria 565 (2023) https://doi.org/10.1016/j.fluid.2022.113627

[15] Z.A. Hamamah, T. Grützner, Measurement and thermodynamic correlation of liquid–liquid equilibria of water + n-propanol + toluene at 20°C and atmospheric pressure, Discover Chemical Engineering 4 (2024) 20. https://doi.org/10.1007/s43938-024-00057-6

[16] M. Neagu, P. Rosca, R.E. Dragomir, O. Mihai. The effect of bioalcohol on the water solubility in reformulated gasoline. in Chemical Engineering Transactions. 2010.

[17] X. Li, L. Hou, C. Chai, S. He, Y. Huang, A New Molecular Insight in Effects of Alcohol Co-Solvents on Miscibility of Anhydrous Ethanol/Diesel Blends, Journal of Energy Resources Technology, Transactions of the ASME 145 (2023) https://doi.org/10.1115/1.4056115

Innovative Strategies in Waste Management and Recycling of Energy Materials: Advancing Circular Economy and Sustainable Energy Systems

Vinod BALMIKI[1,a *], Samrity JALOTA[1,b], Navodit BHANDARI[1,c], Kapil JOSHI[2,d], Ranjay Kumar CHOUDHARY[3,e], ROSY[5,f], M Vijaya SIMHA[4,g], Zahoor Ahmad QUERSHI[4,h]

[1]Department of Civil Engineering, Uttaranchal Institute of Technology, Uttaranchal University, Dehradun-248007, Uttarakhand, India

[2]Department of Computer Science Engineering, Uttaranchal Institute of Technology, Uttaranchal University, Dehradun-248007, Uttarakhand, India

[3]School of Health and Allied Sciences, Sanskaram University, Jhajjar, Haryana, India

[4]Department of Medical Laboratory Technology Institute of Allied Health Sciences, Chandigarh University, India

[5]Department of Medical Lab Technology (UIAHS), Chandigarh University, Mohali-140413, India

[a]vinodbalmiki111@gmail.com, [b]samrityjalota@gmail.com, [c]navodit36@gmail.com, [d]kapilengg0509@gmail.com, [e]r.choudharymt@gmail.com, [f]rosypaul12@gmail.com, [g]mvsimha@gmail.com, [h]zahoor.e16431@cumail.in

Keywords: Renewable Energy Waste, Recycling Technologies, Solar Panels, Lithium-Ion Batteries, Wind Turbine Blades, Circular Economy, Sustainability

Abstract. The large scale deployment of renewable technology has created a major challenge with respect to the disposal of solar panels, lithium-ion batteries, and wind turbine blades, and therefore requires efficient recycling for sustainability purposes. Nevertheless, a lot remains pending to be discovered regarding the economic and environmental viability of recycling solutions, especially in relation to wind turbine blades. Recycling efficiency of existing recycling technologies for these materials is the purpose of undergoing environmental performance and cost -effectiveness. From the review of literature, life cycle evaluation (LCA) and examples from large nations, recycling technologies were assessed along with their compatibility with the global political structure. Solar panels, provide more environmentally friendly alternatives with low carbon emissions, while lithium-ion battery provides the highest recovery rate, but to high environmental value. While contributing the least to the environment, the wind turbine blade has a low recovery rate. From the study, it has been concluded that there is a need to increase material recycling and develop a spherical economy in the renewable energy market, recycling requires strong legislative support and technological innovation.

Introduction

Over the past 20 years, the imperative has to curb greenhouse gas emissions and dependence on fossil fuels accelerated the process of global inflection for renewable energy [1]. It is led by energy storage battery, wind turbine and photovoltaic system (PV) [2]. Although these energy solutions are important for promoting stability, they also address new issues related to resource management and waste generations [3]. Along with the installation of renewable energy equipment [4], the amount of life energy content also increases. For example, solar panels remain as electronic waste with both dangerous and precious components after 25-30 years [5]. The lithium-ion battery that works in energy storage systems and electric vehicles (EV) also gradually comes out over time, and thus recycles or treated properly [6]. Although wind turbines only run 20 to 25 years, they produce a huge amount of total waste because their blades are made of fiberglass and epoxy resin, which cannot be recycled through traditional process [7].

Energy is the most important economic and environmental issues with the increasing amount of waste [8]. Disable recovery of valuable materials is reduced, and incorrect settlement can allow toxins such as cobalt, lead and cadmium to the environment [9]. In addition, the establishment of a circular economy in the energy sector is interrupted by non-standard recycling practices and law [10]. The study of innovative and environmentally friendly methods for managing energy material and recycling is extremely necessary to keep in mind these concerns [11]. The purpose of this study is to study existing processes, technological innovations and legislative movements that can contribute to the recycling of effective waste management and renewable energy materials [12]. The purpose of this research is to contribute to the overall mission to increase resource efficiency and environmental ability in energy systems by highlighting great opportunities and challenges [13].

The long -term stability of the renewable energy system depends on recycling and waste management [14]. Recruitment of useful resources such as cobalt, lithium, rare soil material and other important metals facilitate effective treatment and extraction of unused materials, which also avoid completing environmental risks [15]. Energy storage units, wind turbines and photovoltaic systems are under improvement depending on these materials [16]. Also the reduction in environmental pollution and conservation of natural resources promotes recycling green jobs and flexibility in the supply chain [17]. Even with these benefits, there are many challenges in the energy sector, including inadequate infrastructure for recycling, expensive recovery, absence of standardization and limited state support [18]. Although much research has highlighted the benefits of implementing renewable energy [19]. Through the analysis of both challenges and opportunities within the existing recycling and waste management structures, this research fulfils a significant knowledge interval [20]. Assessment of stability, viability and efficiency of recycling processes for solar panels, batteries and wind turbines is the primary purpose of this study [21]. It also scans for policy, financial and technological obstacle to the improvement of circular energy systems. The study is guided by the following key research aim:

- To examine and match the current recycling technique and technologies used for recycling solar panels, lithium-ion batteries and wind turbine part.
- To Examine the importance of Recycling and waste management in the field of renewable energy with focus on pollution control, resources recovery and environmental sustainability.
- To mention the primary obstacles to the energy sector successful taking into extensive recycling systems, including technological, financial and regulatory issues.
- To give Idea for plan and policies that can improve circular economy and the sustainability of energy material waste management.

The arrangement of rest of the paper is as follows: Recycling renewable energy sources, including solar panel, lithium ion battery and wind turbine blade are presented section 2. It showcases research gaps and talk about different recycling methods, obstacle and legal barrier especially when it comes to mixing recycling in circular economy. A detailed study for the literature, a case study examination of top nations in recycling renewable waste and the use of life cycle Assessment (LCA) to evaluate environmental effects are all included in the research methodology, as detailed in section 3. Section 4 offers a detailed connection of the primary materials, economic effectiveness, environmental effects and recycling efficiency with international standards. Moreover, it gives advices for improved recycling procedure, specifically with regard to wind turbines blades. The importance of technological innovations, improved regulatory framework and the shift to circular economy are highlighted in the Section 5. It presents the main result aimed at improving recycling rates and sustainability in the renewable energy sector.

Literature Review

As the growing use of renewable energy technologies has increased the amount of waste produced from end-of-life energy products,effective recycling procedures must be put in place [21].Many recyling methods have been investigated for items like solar panel,lithium ion batteries and wind turbine blades [22].For example,mechanical,thermal and chemical methods have all been used to recycle wind turbine blades;solvolysis has become a practical alternative because of high material recovery rates and low carbon intensity [23].Improvement in battery recyling also alternative organic solvents for harsh checmical to lessen their enviromental impact [24].Standardizing procedure and guaranteeing regional cooperations remain difficult,despite the fact that policy initiative,like the European Union WEEE Directive,have been crucial in encouraging recyling [25].A lot of research has been done on recyling electrical waste such as solar panel and batteries [26].

As studies have shown that solar panel can recycle valuable components like silicon, silver and rare earth elements, actual recycling efforts have been put by both financial and technological constraints [27]. Research on battery recycling has focused on variety of technique for recovering important minerals such as nickel, cobalt, and lithium [28]. Technologies such as the creation of low carbon battery production technique and the placement of recycling facilities in areas with a wealth of renewable energy sources have been put forward as solution to the environmental issues related to battery disposal [29]. Recycling energy materials is dangerous with difficulties [30]. Economic factors such a low material recoveries and high recycling costs have become major problem [31]. Recovering valuable materials is difficult due to technical issues, such as difficulties of recycling composite material used in wind turbine blades and gradual deterioration of battery component [32]. It is challenging to develop a uniform recycling strategy due to policy issues such as lack of infrastructure, inconsistent laws in different jurisdiction and inconsistent recycling [34]. Step by step studies that analyze the technological, economic and environmental impact of recycling processes for various energy materials are required [35]. Also there are not many empirical analyses of mixing recycling operations in the energy industry with a circular economy approached [36].

This article seeks to fill the gap by reviewing of recycling technologies, their efficacy and sustainable and also recommending approaches to increases the circularity of energy material and waste material [37]. By means of a thorough analysis of recycling technologies, an evaluation of their sustainability and efficiency, and a narration of measures to foster the circularity of energy material waste management, this study seeks to bridge these gaps [37]. The current practice, technology, and policy of recycling and waste management of valuable energy resources such as solar panels, wind turbine blades, and lithium-ion batteries are discussed in this research through a qualitative and analytical approach [38]. Along with particular strategies for driving circularity in the renewable energy industry, the strategy seeks to offer an extensive understanding of the issues and opportunities of this business [39].

Methodology

Efficient recycling practices need to be implemented since the expanding application of renewable energy technologies has led to a higher volume of waste generated from end-of-life energy products [21].

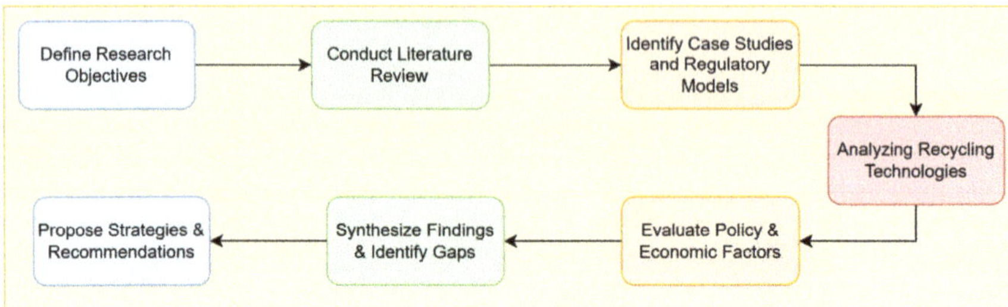

Fig. 1 Flowchart Depicting the Logical Sequence of the Study

The organized approach flowchart for recycling system research is shown in Fig. 1. To develop the background information, a thorough literature review comes after the research goals have been clearly determined. After that, appropriate case studies and regulatory frameworks are found in order to provide context knowledge. The existing recycling technologies are then analyzed, which leads to an assessment of the relevant economic and policy factors. To identify present knowledge gaps, these findings are combined. Strategic suggestions and ideas are created to address the problems and improve recycling systems based on the in-depth knowledge gained. This logical order ensures a thorough and fact-based research procedure.

An outline of the most recent recycling technology, legal frameworks, and the financial and environmental effects of the different strategies were given by this review. Case studies of major recycling nations for renewable waste, including the United States, Japan and Germany, were chosen after analysis of literature. Case Study's was used to determine the purpose of recycling technology and the effect of various policy equipment. According to cost -effectiveness, environmental stability, skills in material recovery and regulatory support, allowed comparative analysis to determine the most appropriate recycling approach [41].

Research also assessed global standards, including the EU instructions for waste electrical and electronic equipment (WEEE) and battery instructions, in an attempt to consider extensive economic and political issues. This provided a reference where technical successes against governance and politics were determined [42]. During the product's life cycle, the conceptual life cycle evaluation (LCA) approach was used and their environmental benefits and recycling procedures were analyzed for business bands [43]. Finally, research on case studies, literature and regulatory subpoena used to determine knowledge gaps and recommend rational strategies. These strategies will be to promote the installation of integrated recycling systems, increase the frequency of materials extraction and advocate for legislative reforms that favor circular economy in renewable energy [44].

Result and Discussions

An analysis of modern recycling technologies, their effectiveness and environmental results, and a comparative evaluation of the financial feasibility of recycling operations in the strength sector served as the inspiration for the study's findings. The following records was amassed via case studies, literature evaluations, and lifestyles cycle assessment (LCA) computations, with a focus on recycling 3 important energy commodities: wind turbine blades, lithium-ion batteries, and sun photovoltaic panels [45,49].

The impact values presented in Table 1 are estimated using a 'cradle-to-gate' Life Cycle Assessment approach, covering the processes of raw material collection, preprocessing, and recycling, up to the recycling facility gate. The processes of product reuse, distribution, and disposal are not included in the study.

Materials Research Proceedings 66 (2026) 115-125 https://doi.org/10.21741/9781644904152-11

Table 1 Recovery Rates of Materials through Recycling Technology

Energy Material	*Recycling Technology*	*Material Recovery Rate*	*Environmental Impact [kg CO_2-eq./ton]*	*Cost Efficiency*
Solar Panels	Mechanical & Thermal Recovery	85%	40 kg CO_2-eq./ton	Moderate
Lithium-Ion Batteries	Chemical Leaching	90%	50 kg CO_2-eq./ton	High
Wind Turbine Blades	Solvolysis	70%	25 kg CO_2-eq./ton	Low

The recovery rate of key material from all energy sectors are matched in this table, along with each material cost effectiveness and environmental impact. Recycling solar panel with mechanical and thermal recovery method provides a high material recovery rate of 85% [49]. On the other hand chemical leaching processes with a 90% recovery rate are advantages for lithium-ion batteries [46,48].These steps are costlier, though because they require the use of complex chemicals. But more recent technique like solvolysis offer potential solutions, wind turbine blades show a lower material recovery rate (70%) primarily because of the difficult of composite materials [47].

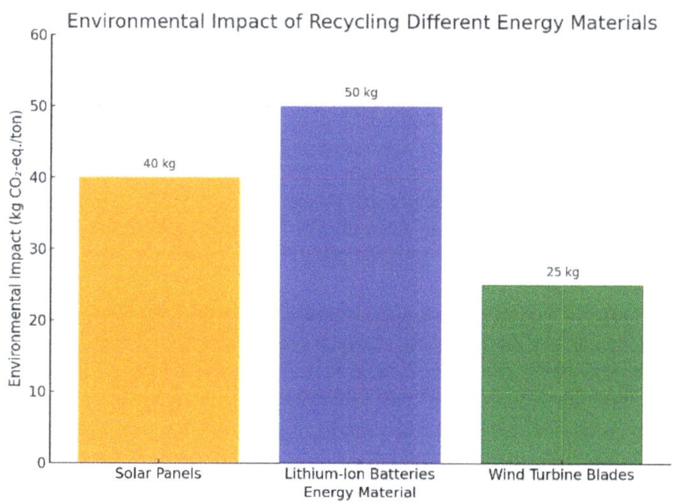

Fig. 2 Recycling Efficiency vs. Environmental Impacts of Energy Materials

Fig. 2 illustrates the connection between recycling usefulness and environmental impact. It is clear that even though lithium-ion batteries have the highest recycling efficiency the big nature of related processes result in a higher carbon footprint. Recycling solar panel is more sustainable per ton, even if it is slightly less successful at recovering material and has a smaller environmental impact [46,49].

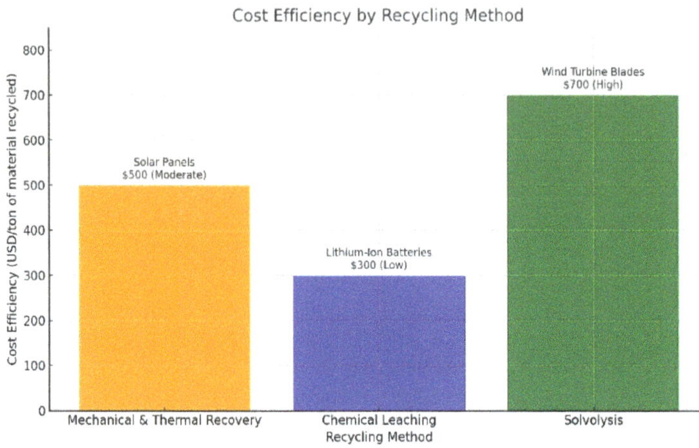

Fig. 3 Economic Feasibility of Recycling Methods

The economic workability of recycling technique for all material is shown in this Fig. 3. As seen lithium ion battery recycling method are the expensive due to advanced equipment required to recover important metals. The main reason for the higher cost of recycling lithium-ion batteries is related to the higher price of reagents used in the hydrometallurgical leaching process, as well as the specialized equipment and conditions needed for efficient metal recovery. Despite their environmental appeal wind turbine blade recycling technique are still less appealing from an economic standpoint due to challenges associated with a processing combined material [46, 49].

It is possible to draw many important results from the data:

- **Recycling Efficiency:** Because lithium ion batteries are widely used in electric vehicle and energy storage, their maximum Recycling productivity is 90%. But the high cost of these technology may limit their widespread adoption until improvement reduce the cost per ton of recycled material.
- **Environmental Impact:** Even with lower ability of work, recycling solar panel is the most environmental friendly way to reduce carbon emission. Also recycling solar panels should be a top priority for nations looking to meet low carbon targets.
- **Economic Viable:** Solar panel and wind turbine blades are the least expensive material to recycle. As the technology advances, recycling lithium ion batteries might become more cost effectiveness over time, making it a more attractive option down the road.

The results of this study were matched with the benchmark in order to judge how well these recycling methods perform in relation to global standards. For example, the European Union's Battery Directive has established strict recovery goal for battery recycling, with a weight recovery of 65% [51]. The research finding are in line with these objective because lithium ion batteries can recover of to 90% in advanced recycling facilities. Alike recovery rate of 80-90% are also proved by German Solar panel recycling systems, which recycle a sizeable number of solar panel(IRENA,2023) [50].

But, recycling wind turbine blades still lags behind other international standard. Material recovery rates for blades recycling procedure, like mechanical grinding, are still low and ineffective when compared to other renewable technologies. Solvolysis is a new alternative technique that has a possible transform wind turbine recycling, despite not being widely used in commercial sector yet (European Wind Energy Association, 2023) [52].

The study finding highlights many important policy recommendations for promoting energy material recycling:

- **Regulatory Support:** Stronger controlling mechanism is needed where policy gaps already exist, especially when it comes to recycling wind turbine blades [6, 20]. Well defined regulatory and rewards for companies to implement recycling technologies will guarantee the proper disposal of these product at the end of their useful lives.
- **Promoting Innovation:** Government may also take the lead in encouraging innovations by funding research and public-private partnerships to advance recycling technologies. The high cost of recycling lithium ion batteries, for instances, might be lessened with government subsidies or tax credits for businesses that take in green plan [25, 48].
- **Circular Economy Framework:** The shift to a circular economy requires both better recycling technologies and design changes for energy material. Policy framework must enhance the creations of products that are simpler to recycle and reuse in order to help a systematic shift towards materials recovery across the renewable energy sector [44].
- **Environmental and Economic Sustainability:** According to the result, recycling practises for wind turbine and solar panel can be improved to drastically reduce carbon emissions and help achieve sustainability objectives. Despite their high initial cost, some recycling technique will become more economically viable as technology advances due to long term advantages of material recovery, resource protection and decreased landfill trash [13, 33].

Conclusion

The economic practicability, environmental impact and recycling productivity of recycling wind turbines blades, lithium ion batteries and solar panel were checked in this study. As per the finding, lithium-ion batteries have the maximum recyling effectiveness (90%) but come at a higher enviromental cost (50kg CO2-eq/ton). With an 85% recovery rate and smallest carbon footprint (40kg CO2-eq./ton) solar panel are the most enviromental friendly. Even though recyling technique are still evolving, wind turbine blades have a lower recovery rate (70%) and a slight smaller enviromental footprint (25kg CO2-eq./ton). Batteries are the most expensive to recycle economically because of their detailed procedures, whereas wind turbine blades are less costly but less efficient. Recycling solar panel is the least expensive option. Policy effort such as the EU WEEE Directive, which encourage the recycling of solar panel and batteries but fall short when it comes to wind turbine blades, are slowing down development in this area. The study end that more legislative support and technological innovation are necessary to increase recyling usefulness and lessen environmental impact,especially for wind turbine blades. The report encourages innovative recyling methods and inclusive policies to increase long lasting in the renewable energy industry.

References

[1] D. Anderson, Renewable energy technology and policy for development, Annual Review of Energy and the Environment 22.1 (1997) 187-215. https://doi.org/10.1146/annurev.energy.22.1.187

[2] K. Moustakas, M. Loizidou, M. Rehan & A. S. Nizami, A review of recent developments in renewable and sustainable energy systems: Key challenges and future perspective, Renewable and Sustainable Energy Reviews 119 (2020) 109418. https://doi.org/10.1016/j.rser.2019.109418

[3] S. A. Muawad & A. A. Omara, Waste to energy as an alternative energy source and waste management solution, International Conference on Computer, Control, Electrical, and Electronics Engineering (ICCCEEE) IEEE (2019). https//10.1109/ICCCEEE46830.2019.9071080

[4] H. F. Yu, M. Hasanuzzaman, N. A. Rahim, N. Amin & N. Nor Adzman, Global challenges and prospects of photovoltaic materials disposal and recycling: a comprehensive review, Sustainability 14(14) (2022) 8567. https://doi.org/10.3390/su14148567

[5] J. Xiao, J. Li & Z. Xu, Challenges to future development of spent lithium ion batteries recovery from environmental and technological perspectives, Environmental Science & Technology 54(1) (2019) 9-25. https://pubs.acs.org/doi/10.1021/acs.est.9b03725

[6] J. P. Jensen & K. Skelton, Wind turbine blade recycling: Experiences, challenges and possibilities in a circular economy, Renewable and sustainable energy reviews 97 (2018) 165-176. https://doi.org/10.1016/j.rser.2018.08.041

[7] R. Prakash & I. K. Bhat, Energy, economics and environmental impacts of renewable energy systems, Renewable and sustainable energy reviews13(9) (2009) 2716-2721. https://doi.org/10.1016/j.rser.2009.05.007

[8] C. M. Motta, R. Cerciello, S. De Bonis, V. Mazzella, P. Cirino, R. Panzuto & B. Avallone, Potential toxicity of improperly discarded exhausted photovoltaic cells, Environmental Pollution 216 (2016) 786-792. https://doi.org/10.1016/j.envpol.2016.06.048

[9] D. Su, Y. Mei, T. Liu & K. Amine, Global Regulations for Sustainable Battery Recycling: Challenges and Opportunities, Sustainability 17(7), (2025) 3045. https://doi.org/10.3390/su17073045

[10] A. Srivastava, S. Pandey, R. Shahwal & A. Sur, Recycling of waste into useful materials and their energy applications, In Microbial Niche Nexus Sustaining Environmental Biological Wastewater and Water-Energy-Environment Nexus (2025) 251-296. Cham: Springer Nature Switzerland. https://doi.org/10.1007/978-3-031-62660-9_11

[11] N. Kamboj & A. Ray, Advanced Recycling Technology. AI Technologies for Enhancing Recycling Processes (2025) 393.

[12] J. Geldermann, L. M. Kolbe, A. Krause, C. Mai, H. Militz, V. S. Osburg & S.Westphal, Improved resource efficiency and cascading utilisation of renewable materials, Journal of Cleaner Production 110 (2016) 1-8. https://doi.org/10.1016/j.jclepro.2015.09.092

[13] S. Bringezu, Carbon recycling for renewable materials and energy supply: recent trends, long-term options, and challenges for research and development, Journal of Industrial Ecology 18(3) (2014) 327-340. https://doi.org/10.1111/jiec.12099

[14] S. M. Jowitt, T. T. Werner, Z. Weng & G. M. Mudd, Recycling of the rare earth elements, Current Opinion in Green and Sustainable Chemistry 13 (2018) 1-7. https://doi.org/10.1016/j.cogsc.2018.02.008

[15] L. Grandell, A. Lehtilä, M. Kivinen, T. Koljonen, S. Kihlman & L. S. Lauri, Role of critical metals in the future markets of clean energy technologies, Renewable energy 95 (2016) 53-62. https://doi.org/10.1016/j.renene.2016.03.102

[16] A. Sulich & L. Sołoducho-Pelc, The circular economy and the Green Jobs creation, Environmental Science and Pollution Research 29(10) (2022) 14231-14247. https://doi.org/10.1007/s11356-021-16562-y

[17] B. Mignacca & G. Locatelli, Modular circular economy in energy infrastructure projects: Enabling factors and barriers, Journal of Management in Engineering, 37(5) (2021) 04021053. https://doi.org/10.1061/(ASCE)ME.1943-5479.0000949

[18] F. Rossi, C. Zuffi, M. L. Parisi, D. Fiaschi & G. Manfrida, Comparative scenario-based LCA of renewable energy technologies focused on the end-of-life evaluation. Journal of Cleaner Production 405 (2023) 136931. https://doi.org/10.1016/j.jclepro.2023.136931

[19] H. Fazlollahtabar, Waste opportunity gap for energy productivity in new product design, Energy Sources, Part B: Economics, Planning, and Policy 19(1) (2024) 2371407. https://doi.org/10.1080/15567249.2024.2371407

[20] N. Rathore & N. L. Panwar, Environmental impact and waste recycling technologies for modern wind turbines: An overview, Waste Management & Research 41(4) (2023) 744-759. https://doi.org/10.1177/0734242X221135527

[21] D. Gayen, R. Chatterjee & S. Roy, A review on environmental impacts of renewable energy for sustainable development, International Journal of Environmental Science and Technology 21(5) (2024) 5285-5310. https://doi.org/10.1007/s13762-023-05380-z

[22] T. V. Ramachandra, Renewable energy transition: Perspective and Challenges, Energy India, (2020) 175-183. https://wgbis.ces.iisc.ac.in/energy/paper/renewable_energy_transition/

[23] M. K. Awasthi, S. Sarsaiya, H. Chen, Q. Wang, M. Wang, S. K. Awasthi & Z. Zhang, Global status of waste-to-energy technology, In Current developments in biotechnology and bioengineering, Elsevier (2019) 31-52. https://doi.org/10.1016/B978-0-444-64083-3.00003-8

[24] G. V. Kaliyannan, R. Gunasekaran, R. Rathanasamy, S. Subramaniam & V. Chinnathambi, Challenges and prospects in photovoltaic waste management: towards sustainable recycling and disposal of end-of-life solar panels, Sustainable Development Goals Towards Environmental Toxicity and Green Chemistry: Environment and Sustainability (2024) 61-82. https://doi.org/10.1007/978-3-031-77327-3_4

[25] X.Yu, W. Li, V. Gupta, H.Gao, D.Tran, S. Sarwar & Z. Chen, Current Challenges in Efficient Lithium-Ion Batteries' Recycling: A Perspective, Global Challenges 6(12) (2022) 2200099. https://doi.org/10.1002/gch2.202200099

[26] L. Mishnaevsky Jr, Sustainable end-of-life management of wind turbine blades: Overview of current and coming solutions, Materials, 14(5) (2021) 1124. https://doi.org/10.3390/ma14051124

[27] C. Cosmi, M. Macchiato, L. Mangiamele, G. Marmo, F. Pietrapertosa & M. Salvia, Environmental and economic effects of renewable energy sources use on a local case study, Energy Policy 31(5) (2003) 443-457. https://doi.org/10.1016/S0301-4215(02)00073-3

[28] A. Babaei & A. Nasr Esfahani, A review of photovoltaic waste management from a sustainable perspective, Electricity 5(4) (2024) 734-750. https://doi.org/10.3390/electricity5040036

[29] S. Xiao, H. Dong, Y. Geng & M. Brander, An overview of China's recyclable waste recycling and recommendations for integrated solutions, Resources, Conservation and Recycling 134 (2018) 112-120. https://doi.org/10.1016/j.resconrec.2018.02.032

[30] M. Massoud, G. Vega, A. Subburaj & J. Partheepan, Review on recycling energy resources and sustainability, Heliyon, 9(4) (2023). https://10.1016/j.heliyon.2023.e15107

[31] Z. Kang, Z. Huang, Q. Peng, Z. Shi, H. Xiao, R. Yin & J. Zhao, Recycling technologies, policies, prospects, and challenges for spent batteries, Iscience 26(11) (2023). https://10.1016/j.isci.2023.108072

[32] F. Flachenecker & J. Rentschler, From barriers to opportunities: Enabling investments in resource efficiency for sustainable development, Public Sector Economics, 43(4) (2019) 345-373. https://doi.org/10.3326/pse.43.4.2

[33] P. H. Brunner & H. Rechberger, Waste to energy–key element for sustainable waste management, Waste management 37 (2015) 3-12. https://doi.org/10.1016/j.wasman.2014.02.003

[34] S. Weckend, A. Wade & G. A. Heath, End of life management: solar photovoltaic panels (No. NREL/TP-6A20-73852; T12-06: 2016), National Renewable Energy Laboratory (NREL) Golden CO (United States) (2016). https://doi.org/10.2172/1561525

[35] V. Kandpal, A. Jaswal, Santibanez E. D. Gonzalez & N. Agarwal, Challenges and opportunities for sustainable energy transition and circular economy, Sustainable Energy Transition: Circular Economy and Sustainable Financing for Environmental, Social and Governance (ESG) Practices (2024) 307-324. https://doi.org/10.1007/978-3-031-52943-6_11

[36] H. K. Jani, S. S. Kachhwaha, G. Nagababu & A. Das, A brief review on recycling and reuse of wind turbine blade materials, Materials Today: Proceedings 62 (2022) 7124-7130. https://doi.org/10.1016/j.matpr.2022.02.049

[37] R. Wendling, Job Creation through a Green Energy Economy, In Green Energy Economies (2017) 49-86. Routledge.

[38] S. Rahimifard, G. Coates, T. Staikos, C. Edwards & M. Abu-Bakar, Barriers, drivers and challenges for sustainable product recovery and recycling, International Journal of Sustainable Engineering 2(2) (2009) 80-90. https://doi.org/10.1080/19397030903019766

[39] Y. Hotta, C. Visvanathan & M. Kojima, Recycling rate and target setting: challenges for standardized measurement, Journal of Material Cycles and Waste Management 18(1) (2016) 14-21. https://doi.org/10.1007/s10163-015-0361-3

[40] G. Cepeda & D. Martin, A review of case studies publishing in Management Decision 2003-2004: Guides and criteria for achieving quality in qualitative research, Management Decision 43(6) (2005) 851-876. https://doi.org/10.1108/00251740510603600

[41] A. Pires, G. Martinho & N. B. Chang, Solid waste management in European countries: A review of systems analysis techniques, Journal of environmental management 92(4) (2011) 1033-1050. https://doi.org/10.1016/j.jenvman.2010.11.024

[42] I. Ulnicane, W. Knight, T. Leach, B. C. Stahl & W. G. Wanjiku, Framing governance for a contested emerging technology: insights from AI policy, Policy and Society 40(2) (2021)158-177. https://doi.org/10.1080/14494035.2020.1855800

[43] K. Richa, C. W. Babbitt, N. G. Nenadic & G. Gaustad, Environmental trade-offs across cascading lithium-ion battery life cycles, The International Journal of Life Cycle Assessment 22(1) (2017) 66-81. https://doi.org/10.1007/s11367-015-0942-3

[44] J. J. Klemeš, P. S. Varbanov, T. G. Walmsley & A. Foley, Process integration and circular economy for renewable and sustainable energy systems, Renewable and Sustainable Energy Reviews 116 (2019) 109435. https://doi.org/10.1016/j.rser.2019.109435

[45] Y. U. Dan, S. U. N. Yongbin, Y. A. N. G.Qi, G. A. O. Li, D. O. N. G. Biqiong & Z. E. N. G. Xiaochao, Current Status and Prospect of Recycling Technology of Wind Turbine Blades, Southern Energy Construction 10(4) (2023) 113-121.

[46] X. Zeng, J. Li & N. Singh, Recycling of spent lithium-ion battery: a critical review, Critical Reviews in Environmental Science and Technology 44(10) (2014) 1129-1165. https://doi.org/10.1080/10643389.2013.763578

[47] P. Liu, , F. Meng & C. Y. Barlow, Wind turbine blade end-of-life options: An eco-audit comparison, Journal of Cleaner Production 212 (2019) 1268-1281. https://doi.org/10.1016/j.jclepro.2018.12.043

[48] K. M. Winslow, S. J. Laux & T. G. Townsend, A review on the growing concern and potential management strategies of waste lithium-ion batteries, Resources, Conservation and Recycling 129 (2018) 263-277. https://doi.org/10.1016/j.resconrec.2017.11.001

[49] S. B. Darling, F. You, T. Veselka & A. Velosa, Assumptions and the levelized cost of energy for photovoltaics, Energy & environmental science 4(9) (2011) 3133-3139. https://doi.org/10.1039/C0EE00698J

[50] P. Nain & A. Anctil, End-of-life solar photovoltaic waste management: A comparison as per European Union and United States regulatory approaches, Resources, Conservation & Recycling Advances 21 (2024) 200212. https://doi.org/10.1016/j.rcradv.2024.200212

[51] Information on https://ec.europa.eu/environment/topics/waste-and-recycling/batteries-and-accumulators_en

[52] J.Yang, F. Meng, L. Zhang, J. McKechnie, Y., Chang, B. Ma & J. M. Cullen, Solutions for recycling emerging wind turbine blade waste in China are not yet effective Communications Earth & Environment 4(1) (2023) 466. https://doi.org/10.1038/s43247-023-01104-w

Carbon Capture, Utilization, and Storage: Technologies, Applications, and Pathways to Net-Zero

S. SRIVASTAVA[1*], A. DHASMANA[1], S. GARG[2], J. GAIROLA[3], S. DOBHAL[4], M. SHAMEEN[5]

[1]Department of Aerospace Engineering, Uttaranchal Institute of Technology, Uttaranchal University, Dehradun-248007, Uttarakhand, India

[2]School of Business, Department of forensic science, Chandigarh Group of Colleges, Jhanjeri, Mohali-140307, Punjab, India

[3]Department of Telemedicine, AIIMS, Rishikesh, Uttarakhand, India

[4]Department of Biochemistry, AIIMS, Rishikesh, Uttarakhand, India

[5]Allied Health Sciences, Desh Bhagat University, Mandi Gobindgarh-147301, Punjab, India

*sachinrash2004@gmail.com

Keywords: Carbon Capture and Storage (CCS), CO_2 Utilization, Net-Zero Emissions, Climate Change Mitigation, Sustainable Energy Technologies

Abstract: Anthropogenic CO_2 discharge has risen as a larger danger to global climate system stability as an effect of industrial and energy activities. Progress made in renewable technologies aside, present emission abatement measures have been found wanting to be taken on challenging-to-abate sectors, including cement, steel, and chemical production. CCUS is rapidly coming in farther along as a solution to the issue by capturing CO_2 prior to liquefaction, usage, or storage. This paper intends to give an extensive overview of CCUS technologies such as pre-combustion, post-combustion, oxy-fuel, and direct air capture. Different utilization opportunities of CO_2 will be addressed with particular interest being paid to enhanced oil recovery (EOR), fuel production, and mineral carbonation possibilities along with geological storage possibilities assessed in terms of safety factors. It will also analyze key performance indicators (KPIs), economic viability, and operational issues based on relevant case studies. Further, potential policy and regulatory tools that can facilitate the upscaling of CCUS globally will be addressed.

1. Introduction

The combined carbon dioxide capture, utilization, and storage (CCUS) operate once the carbon dioxide is captured directly from a large source or the air, sent to a location for utilization into manufacturing processes, or processed to be injected in geological reservoirs for the long-term storage. The technology especially helps hard-to-abate industries such as cement, steel, chemicals, and fossil fuel power generation, which otherwise are difficult to decarbonize [1-3]. Figure 1 illustrates hard-to-abate sector emissions and the projected role of CCUS in reducing these emissions by 2050 [4].

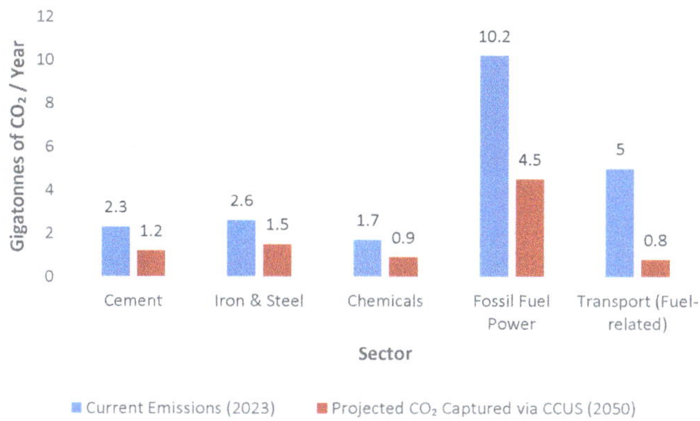

Figure 1. *CCUS data up to 2050*

Recent studies have highlighted CCUS as greater than a mitigation measure; on the contrary, it is regarded as a foundation unit for subsequent low-carbon industrial clusters, particularly for China, where heavy industrialization and emissions dominate. Referring to China, CCUS is increasingly significant to climate change national policies and long-term decarbonization blueprints, with some pilot projects exhibiting technical viability and policy congruence [5-6]. Place deployment CCUS in context from a system standpoint, and inherent to the task are complicated decision-making and optimization issues like site location, infrastructure design, and cost-effectiveness evaluation. Risk analysis and deployment planning enables and ensures the economic viability and public acceptance, especially in economically limited geographical areas where geological storage and transport infrastructure are limited. Recent advancements in machine learning and data-driven modelling enable more efficient design and operation of CCUS systems as from predictive modelling of capture processes to real-time monitoring of storage sites [7-8]. Innovation in materials and process integration has expanded the horizons of CO_2 utilization. For example, industrial by-products, particularly red mud, are being considered as possibly cheaper candidates for absorbents or mineralization agents and contribute further to sustainability and circularity in CCUS applications. In addition, coupling CCUS with emerging technologies, such as green hydrogen and bioenergy with carbon capture and storage (BECCS), provides avenues for achieving negative emissions and even deeper decarbonization [9].

By comparison the multiprocessing machine learning algorithms applied in STLF with data variables, the study demonstrates how accurate future forecasts of electricity output or consumption may be generated using machine learning techniques [10]. To overawed parameters, it talks about edge computing powered by AI and advanced communication protocols. Case studies are used to approximation the contribution of embedded systems to energy efficiency and reliability. Future growths, such as hardware evolution, machine learning for analytical management, and IoT-based smart ecosystems, are also predicted to enhance efficiency by 15–25% over the next ten years [11].

Significant, widespread and successful deployment of CCUS will require a synchronised technological development, policy and financial investment support, and public acceptance. Indeed, with increasing urgency to address climate change, CCUS should serve as a linchpin that connects existing emissions to an eventual net-zero future. The increase in the number of CCUS projects globally is indicative of the increasing uptake and investment, as shown in figure 2.

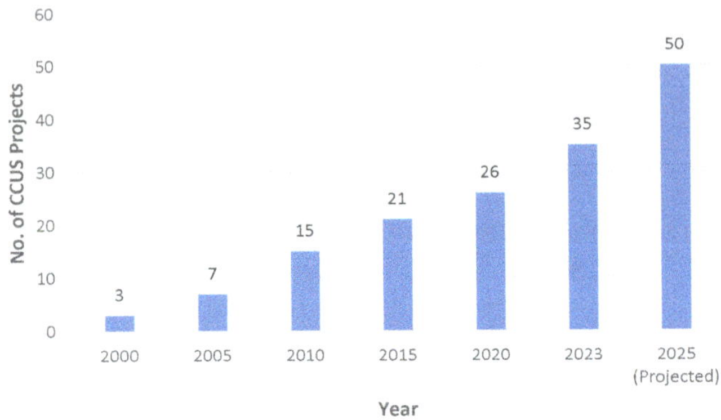

Figure 2. *Growth of Global CCUS Projects Over Time (2000–2025) [12]*

Carbon Capture, Utilization, and Storage (CCUS) is, broadly, a family of technologies to capture carbon dioxide (CO_2) emissions from major point sources, such as fossil fuel power plants, industrial facilities, or even directly from the atmosphere and then utilize or store indefinitely from the captured CO_2 into value-added products or deep geological formations. The heart of CCUS is keeping carbon dioxide from entering the atmosphere and thus avoiding its global warming and climate change effects. This is because CCUS is the only option available to address emissions from hard-to-abate sectors such as steel, cement, and chemicals, all of which are not electrified or decarbonized easily through renewable energy alone. With BECCS or DAC, CCUS is one of the scalable means negative emissions can be achieved. This is noted by DeZern (2021), indicating that CCUS is one of the few determinative criteria for seriousness and effectiveness among global climate-mitigation strategies; it is not merely a transitional or short-term tool but a very strong base technology that can provide sustainability for a very long period [13].

Carbon Capture, Utilization and Storage (CCUS) is advisable to use in space mission also. For any space mission, using the chemical propulsion engine (Solid rocket motor [14-15], Hybrid rocket engine [16-17] or liquid rocket engine [18-19]) then it is compulsory to use this technique in atmosphere as well as in space. The above mentioned references shows the reaction mechanism where it is showing that how much carbon generated.

CCUS has been generally mentioned as a key component of any route to net-zero emissions globally, particularly in situations where complete de-carbonization is not possible using renewable energy technologies and efficiency technology. Greig and Uden (2021) would further argue CCUS are double-faced in climate policies 'firstly by preventing emissions from fossil fuels and industrial infrastructure and secondly by allowing the production of low carbon fuels, notably blue hydrogen from natural gas reforming with carbon capture. In addition, CCUS maintains a fair transition of retaining current industrial capital while lowering emissions so as not to contort the economy in carbon-rich regions too significantly. The second benefit of combining CCUS with net-zero routes is providing time flexibility to the corresponding policymaking and industry: they will phase out fossil fuels on a schedule as they work to develop and deploy renewable alternatives. According to Greig and Uden (2021), net-zero goals in the absence of CCUS would ipso facto mean increased costs and greater technological limitations, particularly in cobalt industrial processes where the emissions are process-based as opposed to energy-based [20].

2. Carbon Capture Technologies

Carbon capture is the first and most important step in the Carbon Capture, Utilization, and Storage (CCUS) technology route. It involves the separation of carbon dioxide (CO_2) from flue gases or ambient air before being let out in the atmosphere. Several technological pathways have been developed for CO_2 capture, oriented toward different applications and various emission sources. The above processes can be conveniently grouped into four main categories of operation: pre-combustion, post-combustion, oxy-fuel combustion, and direct air capture (DAC). A block diagram in Figure 3 gives a visual summary of major carbon capture technologies that are part of CCUS systems. This diagram puts carbon capture in the centre, diffusing out into four main pathways, namely: pre-combustion, post-combustion, oxy-fuel combustion, and DAC. Pre-combustion concerns itself with the gasification of fuel into syngas and CO_2 separation before combustion, while post-combustion processes capture CO_2 from flue gases following combustion, utilizing chemical solvents. Oxy-fuel combustion produces an exhaust enriched in CO_2 through burning the fuel in oxygen and recycled flue gas, facilitating its capture. Contrarily, DAC captures CO_2 directly from the ambient air by means of sorbents. Each of these methods is unique in serving different applications and presenting different advantages, disadvantages, efficiency, scalability, and energy requirement.

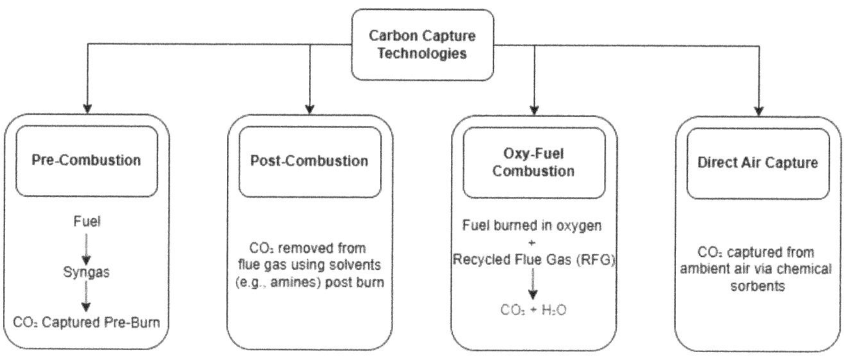

Figure 3. *Overview of Carbon Capture Technologies in CCUS*

2.1 Pre-combustion, post-combustion, and oxy-fuel combustion capture

In pre-combustion capture, the first thing that happens is the conversion of fuel to a mixture of hydrogen and carbon monoxide, which is the syngas, by gasification. Carbon monoxide then reacts with steam in a water-gas shift reactor to give hydrogen and CO_2, which are separable before combustion. This technique is primarily used with integrated gasification combined cycle (IGCC) plants and allows for high-purity CO_2 capture at lower energy penalties. In contrast, post-combustion capture consists of the removal of carbon dioxide from the flue gas after the burning of fossil fuels. This is the most developed and widely applicable method, especially in retrofits to existing power plants and industrial facilities. It uses chemical solvents, mostly amines, to selectively absorb CO_2 from the exhaust gas stream. Oxy-fuel combustion, a newer approach, burns fuel in a mixture of oxygen and recycled flue gas instead of air. This leads to flue gas that is mostly composed of CO_2 and water vapor, so the separation process is simplified. This method also has potential advantages in terms of reduced NO_x emissions and ease of CO_2 purification [21].

2.2 Direct air capture (DAC)

Direct Air Capture (DAC), unlike any aforementioned point-source methods, accomplishes the task of sequestering CO_2 from ambient air. In contrast to other negative emissions technologies, DAC may be viewed differently since it can either offset present or past emissions. It relies on chemical sorbents (either solid or liquid) that bind CO_2 from the air, followed by regeneration processes to release concentrated CO_2 for subsequent storage or utilization. DAC, compared to point- source capture, is currently more energy-consuming and expensive; however, it has strategic advantages such as flexibility in siting and scalability. As noted by Ozkan et al. (2022), DAC is the linchpin of future strategies for carbon removal to reach carbon neutrality or net-negative emissions. Ongoing advancements in sorbent materials, process optimization, and renewable energy utilization will further improve the feasibility and thereby reduce the costs [22].

3. Utilization of Captured Carbon

In the utilization component of Carbon Capture, Utilization, and Storage (CCUS), the use of captured CO2 is turning the captured carbon dioxide into commercially valuable products or applying it in processes that prevent its release into the atmosphere. The entire approach fills into emissions mitigation yet provides economic incentives by revenue generation from waste CO2. Key pathways are EOR (Enhanced Oil Recovery), mineralization, and conversion to chemicals and fuels.

3.1 Enhanced oil recovery (EOR)

CO_2-enhanced oil recovery (CO_2-EOR) is one of the more advanced and widely spread forms of carbon utilization. Such captured carbon dioxide is injected into oil reservoirs that are either depleted or low pressure so that the dormant oil is mobilized, and the oil extraction efficiency is improved. The injected CO_2 combines with the crude oil, thereby reducing its viscosity and enabling more oil to be recovered to the surface. Although intended specifically for long-term sequestration, this kind of EOR also has the dual functions of enhanced oil production alongside CO_2 storage in the subsurface. Large-scale EOR schemes can be seen on the industrial as well as the environmental aspects, particularly in regions that have existing oil infrastructures. As Shiyi et al. (2022) observe, improvements in capture technologies and injection methods have enhanced the scalability of EOR in relation to making it an aspect of early CCUS deployment-mostly in nations such as the United States and China [23].

3.2 Mineralization and concrete curing

In mineralization, CO2 interacts with calcium or magnesium content-bearing materials and produces stable carbonate minerals. Therefore, it's a natural process over a geological timescale, but one can hasten it under controlled conditions in an industrial scenario. One such interesting application in mineralization is curing the concrete, where CO2 is fed into newly mixed concretes to enhance strength and reduce curing time while storing the gas away for good. In their working paper, Hills et al. (2020) elaborate that this technology effectively reduces the carbon footprint attributed to cement production, one of the leading sources of global CO2 emissions, and at the same time transforms captured carbon into construction materials that are durable. Mineralization is capable of providing a low-leakage solution for long-term storage and is continuously being assessed further for applications involving waste materials, such as steel slag or fly ash, thereby assisting in enhancing circularity [24].

3.3 Conversion to chemicals and fuels

Transformation of CO_2 into fuels and chemicals is another major utility route. This can be through catalytic, electrochemical, or biological pathways for generating methanol, formic acid, urea, and synthesized hydrocarbons from CO_2. When driven by renewable energy, these conversion processes can lead to carbon-neutral or even carbon-negative fuels, particularly in the form of synthetic methane or liquid fuels compatible with current infrastructure. Many of these technologies are still being developed or at early stages of commercialization but have great promise for decarbonizing the chemical and transport sectors (Jones 2020). Innovations in catalytic and process intensification can minimize energy demand and improve the economics of CO_2 conversion [25].

4. Storage Technologies

The storage segment of Carbon Capture, Utilization, and Storage (CCUS) is critical for ensuring the permanent withdrawal of captured CO_2 from the atmosphere. Once CO_2 is captured and used, any remaining or excess volume is always stored in long-term geological or mineralogical formations. The legitimate effectiveness of storage not only depends on the capacity and integrity of the storage site but also on the technologies of monitoring and verification implemented to guarantee long-term containment."

4.1 Geological storage (saline aquifers, depleted reservoirs)

The geological storage method refers to the most practically and technically established approach for the long-term storage of carbon dioxide. In this method, CO2 is compressed and then injected into deep underground formations, such as saline aquifers or depleted oil and gas reservoirs, where it will be trapped by physical and geochemistry mechanisms. The saline aquifer has a much greater potential storage capacity at a global scale, given its large volume and widespread geological distribution. A thorough examination was made with respect to existing CO2 storage operations in saline aquifers by Michael et al. (2010), who provided evidence that successful projects such as those in the Sleipner and In Salah fields have entirely proven both technical feasibility and environmental safety. Various conditions such as structural trapping, residual trapping, solubility trapping, and mineral trapping provide CO2 immobilization in these formations ensuring multi-barrier containment over geological time scales [26].

4.2 Ocean and mineral storage

Another consideration beyond geological formations is whether the CO_2 can be sunk in oceans or interacted with minerals. CO_2 injection into the deep ocean is considered a potential mechanism for deep-sea storage at subduction depths for upwards of a few centuries due to extremely high pressure and low temperature. The marine ecosystem impacts, and ocean acidification concerns limit its advancement and public acceptance. Mineral storage is, on the other hand, a viable alternate storage solution with such inherent permanence. The method involves the reaction of CO_2 with ultramafic and mafic rocks, naturally available such as peridotite or basalt, to produce stable carbonate minerals. Since mineral carbonation can be done in situ (within the rock formations) as well as ex situ (in industrial reactors), it is an environmentally safe option for CO_2 leak-proof storage as CO_2 is chemically bound in solid form, as presented by Voormeij and Simandl (2004) [27]. Table 1 This table gives a general summary to compare key storage techniques based on permanence, advantages, and concerns.

Table 1. *Comparison of CO_2 Storage Methods*

Storage Method	Description	Permanence	Main Advantages	Challenges
Saline Aquifers	Injection into deep underground brine-filled formations	High	Large capacity, widespread availability	Requires monitoring, potential leakage
Depleted Reservoirs	Use of exhausted oil/gas fields	High	Existing infrastructure, known geology	Potential leakage through old wells
Ocean Storage	CO_2 injected into deep ocean layers	Moderate–High	Huge potential storage volume	Environmental risks, public resistance
Mineral Storage	Reacting CO_2 with ultramafic/mafic rocks to form stable carbonates	Very High (Permanent)	Inherently stable, no leakage	Slow kinetics, cost of ex situ processing

4.3 Long-term stability and monitoring

For any CO_2 storage project, long-term assurance of stored gas is a significant requirement. A sound monitoring technology is then necessary to ensure that it is contained, free from surface or potable water entry. Seismic imaging, pressure monitoring, geochemical sampling, and satellite surface detection techniques are among the possible methodologies. Furthermore, Chadwick et al. (2009) mentioned that the consideration of a full monitoring strategy needs to cover all phases of a project, namely, injection, post-injection, and closure. Insights gained from early implementers of storage have shown that the geological method can be environmentally safe, and publicly acceptable with appropriate monitoring and risk assessment. Research and regulation continue to enhance the credibility and transparency of long-term CO_2 storage [28].

5. Efficiency and Performance Evaluation

For practical deployment of CCUS, technical viability, safety, and economic feasibility are taken into consideration. Therefore, its analysis must span these areas to be able to reach the likely scale of practicality and the directions upon which to develop for cost minimization of CCUS schemes. These controls utilized are based on capture efficiency, energy penalties, storage volume, leakage risk, and cost feasibility. Figure 4 presents the summary block diagram that postulates the factors at the center of the evaluation of the performance of Carbon Capture, Utilization, and Storage (CCUS) systems The capture CO2 system is in the center of performance assessment. There are three interrelated aspects that make up this assessment: capture efficiency or how well CO2 is captured and the energy costs associated; storage performance or ensuring the integrity of the storage site including its capacity and means of monitoring to avoid leakage, and economic feasibility or cost considerations from the standpoint of viable/set-up and running costs and other financial events such as carbon pricing. The two contribute toward a global understanding of the technical and economic feasibility of CCUS technologies.

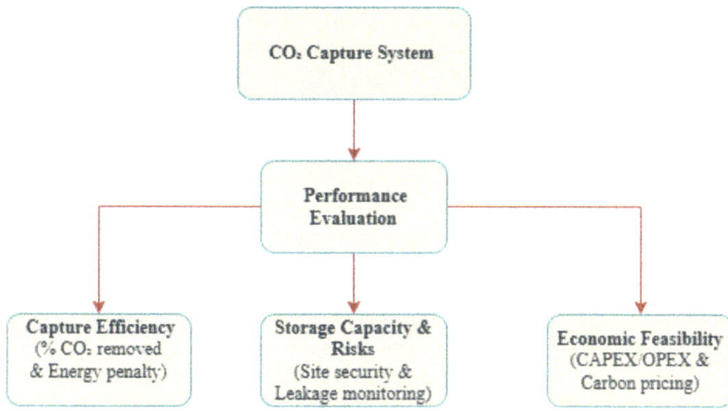

Figure 4. *CCUS Performance Evaluation Framework*

5.1 Capture efficiency and energy penalties

CCUS performance metrics typically include capture efficiencies in the range of 85% to 95% and above, the usual range depending on factors such as technology and source type. The price of having higher capture efficiencies is because there are almost always penalties to energy generation-that is especially true in solvent-based post-combustion cases. Energy penalties, depending on the design, can impose impacts on net power plant efficiencies ranging between 8 and 12 percentage points. Capocelli and De Falco (2022) give an overarching evaluation of energy penalties and benchmarked performance metrics for capture technologies that iron out the inconsistency between capture efficiency being overly high versus operational costing and energy barriers, the retrofits to existing older fossil-fuel installations in particular. Innovative aspects of process integration and heat recovery as well as solvent formulation will act as the key drivers to remove energy losses and increase the performance of the entire system [29].

5.2 Storage capacity and leakage risks

The sustainable long-term future for CCUS depends on the guaranteeing of sufficient and secure storage capacity. Deep saline aquifers and depleted oil-and-gas reservoirs provide enormous potential storage capacity with huge volumes of hundreds of potentially hundreds of gigatonnes in volume globally. Nevertheless, concerns over leakage of CO_2 from storage locations imposes very considerable hurdles to public acceptance and regulatory license. Gholami, Raza, and Iglauer (2021) conducted an extensive review of the evaluation methods concerning the risk of leakage for storage sites of CO_2. Their study focuses on local geological characterization, fault and fracture research, and damage monitoring systems. Site selection based on risk sensing along with good sealing mechanisms (caprocks, wellbore integrity) is essential in preserving environmental safety in long-term CO_2 storage [30].

5.3 Economic feasibility

The environmental advantages of CCUS have always been constrained by high initial capital expenditure and operational costs. The major cost elements are the capture equipment, transport pipes, compression, injection, long-term surveillance, and these cost sites differ with various applications, remembering industrial CCUS is less expensive than power sector applications due to flue gas containing a greater concentration of CO_2. Taghizadeh-Hesary, Vandercamme, and Phoumin (2024) opined that making CCUS projects more financially viable therefore entails a combination of policy measures, market measures (such as carbon pricing), and innovation in technology. They also make observations regarding the necessity of financial mechanisms that will reduce the investment risk like public-private partnerships, tax credits, and carbon trading. Upscaling the CCUS will also be dependent on lowering the unit costs by learning-by-doing, modularity, and economies of scale [31].

6. Modelling Approaches and Technological Advancements in CCUS Systems

6.1 Advanced Process Modelling and System Optimization

Recent CCUS research results show that scientists now use high-resolution process models together with integrated system optimization frameworks to study their research. Scientists now use advanced thermodynamic and kinetic models to create simulations of absorption adsorption and membrane-based capture processes which operate under changing conditions. The models use techno-economic and lifecycle assessment tools to measure their cost efficiency and emissions reduction potential and their ability to scale up operations. Holistic modelling methods which combine social political and economic factors create a comprehensive systems view of CCS deployment paths and carbon neutrality pathways [32]. The multi-layered frameworks enable decision-making processes by showing how technical feasibility connects to both policy matters and market forces.

6.2 Machine Learning and AI-Driven Innovation

Machine learning (ML) and artificial intelligence (AI) are becoming more important for improving CCUS systems. AI-driven digital solutions enable real-time monitoring, predictive maintenance, and risk assessment through advanced data analytics and sensor integration [33]. Intelligent systems help detect anomalies while predicting leakages and maintaining operational safety for capture and storage systems. ML-based material discovery methods help researchers identify and develop advanced sorbents membranes and hybrid materials by predicting their adsorption capacity and selectivity and long-term stability which leads to lower testing expenses and shorter development times [34]. The combination of AI with process modeling enables organizations to create adaptive control methods which help them forecast system performance.

6.3 Digitalization and Smart CCUS Infrastructure

Digital transformation serves as a vital foundation for developing upcoming carbon capture utilization and storage technologies. The implementation of digital twins, cloud-based data management platforms, and IoT-enabled monitoring networks enables ongoing performance assessment and improved visibility across the complete CCUS value chain [36]. Digital ecosystems provide support for combined energy system modeling which allows CCUS facilities to function in harmony with renewable energy systems and industrial clusters [35]. The digitalization model establishes a framework which helps organizations meet regulations while they build stronger operational systems and achieve better financial results.

7. Applications and Case Studies

In the last twenty years, CCUS technology gained a considerable foothold for applications in industry, and pilot programs have been set up to further advance the field. These projects highlight the technical feasibility of CCUS in real-world applications; they also provide important information concerning costs, scalability, and operational issues that will need to be addressed for wider deployment.

7.1 Industrial-scale CCUS projects

Huge scores of CCUS projects have been applied in highly consumed power production, cement, steel, fertilizers, and refining processes. Cumulative application evidenced the capability of CCUS for online industrial application, drastic emission reductions, and decarbonization pathways for hard-to-abate sectors. Olabi et al. (2022) presented a thorough review of CCUS in process industries. It portrays some of the largest facilities such as Boundary Dam (Canada), Sleipner

(Norway), and Gorgon (Australia) in the deployment of CCUS. These have captured millions of tonnes of CO_2 per year, thereby putting a wealth of experience regarding capture technology performance, CO_2 compression and transport coordination, and geologic storage over the long term. Successful as they have proved, the scale-up is further limited by high capital investment and regulatory uncertainties, and those need to be resolved to realize the full climate benefit of CCUS in industry [35].

7.2 Pilot programs and field trials

Field trials and pilot demonstrations are essential to bridge laboratory research to large-scale application. These small-scale demonstrations test capture materials, process design, monitoring methods, and storage facility performance in actual operating conditions. They are used to test models, confirm technical showstoppers and guide policy and investment. Reyes-Lúa and Jordal (2021) examine the status and lessons from a variety of industrial CO2 capture pilot projects from across Europe to Asia to North America. Their findings are that whereas notable technological advance has occurred, many of the demonstration projects have been delayed and overbudget due to onerous permitting processes and policy shortfall. In either event, these pilots have contributed substantially to derisk technologies, increase public awareness, and improve technical readiness for final large-scale implementation [36].

7.3 Recent CCUS projects

The countries of China and Gulf Cooperation Council (GCC) nations have developed multiple partnerships which focus their efforts on large-scale project implementation together with cross-border investment activities while transferring their expertise to solve technical and regulatory issues which will help their energy-intensive industries reach decarbonization targets [37]. The technological field benefits from CCUS-EOR advancements which enable better reservoir modeling and CO_2 injectivity management and long-term storage reliability which results in more economical CCUS deployment throughout mature oil fields according [38]. The Indonesian case study shows that establishing carbon pricing systems together with policy instruments determines the economic viability and emission reduction capacity of carbon capture technology for coal power plants which are vital for emerging economies that depend on coal power [39]. The recent developments show how regional energy transitions now integrate CCUS systems because of improved technology and new carbon governance regulations.

8. Challenges and Future Directions

Despite the consideration that Carbon Capture Utilization and Storage (CCUS) enable one to stride in pursuing decarbonization, there are some challenges it undergoes that prevent its implementation: limitations in technology, economics, regulation, and shifting priorities in research. Further discussing these shall give the chance for CCUS to appear in the world map of low-carbon alternatives.

8.1 Technological and financial barriers

Even at a development stage, CCUS technologies are suffering from technical inefficiencies and operating cost severely restricting their commercial potential. High-energy-intensive CO_2 capture, optimum transport infrastructure, and long-term storage still integrity problems are all significant challenges. Gowd et al. (2023) identify, in a study, aspects related to economy, i.e., unfavorable investment incentives and high capital requirements at the beginning, coempt the technological constraints. Lacking robust policy or carbon price, the majority of CCUS projects have struggled to find financing. Today, they demand massive deployment of CCUS to be woven into sustainable development pathways where economic incentives and scalable business models will be available [40].

8.2 Policy and regulatory frameworks

One of the main barriers to CCUS progress in most places is that most of them still do not have comprehensive and consistent policy frameworks. Practical implementation will involve reconciling rules along the whole value chain-from capture and transportation to use and storage, and a sustained commitment to carbon price, subsidies, liability management, etc. Song, Jiang, and Rao reviewed recent CCUS policy-making in China. They state that while national plans and pilot initiatives have emerged, no common standards, regulatory certainty, or inter-agency coordination exist within the policy. They believe that international cooperation, knowledge sharing, and coalescing CCUS policy with overall climate ambitions can overcome these systemic impediments and speed the deployment of such CCUS policies [41].

8.3 Research trends in CCUS

Sustained R&D and innovation are crucial to overcoming today's limitations towards the next generation of CCUS systems. These are low-energy capture materials, enhanced storage monitoring methods, and CO_2 conversion routes into higher-value products. Ren et al. (2023) conducted a bibliometric analysis of trends in CCUS research in China and the US, following which an extraordinary surge in scientific output has been observed over the past 10 years. The authors report that machine learning is being highlighted more, as are cross-disciplinary and integrated system optimization, which shows the shift in this field from standalone pilot work to system-wide carbon management plans. They propose a boost in international cooperation and investment in basic and applied research to accelerate the CCUS technology breakthroughs required [42].

Conclusion

- In international efforts towards climate, CCUS technology is one of those technologies which have to be developed for those industries which are not suitable for emission removal on a renewable energy basis.
- Of the various methods employed to capture, post combustion capture, and direct air capture are such great potentials as concerns flexibility but are still waiting to be studied for their energy costs to be minimized and their cost-effectiveness maximized so that they may be the wood upon which innovation rests.
- Employing such methods as enhanced oil recovery, synthetic fuel generation, and mineral carbonation assists in generating economic incentives, and hence SC/CCUS gets wide acceptance.
- Though it is possible technically to store CO2 deep within the earth in a secure long-term manner along with rock formations, such as saline aquifers and depleted oil reservoirs, it has to be monitored properly and risk evaluated so that it is safe for the environment.
- Public-private collaborations and global cooperation are part of one of the key ingredients in the mix needed to scale up CCUS infrastructure so that regulatory hurdles could be overcome and innovations could be elicited.
- Multidisciplinary studies and dedicated investment in overcoming current challenges will turn CCUS into a pillar of ecologically friendly future and low-carbon world economy transition.

References

[1] Yan, J., & Zhang, Z. (2019). Carbon capture, utilization and storage (CCUS). Applied Energy, 235, 1289-1299. https://doi.org/10.1016/j.apenergy.2018.11.019

[2] Zhang, Z., Wang, T., Blunt, M. J., Anthony, E. J., Park, A. H. A., Hughes, R. W., ... & Yan, J. (2020).

[3] Chen, S., Liu, J., Zhang, Q., Teng, F., & McLellan, B. C. (2022). A critical review on deployment planning and risk analysis of carbon capture, utilization, and storage (CCUS) toward carbon neutrality. Renewable and Sustainable Energy Reviews, 167, 112537. https://doi.org/10.1016/j.rser.2022.112537

[4] Net, I. E. A. (2021). Zero by 2050: a roadmap for the global energy sector. Int Energ y Agency, 224.

[5] Jiang, K., Ashworth, P., Zhang, S., Liang, X., Sun, Y., & Angus, D. (2020). China's carbon capture, utilization and storage (CCUS) policy: A critical review. Renewable and Sustainable Energy Reviews, 119, 109601. https://doi.org/10.1016/j.rser.2019.109601

[6] Yao, J., Han, H., Yang, Y., Song, Y., & Li, G. (2023). A review of recent progress of carbon capture, utilization, and storage (CCUS) in China. Applied Sciences, 13(2), 1169. https://doi.org/10.3390/app13021169

[7] Yan, Y., Borhani, T. N., Subraveti, S. G., Pai, K. N., Prasad, V., Rajendran, A., ... & Clough, P. T. (2021). Harnessing the power of machine learning for carbon capture, utilisation, and storage (CCUS)-a state-of-the-art review. Energy & Environmental Science, 14(12), 6122-6157. https://doi.org/10.1039/D1EE02395K

[8] Tapia, J. F. D., Lee, J. Y., Ooi, R. E., Foo, D. C., & Tan, R. R. (2018). A review of optimization and decision-making models for the planning of CO2 capture, utilization and storage (CCUS) systems. Sustainable Production and Consumption, 13, 1-15. https://doi.org/10.1016/j.spc.2017.10.001

[9] Liu, X., Liu, X., & Zhang, Z. (2024). Application of red mud in carbon capture, utilization and storage (CCUS) technology. Renewable and Sustainable Energy Reviews, 202, 114683. https://doi.org/10.1016/j.rser.2024.114683

[10] Naudiyal, A., Joshi, K., Singh, A., Chhabra, G., Anandaram, H. and Kumar, A., 2023, July. A Review analysis: comparative study on various machine learning techniques for load forecasting in electric power distribution system with multiprocessing. In 2023 14th International Conference on Computing Communication and Networking Technologies (ICCCNT) (pp. 1-6). IEEE. https://doi.org/10.1109/ICCCNT56998.2023.10306925

[11] Srivastava, S.S., Satyanarayana, G.S., Dhasmana, A., Rawat, V., Rana, A.S. and Bisht, Y.S., 2025. Role of Embedded Systems in Smart Energy Management: Challenges, Innovations, and Future Trends. Solar Energy and Sustainable Development Journal, 14(STR2E), pp.27-50. https://doi.org/10.51646/jsesd.v14iSTR2E.797

[12] Global, C. C. S. (2023). Institute (2022) Global Status of CCS 2022. URL: https://status22. globalccsinstitute. com/2022-status-report/regional-overview/(дата обращения: 01.07. 2024).

[13] DeZern, A. E. (2021). Relevant metrics: how significant CCUS truly is. Blood, The Journal of the American Society of Hematology, 138(11), 918-919. https://doi.org/10.1182/blood.2021013088

[14] Srivastava, S., Rahman, A., Chetty, R. and Haldar, S., 2022. Analytical study of sugar and paraffin/potassium perchlorate as solid rocket propellant. International Journal of Vehicle Structures & Systems, 14(6), pp.729-731. https://doi.org/10.4273/ijvss.14.6.06

[15] Srivastava, S., Hamshitha, G., Jagannath, V., Kothenty, G., Haldar, S., Rajendran, A.R. and Thapliyal, S., 2024, October. Numerical Studies of Solid Rocket Propellant PMMA-PBAN-ALF3-Nitrocellulose-Difluoramine. In Journal of Physics: Conference Series (Vol. 2856, No. 1, p. 012010). IOP Publishing. https://doi.org/10.1088/1742-6596/2856/1/012010

[16] Srivastava, S., Ingenito, A. and Andriani, R., 2019. Numerical and Experimental Study of a 230 N Paraffin/N2O hybrid rocket. In Proc. 8th European Conf. for Aeronautics & Space Sci.

[17] Ommounica, C., Srivastava, S. and Haldar, S., 2025. Numerical study of the hybrid rocket engine. Aerospace Systems, 8(1), pp.237-245. https://doi.org/10.1007/s42401-024-00280-7

[18] Hou, L., Fu, P. and Ba, Y., 2019. Chemical mechanism of MMH/NTO and simulation in a small liquid rocket engine. Combustion Science and Technology, 191(12), pp.2208-2225. https://doi.org/10.1080/00102202.2018.1551214

[19] Hossain, M.A., Morse, A., Hernandez, I., Quintana, J. and Choudhuri, A., 2024. Liquid rocket engine performance characterization using computational modeling: Preliminary analysis and validation. Aerospace, 11(10), p.824. https://doi.org/10.3390/aerospace11100824

[20] Greig, C., & Uden, S. (2021). The value of CCUS in transitions to net-zero emissions. The Electricity Journal, 34(7), 107004. https://doi.org/10.1016/j.tej.2021.107004

[21] Kanniche, M., Gros-Bonnivard, R., Jaud, P., Valle-Marcos, J., Amann, J. M., & Bouallou, C. (2010). Pre-combustion, post-combustion and oxy-combustion in thermal power plant for CO2 capture. Applied Thermal Engineering, 30(1), 53-62. https://doi.org/10.1016/j.applthermaleng.2009.05.005

[22] Ozkan, M., Nayak, S. P., Ruiz, A. D., & Jiang, W. (2022). Current status and pillars o

[23] Shiyi, Y. U. A. N., Desheng, M. A., Junshi, L. I., Tiyao, Z. H. O. U., Zemin, J. I., & Haishui, H. A. N. (2022). Progress and prospects of carbon dioxide capture, EOR-utilization and storage industrialization. Petroleum Exploration and Development, 49(4), 955-962. https://doi.org/10.1016/S1876-3804(22)60324-0

[24] Hills, C. D., Tripathi, N., & Carey, P. J. (2020). Mineralization technology for carbon capture, utilization, and storage. Frontiers in Energy Research, 8, 142. https://doi.org/10.3389/fenrg.2020.00142

[25] Jones, W. D. (2020). Carbon capture and conversion. Journal of the American Chemical Society, 142(11), 4955-4957. https://doi.org/10.1021/jacs.0c02356

[26] Michael, K., Golab, A., Shulakova, V., Ennis-King, J., Allinson, G., Sharma, S., & Aiken, T. (2010). Geological storage of CO2 in saline aquifers-A review of the experience from existing storage operations. International journal of greenhouse gas control, 4(4), 659-667. https://doi.org/10.1016/j.ijggc.2009.12.011

[27] Voormeij, D. A., & Simandl, G. J. (2004). Geological, ocean, and mineral CO 2 sequestration options: a technical review. Geoscience Canada, 31(1), 11-22.

[28] Chadwick, R. A., Arts, R., Bentham, M., Eiken, O., Holloway, S., Kirby, G. A., ... & Zweigel, P. (2009). Review of monitoring issues and technologies associated with the long-term underground storage of carbon dioxide. https://doi.org/10.1144/SP313.15

[29] Capocelli, M., & De Falco, M. (2022). Generalized penalties and standard efficiencies of carbon capture and storage processes. International Journal of Energy Research, 46(4), 4808-4824. https://doi.org/10.1002/er.7474

[30] Gholami, R., Raza, A., & Iglauer, S. (2021). Leakage risk assessment of a CO2 storage site: A review. Earth-Science Reviews, 223, 103849. https://doi.org/10.1016/j.earscirev.2021.103849

[31] Taghizadeh-Hesary, F., Vandercamme, L., & Phoumin, H. (2024). Enhancing the economic feasibility of carbon capture, utilisation, and storage (CCUS) projects. Journal of Environmental

Assessment Policy and Management, 26(01), 2350024. https://doi.org/10.1142/S1464333223500242

[32] Ateş, F., & Javani, N. (2026). Ccs Technologies and Digitalization Perspective on the Road to Carbon Neutrality: A Holistic Research with Social, Political and Economic Dimensions. International Journal of Energy Horizon (IJEH), (1).

[33] Saini, P., Biradar, U., Sonawane, K., Mehta, A., Chauhan, D., Bordoloi, S., & Phani, R. (2026, February). Harnessing Digital Innovation in CCUS: AI-Driven Solutions for Real-Time Monitoring and Risk Management. In SPE Kuwait Oil and Gas Show and Conference (p. D021S015R004). SPE. https://doi.org/10.2118/231066-MS

[34] Mazhar, S., Mumtaz, M. W., El Oirdi, M., Mukhtar, H., Raza, M. A., Farhan, M., ... & Muteeb, G. (2026). Synergizing advanced materials and artificial intelligence for next-generation carbon capture, utilization, and storage (CCUS): a review. RSC advances, 16(3), 2621-2651. https://doi.org/10.1039/D5RA07338C

[35] Olabi, A. G., Wilberforce, T., Elsaid, K., Sayed, E. T., Maghrabie, H. M., & Abdelkareem, M. A. (2022). Large scale application of carbon capture to process industries-a review. Journal of Cleaner Production, 362, 132300. https://doi.org/10.1016/j.jclepro.2022.132300

[36] Reyes-Lúa, A., & Jordal, K. (2021). Industrial CO2 Capture Projects: Status, Lessons Learned and Needs for Progressing Towards Full-Scale Implementation. In TCCS-11. CO2 Capture, Transport and Storage. Trondheim 22nd-23rd June 2021 Short Papers from the 11th International Trondheim CCS Conference. SINTEF Academic Press.

[37] Yang, X., Alkhedhair, A., & Alabduly, A. (2026). Challenges and Opportunities in CCUS: Synergizing Efforts in China and GCC Countries. China and Gulf Cooperation Council Countries, 230-250. https://doi.org/10.4324/9781003610625-12

[38] Yang, Y., Zhao, X., Yang, J., Zhang, Y., & Tian, Y. (2025). Recent developments in CCUS-EOR for depleted reservoirs. Computational Energy Science, 2(1), 17-30. https://doi.org/10.46690/compes.2025.01.04

[39] Sopha, B. M., Ma'mun, S., Wibisono, M. A., & Kim, I. (2026). Exploring Carbon Scheme Scenarios for Retrofitted Coal-Fired Power Plants with Carbon Capture System in Indonesia. Cleaner Energy Systems, 100240. https://doi.org/10.1016/j.cles.2026.100240

[40] Gowd, S. C., Ganeshan, P., Vigneswaran, V. S., Hossain, M. S., Kumar, D., Rajendran, K., ... & Pugazhendhi, A. (2023). Economic perspectives and policy insights on carbon capture, storage, and utilization for sustainable development. Science of the Total Environment, 883, 163656. https://doi.org/10.1016/j.scitotenv.2023.163656

[41] Song, D., Jiang, T., & Rao, C. (2022). Review of policy framework for the development of carbon capture, utilization and storage in China. International Journal of Environmental Research and Public Health, 19(24), 16853. https://doi.org/10.3390/ijerph192416853

[42] Ren, Q., Wei, S., Du, J., & Wu, P. (2023). Research progress and perspectives on carbon capture, utilization, and storage (CCUS) technologies in China and the USA: a bibliometric analysis. Environmental Science and Pollution Research, 30(31), 76437-76454. https://doi.org/10.1007/s11356-023-27749-w

Materials Research Proceedings 66 (2026) 140-154 https://doi.org/10.21741/9781644904152-13

Pathways to Clean and Affordable Energy: Innovations, Policies, and Socio-Economic Impact

S. SRIVASTAVA[1*], S. YADAV[1], S. JASWAL[2], M. PARMAR[3], M. PREET KAUR[2], M. SHAMEEN[4]

[1]Department of Aerospace Engineering, Uttaranchal Institute of Technology, Uttaranchal University, Dehradun-248007, Uttarakhand, India

[2]Department of Medical Lab Sciences (USAHS), Rayat-Bahra University, Mohali-140104, Punjab, India

[3]School of Business, Department of Allied Health Science, Chandigarh Group of Colleges, Jhanjeri, Mohali-140307, Punjab, India

[4]Allied Health Sciences, Desh Bhagat University, Mandi Gobindgrah-147301, Punjab, India

*sachinrash2004@gmail.com

Keywords: Energy Affordability, Clean Energy Technology, Renewable Energy Use, Energy Storage Alternatives, Smart Grid Integration

Abstract: The global energy landscape has totally transformed with enhanced power usage resulting from industry, urbanization, and population increase. The use of fossil fuels is a key driver of greenhouse gas emissions, which in turn lead to climatic change & environmental degradation. Renewables are very much sought after, but hurdles such as their excessive cost, variability, and inadequate infrastructure stop them from being used on a large scale. Since most locations do not have cheap or stable energy sources, economic inequalities are enabled. New technologies for low-cost, clean energy are explored in this research, such as ongoing innovations in hydropower, wind energy, solar energy, and biofuel technology; low-cost energy storage technologies such as thermal systems and battery; resource-efficient methods; and combined smart grid integration. It considers the social implications of making a transition to clean energy, including economic development and the elimination of energy poverty. The article also talks about the policy, economic, and scientific barriers to mass deployment. It juxtaposes promising and emerging trends and future prospects with highlighting successful renewable energy projects in examples and real-world uses. To enable the transition to sustainable, green energy futures, the study accordingly underscores the pressing need for innovative policies, new, attractive investments, and enhanced partnerships.

1. Introduction

One of the most critical challenges that humankind is currently confronted with is shifting to cheap, clean energy, which is inextricably bound up with social, economic, and environmental well-being [1]. All industrialization, advance, and technological advancement run on energy, but the conventional energy sources-primarily fossil fuels-have contributed immensely to depleting resources, contaminating the environment, and causing climatic change. A great deal of attention must be given to SDG 7, which is ensuring that all individuals have access to energy that is renewable and environmentally friendly. The key is that the achievement of SDG 7 requires more than reducing the cost of energy; it also requires the ability to supply reliable electricity to all, promoting equity and promoting human well-being through energy consumption. An economic strategy that is solely aimed at lowering energy costs will not consider the significant socioeconomic and infrastructural barriers that hinder the accessibility and sustainability of energy [2]. The need for clean, cheap energy is an unequivocal call to action for lowering carbon emissions and fighting global warming and its adverse effects. Fossil fuels emit massive quantities of gases into the atmosphere, increase the overall temperature, lead to weather extremes, and are detrimental to human health. With alternatives such as wind, solar energy, hydro, and biofuels available for electricity generation, there is potential for a long life at reduced environmental expense [3]. There are many problems that low-cost energy, especially in developing countries with poor energy infrastructure, suffers from. With increasing demand, economic growth, and environmental pressures, global energy patterns are extremely dynamic and complex. Energy consumption increases as populations and economies grow [4]. The growth in demand adds further pressure to already strained energy infrastructure. These complexities simply make the finicky task of balancing economic growth and environmental objectives all the more intricate. Since cleaner power systems and fuels are better at satisfying demand and doing less harm to the environment, the model has been directed away from traditional power sources. As reported by Zohuri (2023), "Energy manufacturing, consumption, and environmental responsibility are all interconnected" [5]. Also, Smart healthcare technologies are revolutionizing the medical industry by enabling remote patient monitoring, real-time diagnostics, and effective healthcare data administration. However, the expansion of resource-intensive healthcare applications in cloud-based infrastructures has sparked worries about latency and energy usage. By creating a Dew Computing framework that is scalable and energy-efficient and tailored for smart healthcare environments, our study presents a novel approach to overcoming these challenges [6].

Developing new renewable energy infrastructure, enhancing energy storage, and upgrading power grid infrastructure for greater resilience and efficiency are all supply-side and demand-side issues that need to be addressed by an effective energy strategy [7]. In the not-so-distant future, how humanity develops, invests in, and coordinates renewable energy sources will determine the sustainability of the planet. Regardless of their technological and scientific breakthroughs, various limitations inhibit the extensive application of renewable energy technology. Since renewable energy sources such as solar and wind are intermittent, there is a need to construct smart grid systems and the right energy storage technology to ensure a consistent and reliable supply of energy [8]. In addition, national energy policy is subject to economic and geopolitical interests that also often dictate how rapidly these countries will shift towards cleaner alternatives. The timeline of energy shifts in underdeveloped countries is greatly expedited by economic barriers like the very high start-up fees for clean infrastructure. In addition, an economy based currently on fossil fuels will always be convoluted and slow, whether or not the whole world is presently seeking to transition towards clean energy [9]. Thus, the only means of overcoming the hurdles is to possess a properly structured plan that would integrate financing, policies, and technologies. Governments, enterprises, and research organizations need to join forces to design energy systems that are not just accessible and equitable for all, but sustainable. Investing in energy storage technology, better grid management, and the establishment of decentralized energy channels such as microgrids will assist the global energy sector in developing a resilient and fair energy future. Thus, aside from being an ecological necessity, clean energy transition is also a socioeconomic necessity that is receiving immense attention globally as scientists search for new methods of attaining prosperity and sustainable growth. Affordable and clean energy refers to energy that remains reasonably affordable for both people and communities to use. Hence, increased energy efficiency and effective energy storage complement the renewable energy sources such as geothermal, solar, wind, hydroelectricity, and biofuels [8]. The basis of the demand for clean, cheap energy lies in its ability to mitigate global warming, reduce dependence on the consumption of fossil fuels, and ensure that all communities access modern energy services. Affordability is one of the influencing factors of energy accessibility, though by itself, as stated by Hillerbrand (2018), it is nearly useless for garnering substantial gains in sustainability [10]. Rather, a capacity-based strategy needs to be applied in order for clean energy solutions to make it possible for individuals and society to increase their economic opportunities and standard of living. To put it differently, for clean energy technologies to encourage sustainable development and enhance the quality of life for individuals, they need to be affordable and incorporated into society and socioeconomic institutions [2]. Clean energy also comes in space missions, such as using the green propellant in space missions. New technology such as using the hybrid rocket engine [11-12] which will use the green propellant, low cost, eco-friendly and high performance.

The world energy sector is experiencing tremendous change as countries try to balance between their massive energy consumption, economic growth, and environmental goals. As mentioned by Zohuri (2023), the population growth of the world, rapid urbanization, and industrialization have all raised the energy demand [13]. The current energy infrastructures are being strained with electricity demand. Though fossil fuels remain a key component of the global energy mix, they are responsible for most of the pollution, climate change, and emissions of greenhouse gases, among other factors driving the need to transition to cleaner and more friendly fuels [14]. Yet, several challenges need to be overcome to attain essentially sustainable energy systems. A number of recent areas still suffer from energy poverty, which hinders them from accessing all forms of modern energy services. Energy security is further influenced by geopolitical conditions, and since renewable energy sources may be intermittent, there is a need to utilize efficient energy storage technology. An integrated solution using technology innovation, global cooperation, and regulatory arrangements is needed to overcome such hurdles to an equitable and sustainable renewable energy future [15]. With the advent of new technologies, energy efficiency increases, policies for fair energy access are encouraged, clean energy and affordable energy definitions and requirements are comprehended, and the complexities of the global energy scenario are finally revealed, it is more than clear that an oil-free future needs to be attained through various strategies [7].

The research study establishes an integrated framework which simultaneously evaluates renewable energy technologies and low-cost storage solutions and demand-side management and socio-economic equity through the Sustainable Development Goal 7 assessment. The review presents a unique combination of technological progress and cost-based evaluation and socio-economic accessibility which focuses on rural areas and developing countries. The research combines three elements by creating a single transition pathway model which includes cost reduction patterns, community-level microgrid integration, and upcoming hybrid systems that use PV-CSP and gravity storage technology. The paper demonstrates full comprehension of the subject by uniting technological and economic and social justice elements through its comprehensive roadmap which extends beyond evaluating particular technologies.

2. Renewable Energy Technologies

The world shifting towards an increasingly eco-friendly future, renewable energy sources are at the forefront. Today, clean renewable energy technologies harness energy from sunlight, wind, and water into electricity generation without environmental impact. Among the major energy resources that support the greatest extent of renewable power generation are solar energy, wind energy, hydropower, and ocean energy. Each source has its unique advantages and challenges. Figure 1 shows the total installed capacity of these renewable energy sources worldwide. The graph illustrates that hydropower remains the dominant source, while solar and wind are growing rapidly.

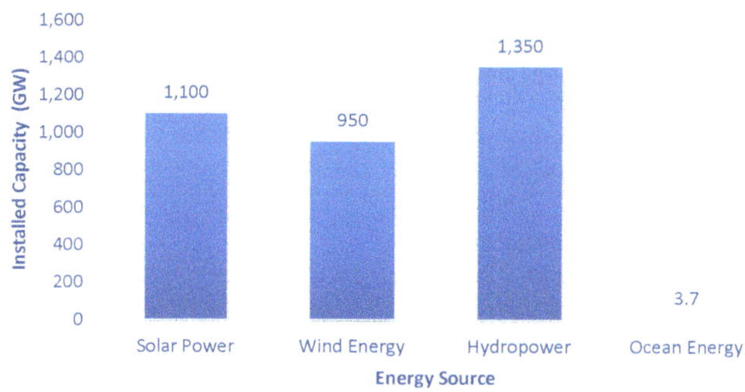

Figure 1. *Global Installed Capacity in 2023 (Source: RENA & IEA Reports)*

2.1 Solar power (PV, CSP)

Current fastest renewable energy source development shows that solar is the fastest. There are two most important technologies in current developments: photovoltaic systems and concentrated solar power systems. Using semiconductor materials- for example, silicon-converts sunlight directly into electricity in a photovoltaic system. Innovations in PV efficiency such as multi-junction cells and perovskite-based solar cells lead to significant improvement in performance and affordability. Because PV systems can be easily installed and modify, they have entered into residential, commercial, and even industrial applications. In comparison, the CSP technology will concentrate sunlight onto a solar receiver by using mirrors or lenses, which will then transform it from solar thermal energy into electricity. This system works very well for generating electricity on a large scale due to the incorporation of thermal energy storage (TES) to supply electricity even when the sun doesn't shine. Hybrid PV-CSP systems are a budding innovation for combining greater efficiency of PV with more extended storage possibilities, increasing reliability of energy while decreasing dependence on fossil fuel backup systems [16]. Figure 2 represents a block diagram of Hybrid PV-CSP System.

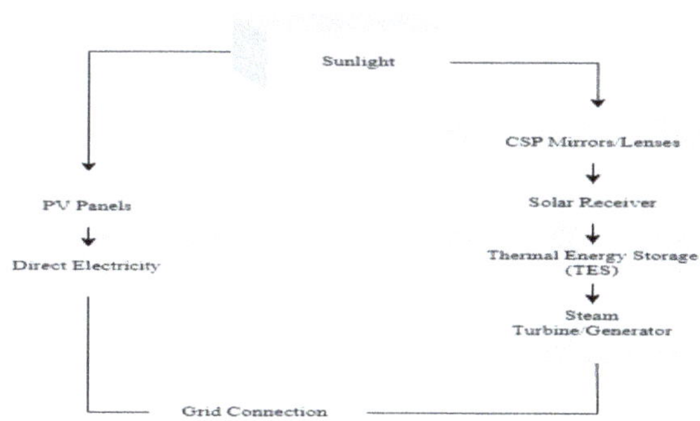

Figure 2. *Hybrid PV-CSP System*

Hybrid PV-CSP systems are a nascent technology that capitalizes on the high efficiency of PV panels and on the long-term power storage possibilities of CSP. Here, sunlight serves as the primary energy input. In this configuration, PV panels are directly converting sunlight into electric power while CSP mirrors or lenses concentrate sunlight onto a solar receiver. The receiver then converts solar radiation into thermal energy. This thermal energy is stored in a Thermal Energy Storage (TES) unit, giving rise to the possibility of supplying power during hours that are not sunny. Heat would then be withdrawn from storage during periods of demand to drive a steam turbine/generator that converts thermal energy to electricity. The electricity produced is then fed to the consumers through a grid connection that can support a continuous and stable power output.

2.2 Wind energy (onshore and offshore)

Wind energy is another major renewable energy source with a rapid growth rate across the globe. Wind turbines convert kinetic energy from wind into mechanical power and then into electricity. Broadly, wind energy is divided into onshore and offshore wind energy. Installation and maintenance of onshore wind farms are less costly than offshore ones. Such farms are generally set up in high wind areas like plains and coastlines. Onshore wind also suffers from challenges like land use conflict, noise pollution, and variability in wind speed. In fact, nowadays offshore wind farms go ahead since they capture much stronger and steadier wind currents offshore. Offshore turbines are built larger and more efficient than those sitting on land, yielding extra energy output. Offshore wind, however, poses certain challenges, such as more expensive installation costs, complicated maintenance systems, and specialized infrastructures. But also, floating wind-turbine technology is evolving and making offshore wind farms more economically viable because they are capable of being installed in deeper waters where winds are steadier and more aligned [17].

2.3 Hydropower and ocean energy

Hydropower is the first and oldest amongst all renewable sources of energy. This form of energy has kinetic energy from flowing water as the prime mover for electric generation. For example, big hydropower plants like dams and reservoirs produce electricity consistently and reliably. Hydropower plants, in addition, store electricity, providing an advantage in absorbing fluctuations of power supply and demands in the energy grid. However, criticisms against conventional hydropower projects frequently arise due to the effects they have on the environment and their social impacts, resulting in many ecosystems disrupts and habitat losses and people displacements. In addition to conventional hydro power, ocean energy technologies today are gaining recognition among sustainable energy sources. These technologies include wave energy, tidal energy, and ocean thermal energy conversion (OTEC). While wave energy captures the movement of the ocean and makes use of it to produce electricity, tidal energy relies primarily on the rise and fall of the tides, which is more predictable than other renewable energy sources. OTEC, on the other hand, produces energy by exploiting the difference in temperatures between warm surface waters and cold, deep waters. Despite all the above, ocean energy has a lot of promise, but development for large-scale deployment has been hindered by technological, economic, and environmental challenges. It still requires continued research and invention to improve efficiency and lower costs [18]. Table 1 summarizes efficiency, advantages, and challenges of different renewable energy sources in a comparative manner.

Table 1. *Efficiency & Challenges of Renewable Energy Sources*

Renewable Energy Source	Efficiency (%)	Key Advantages	Key Challenges
Solar Power (PV, CSP)	15-45% (PV), 30-50% (CSP)	Scalable, low maintenance, abundant	Intermittent generation, storage needed
Wind Energy (Onshore & Offshore)	35-50%	High energy output, zero emissions	Land use (onshore), high costs (offshore)
Hydropower	70-90%	Reliable, long lifespan	Environmental impact, displacement issues
Ocean Energy (Wave, Tidal, OTEC)	30-45% (Wave), 80-90% (Tidal), 3-6% (OTEC)	Predictable (tidal), vast potential	High cost, limited deployment

The technical maturity of renewable technologies has reached a high point yet their ability to scale operations depends more on storage costs and grid system adaptability than their power generation capabilities. Although hydropower operates at its highest efficiency level between 70 and 90 percent its environmental effects prevent any further development. The declining costs of solar photovoltaic systems despite their lower efficiency make these systems essential for upcoming decentralized energy networks. The future market leadership of a product will be determined by its cost path and operational flexibility which will prove more valuable than its efficiency ratings.

3. Low-Cost Energy Storage Solutions

Such storage systems will thus solve the problem of energy storage from renewable energy sources for the large-scale integration of renewables, intermittency, and availability of supply of energy on a more stable basis. These storage alternatives become very important for the complete commercialization of clean energy use, especially in developing regions with poor or absent grid supply. New, cheap forms of batteries are being developed in this area, along with a range of low-cost thermal energy storage systems and various gravity-based or mechanical storage systems. They will all contribute heavily to balancing the energy demand and supply and reduce the reliance on fossil fuels.

3.1 Affordable battery technologies

Batteries are highly important energy storage devices that play a widespread role in grid-scale systems and electric vehicles (EVs). Improvements in battery technology were progressive in the ways they improved energy storage capacities, efficiency, and costs. Despite these advancements, Li-ion batteries remain the most widely adopted technology due to factors such as their high energy density and decreasing costs of production. Concerns about resource availability and problems with recycling have led to extensive research on different chemistries, including sodium-ion, solid-state, and flow batteries. The other significant aspect is that battery electric vehicles (BEVs) and fuel cell electric vehicles (FCEVs) are competing to be the clean transportation. While BEVs rely on Li-ion batteries, FCEVs use hydrogen fuel cells for energy storage and propulsion. According to Vengatesan et al. (2024), both technologies contributed to achieving SDG 7-increased access to affordable and clean energy, but at present, BEVs are more economical because of established infrastructure and lower costs. Technological advances in h2 production and storage may lead FCEVs to an increasingly promising future [19].

Materials Research Proceedings 66 (2026) 140-154 https://doi.org/10.21741/9781644904152-13

3.2 Thermal energy storage

Systems intended for thermal energy storage (TES) are responsible for storing energy in excess in the form of heat for later use in electricity generation, industrial applications, or heating and cooling. TES is highly advantageous as a technology in solar power plants, district heating systems, and industrial waste heat recovery. The three major types of TES are sensible heat storage, latent energy storage, and thermochemical storage. In sensible heat storage, a solid or liquid medium (water, molten salts, or rocks) is heated, and the stored heat is extracted later. It is commonly utilized for thermal storage in concentrated solar power (CSP) plants. Latent heat storage uses phase change materials (PCMs) to absorb and release heat during phase transitions (like solid-to-liquid). This method is more energy-efficient when compared to sensible heat storage. Thermochemical storage works on reversible chemical reactions to release and store energy. This technique holds the potential for long-term storage with very little loss of energy [20]. TES economically balances the requirement and supply of energy, especially when incorporated with renewable energy systems. The other possible viability improvements are the efficiency of materials used, scalability, and retention of energy, which still remains a principal area of investigation.

3.3 Gravity and mechanical storage

They allow the use of physical forces and not batteries charge and discharge energy. These energy systems are especially useful for an energy storage system in the case of the grid scale. It is suitable for long durations and high-capacity storage. Gravity energy storage usually consists of lifting a mass in order to store potential energy to convert it back into electricity as the mass descends. An example is pumped hydro storage, where water is pumped into a higher elevation during times of surplus energy and released through turbines at times of high demand. Other GES techniques include lifting large solid blocks or using large underground mine shafts to store energy. Correspondingly, other energy-saving storage methods are provided by mechanical storage systems like flywheels and compressed air energy storage (CAES). Through energy storage in the rotating mass, flywheels provide frequency stabilization of the grid by fast response and high cycle efficiency. In contrast, CAES machines store air compressed in underground storage caverns using excess electricity, which expands on demand for power generation [21].

4. Energy Efficiency and Demand-Side Management

These are typically energy-saving initiatives. Demand-side management (DSM) combined with energy efficiency would maximize energy utilization, reduce costs, and maintain grid stability. For maximum reduction in waste and tremendous improvement in sustainability, energy-efficient technology and demand-side applications should be given priority as the world becomes increasingly dazzling and energy demand increases. The DSM optimizes energy-use behaviors and improves system function through the use of smart grids, energy-efficient appliances, demand response strategies, and energy-saving measures.

4.1 Smart grids and energy-efficient devices

Therefore, the development of smart grid systems has changed the distribution of energy by integrating automation and digital communication technologies to optimize the flow of electricity. Specifically, a hybrid intelligent grid system integrates new metering infrastructure, power storage facilities, and renewable energy sources to ensure real-time monitoring and management of the distribution of energy from any place. Smart grids contribute to reliability, reduce transmission losses, and improve demand management in city environments, thus playing a pivotal role in the development of tomorrow's smart cities. In parallel with smart grids, energy-efficient devices also contribute to the overall reduction in electricity consumption. Smart thermostats, LED lightings, energy-efficient appliances, and intelligent HVAC systems are just some appliances that keep energy use to a minimum while maintaining comfort and function. Smart appliances integrated with IoT technology will therefore alter their operating schedule basis on energy availability and user input, thus propagating sustainable energy consumption [22].

4.2 Demand response strategies

Demand response is the prime DSM strategy that drives the consumers to modify their energy consumption according to price signals, grid conditions, or incentive schemes. The two main subcategories of demand response programs are as follows:

- **Price-based demand response:** When consumers' energy usage patterns are modified by tariff schedules such as time-of-use (TOU) pricing, real-time pricing (RTP), or critical peak pricing (CPP), they are supposedly encouraged to shift energy-intensive activities to off-peak periods to ease grid constraints.
- **Incentive-based demand response:** During peak electricity demand, they give monetary incentives to their customers who reduce their electricity consumption voluntarily. This includes automating load control, direct load shedding, smart home integration.

A working DR strategy presupposes an advanced metering infrastructure (AMI), real-time monitoring of electricity, and consumer participation. In conjunction with the smart grid, DR programs will also limit energy costs, balance the supply-demand scenario, and basically keep the stability of the grid [23].

4.3 Energy conservation measures

Energy conservation is permanent and it cuts down the total energy they utilize due to efficiency improvements, behavior changes, and regulations. Contrary to demand response, which helps to adjust energy consumption in the short run, energy conservation concentrates on reducing waste of energy as well as optimizing energy use in the long term. Gellings and Parmenter (2016) classify energy conservation measures into three broad categories as follows:

1. **Building energy efficiency:** Upgrading insulation, green building designs, using energy-efficient windows, and so on can easily reduce heating and cooling loads.
2. **Industrial energy management:** Upgrading motors; recovering heat wastes; automating processes for minimal unnecessary use of energy will give an industry the occasions to optimize energy use.
3. **Transportation efficiency:** Step up the use of battery-powered cars (EVs), enhancing public transport, and encouraging non-motorized mobility (e.g., walking and cycling) are all subprojects that decrease overall energy use and dependence on fossil fuels.

In fact, government policies and incentives also have an important effect on energy saving. Initiatives such as energy efficiency labelling, tax incentives, and building performance requirements, for example, are meant to motivate individuals and firms to adopt more environmentally responsible energy strategies [24].

5. Cost-Effective Clean Energy Solutions

It will need efficient, cost-effective, practical, and durable solutions. It was exorbitant to meet initial investments needed for deployment of several clean energy technologies, especially in developing countries. There are now opportunities to enhance the economic feasibility of the energy sector with community-scale microgrids, biomass and biogas production, and lower-cost solar technology. Apart from enhancing the availability of energy, these technologies align with the global community's emphasis on sustainable development. Goal 7 (SDG 7): Affordable of Clean Energy.

5.1 Low-cost solar cells and panels

The most viable source of clean energy has been solar power, but mass domestic consumption has been precluded by the economical manufacturing of photovoltaic (PV) panels. But improved thin-film photovoltaic cells, perovskite-solar cells, and organic solar cells (OPVs) have radically altered this scenario in recent times. Along with enhancing the effectiveness of cheap materials and making manufacturing more efficient, these technologies contributed to a dramatic decrease in cost, opening up even more advanced solar technology to companies and consumers in the future. Perovskite solar cells have resulted in relatively high efficiencies compared to fraction-based costs of making traditional silicon-based PV panels and have become one of the most notable breakthroughs in cost reduction. Such thin-film technologies also gained acceptance and popularity because of their flexible and lightweight designs and low material costs, such as cadmium telluride (CdTe) and copper indium gallium selenide (CIGS). The aforementioned lead innovations have brought solar technology affordability and practicality, especially in off-grid rural areas and developing countries [25]. In figure 3, the graph highlights the steep decline in solar panel costs, along with steady cost reductions in biomass and microgrid technologies, making clean energy solutions more affordable.

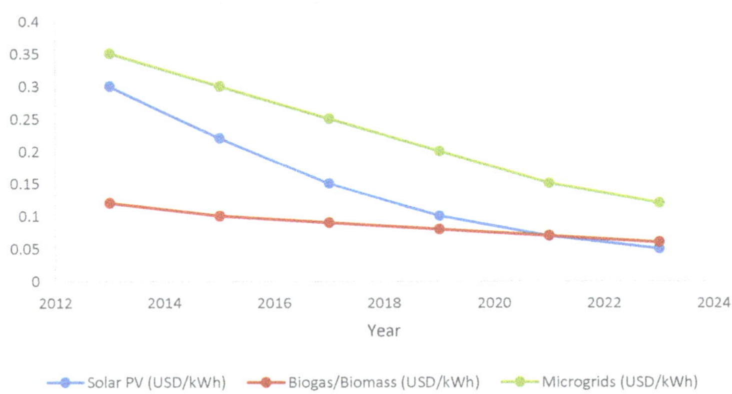

Figure 3. *Average Cost Reduction (in USD per kWh) (Source: IRENA & IEA Reports)*

5.2 Affordable biomass and biogas solutions

Energy generation based on biomass and biogas is an inexpensive and sustainable alternative to fossil fuels, particularly advantageous in regions where there is more agricultural or organic waste. Biomass energy comes from plant materials, agricultural residues, and animal waste; biogas is produced through anaerobic digestion between organic material without oxygen, generating methane-rich gas that can be used for cooking, heating, and electricity generation. One of the main constraints for biogas production has historically been the cost of purification and upgrading technologies for removal from biogas of CO_2 and H_2S. However, recent investigations have focused on exploiting low-cost materials such as activated carbons prepared from agricultural residues, natural zeolites, and biochar's for biogas purification, thus reducing cost and increasing the efficiency of biogas use in the proposed applications for rural households, farms, or small industries [26].

5.3 Community-scale microgrids

Microgrids are being considered increasingly as a viable option for clean and reliable energy supply to remote and underserved areas. The term generally refers to a localized network of electricity, although it can either operate autonomously or be connected to the main grid. Such systems include renewable energy sources: solar and wind, as well as biomass, coupled with battery storage and smart grid technologies to assure stable and affordable electricity provision. The operations of microgrids are particularly effective in an island community or an isolated village where extension of standard power lines is prohibitively expensive. A case study from the Azores islands can show how an integrated energy microgrid provides energy access, independence from fossil fuels, and a reduction in price of electricity. Microgrids improve security of energy supply and local economic growth by integrating renewable energy generation with advanced energy management systems [27]. Figure 4. Shows how community microgrids integrate multiple energy sources with smart management for affordability and reliability.

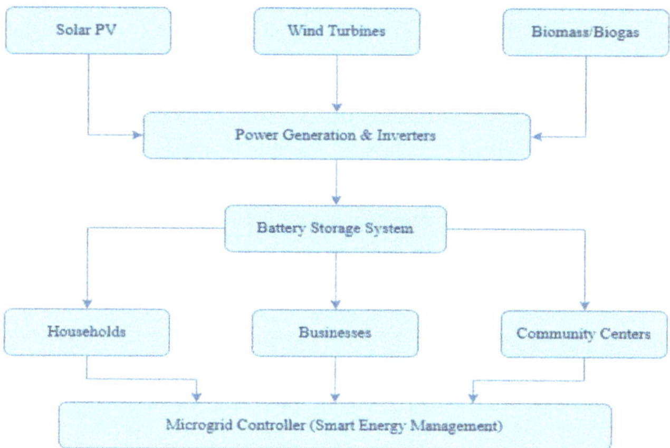

Figure 4. *Community-Scale Microgrid Structure*

A community scale microgrid functions as an autonomous network for energy sustainability by integrating various renewable energy sources such as solar photovoltaic panels, wind turbines, and biomass/biogas generators for the generation of electricity. The sources supply power into the whole power generation and inverter commercial system that converts and stabilizes electricity for local consumption. A battery-based storage system accumulates surplus power for the round-the-clock power supply, which feeds loads during times of low generation. Independent of the external grid connection, the microgrid controller supplies electricity to homes, enterprises, and neighbourhood community centers and executes smart energy management. This guarantees a secure and affordable supply of power. As such, it brings energy security to the remotest and poorest regions.

6. Socio-Economic and Environmental Impact

The transition towards low-cost, clean energy has multiple impacts on social justice, economic development, jobs, and sustainable development, aside from decreasing dependence on fossil fuels. Large-scale use and installation of green energy technologies enhance energy access, offer economic opportunity, and substantially decrease greenhouse gas emissions, especially when coupled with distant and developing areas.

6.1 Affordability and accessibility in rural areas

Unavailability of electrical infrastructure hinders remote off-grid communities in most developing countries from accessing affordably priced and reliable electricity. Kerosene, biomass, and diesel generators are some of the traditional sources of energy that are not only costly but also environmentally unfriendly. Some of the clean, decentralized sources of energy are home solar systems, biogas plants, and microgrids. Clean energy technology is becoming cheaper due to government policies, improved energy storage technology, and reducing prices for solar panels. For instance, pay-as-you-go (PAYG) solar energy systems eschew the hefty initial investments required for solar energy systems since they permit rural dwellers to buy electricity in instalments. In addition, community-based microgrids have also proved effective in filling the energy gap in these areas, usually strengthening local businesses and enhancing standards of living [28].

6.2 Job creation and economic benefits

Another factor that drives job creation and economic growth is the transition to cleaner energy. According to the quantity of energy generated, the renewable energy sector creates more jobs and job opportunities than the fossil fuel industry. Renewable energy jobs cover a variety of economic sectors, including the manufacture of solar panels, maintenance of wind turbines, production of biofuel, and the build-out of energy infrastructure. Numerous career opportunities in research, engineering, development, and energy management exist due to government agencies and private investors making increased funding available for renewable energy projects. In addition to that, the development of regional renewable markets promotes entrepreneurship and small businesses, creating employment. Kumar and Rathore (2023) state that the renewable sector can create millions of jobs across the world, especially in countries shifting away from fossil fuel dependence. Along with creating jobs outright, cheap clean energy reduces domestic and commercial energy bills, making businesses more efficient and allowing consumers to save whatever money they do on other essentials that ultimately build economic resilience [29].

6.3 Reduction in carbon footprint

Reduction in the emission of greenhouse gases is the largest possible environmental advantage of clean energy adoption. Fossil fuel energy is one of the primary sources of greenhouse gas (CO_2) emissions causing climate change. Solar, wind, hydropower, and bioenergy greatly reduce the carbon footprint of a country and enable it to achieve its climate objectives according to agreements such as the Paris Agreement. Biogas also reduces methane emissions from organic waste decomposition and avoids the necessity of burning wood or charcoal; hydroelectric and geothermal power plants can offer stable, long-term low-emission electricity production; and renewable energy technologies emit little or no greenhouse gases after installation [29].

Aside from reducing emissions, the utilization of clean energy also ensures improved air quality, reduced forest cutting, and declining health risks associated with indoor air pollution caused by the combustion of solid fuels. These environmental and health benefits enhance the long-term, sustainable development of societies across the globe.

7. Challenges and Barriers

While the transition to clean and affordable energy seems to be a very successful move towards building a sustainable society, there are several barriers in infrastructure, economics, technology, and politics that do not allow for mass adoption. The key to breaking down all of those barriers and clearing the path to a bright future of the renewable energy revolution is major investments, legislative reforms, and technological innovations.

7.1 Technological and financial constraints

Energy storage, grid modernization, and expansion of renewable energy are still greatly constrained by technology and finance. Most of the sustainable energy technologies, including energy storage systems, wind turbines, and advanced solar panels, are expensive initially, and therefore unavailable to low-income communities and developing nations. But energy storage systems, which are key to a reliable supply of renewable energy, remain expensive and technically challenging. While all of thermal energy storage, gravity-based storage, and lithium-ion batteries are pushing forward, practical application is still limited by cost and efficiency problems. In addition, it is challenging to fund sustainable energy projects, especially in poor countries where numerous gaps in access to cheap loans, government incentives, and private capital remain. The absence of long-term finance models and measures for protecting against risk only exacerbates the problem by deterring scalable investment [30].

7.2 Policy and regulatory hurdles

The efficiency of the employment of clean energy remains mostly dependent on government actions, from laws and rules to incentives in the marketplace. However, almost all countries suffer from policy inconsistency, lack of regulatory framework, and bureaucratic delays, which slow down clean energy development. [31]. Some of the key policy challenges include:

- **Lack of clear renewable energy incentives**: There are other areas where fossil fuel industries are still receiving grants and allowances-tension from clean energy sources.
- **Complex permitting and approval processes**: Time-consuming administrative procedures of projects with regard to renewable energy cause delays and increase costs of these projects.
- **Unstable prices for energy models:** Companies and consumers are reluctant to commit free investments in green power sources because energy pricing is unpredictable and net metering policies.

A clear policy framework for clean energy in the medium to long term, with the support of partnerships between the public and the private sector, can, indeed, help fast-track the widespread implementation of sustainable energies.

7.3 Integration with existing infrastructure

Including renewable sources of energy into the pre-existing arrangement of the energy infrastructures constitutes another major challenge. Most of the electricity grid infrastructure was originally designed to allow centralized fossil fuel-based power generation. Hence, it becomes difficult, in principle, to integrate decentralized renewable sources of energy, such as solar, wind, and microgrid, into these existing infrastructures. Some of the key challenges to integration include:

- **Grid stability and intermittent energy supply:** Wind and solar energies per se are variable sources and as such require grid control and energy storage for stability.
- **Obsolete transmission and distribution systems:** Much renewable energy can at times be too high for existing grids to accommodate, resulting in loss of power during transmission and voltage collapse.
- **Cybersecurity & virtual infrastructure:** Cybersecurity threats are becoming increasingly urgent as smart grids and internet of things energy become increasingly complex, requiring the adoption of strict cybersecurity protocols [32].

Conclusion

- The review presents an integrated clean energy transition model which centers on affordability while it surpasses traditional descriptive survey methods. The study links all elements of renewable generation together with cost-reduction trends and storage innovation and smart grid modernization and socio-economic equity under the unified SDG 7 framework. The future energy transitions will develop through three main factors which include the development of new technologies and the cost structure changes and the establishment of decentralized systems and inclusive policy frameworks.
- To offer reliable, affordable access to clean energy, particularly in those regions where solar and wind power are applied intermittently, new energy storage technologies—with an emphasis on low-cost battery and thermal storage—as well as further upgraded efficiency equipment are needed.
- Clean energy also increases economic development and reduces dependence on fossil fuels by creating employment and eradicating energy poverty. It fosters socioeconomic equity.
- The use of environmentally friendly energy must be supported by proper government policies, subsidies, and financial incentives. A public-private partnership should be put in place to promote technological development and investment.
- It involves continued research, development, and deployment of renewable energy technologies that are scalable and affordable in an effort to achieve a low-carbon, environmentally sustainable energy future. To foster innovative policies and technologies that enhance affordability, accessibility, and environmental sustainability, governments, companies, and researchers will need to work hand in hand.

References

[1] Fiorino, D. J. (2022). The clean energy transition: Policies and politics for a zero-carbon world. John Wiley & Sons.

[2] Hillerbrand, R. (2018). Why affordable clean energy is not enough. A capability perspective on the sustainable development goals. Sustainability, 10(7), 2485. https://doi.org/10.3390/su10072485

[3] Bello, M. (2015). Renewable energy for sustainable socio-economic development in developing countries: A case study of sub-Saharan Africa. Advanced Materials Research, 1116, 33-44. https://doi.org/10.4028/www.scientific.net/AMR.1116.33

[4] Cantarero, M. M. V. (2020). Of renewable energy, energy democracy, and sustainable development: A roadmap to accelerate the energy transition in developing countries. Energy Research & Social Science, 70, 101716. https://doi.org/10.1016/j.erss.2020.101716

[5] Adewumi, A., Olu-lawal, K. A., Okoli, C. E., Usman, F. O., & Usiagu, G. S. (2024). Sustainable energy solutions and climate change: A policy review of emerging trends and global responses. World Journal of Advanced Research and Reviews, 21(2), 408-420. https://doi.org/10.30574/wjarr.2024.21.2.0474

[6] Singh, K.D., Singh, P.D., Maurya, S., Verma, R., Joshi, K. and Khilnani, L., 2024, May. A Framework for Advanced Technology in Energy-Efficient Dew Computing for Smart Healthcare. In 2023 International Conference on Smart Devices (ICSD) (pp. 1-5). IEEE. https://doi.org/10.1109/ICSD60021.2024.10751334

[7] Zohuri, B. (2023). Navigating the global energy landscape balancing growth, demand, and sustainability. J. Mat. Sci. Apl. Eng, 2(7).

[8] Pahle, M., Pachauri, S., & Steinbacher, K. (2016). Can the Green Economy deliver it all? Experiences of renewable energy policies with socio-economic objectives. Applied Energy, 179, 1331-1341. https://doi.org/10.1016/j.apenergy.2016.06.073

[9] Qudrat-Ullah, H. (2024). Transitioning Fossil-Based Economies: Sustainable Strategies for Energy Change. Taylor & Francis. https://doi.org/10.4324/9781003558293

[10] Creutzig, F., Roy, J., Devine-Wright, P., Díaz-José, J., Geels, F. W., Grubler, A., ... & Weber, E. U. (2022). Demand, services and social aspects of mitigation (Chapter 5).

[11] Srivastava, S. and Thakur, A.K., 2022. Review on hybrid rocket engine: past, present and future scenario. International Journal of Vehicle Structures & Systems, 14(5), pp.680-685. https://doi.org/10.4273/ijvss.14.5.24

[12] Srivastava, S., Ingenito, A. and Andriani, R., 2019. Numerical and Experimental Study of a 230 N Paraffin/N2O hybrid rocket. In Proc. 8th European Conf. for Aeronautics & Space Sci.

[13] Okorie, D. M. (2024). The Improvement of Nigeria's Electricity Distribution Services Using Integrated Management Systems.

[14] Haines, A., Smith, K. R., Anderson, D., Epstein, P. R., McMichael, A. J., Roberts, I., ... & Woods, J. (2007). Policies for accelerating access to clean energy, improving health, advancing development, and mitigating climate change. The Lancet, 370(9594), 1264-1281. https://doi.org/10.1016/S0140-6736(07)61257-4

[15] Omotoye, G. B., Bello, B. G., Tula, S. T., Kess-Momoh, A. J., Daraojimba, A. I., & Adefemi, A. (2024). Navigating global energy markets: A review of economic and policy impacts. International Journal of Science and Research Archive, 11(1), 195-203. https://doi.org/10.30574/ijsra.2024.11.1.0029

[16] Ju, X., Xu, C., Hu, Y., Han, X., Wei, G., & Du, X. (2017). A review on the development of photovoltaic/concentrated solar power (PV-CSP) hybrid systems. Solar Energy Materials and Solar Cells, 161, 305-327. https://doi.org/10.1016/j.solmat.2016.12.004

[17] Lynn, P. A. (2011). Onshore and offshore wind energy: an introduction. John Wiley & Sons. https://doi.org/10.1002/9781119954613

[18] Sims, R. E. (2008). Hydropower, geothermal, and ocean energy. MRS bulletin, 33(4), 389-395. https://doi.org/10.1557/mrs2008.79

[19] Vengatesan, S., Jayakumar, A., & Sadasivuni, K. K. (2024). FCEV vs. BEV-a short overview on identifying the key contributors to affordable & clean energy (SDG-7). Energy Strategy Reviews, 53, 101380. https://doi.org/10.1016/j.esr.2024.101380

[20] Alva, G., Lin, Y., & Fang, G. (2018). An overview of thermal energy storage systems. Energy, 144, 341-378. https://doi.org/10.1016/j.energy.2017.12.037

[21] Berrada, A., Loudiyi, K., & Zorkani, I. (2017). System design and economic performance of gravity energy storage. Journal of Cleaner Production, 156, 317-326. https://doi.org/10.1016/j.jclepro.2017.04.043

[22] Khalil, M. I., Jhanjhi, N. Z., Humayun, M., Sivanesan, S., Masud, M., & Hossain, M. S. (2021). Hybrid smart grid with sustainable energy efficient resources for smart cities. sustainable energy technologies and assessments, 46, 101211. https://doi.org/10.1016/j.seta.2021.101211

[23] Kumar, P., Ali, I., & Thanki, D. V. (2018). Demand-side management: Energy efficiency and demand response. In Handbook of Research on Power and Energy System Optimization (pp. 453-479). IGI Global. https://doi.org/10.4018/978-1-5225-3935-3.ch013

[24] Gellings, C. W., & Parmenter, K. E. (2016). Demand-side management. In Energy management and conservation Handbook (pp. 399-420). CRC Press. https://doi.org/10.1201/9781315374178-20

[25] Jacoby, M. (2016). The future of low-cost solar cells. Chem. Eng. News, 94(18), 30-35. https://doi.org/10.1021/cen-v089n018.p030

[26] Jepleting, A., Mecha, A. C., Sombei, D., Moraa, D., & Chollom, M. N. (2025). Potential of low-cost materials for biogas purification, a review of recent developments. Renewable and Sustainable Energy Reviews, 210, 115152. https://doi.org/10.1016/j.rser.2024.115152

[27] Mendes, G., Von Appen, J., & Ioakimidis, C. (2011, May). Integrated energy microgrids for community-scale systems: Case study research in the Azores islands. In 2011 8th International Conference on the European Energy Market (EEM) (pp. 382-387). IEEE. https://doi.org/10.1109/EEM.2011.5953041

[28] Babayomi, O. O., Dahoro, D. A., & Zhang, Z. (2022). Affordable clean energy transition in developing countries: Pathways and technologies. Iscience, 25(5). https://doi.org/10.1016/j.isci.2022.104178

[29] Kumar, S., & Rathore, K. (2023). Renewable energy for sustainable development goal of clean and affordable energy. International Journal of Materials Manufacturing and Sustainable Technologies, 2(1), 1-15. https://doi.org/10.56896/IJMMST.2023.2.1.001

[30] Jayachandran, M., Gatla, R. K., Rao, K. P., Rao, G. S., Mohammed, S., Milyani, A. H., ... & Geetha, S. (2022). Challenges in achieving sustainable development goal 7: Affordable and clean energy in light of nascent technologies. Sustainable Energy Technologies and Assessments, 53, 102692. https://doi.org/10.1016/j.seta.2022.102692

[31] Doukas, A., & Ballesteros, A. (2015). Clean energy access in developing countries: perspectives on policy and regulation.

[32] Hoang, A. T., & Nguyen, X. P. (2021). Integrating renewable sources into energy system for smart city as a sagacious strategy towards clean and sustainable process. Journal of cleaner production, 305, 127161. https://doi.org/10.1016/j.jclepro.2021.127161

Next-Generation Solar Energy: Emerging Materials, Thermal Systems, and Photovoltaic Advancements

A. VERMA[1], A. DHASMANA[1*], S. KUMAR[2], J. GAIROLA[3], UMANG[4], SHWETA[5], N. SHAMEEN[6]

[1]Department of Aerospace Engineering, Uttaranchal Institute of Technology, Uttaranchal University, Dehradun-248007, Uttarakhand, India

[2]Department of Mechanical Engineering, J B Institute of Technology, Dehradun, UK, India

[3]Department of Telemedicine, AIIMS, Rishikesh-249203, UK, India

[4]Department of Computer Applications, DSB Campus, Kumaun University, Nainital

[5]Department of Medical Lab Technology, University Institute of Allied Health Sciences, Chandigarh University, Mohali-140413, Punjab, India

[6]Chandigarh Group of College, Jhanjeri, Mohali-140307, Punjab, India

*abhaydhasmana735@gmail.com

Keywords: Solar Energy Technologies, Photovoltaic Advancements, Perovskite Solar Cells, Thermal Energy Storage, Nanomaterials in Solar Energy

Abstract: The demand for energy in the world is increasing and the input for the necessity of more sustainable substitutes for fossil source energy has created an 'emphasis' on solar energy technologies. Solar photovoltaic and solar thermal systems are along with these technologies 'in the shadows', or 'in the background' to a good extent. Nevertheless, they are not without problems that always hamper them as a technology that has matured. Performance issues, manufacturing costs, material degradation, and non-uniform energy generation have hindered utility on a large-scale, or made sustainability of a low-risk of expect future investments in them, or development of more moder technologies. Moreover, solar thermal system advancements require energy storage to deliver uninterrupted power. This paper evaluates new materials, thermal energy technologies, and evolving photovoltaic technologies, that comprise the future of solar energy systems. Innovative alternatives such as perovskite solar cells, tandem solar structures, and organic photovoltaics will be examined to determine their potential to improve efficiency and reduce costs. We will investigate solar thermal systems and phase-change energy storage to determine their contribution to energy retention and grid stability. Major issues associated with material stability, energy conversion efficiency, and material viability for large-scale deployment are indicated as potential opportunities related to nanomaterials, transparent conductive oxides, and hybrid solar energy approaches. Following an elaborate examination of recent trends, this work aims to provide an all-inclusive overview of the technological innovations that are shaping the future of solar energy toward an efficient and sustainable global energy system.

1. Introduction

Due to concerns about climate change, energy security, and limited fossil fuel resources, the global energy market is changing toward the use of renewable energy resources. Solar energy is regarded as one form of clean energy that is abundant in nature and, if free, would be the principal source of energy for everybody [1]. Along with the rapid technological advancement in both solar PV and solar thermal systems have come considerable improvements in conversion efficiency and cost. Figure 1 depicts the rise and importance of solar energy around the world.

One more solution is the element of highways and streets used for evening illumination is the street light. Moreover, streetlight is crucial to accomplishing the UN's sustainable development objectives by 2030. At the moment, industry 4.0 enabling technologies have the potential to improve the streetlight infrastructure in terms of sustainability with the support of solar panel [2].

The authors focused on solar energy used in UAV, as unmanned aerial vehicle's low-altitude, long-endurance original concept for multi-day breakouts is discussed in the research. The paper for the performance of deliverance, hunting, and monitoring tasks as well as the transportation of food, medicine, and other supplies. In order for the aircraft to be light, the initial design focused on energy operation and flying for longer hours. Unmanned aerial vehicle abidance can be increased using solar energy, one of the limitless renewable energy sources, without significantly increasing the energy system's size or mass [3].

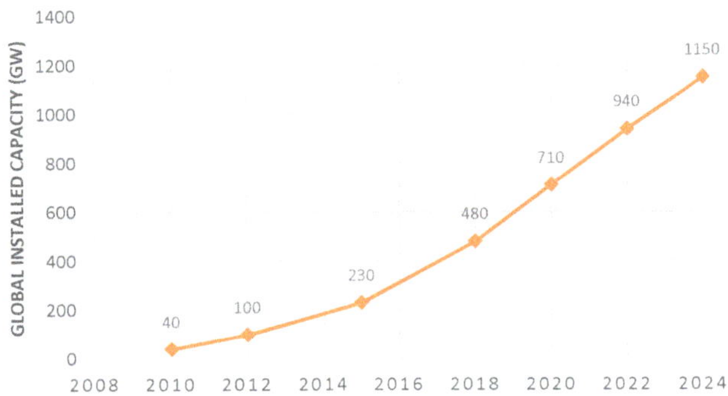

Figure 1. *Global Growth of Solar Power Capacity (Source: IEA (International Energy Agency) – World Energy Outlook 2023) [28]*

This discussion is not recent and the challenges of high capital cost, efficiency drop, and intermittency in energy has been seen with the large-scale application of solar energy without the benefit of solar power factor for long [4]. To mitigate these disadvantages or indeed remove them entirely, research is focusing on new classes of photovoltaic materials, next-generation storage systems, and energy conversion technologies that will complement solar technology in terms of reliability and practicality. The silicon solar cell continues to dominate all other solar energy technologies while simultaneously having the highest commercialization and technology readiness and operate reliably; however, silicon solar cells have reached the theoretical efficiency limits and are therefore focused on new production materials that not only cheap but also operate faster and are active than silicon [5].

Perovskite solar cells, tandem configurations, and organic photovoltaics, among others, are currently the top contenders because of their high light-absorption properties flexibility, and ease of manufacturing. Bifacial PV systems, multi-junction solar cells, and transparent photovoltaic panels are some of the alternative technologies still under development for the purpose of improving electrical conversion efficiency and utilization. Along with photovoltaics, the solar thermal energy systems remain a key competitor among renewable energy alternatives because the solar thermal systems can readily convert sunlight into electricity and heat. CSP has exhibited considerable merit in locations with high solar irradiance with respect to operation under normal operational mode. Nevertheless, the issues of thermal losses, storage potential, and efficiency are still the main issues the troubleshooting engineering efforts are focused on solving. The application of phase change materials (PCMs), molten salt storage systems, and nanostructured thermal absorbers to energized to heat-stored energy, dispatchability of energy, etc. [6].

The latest CSP systems are capable of completely obviating traditional fossil fuel power generation. A big part of the balance is a power battery or H_2-based energy storage for use when the solar energy is not in generation. While lithium-ion and solid-state battery storage are likely to be the most common in households, thermal energy storage (TES) combined with H_2-based would represent a viable option for long-term storage [7]. The TES systems are made of particular materials that allow for a very efficient process of storing and releasing heat. It means that the solar plant at night will still be able to generate power in the night just like it did in the daylight. Strangely in another way, H_2 production via solar-powered electrolysis is a method that leverages sunlight and turns it into a storable and transportable fuel. The evolution of storage technology will be what eventually guarantees that solar energy will be dependable and scalable. In addition, researchers are looking at nanomaterials pairing with transparent conductive oxides and QD solar cells that will not only serve to increase the efficiency of solar energy systems but also the lifetime; additionally, the fusion of ML and AI into areas such as solar tracking, efficiency optimization, and predictive maintenance helps to ensure the reliability of the solar system and energy output operational range [8]. Therefore, solar power is on the cusp to be one of the main drivers of the worldwide clean energy transition as a result of emerging photovoltaic materials and storage technology.

It discusses innovative communication protocols and AI-powered edge computing to overawed parameters. The impact of embedded systems on energy efficiency and dependability is estimated through case studies. Over the next 10 years, further developments including solar panel fit in electric vehicles, hardware evolution, machine learning for analytical management, and IoT-based smart ecosystems are also expected to increase efficiency by 15% to 25% [9].

This article presents a thorough overview of advanced solar energy technologies, including new photovoltaic materials, solar thermal energy, and energy-storage technology. The authors attempt to elucidate the key challenges associated with the desirability of solar energy and explorations relating to solar energy options that may be effective, efficient, economical, and sustainable based on new science and a baseline of technological development.

2. Photovoltaic (PV) Technologies

Photovoltaic (PV) technologies gained tremendous progress throughout the years, and many types of solar cells have been developed for better performances, lower costs, and better scalability. The most important PV technologies are separated into three groups: silicon-based solar cells, thin-film and perovskite solar cells, and organic PVs and multi-junction solar cells. Each of these technologies is very different from the other and offers equally diverse applications to the solar-energy solutions. Figure 2 shows the procedure of a typical solar PV system. Silicon PV cells remain considered to be the dominant solar PV technology in the world today in terms of installed power generation capacity.

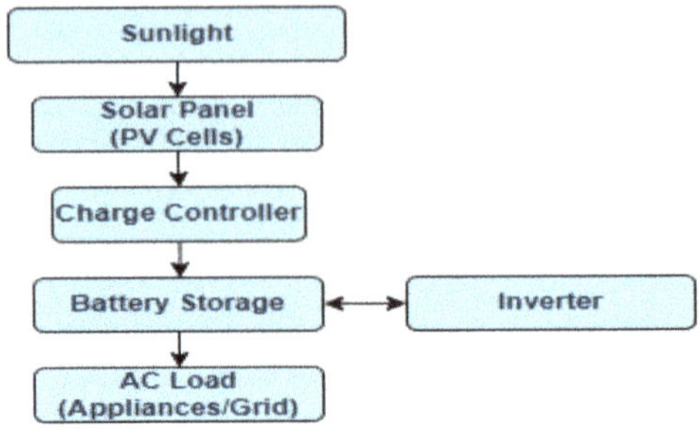

Figure 2. *Solar PV System Workflow*

2.1 Silicon-based solar cells

Solar cells are basically divided on monocrystalline, polycrystalline, and amorphous silicon, or a-Si cells. Since the purity level and order within a crystalline structure are high, monocrystalline-style silicon cells present the most efficiency, while polycrystalline cells represent a cheaper option with slightly less efficiency. Amorphous silicon solar cells have a non-crystalline silicon structure which results in light and flexible designs suitable for use in portable electronics and building-integrated photovoltaics. While these are less efficient than crystalline silicon cells, a-Si cells do have the advantage of relatively low production costs and better performance under low light [10].

2.2 Thin-film and perovskite solar cells

Thin-film solar cells stand as an alternative to the older silicon-based PV technology but are proposed for their cheaper material cost, lighter frames, and flexibility. Thin-film solar cells consist of thin semiconductor layers, like cadmium telluride (CdTe) or copper indium gallium selenide (CIGS), deposited on a glass substrate. Perovskite solar cells are indeed the newest incarnation of thin-film technology, thus fetching them more publicity in recent times for their high efficiency and low cost of manufacture. Perovskite compounds boast excellent light-absorption characteristics, thus allowing for efficiencies on a par with silicon-based solar cells. However, they currently suffer from stability issues; their large-scale manufacturing is also hindered by the lack of scalable processes and by the presence of toxic materials in the composition [11].

2.3 Organic PV and multi-junction cells

Organic photovoltaic (OPV) cells use carbon-based organic materials to change sunlight to electricity. These solar cells have benefits like flexibility, low weight, and inexpensive production methods. Yet, because of their lower efficiency and shorter lifespan of operation in comparison to silicon and perovskite solar cells, their large-scale utilization has been limited. Scientists continue to investigate novel organic materials and device structures for improving OPV performance. Solar cells with multiple junctions, or tandem solar cells, have been developed to achieve highest energy conversion efficiency by stacking layers of semiconductor materials with various bandgaps. Each layer absorbs a particular range of the solar spectrum, giving greater overall efficiency than single-junction solar cells. Multi-junction cells find extensive application in space technology and concentrated photovoltaic systems, where efficiency has to be high. With developments in the field of material science and fabrication processes, multi-junction solar cells will dominate the horizon of high-efficiency solar power generation in the coming times [12]. Figure 2 shows the efficiencies of different types of solar cells.

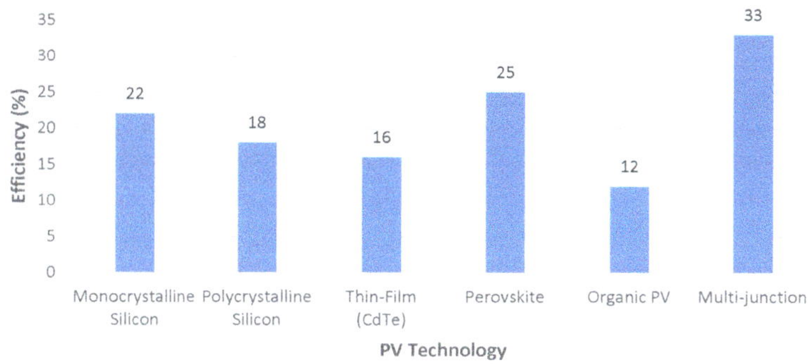

Figure 3. *Efficiency comparison of photovoltaic technologies (adapted from NREL efficiency chart) [28]*

3. Solar Thermal Systems

3.1 Solar water heaters and concentrators

In these systems, solar collectors are utilized to absorb sun's radiation and convey heat to a storage tank, where hot water is accumulated for household, business, and industrial use. Integrated collector storage (ICS) systems, where both the collector and storage unit are integrated into a single part, have been created to improve efficiency and minimize heat losses. To boost the performance of solar water heaters, compound parabolic concentrators (CPCs) have been incorporated into ICS systems. CPCs improve the capture of solar radiation by concentrating sunlight on the collector surface, thus improving temperature and efficiency in the process of heating. Recent technological developments in CPC-based solar water heating systems have shown improved thermal efficiency significantly and hence are a potential solution for high-solar-regions [13].

3.2 Solar air conditioning and heating

Solar air conditioning and heating systems utilize solar thermal energy to regulate indoor temperatures, offering an environmentally friendly alternative to conventional heating, ventilation, and air conditioning (HVAC) systems. These systems operate by using solar heat to drive absorption or adsorption cooling cycles, which can provide space cooling without relying on electricity-intensive compression-based cooling systems. In Europe, solar air conditioning systems have gained traction due to their potential to reduce electricity demand and greenhouse gas emissions. Various configurations, including direct and indirect solar-assisted cooling technologies, have been explored to enhance system efficiency and feasibility. Despite their benefits, widespread adoption is hindered by challenges such as high initial costs, intermittent solar availability, and the need for thermal storage solutions to ensure consistent performance [14].

3.3 Thermal energy storage systems

Thermal energy storage (TES) systems play a crucial role in improving the efficiency and reliability of solar thermal applications by storing excess heat for use when solar radiation is unavailable. TES technologies are categorized into sensible heat storage, latent heat storage, and thermochemical storage.

- **Sensible heat storage** systems store thermal energy by raising the temperature of materials such as water, rock, or molten salts. These systems are widely used in large-scale solar thermal plants.
- **Latent heat storage** utilizes phase change materials (PCMs) to store and release energy during phase transitions, providing higher energy density and improved thermal management.
- **Thermochemical storage** involves reversible chemical reactions to store heat, offering high energy densities and long-duration storage capabilities.

The integration of TES systems with solar thermal technologies enhances energy utilization and enables continuous operation, even during nighttime or cloudy conditions. Ongoing research focuses on developing advanced storage materials and optimizing TES system designs to enhance efficiency and cost-effectiveness [15]. Table 1 shows the comparison between different solar thermal systems.

Table 1. *Comparison of Solar Thermal Systems*

System Type	Operating Temp (°C)	Efficiency (%)	Common Applications
Flat Plate Collectors	30–80	50–70	Domestic hot water
Evacuated Tube	50–150	60–75	Space heating, industrial usage
Parabolic Trough	150–400	70–85	Electricity generation

4. Advanced Energy Materials

The creation of advanced energy materials is also important in enhancing the efficiency, stability, and affordability of solar energy technologies. Advanced energy materials support increased light absorption, charge transport, and overall energy conversion efficiency in PV devices. Three prominent categories of advanced energy materials are nanomaterials in solar cells, light-harvesting materials, and transparent conducting oxides (TCOs), all of which have made important contributions to solar energy applications.

4.1 Nanomaterials in solar cells

Nanomaterials have transformed solar cell technology through the ability to improve light absorption, charge transport, and surface passivation. Their distinct optical and electronic properties have led them to be a potential choice for improving solar cell efficiency. Quantum dots, carbon nanotubes, graphene, and plasmonic nanoparticles are some of the most commonly researched nanomaterials for PV applications. Quantum dots, for example, provide tuneable bandgaps that can provide improved spectral absorption and multi-exciton generation, greatly enhancing efficiency. With high electrical conductivity and transparency, carbon nanotubes and graphene are considered good candidates for charge transport layers in thin-film and organic solar cells. Plasmonic nanoparticles of gold and silver serve as light-trapping agents by scattering and confining light within the active layer of solar cells, thereby enhancing its absorption. Incorporating nanomaterials into PV devices will undoubtedly remain an exciting frontier for the next generation of solar technologies [16].

4.2 Light-harvesting materials

Optimum light harvesting is needed to ensure optimum solar cell performance because light harvesting would directly influence photon absorption and the production of charge carriers. Organic solar cells, conversely, rely on new light adsorption materials to improve its efficiency. Optimized optical properties of the materials such as conjugated polymers and small molecules are tailored to absorb a broader spectrum of solar illumination and transport charge efficiently. With the aid of photonic nanostructures, such as anti-reflective coatings and plasmonic structures, light absorption is maximized by reducing loss via reflection and increasing light trapping. By nano structuring the materials, one aims to fabricate inexpensive and flexible organic and hybrid solar cells with improved performance [17].

4.3 Transparent conducting oxides (TCOs)

Thin film solar Application of transparent conducting oxides (TCOs) has been used in the last decade as a thin coating layer in the electrodes found in solar cells to conduct electricity and transmit light simultaneously. Among the many types of TCOs used, indium tin oxide (ITO), fluorine-doped tin oxide (FTO), and aluminum-doped zinc oxide (AZO) are most popular materials for commercial installations of photovoltaic cells. Their basic property is to allow transmission of light and conduction of electricity, which directly affects the conversion efficiency. But the high price and limited availability of indium have motivated scientists to look for substitute TCO materials like doped zinc oxides and metal oxides with better stability and performance. The creation of new TCOs with enhanced optoelectronic properties is a top priority for the advancement of next-generation photovoltaic technology [18]. Table 2 illustrates new emerging solar cell materials along with their benefits and their used cases.

Table 2. *Emerging Materials in Solar Cells*

Material Type	Benefit	Use Case
Perovskites	High efficiency, low cost	Next-gen PV
Quantum Dots	Tunable bandgap	Multi-junction solar cells
Transparent Conducting Oxides (TCOs)	High conductivity and transparency	Front electrode in PVs
Organic Semiconductors	Lightweight and flexible	Wearable and portable solar tech

5. Efficiency and Performance Factors

The efficiency and performance of solar energy systems are influenced by various factors, including material engineering, optical enhancements, and environmental conditions. As solar technologies continue to evolve, researchers focus on improving energy conversion rates and system reliability through innovative solutions. Key aspects affecting solar cell performance include efficiency enhancements through material engineering, anti-reflective coatings and surface texturing, and performance under different weather conditions. Figure illustrates how temperature affects PV output efficiency.

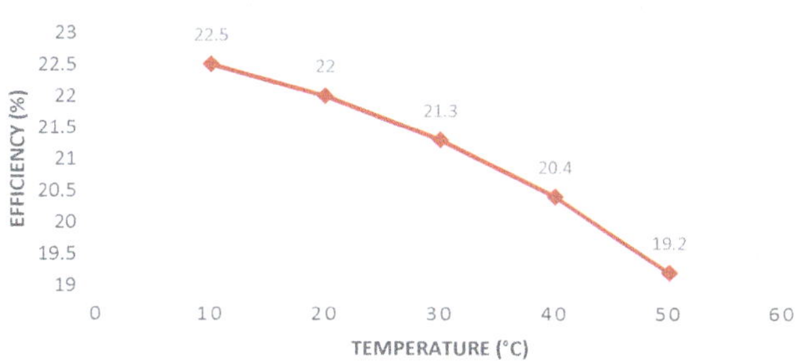

Figure 4. *Effect of temperature on photovoltaic module efficiency (adapted from NREL technical reports) [21]*

5.1 Efficiency enhancements through material engineering

Material engineering plays a crucial role in improving the efficiency of photovoltaic (PV) systems. Hybrid photovoltaic/thermal (PV/T) systems have been explored as an effective method for enhancing energy conversion efficiency. These systems simultaneously generate electricity and utilize the excess heat, improving the overall energy yield of solar panels. Advanced semiconductor materials, multi-junction solar cells, and surface modifications further contribute to efficiency improvements. By incorporating new materials such as perovskites, nanomaterials, and transparent conducting oxides, researchers aim to minimize energy losses and enhance charge transport properties. Additionally, better thermal management strategies in PV/T systems help maintain stable operating temperatures, preventing efficiency degradation due to overheating [19].

5.2 Anti-reflective coatings and surface texturing

Losses in efficiency occur due to sunlight reflection against the surface of the solar cell, and to compensate for this loss, ARCs and surface texturing treatments are applied to enhance light absorption. ARCs, in their role to stop reflection, create an optic interface that prohibits refractive index mismatches so that more light can travel into the active layer of the solar cell. Different materials such as Si_3N_4 for silicon nitride, TiO_2 for titanium dioxide, and MgF_2 for magnesium fluoride have been applied as anti-reflective coatings to increase solar cell efficiency. Apart from this, nanostructured surfaces in the form of pyramid-shaped and moth-eye textured structures have been developed for enhanced light trapping and energy absorption over a wider range. Ultimately, this led to an improvement in the photovoltaic conversion efficiencies, mainly for silicon-based and thin-film solar cells [20].

5.3 Performance under different weather conditions

Solar photovoltaic module efficiency and performance are highly dependent on environmental conditions with temperature, humidity, dust deposits, and the intensity of solar radiation being main parameters through which environmental conditions influence PV modules. In hot arid and desert conditions, excessive temperature can cause loss of efficiency through charge carrier recombination and degradation of PV materials due to heat. Dust deposition, on the other hand, is another significant reason impeding solar panel performances since it decreases the fraction of light incident on the active layer. Experimental evidence has shown how frequent cleaning and self-cleaning thin films can mitigate efficiency loss caused by dust deposition. Besides, these panels are exposed to ever-changing weather conditions from cloudy to rainy, requiring reliable energy storage and hybrid energy systems to stabilize the power output [21].

6. Applications and Case Studies

The energy of the sun has been used for a variety of uses from large utility solar farms to home solar installations on buildings. Case studies based on real applications demonstrate how solar energy systems can operate effectively depending on geography and economic conditions. This chapter discusses utility-scale solar farms, solar rooftops, and case studies to describe the real-time benefits and constraints in utilizing solar energy.

6.1 Utility-scale solar farms

Utility-scale solar farms are big PV power plants that generate electricity on a commercial scale, typically entering into national or regional power grids. Thus in the huge transition into greener energies, they bind thousands of homes and businesses, offering clean and sustainable electricity with which to grow. The facility's design and construction thus need to be carefully planned, considering site selection, panel orientation, grid integration, and maintenance. Modern solar farms utilize advanced tracking systems that optimize sunlight capture by adjusting the angle of solar panels throughout the day. Additionally, large-scale projects often incorporate battery storage solutions to mitigate the intermittent nature of solar power, ensuring a stable electricity supply.

While solar farms offer substantial environmental benefits, challenges such as land use conflicts, grid connectivity, and high initial capital costs must be addressed to enhance their widespread adoption [22]

6.2 Rooftop solar systems

Rooftop solar systems have emerged as a key solution for decentralized energy production, allowing residential, commercial, and industrial buildings to generate their own electricity. The widespread adoption of rooftop solar is driven by factors such as declining PV module costs, government incentives, and increasing consumer awareness of renewable energy benefits. In countries like Australia, rooftop PV systems have seen rapid growth, significantly contributing to the national electricity supply. However, there are technical and regulatory challenges to integrating such solar power distributed generation into existing grid infrastructure. Voltage fluctuations, grid stability, and solar generation variability all have to be tackled through grid technologies and energy management systems. Though there are these challenges, one should keep in mind that rooftop solar is the most cost-efficient and environmentally friendly energy solution, especially in an urban context, where electricity demand peaks [23].

6.3 Real-world case studies

Several case studies from real-life examples show the effective application of solar energy technologies for various uses. One of them is the extensive use of solar power in Germany, where considerable declines in carbon emissions and energy expenses have been witnessed. Solar energy policies such as feed-in tariffs and subsidies have been fundamental in making PV systems affordable and financially sustainable in Germany. An example is the application of solar energy in rural electrification schemes of developing nations. Where there is limited access to grid-based centralized electricity, solar microgrids and stand-alone PV systems offer secure and economic power supply options. The schemes have enhanced living standards through enhanced access to lighting, clean cooking technologies, and digital connectivity. Furthermore, improved energy storage technologies have increased the reliability of solar energy in areas with intermittent weather patterns. Hybrid solar-wind systems and solar installations integrated with batteries are being used more and more to provide a consistent power supply, especially in off-grid and disaster-susceptible locations [24].

7. Future Trends and Innovations

Solar energy technologies have been adopted on a wide range of applications, ranging from large-scale utility solar farms and small-scale solar installations on rooftops. Real-world application case studies introduce the success of solar energy systems across different geographic and economic conditions. This section investigates the adoption of utility-scale solar farms, rooftop solar systems, and real-world case studies, illustrating the practical benefits and challenges of solar energy adoption.

7.1 Next-gen PV technologies

Utility-scale solar farms are large photovoltaic (PV) power plants designed to generate electricity on a commercial scale, typically feeding into national or regional power grids. These solar farms play a significant role in the transition to renewable energy, providing clean and sustainable electricity to thousands of homes and businesses. The design and construction of such facilities require careful planning regarding site selection, panel orientation, grid integration, and maintenance. Modern solar farms utilize advanced tracking systems that optimize sunlight capture by adjusting the angle of solar panels throughout the day. Additionally, large-scale projects often incorporate battery storage solutions to mitigate the intermittent nature of solar power, ensuring a stable electricity supply. While solar farms offer substantial environmental benefits, challenges such as land use conflicts, grid connectivity, and high initial capital costs must be addressed to enhance their widespread adoption [25]

7.2 Tandem solar cells and flexible PV

Rooftop solar systems have emerged as a key solution for decentralized energy production, allowing residential, commercial, and industrial buildings to generate their own electricity. The widespread adoption of rooftop solar is driven by factors such as declining PV module costs, government incentives, and increasing consumer awareness of renewable energy benefits. In countries like Australia, rooftop PV systems have seen rapid growth, significantly contributing to the national electricity supply. However, integrating distributed solar power into existing grid infrastructure presents technical and regulatory challenges. Issues such as voltage fluctuations, grid stability, and variability in solar generation need to be managed through advanced grid technologies and smart energy management systems. Despite these challenges, rooftop solar remains a cost-effective and environmentally friendly energy solution, particularly in urban areas with high electricity demand [26].

7.3 Integration with AI and IoT

Solar energy technologies are being applied effectively across a variety of real-world contexts. In Germany, the integration of solar energy on a large scale has also been phenomenal in conjunction with the existing trajectories of carbon reduction and electricity bill-reduction. With respect to solar energy policy, especially for wonderful public relations and potential profitability to renewable energy providers, the mechanism of feed-in tariffs and subsidies plays an important role in expanding the accessibility and affordability of photovoltaic (PV) systems. Another context is solar energy systems in rural electrification efforts in developing countries. Solar microgrids and stand-alone PV systems provide a more reliable and economically feasible energy option for communities and businesses that are distant from the central grid. These programs directly impact the improvement of living standards by providing light, safe cooking technologies, and digital connectivity. Furthermore, the growing diversity of energy storage solutions has provided greater continuity of solar energy in areas where the weather is less stable compared to places that have more consistent weather. Solar-wind hybrid systems and solar systems with batteries are becoming an increasingly common energy strategy to address issues associated with outages and climate change impacts to the energy infrastructure [27].

8. Conclusion

The development of solar energy technologies has recently achieved major progress because of the increasing demand for sustainable energy systems which do not harm the environment. The development of new photovoltaic materials like perovskites, tandem solar cells and organic photovoltaics has created better energy conversion efficiency combined with lower production expenses. Solar power generation for large-scale applications receives improved performance through the combination of photovoltaic systems with solar thermal technologies and advanced thermal energy storage systems. Advanced materials development, which includes nanomaterials and transparent conducting oxides and light-harvesting structures, has resulted in better device performance and longer stability maintenance. The combination of artificial intelligence with machine learning and smart energy management systems operates as a critical component that enhances solar power generation and predictive maintenance and grid integration processes. The field faces multiple obstacles which include intermittent power supply problems and material degradation issues and expensive equipment installation costs, despite the presence of technological progress. Researchers must continue their work on advanced materials and energy storage technologies and smart energy systems to find solutions for these problems. The ongoing development of photovoltaic technologies together with energy storage advancements and smart energy management systems will bring about a faster shift towards a dependable and sustainable worldwide energy network.

References

[1] Kannan, N., & Vakeesan, D. (2016). Solar energy for future world: A review. Renewable and Sustainable Energy Reviews, 62, 1092-1105. https://doi.org/10.1016/j.rser.2016.05.022

[2] Tyagi, S., Krishna, K.H., Joshi, K., Ghodke, T.A., Kumar, A. and Gupta, A., 2023, November. Integration of PLCC modem and Wi-Fi for Campus Street Light Monitoring. In 2023 International Conference on Computing, Communication, and Intelligent Systems (ICCCIS) (pp. 1113-1116). IEEE. https://doi.org/10.1109/ICCCIS60361.2023.10425715

[3] Vijaya, P.S.B., Srivastava, S., Haldar, S., Sai, N.V., Pooja, K., Shivani, S. and Rajendran, A.R., 2025. Design and experimental testing of solar UAVS for low altitude and long endurance. International Journal of Aerospace System Science and Engineering, 1(2), pp.143-156. https://doi.org/10.1504/IJASSE.2025.145201

[4] Sakthivadivel, D., Balaji, K., Rufuss, D. D. W., Iniyan, S., & Suganthi, L. (2021). Solar energy technologies: principles and applications. In Renewable-Energy-Driven Future (pp. 3-42). Academic Press. https://doi.org/10.1016/B978-0-12-820539-6.00001-7

[5] Ogundipe, O. B., Okwandu, A. C., & Abdulwaheed, S. A. (2024). Recent advances in solar photovoltaic technologies: Efficiency, materials, and applications. GSC Advanced Research and Reviews, 20(1), 159-175. https://doi.org/10.30574/gscarr.2024.20.1.0259

[6] Panagoda, L. P. S. S., Sandeepa, R. A. H. T., Perera, W. A. V. T., Sandunika, D. M. I., Siriwardhana, S. M. G. T., Alwis, M. K. S. D., & Dilka, S. H. S. (2023). Advancements in photovoltaic (PV) technology for solar energy generation. Journal of Research Technology & Engineering, 4(30), 30-72.

[7] Kant, N., & Singh, P. (2022). Review of next-generation photovoltaic solar cell technology and comparative materialistic development. Materials Today: Proceedings, 56, 3460-3470. https://doi.org/10.1016/j.matpr.2021.11.116

[8] Elkelawy, M., M El-Ashmawy, W., & Ahmed, S. M. (2024). State of the Art in Concentrated Solar Power: Latest Technological Advancements and Innovations in Efficiency and Energy Storage. Pharos Engineering Science Journal, 1(1), 17-28. https://doi.org/10.21608/pesj.2025.344983.1007

[9] Srivastava, S.S., Satyanarayana, G.S., Dhasmana, A., Rawat, V., Rana, A.S. and Bisht, Y.S., 2025. Role of Embedded Systems in Smart Energy Management: Challenges, Innovations, and Future Trends. Solar Energy and Sustainable Development Journal, 14(STR2E), pp.27-50. https://doi.org/10.51646/jsesd.v14iSTR2E.797

[10] Deng, X., & Schiff, E. A. (2003). Amorphous silicon based solar cells. https://doi.org/10.1002/0470014008.ch12

[11] Ball, J. M., Stranks, S. D., Hörantner, M. T., Hüttner, S., Zhang, W., Crossland, E. J., ... & Snaith, H. J. (2015). Optical properties and limiting photocurrent of thin-film perovskite solar cells. Energy & Environmental Science, 8(2), 602-609. https://doi.org/10.1039/C4EE03224A

[12] Hadipour, A., de Boer, B., & Blom, P. W. (2008). Organic tandem and multi-junction solar cells. Advanced functional materials, 18(2), 169-181. https://doi.org/10.1002/adfm.200700517

[13] Devanarayanan, K., & Murugavel, K. K. (2014). Integrated collector storage solar water heater with compound parabolic concentrator-development and progress. Renewable and Sustainable Energy Reviews, 39, 51-64. https://doi.org/10.1016/j.rser.2014.07.076

[14] Balaras, C. A., Grossman, G., Henning, H. M., Ferreira, C. A. I., Podesser, E., Wang, L., & Wiemken, E. (2007). Solar air conditioning in Europe-an overview. Renewable and sustainable energy reviews, 11(2), 299-314. https://doi.org/10.1016/j.rser.2005.02.003

[15] Alva, G., Lin, Y., & Fang, G. (2018). An overview of thermal energy storage systems. Energy, 144, 341-378. https://doi.org/10.1016/j.energy.2017.12.037

[16] Tala-Ighil, R. (2015). Nanomaterials in solar cells. Handbook of nanoelectrochemistry, 1-18. https://doi.org/10.1007/978-3-319-15207-3_26-1

[17] Duché, D., Drouard, E., Simon, J. J., Escoubas, L., Torchio, P., Le Rouzo, J., & Vedraine, S. (2011). Light harvesting in organic solar cells. Solar energy materials and solar cells, 95, S18-S25. https://doi.org/10.1016/j.solmat.2010.12.008

[18] Fortunato, E., Ginley, D., Hosono, H., & Paine, D. C. (2007). Transparent conducting oxides for photovoltaics. MRS bulletin, 32(3), 242-247. https://doi.org/10.1557/mrs2007.29

[19] Tripanagnostopoulos, Y. (2007). Aspects and improvements of hybrid photovoltaic/thermal solar energy systems. Solar energy, 81(9), 1117-1131. https://doi.org/10.1016/j.solener.2007.04.002

[20] Shanmugam, N., Pugazhendhi, R., Madurai Elavarasan, R., Kasiviswanathan, P., & Das, N. (2020). Anti-reflective coating materials: A holistic review from PV perspective. Energies, 13(10), 2631. https://doi.org/10.3390/en13102631

[21] Mussard, M., & Amara, M. (2018). Performance of solar photovoltaic modules under arid climatic conditions: A review. Solar Energy, 174, 409-421. https://doi.org/10.1016/j.solener.2018.08.071

[22] Neill, S., Stapleton, G., & Martell, C. (2017). Solar farms: the earthscan expert guide to design and construction of utility-scale photovoltaic systems. Routledge. https://doi.org/10.4324/9781315651002

[23] Mountain, B., & Szuster, P. (2015). Solar, solar everywhere: opportunities and challenges for Australia's rooftop PV systems. IEEE power and energy magazine, 13(4), 53-60. https://doi.org/10.1109/MPE.2015.2416113

[24] Hasan, M. M., Hossain, S., Mofijur, M., Kabir, Z., Badruddin, I. A., Yunus Khan, T. M., & Jassim, E. (2023). Harnessing solar power: a review of photovoltaic innovations, solar thermal systems, and the dawn of energy storage solutions. Energies, 16(18), 6456. https://doi.org/10.3390/en16186456

[25] Madani, M., Aini, Q., Supriyono, I. A., Sihotang, S. V., & Azuddin, M. (2024, August). Critical Elements for Urban Next-Gen Solar Panel Efficiency and Adoption. In 2024 3rd International Conference on Creative Communication and Innovative Technology (ICCIT) (pp. 1-7). IEEE. https://doi.org/10.1109/ICCIT62134.2024.10701198

[26] Jiang, Y., & Qi, Y. (2021). Metal halide perovskite-based flexible tandem solar cells: next-generation flexible photovoltaic technology. Materials Chemistry Frontiers, 5(13), 4833-4850. https://doi.org/10.1039/D1QM00279A

[27] Harry, A. (2023). Unlocking Synergy: Solar Cell Integration with Radio Waves for AI, IT, and Business Advancements. International Journal of Multidisciplinary Sciences and Arts, 2(4), 188-196. https://doi.org/10.47709/ijmdsa.v2i2.3211

Sustainable Waste Management and Energy Material Recycling: Innovations and Circular Economy Strategies

D. ARORA[1*], A. BHARDWAJ[1], N. PARVEEN[2], D. JAYARAJAN[3], R. P. JAYASWAL[4], A. H. SHEIKH[5]

[1]Department of Aerospace Engineering, Uttaranchal Institute of Technology, Uttaranchal University, Dehradun-248007, Uttarakhand, India

[2]Department of Medical Radiology and Imaging Technology, Rayat-Bahra University, Mohali, Punjab, India

[3]Department of MLT, Faculty of Paramedical Sciences, Assam Downtown University, Guwahati, Assam, India

[4]Department of Medical Lab Technology (UIAHS), Chandigarh University, Mohali-140413, India

[5]Department of Operation Theatre and Anaesthesia Technology (USAHS) Rayat-Bahra University, Mohali-140104, India

*divijarora2000@gmail.com

Keywords: Waste-to-Energy, Circular Economy, Energy Material Recycling, Renewable Energy Waste, Sustainable Waste Management

Abstract: The generation of end-of-life waste from energy materials such as solar panels, wind turbine blades, and lithium-ion batteries is an ecological problem that is presently under-addressed, even though renewable energy sources play a very critical role in minimizing the impacts of climate change. Besides initiating the loss of essential raw materials needed for the green energy shift, the elimination of this waste into the environment and the absence of efficient recycling methods pollute the environment and threaten public health. Hence, a shift from a linear model of consumption to an economic paradigm for sustainable development would be required to manage this new challenge. The challenges and potentials for green waste management and recycling of energy material are treated in this article. It assesses the economic viability and scalability of the latest chemical, mechanical, and heat recycling technologies. The application of closed-loop recycling and waste-to-energy (WTE) systems are also discussed in the paper as two feasible measures towards minimizing environmental hazards and material shortages. The assessment of policy context and concomitant global case studies emphasizes adoption challenges and proposes solutions for integrating digital technologies, e.g., AI and Internet of Things, into current waste management systems. The findings will be utilized in creating circular, efficient, and sustainable approaches to managing the energy-generating produced waste so as to enhance resilience and the generation of green energy.

1. Introduction

As the planet begins to move towards cleaner energy, innovation in advanced energy technologies such as solar energy, fuel cell technology, and batteries is increasing. While these technologies reduce the carbon footprint of the energy industry, they also complicate the problem of disposing of waste and obsolescent materials generated and discarded. Energy waste and recycling have become important elements of those sustainable energy systems which emphasize recovery of resources and minimizing the environmental footprint over dependency on virgin material. Figure 1 forecasts the eventual global waste of solar photovoltaics.

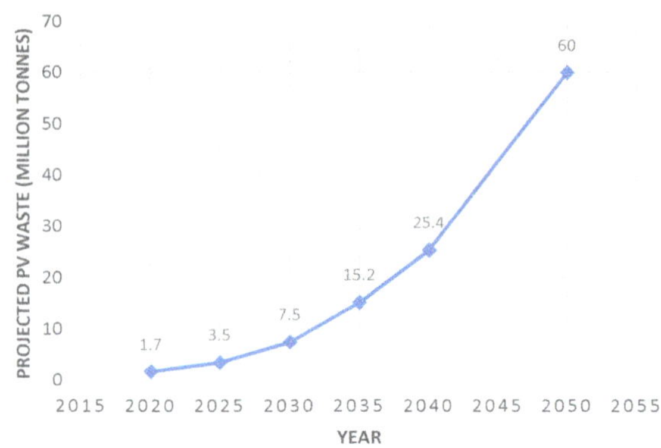

Figure 1. *Global Solar PV Waste Projections [1]*

As Giugliano et al. (2011) put it, combined systems of waste management are most effective when they emphasize material and energy recovery and employ stringent lifetime assessment (LCA) to guide their environmental performance analysis. garbage-to-energy, or WTE, refers to the recovery of energy from trash. It minimizes landfill dependency, facilitates circular use of resources, and more. Consonni & Viganò (2011) opine that such a system would achieve enhanced energy efficiency [2-3] by means of more efficient extraction of hidden energy from complicated material streams. Further, Brunner and Rechberger (2015) cite that WTE is an essential part of sustainable waste management, especially in the context of rubbish recycling methods that lead to the recovery of essential raw materials utilized in energy systems. As per Sadef et al. (2016), waste-to-energy and material recycling can boost the economic and environmental benefits of emerging economies like Lahore. But this holds only in case of integrated waste management plan [4-5].

'Smart frameworks' have recently been suggested to employ intelligent modeling for maximum energy recovery and material recycling. In WTE systems, Kaya et al. (2021) have suggested a new artificial intelligence-based methodology. Given the increasingly complicated energy waste material streams they must handle, this reflects the growing importance of smart waste management systems [6]. Improved and more intelligent waste management technologies, focusing on recycle and energy recovery, will be the only way forward as the demand for clean energy continues to increase. With the assistance of modern energy technology, these systems aim to deliver sound, efficient use of resources along with reduced environmental expense. Waste management of solids is an important element of clean energy systems since it directly assists in meeting those long-term energy independence, reduced carbon emissions, and efficiency in the use of resources objectives. Utilization of cleaner technologies like fuel cells, solar cells, and batteries leads to an increase in the amount of outmoded energy materials. Unless effectively managed, this trash alone would have severe environmental impacts, such as greenhouse gas emissions, toxicity of leachate, loss of valuable raw materials, and so forth. Waste management, as stated by Eriksson (2017), is necessary for reducing the environmental impacts and maximizing energy recovery, which subsequently reduces the use of fossil fuels. Recycling the energy content of non-recyclables is an extremely advanced application of waste-to-energy (WTE), noting the applications' promotion of environmental sustainability and sustainable development goals. Based on this definition, managing waste is a valuable interface between energy generation and environmental conservation, making renewable energy sources sustainable for the long term [7].

Recycling energy materials is crucial but is confronted with various financial, technological, and systemic challenges. Contemporary energy technology employs advanced materials such as cobalt, lithium, rare earth metals, and composite polymers. Since they are heterogeneous and toxic, most of these compounds are very difficult to remove, separate, or process towards the end of life. They complicate common recycling practices by exacting more complex demands, which often demand costly methods. Bringezu (2014) states that recycling systems, such as for material and carbon, which are not yet sufficiently developed for energy purposes, pose one of the largest issues. To ensure long-term sustainability, there has to be improvement in the method of material recovery, support by legislation, and product redesign for recyclability. In addition, the threat of the supply chain choke and pollution resulting from wasted energy resources will continue to increase in the absence of strong recycling infrastructure and international cooperative action. For energy materials to be included in a sustainable, low-carbon society, these issues require continued research and development in eco-design, smart disassembly, and environmentally friendly recycling technologies [8].

The existing research on waste-to-energy systems and energy material recycling shows multiple limitations despite the increasing amount of academic studies that explain these systems. The majority of research examines single recycling methods which include mechanical processing and thermal conversion, while only a small number of studies investigate complete waste management systems that unite recycling and energy recovery and circular economy systems. Existing research studies do not present a dependable framework for industrial use because their current focus on technical capabilities does not consider the complete economic system and the regulatory framework and supply network difficulties which come with large scale deployment. The recycling process for lithium-ion batteries and composite wind turbine blades needs to be studied because the current methods for recovering those materials and their environmental effects remain unoptimized. The use of digital technologies needs to receive more attention because organizations have started to adopt artificial intelligence and Internet of Things technologies for the purpose of monitoring waste and optimizing processes. The research study will examine the existing waste management and recycling methods used for energy materials. The research will assess these methods through their effectiveness and their limitations. The study will discover methods to connect circular economy concepts with future energy system development.

2. Types of Energy Waste

The level of trash generated increases with the world's growing dependence on new and renewable energy systems. Types of energy-related wastes differ, depending on the technology and the sector involved, with solar panels, batteries, electronics, and industrial systems contributing fairly large volumes into the entire waste stream. It is imperative to know the composition and effects of each waste type for designing sound management and recycling schemes.

2.1 Solar panel waste

Solar energy technologies-PV panels in particular-have been deployed rapidly worldwide. After their average lifespan of 25-30 years, however, these panels become a huge source of waste. Solar panel waste consists of glass, silicon wafers, polymers, aluminium frames, and trace amounts of toxic elements, such as cadmium and lead, mostly in thin-film technologies. Xu et al. (2018) maintain that while solar-panel recycling systems globally are still in their infancy, the amount of waste is set to rise tremendously, potentially reaching millions of tons over the coming decades. These thermal, mechanical, and chemical methods employed in recycling currently are meant to salvage materials but have not yet spread widely due to technical and economic barriers. To respond to this looming challenge, the research emphasizes the need for extended producer responsibility and uniform recycling procedures [9].

Materials Research Proceedings 66 (2026) 167-178 https://doi.org/10.21741/9781644904152-15

2.2 Battery and electronic waste

The boom of batteries, especially lithium-ion batteries for electric vehicles, energy storage systems, or those powering consumer electronics, has created a rapidly growing stream of complex electronic waste-hazardous e-waste. Lithium, nickel, cobalt, and other unstable electrolytes are some of the hazardous substances found in such batteries, but they might also contain some valuable items. Veit and Bernardes (2015) assert that owing to the diversity of materials and possibility of chemical or thermal hazards, battery and e-waste recycling require particular treatment procedures. Direct component recycling, hydrometallurgy, and pyrometallurgy are some of the common recycling methods. Moreover, certain unregulated recycling processes evidently impact the environment's safety and human health, which underlines the significance of having a regulated and safe recycling facility [10].

2.3 Industrial waste

Industrial processes create a lot of waste associated with energy when they manufacture or utilize energy-intensive products. Examples of industrial process wastes derived from heating or temperature conversion systems include ash, wastewater, slag, process materials, and others. Waste-to-energy (WTE) processes, including gasification and incineration, could potentially provide a solution for managing industrial waste streams as well as recovering usable energy, according to Tabasová et al. (2012). The associated negative environmental impacts, for example, air pollutants and leftover ash, require monitoring of the processes. The discussion proposes an advanced emission control technology potentially capable of evaluating energy recovery against ecologic concerns [11].

3. Recycling Techniques

The energy resource recycling is done through a combination of process mechanical and chemical and thermal methods. This is dependent on the nature and complexity of the waste. These processes are not only for reducing environmental impacts but also for recovering precious components, including metals, polymers, and rare earth elements, essential for modern energy technologies. The recycling process of lithium-ion batteries is depicted in Figure 2.

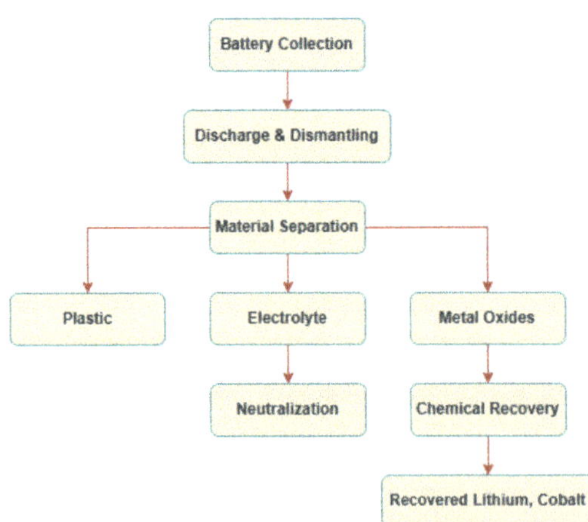

Figure 2. *Recycling Flow of Lithium-Ion Batteries [10]*

Researchers have implemented three recycling methods which include mechanical and chemical and thermal techniques to retrieve valuable materials from waste. The mechanical recycling method achieves low operational costs but faces challenges because materials lose quality and become contaminated. Chemical recycling enables higher material purity outcomes but demands development of advanced processing systems which consume considerable energy. The thermal methods of pyrolysis and gasification enable waste facilities to process multiple waste types although these methods produce emissions that need special monitoring systems. The different types of energy material waste require multiple recycling methods because no single method can successfully process all waste. The researchers should concentrate on developing hybrid recycling systems which use various techniques to achieve maximum material recycling while reducing their environmental effects.

3.1 Mechanical, chemical, and thermal recycling

Mechanical, chemical, or thermal recycling have been common techniques to manage solid waste, particularly plastic and composite waste, originating from energy devices. Whereas mechanical recycling, shredding, grinding, and melting the material to acquire raw inputs for new products, presents some limits due to material degradation and contamination issues, chemical recycling decomposes polymers and compounds into their chemical constituents so that they may be reused in manufacturing. However, thermal techniques such as pyrolysis, gasification, or incineration are best suited to handle complex or contaminated waste streams. These latter processes convert waste into energy and useful by-products, i.e., syngas, oil, or char. Kijo-Kleczkowska and Gnatowski (2022) highlight the increasing relevance of thermal recycling, especially in combating plastic waste originating from composite energy materials which include battery housings and PV encapsulants. They mention that these thermal methods have high conversion efficiency and treat substantial volumes of mixed or degraded waste, with emission allowances and energy input considered for sustainability [12].

3.2 Recovery of rare earth materials

Such neodymium, dysprosium, and terbium are rare earth elements (REEs), essential for making permanent magnets, batteries, and advanced electronics. These materials are hence limited, and their extraction poses an ecological as well as geopolitical challenge. Recycling offers an alternative and sustainable route to recovering REEs from electronic and industrial waste. Agrahari et al. (2024) discuss the developments of new devices and separation techniques to efficiently recover REEs from waste streams. Solvent extraction, ion exchange, membrane separation, and microfluidic separation are some of such processes which present extremely high selectivity and efficiency. Most importantly, they will prove useful in recycling of magnet scrap from wind turbine or electric vehicle motors, which has high though usually not easily recoverable rare earth content [13]. Development in REE recycling strengthens the supply chains of valuable strategic materials while complementing global initiatives to reduce environmental burden from primary mining. The efficiency of different energy materials in recycling is depicted in Figure 3.

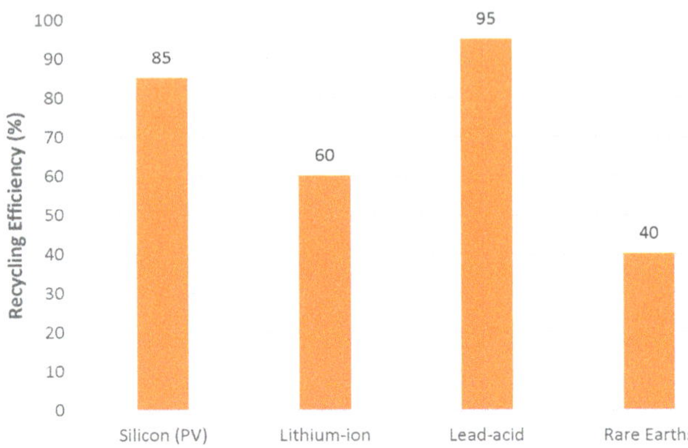

Figure 3. *Recycling Efficiency of Different Energy Materials [13]*

4. Waste-to-Energy (WTE) Technologies

Waste to Energy (WTE) technologies play a major role in integrated waste management systems as they convert waste materials into usable energy forms like electricity, heat, and fuel. These technologies aid in volume reduction of wastes to landfills and also help in the generation of energy from non-recyclable fractions. There are currently different treatment methods employed in waste-to-energy processing: thermal and from biological processes-proposing advantages and disadvantages depending on the type of waste treated and to the desired energy output. The primary Waste-to-Energy (WTE) methods and their conversion procedures—gasification, pyrolysis, incineration, and anaerobic digestion—are illustrated in the figure. These technologies convert waste materials with energy content into forms that final consumers can utilize to generate biogas, heat, electricity, and fuel.

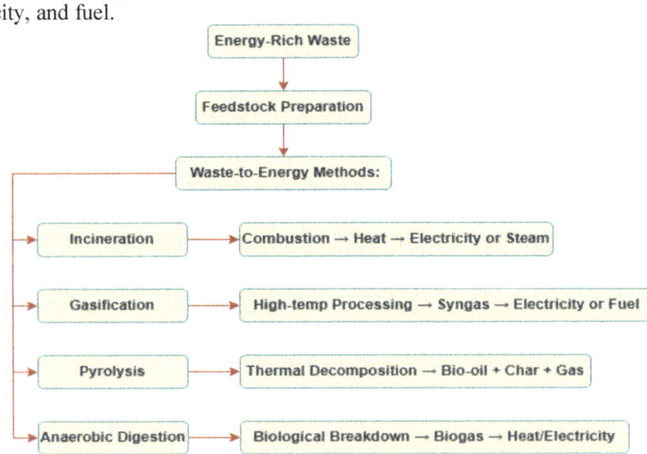

Figure 4. *Waste-to-Energy (WTE) Pathways*

Waste-to-energy technologies create a valuable solution which helps decrease landfill waste while generating energy from waste that cannot be recycled. The environmental and economic sustainability of these systems continues to be subject to ongoing debates. Incineration systems can achieve significant waste volume reduction but often face public opposition due to emission concerns. Gasification and pyrolysis technologies provide cleaner waste disposal methods which need greater financial resources and specialized knowledge for implementation. The operational efficiency of these systems operates based on two factors which include the trash materials and how well they get separated at the starting point. Successful WTE system implementation needs three essential components which include efficient recycling systems and active government monitoring and ongoing technological progress.

4.1 Incineration and gasification

Incineration is the most sophisticated WTE method; it generates heat by combusting refuse at high temperature. Power and steam are thereafter generated out of this heat. More particularly, this process reduces the quantity of trash in MSW uses. This process can release toxic emissions, including dioxins, and requires sophisticated emissions control systems to avoid environmental impacts. Conversely, incineration is oxygen-depleted, high-temperature processing that generates syngas, an integration of hydrogen and carbon monoxide, from organic or carbonaceous refuse materials. Syngas can be utilized as fuel for energy production or as a chemical synthesis feedstock. Gasification is much cleaner and efficient compared to incineration because it yields fewer pollutants and has greater flexibility in controlling different waste streams. The writers Beyene, Werkneh, and Ambaye (2018) assert that advances in these technologies over recent years have also greatly enhanced their effectiveness, environmental management, and application in different waste management systems. These options are highly applicable considering the reduced landfill availability and the high demand for energy in cities [14].

4.2 Pyrolysis and anaerobic digestion

Pyrolysis refers to the thermal decomposition of organic substances in the absence of air and produces predominantly in the form of bio- oil, syngas, and charcoal. Plastics, biomass, and certain types of industry and electronic waste can all be processed using this method. Pyrolysis provides a lot of flexibility for energy recovery uses and can be designed for a variety of outputs depending on feedstock and temperature. On the contrary, anaerobic digestion (AD) is a natural process by which microorganisms digest organic waste in the absence of oxygen, generating nutrient-rich digestate that can be used to fertilize soil and biogas, mostly carbon dioxide and methane. As Caiardi et al. (2022) reports, hybrid systems incorporating pyrolysis and anaerobic digestion have been viewed as potential means to maximize energy recovery and minimize waste. Hence, hybrid systems offer more efficiency and better sustainability for the overall system through making the most of both methods for different types of wastes—thermal conversion to dry or recyclable fractions and biological conversion for wet, recyclable waste [15]. The emissions and electric power output of various WTE technologies are compared in the figure.

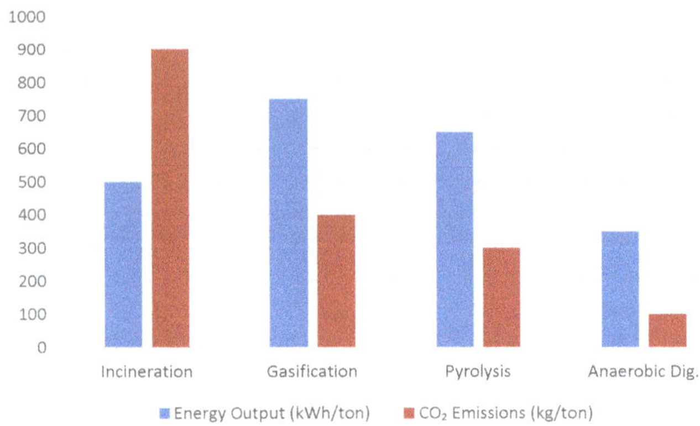

Figure 5. *Comparison of WTE Technologies – Energy Output vs. Emissions (Source: Waste Management Journal, 2022)*

5. Circular Economy Approaches

At the heart of sustainable development, particularly in matters of energy materials, are strategies that advocate a circular economy (CE). Critical issues such as resource scarcity and the environmental impact concerning the energy materials are all addressed by these strategies: they keep materials in use as much as possible by reusing, refurbishing, and undergoing closed-loop recycling processes, while lowering waste and minimizing dependence on virgin resources. Such principles within energy can be utilized to make increased developments towards material efficiency, lower environmental impacts, and sustain energy over long time periods.

5.1 Reuse and refurbishment of energy materials

One of the most promising tools for prolonging the life cycle of energy products, including batteries, photovoltaic modules, and electronic components, is reuse and refurbishment. These methods, in their essence, are subjecting energy devices or their components to scrutiny, repair, and finally reuse before discarding them altogether. For instance, electric vehicle batteries that cannot meet high-performance standards can be refashioned for usage in stationary energy storage systems. Walker et al. (2018) claimed that reuse and refurbishment were some of the highly needed strategies for material efficiency within the circular economy. The study emphasized that in comparison to traditional recycling or disposal practices, reuse and refurbishment had considerable impacts on environmental benefits since their reuse mode significantly cuts down on the new materials extraction and manufacturing processes. Refurbishment tends to have environmental benefits at any time when the energy and emissions involved in producing another new component are high, especially for lithium-ion batteries and other energy-intensive products [16].

5.2 Industrial symbiosis and closed-loop systems

According to Kara et al., 2022, linear production systems must evolve towards closed-loop system models to fulfil the transition process of circular economy connotation. Closed loop as the terminology itself should offer features to a system or user within the design of its products and processes such that they can have these parts fed back into the production systems through one of the methods: direct reuse, high-quality recycling, and energy recovery with minimal losses. In energy systems, this could mean surplus heat sharing, repurposing by-products from processes, or interlinking waste materials streams from power generation, manufacturing, or waste management sectors (describing industrial symbiosis). Their report highlights the manner in which lifecycle thinking, information and communication technologies, and system-level integration can foster efficient circular cycles and optimize resource utilization within industrial networks [17]. Industrial symbiosis, together with reuse & refurbishment initiatives, will enhance circular economy strategies and provide a solid footing for sustainable management of energy from materials that diminishes the stress on the environment and encourages innovation in the energy and waste industries.

6. Applications and Case Studies

Waste management is a reality used as the basis for illustrating ideas on circular economies and sustainability. Cases from across the world illustrate how carefully planned recycling programs and well-running waste management systems can lead to resource efficiency, environmental preservation, and increased renewable energy. Such types of cases demythologize this field and act as a guide for future technical and regulatory progress.

6.1 Successful recycling programs

Experimental recycling programs are being piloted in the management of municipal solid waste (MSW). Effective programs include prompt source separation, infrastructure, enabling policies, and public engagement. Different programs in several European communities demonstrated that community participation and integrated systems could result in high recycling rates and low levels of landfill dependence. According to Xevgenos et al. (2015), detailed case-by-case analysis examines successful municipal reuse and offers examples from San Francisco, Sweden, Malmö, and Ljubljana, Slovenia, where recycling levels over 60% are now achieved by means of public awareness campaigns, source separation, composting, and policies with incentives. The case studies indicate the necessity of continuous monitoring and stakeholder engagement in ensuring system performance in terms of long-term sustainability [18].

6.2 Waste management in renewable energy

Of course, however, combined waste management appears to be the key to the longevity of renewable energy systems. This particular connection between garbage processing and renewal is largely due to waste-to-energy plants and biogas projects utilizing organic waste. Through presenting an alternative method of closing resource loops, these technologies are decreasing the demand for fossil fuels. Through the process of converting organic and combustible waste fractions into energy using anaerobic decomposition, incineration, and gasification, Eriksson et al. (2014) demonstrate how integrated GWM enhances the use of renewable energy. They proved that this process would feed renewable energy to adjacent grids or boilers as well as decrease environmental footprints. They asserted that energy efficiency would greatly rise with waste management and would have a beneficial impact on climate mitigation if it was tied to energy policy [19].

7. Challenges and Future Directions

If we are to progress in a full-scale shift towards a circular economy and more sustainable waste-to-energy future, we must surmount the different technological, coordination, financial, and policy barriers that still hinder the large-scale adoption and integration of sustainability.

7.1 Technological and logistical barriers

The absence of advanced recycling facilities and the technology bottleneck of handling heterogeneous waste streams pose significant barriers to circular economy adoption. For example, innovative separation and treatment technologies must be developed in order to recycle composite materials such as solar panels or electric car battery. Moreover, logistical difficulties generated by a lack of waste collection system, inefficient sorting capacities, and regional coordination deficiencies hinder the effective management of municipal or national garbage. Fedotkina,

Gorbashko, and Vatolkina (2019) enumerate the particular hindrances found in the Russian experience, which include the lack of financial provision for recycling technology, a deficit of qualified staff able to work recyclable technologies, and an insufficient amount of operating recycling infrastructure as substantial barriers. They also note that taking on a circular economy approach is severely undermined by the fragmentation of the waste management system and lack of collaboration among stakeholders. These are not special problems for Russia; instead, they are symptomatic of more ingrained systemic problems that afflict much of the world [20].

7.2 Policy and financial incentives

Intensive regulatory measures and financial incentives are badly needed to drive and reward sustainable waste management. Besides acknowledging the abstract ideals, so too is this the case when adapting them into a regional action strategy. A good number of regions have disjointed or weakly implemented legislation, and initial costs persistently hinder attempts to realize control in the often-ambiguous feasibility of recycle and waste-to-energy projects. Private investment in circular technology will be less likely without some subsidies, tax incentives, or market-based instruments. Some of the integral policy tools supporting systemic change include producer responsibility, landfilling charges, recycling targets, and public procurement directives. Finnveden et al. (2013) examined whether context and procedure in the system would change due to this kind of policy tool. The result proves that employing a mix of information, financial, and legal instruments is the most effective way to encourage innovations and accelerate the implementation of environmentally friendly waste management strategies. In addition, waste management activities could be integrated into more comprehensive environmental and energy projects to further combine economic and ecological objectives [21].

8. Conclusion

The shift to renewable energy technologies has created more waste materials which include solar panels and batteries and industrial parts that require disposal. The upcoming waste stream demands effective management because it impacts the sustainable development of upcoming energy systems. The study examined existing recycling technologies and waste-to-energy systems and circular economy methods which aim to enhance resource recovery while reducing environmental damage. The analysis shows that mechanical recycling and chemical recycling and thermal recycling methods have achieved substantial development yet they still encounter difficulties that impact their ability to expand and their cost efficiency and their operational performance. Waste-to-energy technologies create valuable energy from non-recyclable waste materials but their environmental impact relies on the efficiency of their waste management systems and their emission control technologies. Energy systems can achieve better sustainability through circular economy practices which include reuse and refurbishment and industrial symbiosis and closed-loop material flows. Large-scale implementation of this system needs active collaboration between government bodies and business organizations and academic research centers. Digital technology developments in artificial intelligence and IoT-based monitoring systems will enhance waste tracking and sorting and recycling optimization in future scenarios. The combination of advanced recycling technologies with circular economy principles and supportive policy frameworks will create sustainable waste management systems that enable the world to transition to clean energy sources.

References

[1] IEA-PVPS, I. R. E. N. A. (2016). End-of-life management: Solar photovoltaic panels. In International renewable energy agency and international energy agency photovoltaic power systems.

[2] Giugliano, M., Cernuschi, S., Grosso, M., & Rigamonti, L. (2011). Material and energy recovery in integrated waste management systems. An evaluation based on life cycle assessment. Waste management, 31(9-10), 2092-2101. https://doi.org/10.1016/j.wasman.2011.02.029

[3] Consonni, S., & Viganò, F. (2011). Material and energy recovery in integrated waste management systems: The potential for energy recovery. Waste Management, 31(9-10), 2074-2084. https://doi.org/10.1016/j.wasman.2011.05.013

[4] Brunner, P. H., & Rechberger, H. (2015). Waste to energy-key element for sustainable waste management. Waste management, 37, 3-12. https://doi.org/10.1016/j.wasman.2014.02.003

[5] Sadef, Y., Nizami, A. S., Batool, S. A., Chaudary, M. N., Ouda, O. K. M., Asam, Z. U. Z., ... & Demirbas, A. (2016). Waste-to-energy and recycling value for developing integrated solid waste management plan in Lahore. Energy Sources, Part B: Economics, Planning, and Policy, 11(7), 569-579. https://doi.org/10.1080/15567249.2015.1052595

[6] Kaya, K., Ak, E., Yaslan, Y., & Oktug, S. F. (2021). Waste-to-Energy Framework: An intelligent energy recycling management. Sustainable Computing: Informatics and Systems, 30, 100548. https://doi.org/10.1016/j.suscom.2021.100548

[7] Eriksson, O. (2017). Energy and waste management. Energies, 10(7), 1072. https://doi.org/10.3390/en10071072

[8] Bringezu, S. (2014). Carbon recycling for renewable materials and energy supply: recent trends, long-term options, and challenges for research and development. Journal of Industrial Ecology, 18(3), 327-340. https://doi.org/10.1111/jiec.12099

[9] Xu, Y., Li, J., Tan, Q., Peters, A. L., & Yang, C. (2018). Global status of recycling waste solar panels: A review. Waste management, 75, 450-458. https://doi.org/10.1016/j.wasman.2018.01.036

[10] Veit, H. M., & Bernardes, A. M. (2015). Electronic waste. Recycling Techniques, 165, 3-12. https://doi.org/10.1007/978-3-319-15714-6_2

[11] Tabasová, A., Kropáč, J., Kermes, V., Nemet, A., & Stehlík, P. (2012). Waste-to-energy technologies: Impact on environment. Energy, 44(1), 146-155. https://doi.org/10.1016/j.energy.2012.01.014

[12] Kijo-Kleczkowska, A., & Gnatowski, A. (2022). Recycling of plastic waste, with particular emphasis on thermal methods. Energies, 15(6), 2114. https://doi.org/10.3390/en15062114

[13] Agrahari, G. K., Vignesh, M. S., & Nigam, K. D. P. (2024). Novel devices for the extraction and recovery of rare-earth metals through recycling of waste. Journal of Material Cycles and Waste Management, 26(1), 109-137. https://doi.org/10.1007/s10163-023-01862-x

[14] Beyene, H. D., Werkneh, A. A., & Ambaye, T. G. (2018). Current updates on waste to energy (WTE) technologies: a review. Renewable Energy Focus, 24, 1-11. https://doi.org/10.1016/j.ref.2017.11.001

[15] Caiardi, F., Belaud, J. P., Vialle, C., Monlau, F., Tayibi, S., Barakat, A., ... & Sablayrolles, C. (2022). Waste-to-energy innovative system: Assessment of integrating anaerobic digestion and pyrolysis technologies. Sustainable Production and Consumption, 31, 657-669. https://doi.org/10.1016/j.spc.2022.03.021

[16] Walker, S., Coleman, N., Hodgson, P., Collins, N., & Brimacombe, L. (2018). Evaluating the environmental dimension of material efficiency strategies relating to the circular economy. Sustainability, 10(3), 666. https://doi.org/10.3390/su10030666

[17] Kara, S., Hauschild, M., Sutherland, J., & McAloone, T. (2022). Closed-loop systems to circular economy: A pathway to environmental sustainability?. CIRP Annals, 71(2), 505-528. https://doi.org/10.1016/j.cirp.2022.05.008

[18] Xevgenos, D., Papadaskalopoulou, C., Panaretou, V., Moustakas, K., & Malamis, D. (2015). Success stories for recycling of MSW at municipal level: a review. Waste and biomass valorization, 6, 657-684. https://doi.org/10.1007/s12649-015-9389-9

[19] Eriksson, O., Bisaillon, M., Haraldsson, M., & Sundberg, J. (2014). Integrated waste management as a mean to promote renewable energy. Renewable Energy, 61, 38-42. https://doi.org/10.1016/j.renene.2012.04.024

[20] Fedotkina, O., Gorbashko, E., & Vatolkina, N. (2019). Circular economy in Russia: Drivers and barriers for waste management development. Sustainability, 11(20), 5837. https://doi.org/10.3390/su11205837

[21] Finnveden, G., Ekvall, T., Arushanyan, Y., Bisaillon, M., Henriksson, G., Östling, U. G., ... & Guath, M. (2013). Policy instruments towards a sustainable waste management. Sustainability, 5(3), 841-881. https://doi.org/10.3390/su5030841

Advanced Materials and Sustainable Energy Technologies – AMSET2025 Materials Research Forum LLC
Materials Research Proceedings 66 (2026) 179-189 https://doi.org/10.21741/9781644904152-16

Sustainable Infrastructure: Green Buildings and Low-Carbon Transportation for A Greener Future

M. PRAMODH KUMAR[1*], A. S. RANA[1], R. SINGH[2], K. SARASWATI[3], R. RAJPUT[4], D. SRINIVASARAO[5]

[1]Department of Aerospace Engineering, Uttaranchal Institute of Technology, Uttaranchal University, Dehradun-248007, Uttarakhand, India

[2]Uttaranchal Institute of Technology, Uttaranchal University, Dehradun-248007, Uttarakhand, India

[3]Department of Sciences, Chandigarh School of Business, Chandigarh Group of Colleges Jhanjeri, Mohali-140307, Punjab, India

[4]Medical Radiology & Imaging Technology, Department of Paramedical Sciences, Faculty of Health & Allied Sciences, Usha Martin University, Angara, Ranchi-835103, Jharkhand, India

[5]Department of Medical Laboratory Technology (UIAHS), Chandigarh University, India

*mpramod2323@gmail.com

Keywords: Sustainable Infrastructure, Green Building Design, Low-Carbon Transportation, Energy Efficiency, Urban Sustainability

Abstract: The building and transportation sectors together account for a substantial share of global greenhouse gas emissions which makes these sectors essential targets for urban decarbonization efforts. The paper explores sustainable infrastructure by studying green building technologies and low-carbon transportation systems through an integrated analysis method. The research assesses key technologies which include energy-efficient HVAC systems and passive design strategies and electric mobility and hydrogen transport and sustainable public transit by comparing their energy savings and emissions reduction potential and economic viability. The study demonstrates that coordinated planning methods which include renewable-powered electric mobility and transit-oriented green development are essential for successful planning efforts. The findings show that cities need to use both building and transport solutions together with smart energy systems and policy incentives to achieve sustainability and adapt to climate change over the long term.

1. Introduction

The first decades of urbanization and climate change gave rise to an acknowledgment that the urban environment could not do without measures for sustainable development henceforth. One of the most important ways to achieve low-carbon cities is through the promotion of green buildings and low-carbon transportation systems. The buildings and transportation sectors are among the top contributors of global greenhouse gas emissions and they also account for a large part of the consumption of energy and degradation of the environment [1].

A green building or eco-friendly building is one that is designed to use renewable resources and has very little impact on the environment throughout its lifetime — from energy-efficient measures like resource conservation to improved indoor environmental quality. There exist various green building rating systems such as LEED, BREEAM, and China's Three Star, which are implemented in various regions around the globe to guide construction and retrofitting towards the low carbon path. However, retrofitting of already existing public buildings is complicated and presents unbearable challenges on account of variations in climate, economy, and technical conditions. The policy stimuli, emerging technologies, and appraisal instruments will accelerate the change to low-carbon buildings and districts [2-3]. Certain barriers, however, including initial high costs, regulatory loopholes, and stakeholder awareness limitations have slowed the extensive use of green construction, particularly in developing countries. This makes it necessary to employ a multi-faceted strategy consisting of [4] financial incentives, institutional reform, and capacity-building. The development of low-carbon transport that would be along the same vein as building design and construction industries is not only a critical component but the primary characteristic of sustainable and eco-friendly urban planning. Emissions will be minimized and urban livability enhanced through the use of energy-efficient means like electric vehicles (EV), public transport, bicycles, and walking. Moving towards the greener modes of transport is not just a means of cutting carbon footprint but also assists in addressing urban issues such as congestion, air pollution, and public health risks [5]. The utmost substitute is to drive an electric vehicle (EV). It doesn't produce gas and is safe. Moreover, there isn't any loud noise. Additionally, the Indian government has taken some action to reduce pollution [6]. The integration of low-carbon transport and renewable buildings offers the foundation for sustainable urban design. Their interconnection with harmonious policies and new technologies can be a significant contribution to global climate goals and the establishment of low-carbon cities [7].

It discourses edge computing driven by AI and sophisticated communication protocols to exceeded parameters. The contribution of embedded systems on energy efficiency and dependability is approximated through case studies. Efficiency is expected to increase by 15–25% over the next 10 years due to future developments including IoT-based smart ecosystems, machine learning for analytical management, and hardware evolution [8].

The worldwide urgency for climate action has resulted in the use of green buildings and sustainable transport as major strategies to reduce carbon emissions in urban areas. Green buildings whether high performance, sustainable or eco-friendly development are not mere terms; they point to a new method of designing and building that concentrates on resource-efficient, environmentally friendly, and health-promoting spaces. The very same intent would be with whatever term is used: to reduce the environmental impact while at the same time improving the well-being of the occupants and the operational performance of these buildings. The same is true for sustainable transport that hinges on the mobility solutions that both cut down GHG emissions and guarantee just access to mobility [9]. Consequently, green building and low-carbon transport are considered critical pillars of sustainable urban development as both indirectly and directly contribute to various areas such as energy security, health, environmental protection, and adaptation to climate change. These are the first steps towards the sustainability agenda for cities that are major contributors to global emissions and thus might play an important role in achieving the net-zero target.

Energy demand and carbon emissions from the built environment and urban mobility systems are the main areas to intervene in the environmental conservation. At least the energy and carbon-intensive buildings to design, and construction standards, and operations so that climate change and air pollution do not cause problems. Among the technologies developed, high-performance insulation, active design, and renewable energy integration have proven to have no emissions at the same time heat comfort is achieved. Green buildings and eco-friendly transportation financially contribute through long-term savings, increased asset values, and the creation of green jobs. The cost of operating energy-efficient buildings is lower, while the cities with very good low-carbon transport systems will have healthier and thus less costly population and also able to reap productivity benefits due to better air quality and accessibility. Furthermore, investments in sustainable infrastructure trigger innovations that open pathways to new business models around circularity and carbon neutrality [10].

The research establishes an integrated infrastructure approach which evaluates energy consumption, greenhouse gas emissions, and economic results of both green building practices and low-carbon transportation methods. The research study combines three components which include performance-based comparison tables and emission reduction pathways across multiple sectors and case-based evidence that demonstrates how building energy efficiency impacts transportation decarbonization results. The integrated framework enables urban planners and policymakers to develop coordinated low-carbon strategies which extend beyond individual sector solutions.

2. Green Building Technologies

In the building industry, the development and integration of advanced technologies have attained the status of primary concerns for the attainment of energy efficiency, operational sustainability, and lower carbon emissions. Green buildings embrace a wide range of strategies and systems contributing to the environmental performance of buildings in a life cycle context. The most influential include energy-efficient HVAC systems, passive solar design and insulation, and building management systems. Energy-efficient systems and smart technologies are thus integrated with green buildings, as demonstrated in Figure 1.

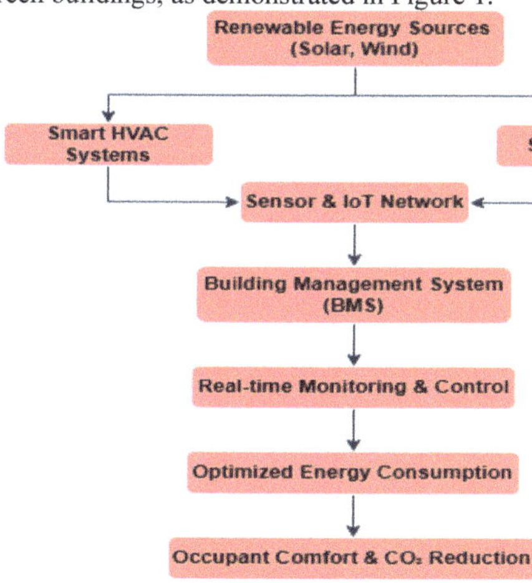

Figure 1. *Smart Green Building System*

2.1 Energy-efficient HVAC systems

Within buildings, heating, ventilation, and air conditioning (HVAC) systems emerge as one of the world's largest consumers of energy, whereby efficiency becomes a prime concern for green construction. Improvement in HVAC performance reduces energy consumption, therefore saving costs and further decreasing environmental deterioration. Among other various best practices to enhance HVAC efficiency discussed by Simpeh et al. (2022) are variable refrigerant flow (VRF) systems, energy recovery ventilators, smart thermostats, and maintenance protocols. Their further promotion along with the joining up of renewable energy systems and installations with automated controls ensures a fast shift to low-carbon building operation [11]. Table 1 summarizes the cost-benefit and performance of key green building innovations.

Table 1. *Performance Metrics of Green Building Technologies*

Technology	Energy Savings (%)	Cost Increase (%)	Payback Period (Years)	CO_2 Reduction Potential
Smart HVAC	25–40	10–15	3–5	High
Passive Solar	20–30	5–10	2–4	Medium
Smart Lighting	30–50	8–12	1–3	High

Source: Compiled by authors based on Simpeh et al. (2022), Chen et al. (2015), and USGBC reports. This table highlights that smart HVAC and lighting systems offer the highest energy savings with relatively short payback periods, making them particularly suitable for urban retrofitting applications.

2.2 Passive solar design and insulation

Passive solar design and thermal insulation represent the two main pillars of sustainable architecture, promoting the use of natural light and heat to minimize reliance on mechanical systems. Some of the most useful techniques are building orientation, window placement, thermal mass utilization, and shading devices. High-performance insulation materials and tight construction reduce heating and cooling loads. Passive design imparts energy savings while providing improved levels of indoor comfort and resilience against climate variations. Also, Chen et al. (2015) shows many international green building rating tools nowadays seriously promote passive strategies in their assessment systems [12].

2.3 Smart building management systems

Through the incorporation of digital technologies in Smart Building Management Systems (SBMS), building performance is monitored and optimized in real-time. SBMS integrate sensors, data analytics, and automation, using them to control lighting, HVAC, water use, and occupancy-based systems. All these contribute to increasing efficiency and user experience. Emerging trends, according to Patil, Boraste, and Minde (2022), with respect to smart green buildings, are the use of Internet of Things (IoT) devices, AI-enabled energy optimization, and predictive maintenance tools. The systems thus enhance sustainability by providing an adaptive response to energy demand, occupant behaviour, and environmental conditions [13]. The contrast in CO_2 emissions between the traditional and various certified green buildings is revealed in Figure 2. The reduction of CO2 emissions from buildings by LEED (Leadership in Energy and Environmental Design) standards is highly pronounced.

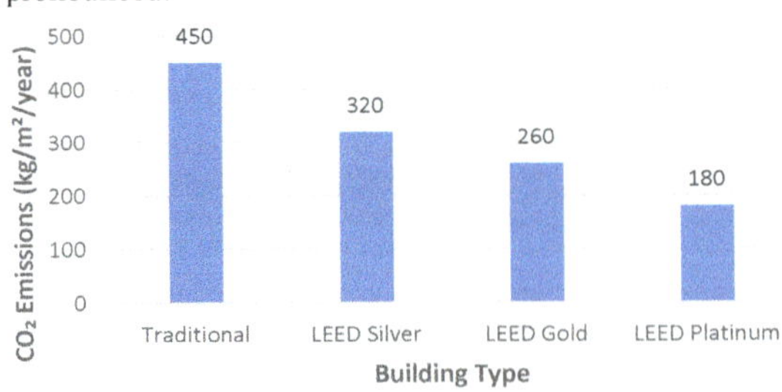

Figure 2. *Carbon Emission Comparison – Traditional vs Green Buildings (Source: US Green Building Council (USGBC), EPA Energy Star Reports)*

3. Low-Carbon Transportation

The changeover to low-carbon mobility solutions happens to be under focus as a top priority for sustainable development and climate mitigation, as the transport sector remains a major culprit of greenhouse gas emissions throughout the world. Key transitioning strategies include creating hydrogen-powered transport systems and implementing sustainable public transit networks, which go hand in hand with the adoption of electric and hybrid vehicles. These technologies and systems would therefore be fundamental to lowering fossil fuel reliance and urban pollution and supporting carbon neutrality goals. Figure 3 shows the links between EVs; hydrogen; public transport; and smart mobility platforms.

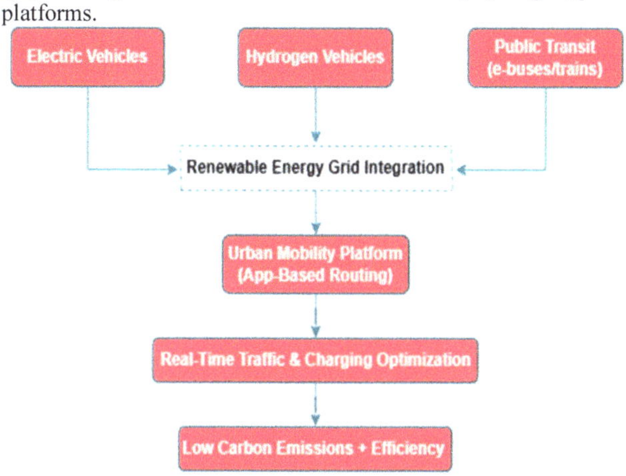

Figure 3. *Integrated Low-Carbon Transportation Ecosystem*

The sustainable urban infrastructure effectiveness requires both sectoral advancements and successful implementation of green buildings and low-emission transportation systems. The installation of rooftop solar photovoltaic systems in buildings enables direct support for electric vehicle charging stations while decreasing dependence on fossil fuel power grids. Transit-oriented development creates additional benefits through its combination of energy-efficient buildings and public transportation systems which decrease the need for private car usage.

3.1 Electric vehicles (EVs) and hybrid vehicles

Electric vehicles (EVs) and hybrid types represent a major shift into one form of adoption of electric propulsion in transportation, either fully or partially replacing internal combustion engines. Having them in the area with zero tailpipe emissions can minimize greenhouse gas emissions over the vehicle's entire life cycle, especially when source-generated electricity is taken into consideration. This is emphasized in the paper by Gibbins et al. (2007), explaining that electric vehicles help in the reduction of carbon emissions while at the same time being efficient and environmentally friendly by reducing noise pollution in some cities. Hybrid vehicles which use electric motor propulsion together with an internal combustion engine could be viewed as a transitional powertrain system that is less polluting and less consuming of fuel when compared with conventional vehicles [14].

3.2 Hydrogen-powered transportation

Hydrogen-fuelled vehicles provide another approach that shows promise for low-carbon transport, especially for heavy-duty and long-distance applications. Here, hydrogen fuel cells convert hydrogen to electricity, with water vapor being the only emission. Sandoval et al. (2009) review the long-term outlooks for hydrogen transportation and state that hydrogen could decarbonize sectors for which battery-electric solutions are less fit. Limited infrastructure and hydrogen production capabilities have hindered acceptance today; however, technology breakthroughs and a favourable policy environment could position hydrogen well for substantial insertion into future low-carbon mobility systems [15]. Table 2 outlines the comparison of EVs, hydrogen, and hybrid vehicles regarding their efficiencies and infrastructural needs.

Table 2. *Comparison of Low-Carbon Transportation Technologies*

Technology	Fuel Type	Range (km)	Charging/Fuelling Time	CO_2 Emissions (g/km)	Infrastructure Needs
EV	Electricity	300–500	30 mins – 8 hours	0 (tailpipe)	Moderate
Hydrogen	Hydrogen	500–600	3–5 minutes	0 (tailpipe)	High
Hybrid	Gas + Electric	800–1000	N/A (refuelling)	70–100	Low

3.3 Sustainable public transport systems

Central to cutting carbon emissions, reducing traffic congestion, and consuming energy is public transportation systems. Sustainable public transport systems comprise electric buses, rail systems, and integrated transit networks that emphasize user access, efficiency, and performance within green benchmarks. Those cities that have evaluated new public transport models-such as bus rapid transit (BRT), light rail, and multimodal integration-have shown significant potential for supporting low-carbon transitions, as highlighted by Wimbadi et al. (2021). Such systems have potential not only to reduce per capita emissions but also to improve social equity and urban resilience [16].

Smart grids along with energy storage systems create a two-way energy exchange between buildings and electric mobility systems which enhances system performance and reliability. The integrated planning methods enhance emission reduction results while optimizing land distribution and creating better living conditions in cities which proves that cities need building and transportation systems to work together for their carbon neutrality goals.

Figure illustrates emissions per passenger-km for various transportation modes.

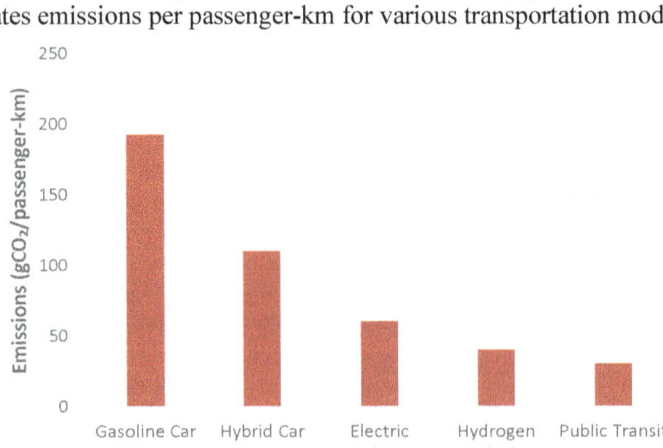

Figure 4. *Transportation CO_2 Emissions by Vehicle Type (Source: International Energy Agency (IEA), IPCC Transport Emissions Report 2022)*

4. Materials and Construction Techniques

The selection of materials and methods of construction forms the backbone with which the carbon footprint of a building has reduced and contributes towards achieving the sustainable development goals. Low-carbon innovation, particularly in terms of materials, and internationally recognized accrediting systems steer the construction industry further down the path of greener and resource-efficient practices. The comparative assessment of various principal green building certification frameworks against some core criteria is shown in Table 3.

Table 3. *Building Certification Standards Comparison*

Certification	Energy Efficiency Focus	Materials & Resources	Water Efficiency	Global Adoption
LEED	High	High	High	Widely used
BREEAM	Moderate to High	Very High	Moderate	Europe-centric
WELL	Moderate	Focus on Health	Low	Emerging

4.1 Low-carbon concrete and recycled materials

Globally, concrete remains the most widely used material in construction, but it is also a significant contributor to carbon dioxide emanations through energy exertion in the process of cement production. In order to somehow lessen its infringement on the environment, researchers and industrialists are now looking into low-carbon alternatives-a common practice which is, for instance, the incorporation of recycled aggregates into concrete mixtures. Bostanci, Limbachiya, and Kew (2018) maintain that the use of recycled aggregates reduces the demand for virgin materials and lowers overall embodied carbon as well as construction costs. This approach contributes to the principles of a circular economy, promotes sustainability in construction, and has shown considerable potential in terms of performance and structural integrity [17].

4.2 Sustainable building certifications (LEED, BREEAM)

Sustainable building certification systems, such as LEED (Leadership in Energy and Environmental Design) and BREEAM (Building Research Establishment Environmental Assessment Method), have become essential strategies for promoting and benchmarking environmentally friendly construction. These framework decision processes evaluate buildings against criteria related to energy efficiency, material use, indoor environmental quality, and overall sustainability. According to Maqbool, Thompson, and Ashfaq (2023), LEED and BREEAM are key in reducing environmental harms and assisting the development of cost-effective urban housing solutions which incorporate energy efficiency. By defining standards for good practice and granting recognition in the marketplace, the approach encourages the take-up of best practices and continuous improvement in building performance [18].

5. Energy Efficiency and Performance

The increased energy efficiency is one of the core strategies through which emission reductions and sustainability are sought in the building and transportation sectors. Advanced modelling tools, simulation techniques, and standards of regulation are not only complementary, but also provide the basis for energy performance improvements in infrastructure systems. This section addresses two key components: building energy modelling and simulation and transportation fuel efficiency standards.

5.1 Building energy modelling and simulation

Building energy modelling (BEM) has become a useful tool in designing, evaluating, and running energy-efficient buildings. It provides stakeholders with a chance to simulate and analyze several energy-related factors: heating, cooling, lighting, and ventilation-before construction has started or in the course of retrofitting. But accuracy and reliability can only be guaranteed if calibration of the model is done. This is supported by Mustafaraj et al. (2014), who state, "It is essential to properly calibrate simulation models to real-world building performance data." Their findings demonstrate that calibration becomes an important aspect in enhancing the accuracy of the forecasting of energy consumption so that well-informed arguments could then be made for energy conservation measures and cost-effective retrofits [19].

5.2 Transportation fuel efficiency standards

Fuel efficiency is perhaps the most direct action in the transport industry to offset energy use and emissions. Fuel efficiency regulations mandate a minimum level of performance for cars, prompting manufacturers to create and sell technologies that consume less energy. However, as Litman observes, concentrating on just vehicle efficiency may be jeopardizing more substantial possibilities for system-wide energy efficiency. It is a stronger strategy that involves increased efficiency in the mechanisms of the transport system-often through land-use planning, through mode shifting to public transport, and through demand management techniques. Litman highlights that vehicle improvements together with other changes at the system level can therefore increase energy savings and sustainability gains greater than those achievable with either option individually [20].

6. Case Studies and Applications

The examples of green building and low-carbon transport strategy implementation in real world provide useful lessons around their effectiveness, limitations, and long-term benefits. Case studies demonstrate the applicability of theoretical approaches in practical contexts to realize actual environmental and economic benefits. Here, we studied two important examples-net-zero energy buildings and low-carbon transport initiatives.

6.1 Net-zero energy buildings

NZEB or Net Zero Energy Building is a building that is designed such that it produces at least as much energy as it consumes in any given year and thus are considered the paradigm of sustainability in the built environment. Such buildings would employ energy-efficient designs and renewable energy generation in order to balance their energy accounts through time. An example study in Italy, developed by Cellura et al. (2014), applied an Energy Life Cycle Assessment to an NZEB with the aim to assess operational and embodied energy. The study concludes that operational energy consumption can be minimized through design and technology with some degree of success but that embodied energy—energy spent producing construction materials and during construction processes-also plays a relevant role in the overall energy balance of the building. In a broader perspective, this also implies the need to consider both operational and embodied impact effect for the sustainability assessment of buildings [21].

6.2 Low-carbon transport initiatives

Low-carbon transport programs have increasingly grown imperative in addressing the environmental impact of the mobility systems in transport that cities employ. In India, where urbanization as well as motorization occurs, numerous policy and infrastructure measures addressed towards low carbon transport. As put forth by Shukla and Pathak (2016), measures towards increasing public transport systems, promoting non-motorized transport (e.g., walking and cycling), and promoting cleaner vehicle technology are at the core of India's green financing initiative. Furthermore, the study also emphasized the need for integrated planning and institutional backing with co-benefits comprising lower air pollution, energy security, and enhanced public health in facilitating this shift [22].

7. Challenges and Future Prospects

Even with significant advancements in green buildings and low-carbon transport, several challenges remain to be resolved for widespread adoption and optimal implementation. These challenges can be categorized into technological, financial, and policy-related barriers. They need to be understood and overcome for the scaling up of sustainable solutions in achieving long-term climate objectives.

7.1 Technological limitations

The penetration of low-carbon technologies in building design and construction still shows significant gaps. As per Xi and Cao (2022), the major technology barriers included the complicated integration of renewable energy systems, the limitations on the actual building materials, and the insufficient data for proper performance modelling. Most of the time, building professionals tend to come across areas where energy performance is cost, aesthetic, or functional. Besides, the speed of technology advancement often puts it ahead of design teams and contractors in terms of exposure and utilization of new tools. Xi and Cao conclude that "the above barriers for low-carbon design could be broken through the interdisciplinary collaboration, more advanced simulation techniques and better training of architects and engineers [23].

7.2 Policy and financial incentives

Financial incentives and policy frameworks have essential roles in enabling the stakeholders to be invested in low-carbon technologies and practices. However, hurdles still exist due to lack of long-term vision, inconsistent regulations, and inadequate funding mechanisms. These issues have been covered by Davies and Osmani (2011) from the point of view of the architects involved in low-carbon housing refurbishments, demonstrating that incentives such as tax rebates, subsidies, and green certifications create motivation, while other financial and regulatory ambiguities are still considered by a large number of professionals as serious barriers. More often, the client is not well-informed or motivated to do energy-efficient upgrades without visible financial gains or policy compulsion. Addressing these issues requires much stronger manoeuvres of government support through well-drawn comprehensive policies, improved access to finance and public awareness. Precise and consistent standards as well as performance-based incentives can speed up the implementation of low-carbon practices on both new constructions and retrofits [24].

8. Conclusion

The research shows that sustainable urban infrastructure depends on green buildings and low-carbon transportation systems as its main components. The combined use of energy-saving building technologies and renewable energy systems and advanced mobility solutions results in both significant carbon emission reductions and better economic and social results. The research shows that separate improvements in different sectors do not provide adequate solutions because effective sustainability benefits need planning that connects building energy systems with low-carbon transportation and public transit systems. The introduction of policy support and financial incentives and public education initiatives creates the necessary conditions that enable people to overcome both initial adoption costs and all related barriers. The future research direction should investigate three areas which include quantitative modeling of integrated infrastructure systems and monitoring their actual performance and developing region-specific policy frameworks to create scalable solutions for resilient and low-carbon urban growth.

References

[1] Chen, L., Huang, L., Hua, J., Chen, Z., Wei, L., Osman, A. I., ... & Yap, P. S. (2023). Green construction for low-carbon cities: a review. Environmental chemistry letters, 21(3), 1627-1657. https://doi.org/10.1007/s10311-022-01544-4

[2] Liu, K., Tian, J., Chen, J., & Wen, Y. (2022). Low-Carbon retrofitting path of existing public buildings: A comparative study based on green building rating systems. Energies, 15(22), 8724. https://doi.org/10.3390/en15228724

[3] Li, Y., Li, S., Xia, S., Li, B., Zhang, X., Wang, B., ... & Zheng, W. (2023). A review on the policy, technology and evaluation method of low-carbon buildings and communities. Energies, 16(4), 1773. https://doi.org/10.3390/en16041773

[4] Azhgaliyeva, D., & Rahut, D. B. (2022). Promoting green buildings: barriers, solutions, and policies (No. 1331). ADBI Working Paper. https://doi.org/10.56506/CVYC4388

[5] Senin, S. N., Fahmy-Abdullah, M., & Masrom, M. A. N. (2021, April). The implementation of green transportation towards low carbon city. In IOP conference series: Earth and environmental science (Vol. 736, No. 1, p. 012063). IOP Publishing. https://doi.org/10.1088/1755-1315/736/1/012063

[6] Kannapiran, E., Joshi, K., Chougale, R.K., Rana, N. and Kaur, C., 2022, July. Smart electric vehicle charging station for residential complex. In 2022 International Conference on Innovative Computing, Intelligent Communication and Smart Electrical Systems (ICSES) (pp. 1-4). IEEE. https://doi.org/10.1109/ICSES55317.2022.9914182

[7] Shi, Q., Yu, T., & Zuo, J. (2015). What leads to low-carbon buildings? A China study. Renewable and Sustainable Energy Reviews, 50, 726-734. https://doi.org/10.1016/j.rser.2015.05.037

[8] Srivastava, S.S., Satyanarayana, G.S., Dhasmana, A., Rawat, V., Rana, A.S. and Bisht, Y.S., 2025. Role of Embedded Systems in Smart Energy Management: Challenges, Innovations, and Future Trends. Solar Energy and Sustainable Development Journal, 14(STR2E), pp.27-50. https://doi.org/10.51646/jsesd.v14iSTR2E.797

[9] Prum, D. A. (2010). Green buildings, high performance buildings, and sustainable construction: does it really matter what we call them. Vill. Envtl. LJ, 21, 1.

[10] Alhorr, Y., Eliskandarani, E., & Elsarrag, E. (2014). Approaches to reducing carbon dioxide emissions in the built environment: Low carbon cities. International Journal of Sustainable Built Environment, 3(2), 167-178. https://doi.org/10.1016/j.ijsbe.2014.11.003

[11] Simpeh, E. K., Pillay, J. P. G., Ndihokubwayo, R., & Nalumu, D. J. (2022). Improving energy efficiency of HVAC systems in buildings: a review of best practices. International Journal of Building Pathology and Adaptation, 40(2), 165-182. https://doi.org/10.1108/IJBPA-02-2021-0019

[12] Chen, X., Yang, H., & Lu, L. (2015). A comprehensive review on passive design approaches in green building rating tools. Renewable and Sustainable Energy Reviews, 50, 1425-1436. https://doi.org/10.1016/j.rser.2015.06.003

[13] Patil, M., Boraste, S., & Minde, P. (2022). A comprehensive review on emerging trends in smart green building technologies and sustainable materials. Materials Today: Proceedings, 65, 1813-1822. https://doi.org/10.1016/j.matpr.2022.04.866

[14] Gibbins, J., Beaudet, A., Chalmers, H., & Lamperth, M. (2007). Electric vehicles for low-carbon transport. Proceedings of the Institution of Civil Engineers-Energy, 160(4), 165-173. https://doi.org/10.1680/ener.2007.160.4.165

[15] Sandoval, R., Karplus, V. J., Paltsev, S., & Reilly, J. M. (2009). Modelling prospects for hydrogen-powered transportation until 2100. Journal of Transport Economics and Policy (JTEP), 43(3), 291-316. https://doi.org/10.3828/jtep.2009.43.3.291

[16] Wimbadi, R. W., Djalante, R., & Mori, A. (2021). Urban experiments with public transport for low carbon mobility transitions in cities: A systematic literature review (1990-2020). Sustainable Cities and Society, 72, 103023. https://doi.org/10.1016/j.scs.2021.103023

[17] Bostanci, S. C., Limbachiya, M., & Kew, H. (2018). Use of recycled aggregates for low carbon and cost effective concrete construction. Journal of Cleaner Production, 189, 176-196. https://doi.org/10.1016/j.jclepro.2018.04.090

[18] Maqbool, R., Thompson, C., & Ashfaq, S. (2023). LEED and BREEAM green building certification systems as possible game changers in attaining low-cost energy-efficient urban housing projects. Journal of Urban Planning and Development, 149(3), 04023024. https://doi.org/10.1061/JUPDDM.UPENG-4292

[19] Mustafaraj, G., Marini, D., Costa, A., & Keane, M. (2014). Model calibration for building energy efficiency simulation. Applied Energy, 130, 72-85. https://doi.org/10.1016/j.apenergy.2014.05.019

[20] Litman, T. (2005). Efficient vehicles versus efficient transportation. Comparing transportation energy conservation strategies. Transport Policy, 12(2), 121-129. https://doi.org/10.1016/j.tranpol.2004.12.002

[21] Cellura, M., Guarino, F., Longo, S., & Mistretta, M. (2014). Energy life-cycle approach in Net zero energy buildings balance: Operation and embodied energy of an Italian case study. Energy and Buildings, 72, 371-381. https://doi.org/10.1016/j.enbuild.2013.12.046

[22] Shukla, P. R., & Pathak, M. (2016). Low-Carbon Transport in India. Enabling Asia to Stabilise the Climate, 153. https://doi.org/10.1007/978-981-287-826-7_8

[23] Xi, C., & Cao, S. J. (2022). Challenges and future development paths of low carbon building design: a review. Buildings, 12(2), 163. https://doi.org/10.3390/buildings12020163

[24] Davies, P., & Osmani, M. (2011). Low carbon housing refurbishment challenges and incentives: Architects' perspectives. Building and environment, 46(8), 1691-1698. https://doi.org/10.1016/j.buildenv.2011.02.011

Advanced Materials and Sustainable Energy Technologies – AMSET2025 Materials Research Forum LLC
Materials Research Proceedings 66 (2026) 190-202 https://doi.org/10.21741/9781644904152-17

Fuel Cells and Hydrogen Technology: Pioneering the Clean Energy Revolution

G. SAI SATHYANARAYANA[1,2*], V. RAWAT[1], P. A. KUMAR[1], A. S. RANA[1], AJITH R. RAJENDRAN[3], A. C. THAKURI[1]

[1]Department of Aerospace Engineering, Uttaranchal Institute of Technology, Uttaranchal University, Dehradun-248007, Uttarakhand, India

[2]Department of Aeronautical Engineering, MRCET, Hyderabad, Telangana, India

[3]Division of Aerospace Engineering, Karunya Institute of Technology and Sciences, Coimbatore, India

*sathyanarayanas305@gmail.com

Keywords: Hydrogen Fuel Cells, Clean Energy Technologies, Hydrogen Production, Zero Emission Systems, Proton Exchange Membrane Fuel Cell (PEMFC)

Abstract: Hydrogen fuel cells are effective, produce no emissions, and have many applications, yet they are becoming a front-runner clean energy choice in the world's sustainable energy transition. There are, however, several significant barriers that need to be alleviated before hydrogen fuel cells can become a mass-use technology, such as high costs, insufficient infrastructure, and technical challenges with producing and storing hydrogen. With focus on PEMFCs and SOFCs, the present study analyzes the status of hydrogen & fuel cell technology and evaluates the major methods of hydrogen production, such as electrolysis, steam methane restructuring, and biomass gasification. The report further analyzes the issues in storage and transit, material developments in fuel cells, and transportation and generation case studies. Priority is given to measures to reduce expenses, enhance system resilience, and create infrastructure to support the transition to a global hydrogen economy.

1. Introduction

As part of the decline in fossil fuels, pollution has joined it as a climate change issue, making the imperative to have clean, green, and very energy-efficient systems even more urgent. Hydrogen fuel cells represent one of the most promising new alternatives. It is an emerging key solution to the critical energy and environmental issues that the world is now facing in the twenty-first century. Fuel cells are devices that employ electrochemical reactions to convert hydrogen into electricity. They possess a number of advantages, including efficiency, no emissions of greenhouse gases at the point of use, flexibility, and the possibility of large-scale integration in a system across industries, power generation, and transportation [1-2]. Governments and industries globally are starting to recognize hydrogen as a clean energy resource. By creating efficient technologies for hydrogen production, storage, and distribution and for installing fuel cells in homes and businesses, the world is working hard to make the power grid cleaner [3]. The increasing role of hydrogen in the transition to global energy sources and sector integration is illustrated by Figure 1, which shows the global rise in hydrogen demand from 2020 to 2030 in the principal sectors of industry, transport, power generation, and buildings.

1.1 Research Gap & Objective

The research provides an integrated assessment of hydrogen technologies through its three major components which include (i) efficiency metrics assessment and (ii) production method economic evaluation and (iii) environmental impact study of gray and blue and green hydrogen systems.
The research study intends to achieve its goals through the following objectives:

i. The study offers a quantitative assessment that compares the performance of PEMFC systems with SOFC systems.
ii. The study evaluates different hydrogen production methods by assessing their efficiency and costs and carbon emissions.
iii. The study examines storage and infrastructure issues through the application of techno-economic evaluation methods.
iv. The research provides evidence-based recommendations which guide future hydrogen deployment activities.

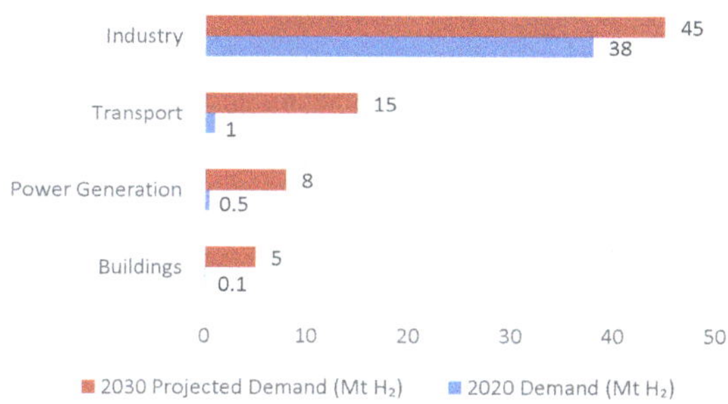

Figure 1. *Global Hydrogen Demand by Sector (2020–2030 Projected) (Source: IEA Net Zero by 2050 scenario projections (2021))*

The figure 1 demonstrates that industrial hydrogen demand continues to lead the market by supplying almost 60 percent of expected hydrogen consumption in 2030. The research projects that transportation sector development will exceed 200 percent which establishes mobility as a primary factor driving future hydrogen usage.

The United Nations Agenda for Sustainable Development Goals and the net-zero emission goals [4]. Fuel cells are now feasible to be employed in practical applications as a result of developments in fuel cell designs and components that have enhanced performance, longevity, and cost [5]. Because of the compatibility with different operating conditions and power generating scales, PEMFCs and SOFCs are under strong research and development [6]. Against this backdrop, international cooperation and national policies are driving large-scale investments in hydrogen production and infrastructure, and several countries have incorporated hydrogen into their energy plans prominently [7]. It deliberates original communication procedures and AI-powered edge computing to intimidated constraints. The effect of embedded systems on energy competence and reliability is estimated through case studies. Over the following 10 years, more growths including hardware evolution, machine learning for analytical management, and IoT-based smart ecosystems are also predictable to increase efficiency by 15% to 25% [8].

There are still major barriers to the large-scale deployment of hydrogen and fuel cell technologies. Some of these barriers are the high costs of hydrogen production and storage, inadequate infrastructure, and the need for sustainable and scalable supply chains. This paper aims to provide a state-of-the-art overview of the present landscape-state, barriers, and future outlooks of fuel cells and hydrogen technologies on technological innovations, production methods, applications in the real world, and urgent policies shaping the global hydrogen economy.

2. Types of Fuel Cells

Fuel cells are classified on the basis of the electrolyte involved and operating temperature, depending on where it is suitable for an application. About the majority of various fuel cell types, proton exchange membrane fuel cells (PEMFCs) and solid oxide fuel cells (SOFCs) have also been studied extensively and applied intensively in various areas. Alkaline and direct methanol fuel cells are some of the other types. However, each one has its positives to be applied in particular applications.

2.1 Proton exchange membrane fuel cells (PEMFC)

Proton exchange membrane fuel cells (PEMFCs) are among those fuel cell technologies that boast of being one of the most widely used across applications ranging from those in vehicular use to portable power uses. These fuel cells function at lower temperatures (60–80°C) and employ polymer electrolyte membranes allowing proton passage while preventing electrons from doing so, thus compelling them to generate an external current. This fuel cell type is characterized by high power density, short start-up time, and use of automotive applications, all yielding advantages in its use. These fuel cells can be found in, for example, hydrogen-powered vehicles, including fuel cell electric cars and hydrogen-powered buses, because of their small size and quick response to load changes. Unfortunately, barriers to the use of this fuel cell for large-scale commercialization include water management, the use of platinum-based catalysts, and durability problems [9].

2.2 Solid oxide fuel cells (SOFC)

Solid oxide fuel cells (SOFCs), a type of high-temperature fuel cells, usually operate in the range of 500–1000°C. They use a ceramic electrolyte to transport oxygen ions. SOFCs stand in stark contrast to PEMFCs in that they can use a wide variety of fuels, such as hydrogen, natural gas, and biogas, without requiring external reforming. SOFCs' principal advantages are their high efficiency (up to 60% efficiency alone or 85% when cogenerating heat and electricity), fuel flexibility, and durability over decades of operation. These very features have made SOFCs particularly suitable for stationary power generation in distributed energy systems, residential cogeneration, and industrial power plants. However, high operating temperatures create challenges with thermal stresses, material degradation, and long startup time of systems. Current efforts are focused on reducing operating temperatures in order to prolong system life while permitting many more applications [10].

2.3 Alkaline and direct methanol fuel cells

The alkaline fuel cell (AFC) technology is one of the earliest fuel cell technologies used in the space applications of the Apollo space program and space shuttle. The alkaline fuel cell functions due to the potassium hydroxide (KOH) electrolyte, which enables it to work with high electrical efficiency. Due to fast electrochemical kinetics, simple construction, and good performance in pure hydrogen-oxygen environments, alkaline fuel cells are well established in fuel cell history. However, the major drawback of these alkaline fuel cells is that they are very much sensitive to carbon dioxide (CO_2) contamination, which reduces their performance with time, limiting their popularity in terrestrial applications without extensive purification systems. Direct methanol fuel cells (DMFC) use liquid methanol directly as fuel, so there is no need for hydrogen storage, which allows a conveniently available feedstock. DMFCs are a suitable candidate for portable electrical use, including laptops, mobile phones, and military field use, since they weigh much less and possess a greater energy density. A limitation of DMFCs is methanol crossover, which results from methanol diffusing across the membrane and decreasing efficiency and consuming fuel. DMFCs also generate much less power compared to PEMFCs, so they have limited use in applications that need more power [11].

3. Hydrogen Production Methods

Every one of the different methods of hydrogen production has benefits, drawbacks, and impacts on the environment. The three primary processes of hydrogen production are gasification of biomass, methane reforming with steam (SMR), and electrolysis. With the planned low- or carbon-free energy infrastructures, they have to enable hydrogen to be an environmentally beneficial energy carrier. Multiple sources' contribution to hydrogen generation is illustrated in this graph.

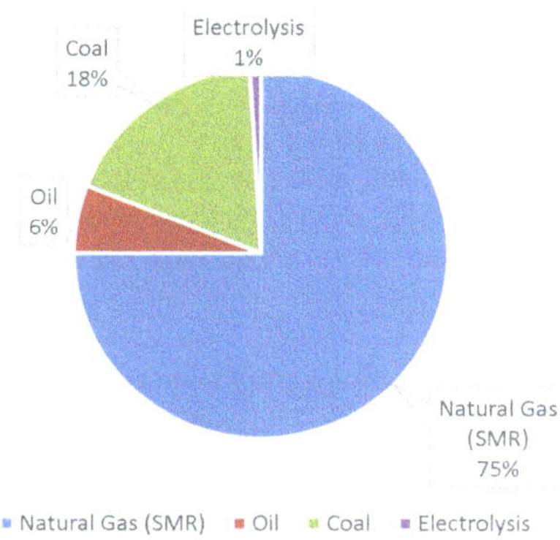

Figure 2. *Global Hydrogen Production by Source (2022) (Source: IEA (International Energy Agency))*

Figure 2 demonstrates that current hydrogen production relies primarily on fossil fuels while green hydrogen accounts for less than 5 percent of total output. The situation demands urgent implementation of renewable energy sources for electrolysis because it serves as a vital solution to achieve decarbonization objectives.

3.1 Electrolysis (PEM, alkaline, SOEC)

An electrochemical process, electrolysis, splits water (H2O) into oxygen (O_2) and hydrogen (H_2). When the electrolytic process is powered by renewable energies such as wind or sunlight, green hydrogen results and is an essential part of a sustainable energy system. PEM, AWE, and SOECs are the three broad types of electrolysis technologies [12].

- **Proton interchange membrane (PEM) electrolysis:** This process allows protons but blocks electrons using a polymer membrane as an electrolyte. Due to its high efficiency, rapid response to changes in input power, and compact design, PEM electrolysis is well suited for coupling with renewable energy sources. Platinum-based catalysts are very costly, though, making this method not widely applied.
- **AWE, or alkaline water electrolysis:** The electrolyte of this legitimate, established, and cost-effective technology is a liquid alkaline substance, typically sodium hydroxide or potassium hydroxide. Alkaline electrolysis is much cheaper and has been used to produce hydrogen in large industrial scale for many decades already. However, lower efficiency and poor dynamic response concerning PEM electrolysis render it less suitable for variable renewable energy sources.
- **Solid Oxide Electrolysis Cells (SOECs):** This work with high temperatures (700–1,000 °C) utilizing ceramic electrolytes to split water. High efficiency is reached with this process that utilizes the waste heat produced by the industrial processes or power plants. Unlike the PEM

Materials Research Proceedings 66 (2026) 190-202 https://doi.org/10.21741/9781644904152-17

and alkaline electrolysis, SOECs can also use carbon dioxide (CO_2) as feedstock and produce syngas (H_2 + CO) for production of synthetic fuels. In spite of these advantages, SOEC technology is still in the early phases of commercialization because of material degradation at prolonged high temperatures.

3.2 Steam methane reforming (SMR)

Steam methane reforming (SMR) is currently the dominant method for hydrogen production, accounting for Almost 80 percent of hydrogen on Earth is produced by a process that adds steam to natural gas (mainly methane, CH_4) at high temperature (700-1,000°C) and pressure in the presence of a catalyst, converting into hydrogen (H_2) and carbon monoxide (CO). The water-gas shift reaction converts the CO into more H_2 and CO_2. SMR is convenient and is the most economical with an extremely high yield of hydrogen. On the other hand, this process emits a lot of CO_2, hence making it a "Gray method" of hydrogen production. SMR could produce low carbon "blue hydrogen" when combined with carbon capture and storage (CCS), which further mitigates environmental impact [13].

3.3 Biomass gasification

Biomass gasification is a novel way of producing hydrogen, wherein organic materials (wood, agricultural waste, and municipal solid waste) are converted to hydrogen syngas. The process of gasification operates at high temperatures, between 700°C and 1,200°C, with a limited supply of oxygen to facilitate the decomposition of biomass into hydrogen (H_2), carbon monoxide (CO), carbon dioxide (CO_2), and methane (CH_4). Some of the prime attributes of biomass gasification are as follows:

- Carbon neutrality, since the CO_2 emission is compensated by CO_2 absorption during the growth of biomass.
- Valuable waste materials can be used as feedstock for gasification, which decreases the dependency on fossil fuels.
- Decentralized production is suitable for hydrogen, thereby helping in local energy independence.

Still, problems like tar production, efficiencies in processes, and variability in feedstocks need solutions to develop biomass gasification into a viable hydrogen source on a larger scale [14].

4. Hydrogen Storage and Transportation

To be a clean energy carrier to be widely accepted, hydrogen should be efficiently stored and transported. Due to its low volumetric energy density, hydrogen is thus extremely difficult to store and transport in an efficient manner. The major methods for storing and transporting hydrogen include compressed and liquid hydrogen storage, metal hydrides and chemical carriers, and other dedicated hydrogen infrastructure, which include pipelines and distribution networks. Storage classifications of hydrogen are shown in Figure 3.

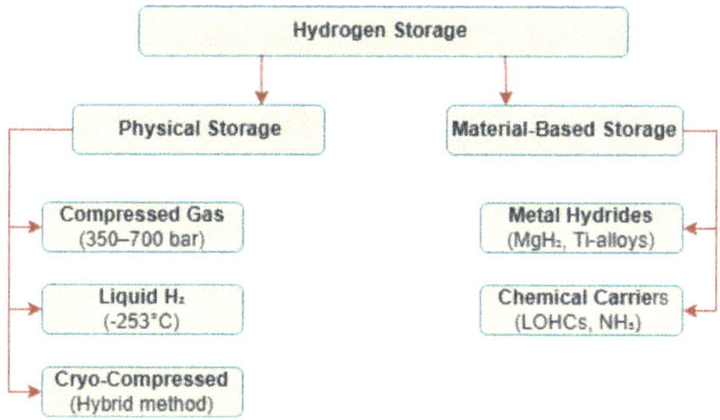

Figure 3. *Classification of hydrogen storage*

The efficiency analysis shows that SOFC systems achieve better total system efficiency for stationary CHP systems while PEMFCs deliver their best performance through their ability to start up quickly and produce high power output.

4.1 Physical storage

Hydrogen can be stored as gas or liquid, and each has its distinct advantages and problems [15].

- **Compressed hydrogen storage:** in this view, hydrogen in the form of gas is stored at high pressure, typically 350-700 bar, in reinforced tanks. This is a common storage technique in fuel cell vehicles (FCVs), industrial uses, and stationary storage because it has been matured enough and requires low energy for compression. High pressure also demands use of strong material with good safety precautions to avoid leaks and explosions.
- **Liquid Hydrogen Storage:** Hydrogen can be liquefied at cryogenic temperatures (-253°C), raising the energy density per unit volume compared to that of compressed gas storage. Liquid hydrogen is especially suitable for aerospace and long-distance haulage applications. However, liquefaction is energy consuming and utilizes about 30-40 percent of the total energy content of hydrogen. In addition, evaporation can ultimately cause losses via boil-off.
- **Cryo-compressed hydrogen storage:** This form of storage employs cryogenic freezing alongside mild pressure (up to 300 bar) in an effort to enhance the efficiency of containers and minimize energy losses. To allow very dense and efficient operations, cryo-compressed storage is under investigation with regard to future hydrogen vehicles and fueling stations. Material-Based Storages.

Scientists are looking into solid-state hydrogen storage in the form of metal hydrides and biochemical carriers to bypass the problems involved with compression and storage as liquid hydrogen [16].

- **Metal Hydrides:** They are safe and mobile storage materials that absorb and deliver hydrogen through reversible chemical processes. Titanium alloy and high-capacity metal hydrates such as MgH2 operate under low pressure, making them suitable for fuel cell and stationary energy storage.
- **Chemical Carriers:** Liquid organic hydrogen carriers (LOHCs) and ammonia are being developed as alternative solutions for hydrogen storage and transport. LOHCs, like toluene-methylcyclohexane systems, chemically bond hydrogen and release it on-demand,

ensuring safe and efficient transportation. Ammonia (NH_3) is another very promising carrier because it has high hydrogen content, an existing distribution infrastructure, and can be cracked to release hydrogen when necessary.

4.2 Hydrogen pipelines and infrastructure

Infrastructure is dedicated to hydrogen pipelines, refuelling stations, and distribution networks for large-scale deployment. [17]

- **Hydrogen pipelines:** Hydrogen has dedicated pipelines through which it can be transported or blended into existing networks with natural gas. However, hydrogen presents a great challenge in its use due to its low density and high diffusivity, causing risks like material embrittlement and leakage. Research is under way on the advanced pipeline materials and coatings for enhancing durability and safety.
- **Refuelling stations and storage hubs:** Building hydrogen refuelling networks is paramount for fuel-cell electric vehicles (FCEVs) and heavy-duty transport. Also, hydrogen storage hubs should be situated near renewable energy produce or industrial centres so that transport costs would be low for distribution efficiency.
- **International Hydrogen trade and transport:** Hydrogen is traded internationally through liquefied hydrogen carriers and ammonia carriers as well as shipping based on LOHCs. Countries rich in renewable energy resources invested on hydrogen export infrastructures thus paving the way toward hydrogen as a global traded energy commodity.

5. Fuel Cell Performance and Efficiency

Fuel cells have become overly critical regarding their performance and efficiency since they determine the commercial future of the cells, whether it is for transportation, stationary power generation, or portable applications. Some of the most important parameters that influence fuel cell performance are efficiency and power density, durability and stability issues, and cost and lifecycle analysis. The approximate efficiencies of different fuel cell systems are illustrated by Figure 3.

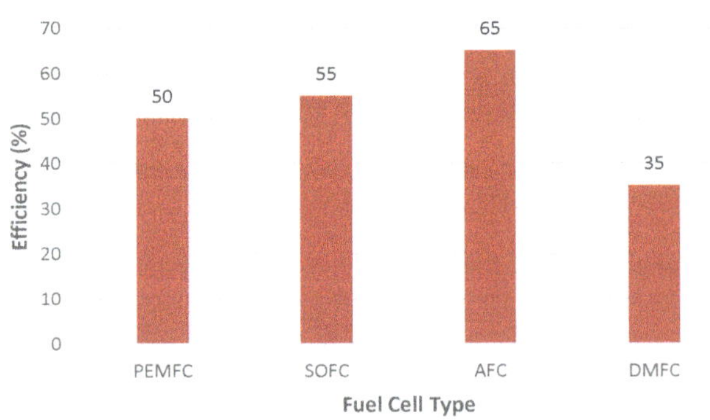

Figure 3. *Fuel Cell System Efficiency Comparison (Approx.) (Source: US department of Energy)*

5.1 Efficiency and power density

A fuel cell's efficiency is determined as the ratio between the useful electrical energy to the energy stored within the fuel input. There are fuel cells of different efficiencies dependent upon the type of fuel cell type, operating temperature and system design. The Proton Exchange Membrane Fuel Cells (PEMFC) operates between an efficiency range of 40–60 whereas the Solid Oxide Fuel Cell (SOFC) achieves maximum efficiency of around 60% under stand-alone mode, and 85% when used in a combined heat and power (CHP) system. Another performance metric is power density, which is expressed in watts/unit area (W/cm^2). Power density should be high for miniaturized and compact fuel cell systems to make them more applicable to automotive and portable applications. Studies have found that perfecting electrode materials, distributing catalysts, as well as configuring cells optimise the power density and could enhance the computational efficiency of fuel cells with respect to the fraction of electrons transferred in a reaction that contributes to power generation [18].

5.2 Durability and stability issues

Durability has remained one of the biggest challenges on the path to fuel cell commercialization, especially in applications requiring long hours of operation such as automotive and stationary power generation. Fuel cells degrade for many reasons, including catalyst degradation, membrane degradation, and thermal cycling issues. Address these challenges, studies are devoted to the development of more durable catalyst materials; improvement of membrane stability; and optimization of system integration for reliability and longevity [19].

5.3 Cost and lifecycle analysis

The extensive cost of fuel cells is a major obstacle to their large-scale application. The large contributions to the cost are due to the cost of Catalyst materials, especially platinum, Manufacturing complexity, and Balance-of-plant components, such as pumps, compressors, and cooling systems. Lifecycle analysis of fuel cell systems examines both the economic and environmental aspects by including such as initial production, operation, maintenance, and end-of-life recycling. Fuel cell hybrid vehicles studies prove that although the initial cost of fuel cells is higher than for internal combustion engines, their operating costs over a vehicle's lifetime would be lower, because of the higher fuel efficiency and much smaller emissions. Cost competitiveness with conventional energy technologies can only be achieved by improving economies of scale, reducing platinum loading, and increasing material durability [20]. Digital twins (DT) have been used to make noteworthy progressions not only from the start of the vehicle expansion process to the end of the vehicle's construction, but also while the vehicle is being used, gathering vigorous data from its regular operations to make travel safer, more enjoyable, and more stress-free [21].

6. Applications and Case Studies

Hydrogen fuel cells have received considerable interest across various sectors as a clean and efficient energy source. These energy systems are an excellent replacement for fossil fuel-based systems since they are able to generate electricity with no emissions. The primary applications of fuel cells include transportation, stationary power generation, and portable electronics; all three applications take advantage of certain aspects of fuel cell hydrogen technologies.

6.1 Transportation (cars, buses, trains)

Fuel cells are among the promising solutions to decarbonize mobility because the transport sector is among the largest emitters of greenhouses gases (GHGs). In hydrogen FCEVs, hydrogen is transformed into electricity through a PEMFC in vehicles, trains, buses, and even boats, with water being the only byproduct. The following are a number of advantages of fuel cell transportation: Longer distances than for battery-electric vehicles; fast refuelling (three to five minutes for passenger vehicles); and energy conservation efficiency with even less environmental damage than with internal combustion engines. However, several shortcomings exist to be tackled for widespread commercialization, including constraints on hydrogen infrastructure, the high production costs of fuel cells, and durability of the fuel cell. Updating the hydrogen refuelling infrastructure and improved fuel cell efficiencies will be important steps in scaling-up the deployment of FCEVs [22].

6.2 Stationary power generation

In fact, fuel cells are also being increasingly introduced in the field of stationary power generation, and providing dependable, clean, and resource-efficient electricity for residential, commercial, and industrial applications. Solid oxide fuel cells and phosphoric acid fuel cells are two types of fuel cells that can be well suited for stationary applications because of their high efficiencies and availability for all of the different fuel types, including natural gas and biogas. Some of the applications of stationary fuel cells are as follows: backup for critical infrastructure (such as hospitals, data centres, telecom towers), independent power provider for remote and off-grid places, and combined heat and power (CHP) applications, where waste heat can supplement the overall energy efficiency.

Reducing greenhouse gas emissions needs us to utilize fossil fuels as low as possible. In this state, electric energy storage systems are vital because they permit them to trade energy with the house or the grid based on the quantity of electricity required [23].

Challenges such as high initial costs, limited fuel supply, degradation of the systems, etc., still need to be overcome. Research in such areas is aimed at improved materials for the fuel cells, increased life of the fuel cells, and reducing costs by integrating hydrogen use in renewable energy systems to make stationary fuel cells economically attractive [24].

6.3 Portable electronics

Hydrogen fuel cells are also examined for use in portable power applications, including consumer electronics, military installations, and remote sensing devices. Different from conventional lithium-ion batteries, which need to be frequently recharged, micro fuel cells provide longer operate times and rapid refuelling options. The salient benefits of hydrogen fuel cells for portable electronics are light and compact design fit for mobile devices, rapid refuelling through hydrogen cartridges to eliminate long charging times, and high energy density allowing for prolonged operational use. Fuel cells are being contemplated in applications such as laptop, smartphone, drone, and wearable device applications requiring long power supply. Military applications benefit from the silent operation of fuel cells and reduced thermal signature, enhancing their viability in field use [25].

6.4 Comparative Evaluation of Fuel Cell Technologies

The comparison between PEMFC and SOFC systems shows that PEMFCs perform better in mobile use because they operate between 60 and 80 degrees Celsius and generate power above 1 watt per square centimetre. SOFCs perform better in stationary applications because they can use different fuels and achieve higher efficiency. SOFC technology experiences operational problems because thermal stresses reduce its lifespan while PEMFC technology suffers from catalyst degradation problems.

7. Challenges and Future Prospects

Although significant progress has been made in the sphere of fuel cell and hydrogen technologies, barriers to large-scale commercialization still exist. These include cost factors, barriers to scaling up, hydrogen-economy policy support, and development of fuel cell materials. It is essential to address these challenges to fast-track commercialization, as well as ensure long-term sustainability of fuel cells as a clean energy solution.

7.1 Cost reduction and scalability

Mass-scale market penetration for fuel cells could be hindered due to costs. Even though platinum-based catalysts are costly and difficult to obtain, they are the costliest components of a fuel cell utilized for the electrochemical reaction. Solid oxide fuel cells and proton exchange membrane fuel cells are instances of production difficulties which require a well-engineered environment for high-grade materials. Compressors, pumps, and cooling networks are a few balance-of-plant (BoP) components that will make up fuel cell system expenses. Unlike platinum load or non-precious metals catalysts, the transition will be more affordable for scientists and industry stakeholders. Mass production techniques shall be employed to reduce prices per unit and enhance durability and reliability to reduce maintenance and replacement expenses. Importantly, economies of scale would help lower cost through an increase in demand and production volume, which would result in cheaper material costs and more efficient manufacturing processes [26].

7.2 Hydrogen economy and policy support

Strong policy, investment, and financial incentives are required within the hydrogen economy for the effective transition to it. Countries worldwide are developing hydrogen roadmaps for the application of fuel cell technologies in the transport, industrial, and energy sectors. Major policy measures on hydrogen involve Directing energy investments into hydrogen distribution, production, and fuel cell adoption; investing in hydrogen refuelling infrastructure; facilitating and enhancing widespread availability for fuel cell vehicles; and introducing regulatory frameworks to Favor low-carbon production pathways for hydrogen in public and private sectors such as those through renewable energy-promoted electrolysis. Due to the fact that they put significant focus on funding hydrogen innovation projects and public-private collaborations for development enhancement, they exist in locations such as the US, Japan, and the EU. However, international collaboration and long-term national policy continuity are necessary for the building of a successful hydrogen market [27].

7.3 Advances in fuel cell materials

Efficiency, durability, and price of fuel cells can be enhanced through material science advancements. There are many research projects underway to increase the performance of electrode catalysts, catalyst membranes, and electrolytes—three key fuel cell components. In these innovative developments, Alternative catalyst materials like nickel and perovskite catalysts replace expensive platinum-group metals: Advanced electrolyte materials, such as proton conducting ceramics for SOFCs associated with higher efficiency in lesser operating temperatures: Durable polymer membranes improve the chemical and thermal stability of fuel cell lifetimes Nanotechnology and materials new phenomena are expected to speed up the performance of fuel cells, thereby enhancing their competition with conventional forms of energy **[28]**.

Conclusion

- With possible superior efficiency, no emissions during use, and widespread application to various industries such as transportation and electricity generation, hydrogen fuel cell technology is the most viable clean energy source.
- Technological improvement in fuel cell materials and also hydrogen production methods have collided, mainly in green hydrogen production via electrolysis, system integration, and other aspects that speed up the phasing out of hydrogen as the basic element of energy economy or even the energy economy itself.
- Though promising developments exist, the high costs of manufacturing and storage, the lack of infrastructure, and dependence on non-renewable hydrogen sources remain obstacles to their wider use.
- The establishment of an economically sustainable hydrogen infrastructure and the upscaling of hydrogen technologies will rely on facilitating policies by governments, global cooperation, and directed investments.
- To address existing challenges and realize the complete potential of hydrogen & fuel cell technologies for a renewable energy-based future, research and innovation and partnerships between the public and private sectors will be crucial.

- The study shows that hydrogen fuel cells achieve high efficiency with zero tailpipe emissions but require two conditions for large-scale adoption. The first condition requires green hydrogen costs to drop below $2 per kilogram while the second condition demands fuel cell durability to exceed 40,000 operating hours and new low-carbon infrastructure to become available. Among technologies, PEMFCs are optimal for mobility, whereas SOFCs are more suitable for stationary CHP systems. Future research should focus on non-precious metal catalysts, high-temperature membranes, integrated renewable electrolysis systems, and comprehensive lifecycle carbon assessment frameworks.

References

[1] Cook, B. (2002). Introduction to fuel cells and hydrogen technology. Engineering science and education journal, 11(6), 205-216. https://doi.org/10.1049/esej:20020601

[2] Hoffmann, P. (2012). Tomorrow's energy, revised and expanded edition: Hydrogen, fuel cells, and the prospects for a cleaner planet. MIT press. https://doi.org/10.7551/mitpress/8625.001.0001

[3] Thomas, J. M., Edwards, P. P., Dobson, P. J., & Owen, G. P. (2020). Decarbonising energy: The developing international activity in hydrogen technologies and fuel cells. Journal of Energy Chemistry, 51, 405-415. https://doi.org/10.1016/j.jechem.2020.03.087

[4] Islam, A., Islam, T., Mahmud, H., Raihan, O., Islam, M. S., Marwani, H. M., ... & Awual, M. R. (2024). Accelerating the green hydrogen revolution: A comprehensive analysis of technological advancements and policy interventions. International Journal of Hydrogen Energy, 67, 458-486. https://doi.org/10.1016/j.ijhydene.2024.04.142

[5] Singla, M. K., Nijhawan, P., & Oberoi, A. S. (2021). Hydrogen fuel and fuel cell technology for cleaner future: a review. Environmental Science and Pollution Research, 28(13), 15607-15626. https://doi.org/10.1007/s11356-020-12231-8

[6] Sorensen, B., & Spazzafumo, G. (2018). Hydrogen and fuel cells: emerging technologies and applications. https://doi.org/10.1016/B978-0-08-100708-2.00003-5

[7] Capurso, T., Stefanizzi, M., Torresi, M., & Camporeale, S. M. (2022). Perspective of the role of hydrogen in the 21st century energy transition. Energy Conversion and Management, 251, 114898. https://doi.org/10.1016/j.enconman.2021.114898

[8] Srivastava, S.S., Satyanarayana, G.S., Dhasmana, A., Rawat, V., Rana, A.S. and Bisht, Y.S., 2025. Role of Embedded Systems in Smart Energy Management: Challenges, Innovations, and Future Trends. Solar Energy and Sustainable Development Journal, 14(STR2E), pp.27-50. https://doi.org/10.51646/jsesd.v14iSTR2E.797

[9] Wee, J. H. (2007). Applications of proton exchange membrane fuel cell systems. Renewable and sustainable energy reviews, 11(8), 1720-1738. https://doi.org/10.1016/j.rser.2006.01.005

[10] Stambouli, A. B., & Traversa, E. (2002). Solid oxide fuel cells (SOFCs): a review of an environmentally clean and efficient source of energy. Renewable and sustainable energy reviews, 6(5), 433-455. https://doi.org/10.1016/S1364-0321(02)00014-X

[11] Wang, Y., Li, L., Hu, L., Zhuang, L., Lu, J., & Xu, B. (2003). A feasibility analysis for alkaline membrane direct methanol fuel cell: thermodynamic disadvantages versus kinetic advantages. Electrochemistry Communications, 5(8), 662-666. https://doi.org/10.1016/S1388-2481(03)00148-6

[12] Hu, K., Fang, J., Ai, X., Huang, D., Zhong, Z., Yang, X., & Wang, L. (2022). Comparative study of alkaline water electrolysis, proton exchange membrane water electrolysis and solid oxide electrolysis through multiphysics modeling. Applied Energy, 312, 118788. https://doi.org/10.1016/j.apenergy.2022.118788

[13] Bhat, S. A., & Sadhukhan, J. (2009). Process intensification aspects for steam methane reforming: An overview. AIChE Journal, 55(2), 408-422. https://doi.org/10.1002/aic.11687

[14] Molino, A., Chianese, S., & Musmarra, D. (2016). Biomass gasification technology: The state of the art overview. Journal of energy chemistry, 25(1), 10-25. https://doi.org/10.1016/j.jechem.2015.11.005

[15] 1Yanxing, Z., Maoqiong, G., Yuan, Z., Xueqiang, D., & Jun, S. (2019). Thermodynamics analysis of hydrogen storage based on compressed gaseous hydrogen, liquid hydrogen and cryo-compressed hydrogen. International Journal of Hydrogen Energy, 44(31), 16833-16840. https://doi.org/10.1016/j.ijhydene.2019.04.207

[16] 1El Kharbachi, A., Dematteis, E. M., Shinzato, K., Stevenson, S. C., Bannenberg, L. J., Heere, M., ... & Hauback, B. C. (2020). Metal hydrides and related materials. Energy carriers for novel hydrogen and electrochemical storage. The Journal of Physical Chemistry C, 124(14), 7599-7607. https://doi.org/10.1021/acs.jpcc.0c01806

[17] Kim, C., Cho, S. H., Cho, S. M., Na, Y., Kim, S., & Kim, D. K. (2023). Review of hydrogen infrastructure: the current status and roll-out strategy. International Journal of Hydrogen Energy, 48(5), 1701-1716. https://doi.org/10.1016/j.ijhydene.2022.10.053

[18] Fan, Y., Hu, H., & Liu, H. (2007). Enhanced Coulombic efficiency and power density of air-cathode microbial fuel cells with an improved cell configuration. Journal of Power Sources, 171(2), 348-354. https://doi.org/10.1016/j.jpowsour.2007.06.220

[19] Wang, J. (2017). System integration, durability and reliability of fuel cells: Challenges and solutions. Applied Energy, 189, 460-479. https://doi.org/10.1016/j.apenergy.2016.12.083

[20] Jeong, K. S., & Oh, B. S. (2002). Fuel economy and life-cycle cost analysis of a fuel cell hybrid vehicle. Journal of Power Sources, 105(1), 58-65. https://doi.org/10.1016/S0378-7753(01)00965-X

[21] Joshi, K., Khurana, V., Krishna, C.S., Azam, M., Singh, Y. and Johari, L., 2022, July. Smart routing system in wireless sensor networks. In 2022 International Conference on Innovative Computing, Intelligent Communication and Smart Electrical Systems (ICSES) (pp. 1-6). IEEE. https://doi.org/10.1109/ICSES55317.2022.9914036

[22] Fakhreddine, O., Gharbia, Y., Derakhshandeh, J. F., & Amer, A. M. (2023). Challenges and solutions of hydrogen fuel cells in transportation systems: A review and prospects. World Electric Vehicle Journal, 14(6), 156. https://doi.org/10.3390/wevj14060156

[23] Joshi, K., Gupta, A., Kaushik, V., Anandaram, H., Gupta, H. and Kumar, A., 2023, July. Intervention of digital twin for smart monitoring of battery management system. In 2023 14th International Conference on Computing Communication and Networking Technologies (ICCCNT) (pp. 1-6). IEEE. https://doi.org/10.1109/ICCCNT56998.2023.10306563

[24] Felseghi, R. A., Carcadea, E., Raboaca, M. S., Trufin, C. N., & Filote, C. (2019). Hydrogen fuel cell technology for the sustainable future of stationary applications. Energies, 12(23), 4593. https://doi.org/10.3390/en12234593

[25] Kundu, P. P., & Dutta, K. (2016). Hydrogen fuel cells for portable applications. In Compendium of hydrogen energy (pp. 111-131). Woodhead Publishing. https://doi.org/10.1016/B978-1-78242-364-5.00006-3

[26] Wang, J. (2015). Barriers of scaling-up fuel cells: Cost, durability and reliability. Energy, 80, 509-521. https://doi.org/10.1016/j.energy.2014.12.007

[27] Bleischwitz, R., & Bader, N. (2010). Policies for the transition towards a hydrogen economy: the EU case. Energy Policy, 38(10), 5388-5398. https://doi.org/10.1016/j.enpol.2009.03.041

[28] Brandon, N. P., Skinner, S., & Steele, B. C. (2003). Recent advances in materials for fuel cells. Annual review of materials research, 33(1), 183-213. https://doi.org/10.1146/annurev.matsci.33.022802.094122

Design and Development of Coconut Dehusking Machine for Enhanced Efficiency and Safety of Coconut Farmers

Benjamin Roy T. IDIAN[1,a], Bryant Keanu T. MAESTRE[1,b], Kevin Francis F. EVANGELISTA[1,c], Raynold M. MARAVILLA[1,d], Jeson Elijah M. KALONG[1,e], Luvisminda MARCELO[1,f *]

[1]Mechanical Engineering Department, National University, Manila, Philippines

[a]idian.benjamin@gmail.com, [b]bryant.maestre@gmail.com, [c]kevinfrancisevangelista@gmail.com, [d]maravilla_raynold28@gmail.com, [e]kalong.elijah@gmail.com, [f]ldmarcelo@national-u.edu.ph

Keywords: Automated, Dehusking, Efficiency, Fabrication, Innovation, Labor-Intensive, Malfunction, Manual Labor, Semi-Automatic Coconut Dehusking Machine, Coconut Farming, Dehusking Efficiency, Manual Dehusking, Performance Testing, Operational Efficiency, Small Farmers

Abstract. The goal of this study is to improve the productivity, safety, and efficiency of coconut farming by designing, developing, and testing a semi-automated coconut dehusking machine. The study, which was carried out in Barangay Tumbaga II, Sariaya, Quezon Province, Philippines, examines the difficulties faced by regional small-scale farmers who use manual dehusking techniques. The machine's material loss, production, and operational effectiveness were assessed. The total efficiency rate was 82%, and the loss rate was 18%, with an average of 8.2 successfully dehusked coconuts and 1.8 damaged coconuts each trial. Although the rate of damaged coconuts indicates scope for improvement, these results show the machine's potential to increase processing capacity. Improvements including better feed systems, pressure management, and blade alignment could reduce product loss. All things taken into account, the equipment offers a viable substitute for conventional techniques, providing rural Filipino farmers with a safer and more environmentally friendly way to dehusk coconuts.

Introduction

Coconuts (Cocos nucifera), commonly referred to as the "tree of life," play an important role in tropical regions, providing food, fuel, and medicine [1]. In the Philippines, coconut farming contributes 3.6% to agricultural gross value added and ranks among the top global producers despite setbacks from typhoons and pests [2,3,4]. Dehusking, a key post-harvest step, remains labor-intensive and hazardous, typically performed manually with machetes and crowbars [5]. To address this, researchers aim to develop a mechanized dehusking machine for farmers in Sariaya, Quezon Province, reducing labor and injury while improving productivity. Interviews confirm the challenges of manual dehusking and the need for safer, more efficient solutions.

Traditionally, the process of dehusking a coconut has been a time-consuming, laborious, wasteful, and unproductive job for farmers, often involving dangerous conditions. It is a complex engineering problem that must be solved by the community. Designing and fabricating a coconut dehusking machine that would improve the efficiency and safety of the farmers would solve the problem. Through the needs analysis assessment, farmers need help with traditional dehusking methods involving labor- intensive, inefficient, safety concerns, and technological limitations that limit effective dehusking processes. The researchers will address this problem by ideating, creating, and experimenting using various engineering techniques that we have learned. This study enables researchers to find effective ways and development areas in designing a dehusking machine.

General Objective. This study focused on developing and evaluating a gasoline powered. This research study aims to design and fabricate a coconut dehusking machine using a gas-fed motor that enhances efficiency, productivity, and safety in the province of Barangay Tumbaga II, Sariaya, Quezon.
Specific Objectives. To design a dehusking machine that is adaptable for the coconut farmers, Provide costing and comparisons with conventional methods of design, To select appropriate materials and components for the fabrications to be made to satisfy operation and design conditions, and To develop, evaluate and compare the performance of the gas-powered coconut dehusking machine prototype with the PCA-developed machine in terms of dehusking time, efficiency, and ease of use.

Methodology

The study employed a mixed-method approach combining farmer interviews, mechanical design calculations, prototype fabrication, and performance testing. Interviews with local coconut farmers in Barangay Tumbaga II, Sariaya, Quezon Province were conducted to identify challenges in traditional dehusking practices and establish design requirements. Mechanical design methods were applied to size critical components, including shafts, rollers, sprockets, pulleys, bearings, and the motor, using standard engineering equations for mass, torque, power, gear ratio, belt drive, and bearing capacity.

Following the design phase, the prototype was fabricated using stainless steel and industrial-grade components to ensure durability and safety. Experimental testing was then carried out through multiple trials, each involving ten coconuts, to measure processing time, efficiency, and product loss. Comparative analysis was performed against manual dehusking methods and the Philippine Coconut Authority (PCA) machine to evaluate improvements in speed and productivity. Finally, an economic assessment was conducted to estimate fabrication cost, payback period, and return on investment, thereby determining the financial viability of the prototype for small-scale farmers.

Table 1. List of Components

Name	Quantity
Frame	1
Coconut Roller	2
Motor	1
Shaft	2
Sprocket	6
Pillow block bearing	10
Pulley	2
Hopper	1
Cover	1

Materials Research Proceedings 66 (2026) 203-212 https://doi.org/10.21741/9781644904152-18

Table 2. Bill of Materials

ITEM NO.	NAME	TYPE	QUANTITY
1	Frame	Stainless steel	1
2	Pillow block bearing	Stainless steel	10
3	Solid shaft	AISI 304	2
4	Motor	1hp gas	1
5	Sprocket	Industrial grade	6
6	Chain	roller chain #25	-
7	Belt	3 ply	-
8	Pulley	Industrial grade	2
9	Bolts	Stainless steel	-
10	Roller	Stainless steel	2

Equation

Mass of Shaft and Roller. The mass of the shaft and roller was determined using the volume of a cylinder, see eq. 1, and the relation m=Vρ, where ρ is the density of steel. This calculation established the total mass of the rotating components, which is critical for evaluating the forces acting on the system during operation.

$$V = \pi r^2 L \;;\; m = V\rho \tag{1}$$

Where: V= volume

L= length of shaft

r = radius of shaft

ρ = density

$$V_{shaft} = \pi(0.0127m)^2(0.75m) = 0.00038m^3$$
$$V_{roller} = \pi(0.1m)^2(0.5m) = 0.01571m^3$$

According to Density of Steel, the average density of steel is 7,850kg/m3

$$mass_{shaft} = V\rho \tag{2}$$
$$mass_{shaft} = V\rho = (0.00038m^3)\left(7850\frac{kg}{m^3}\right) = 2.98kg$$
$$mass_{roller} = V\rho = (0.01571m^3)\left(7850\frac{kg}{m^3}\right) = 123.31kg$$
$$mass_{total} = (mass_{roller} + mass_{shaft}) \times 2$$
$$mass_{total} = (123.31kg + 2.98kg) \times 2 = 252.58kg$$

Applied Force. The total force exerted by the rollers was calculated using Newton's second law, see eq. 3, with a equal to gravitational acceleration. An additional husk-separating force was included to account for operational resistance. This ensured that the drive system was designed to withstand realistic working loads.

$$F = m \times a \tag{3}$$

Where: F = force

m=mass

a = acceleration due to gravity, 9.81 m/s^2

$F = 252.58\,kg \times 9.81 m/s^2 = 2477.81\,N$
$F = 2477.81\,N + 520\,N\ (husk\ separating\ force$
$F = 2997.81\,N$

Torque. Torque was obtained using eq. 4, where r is the roller radius. This value represents the twisting moment applied to the shaft and is essential for determining both shaft strength and motor requirements.

$$T = F\ \times r \qquad (4)$$

Where: T = torque
F = Force
r = radius of roller, 0.1 m

$T = 2997.81\,N \times 0.1\,m$
$T = 299.781\,N - m \approx 299781\,N - mm$

Factor of Safety and Torsional Stress. The shaft was evaluated using the factor of safety equation with a conservative FoS of 2. Torsional shear stress was then calculated using eq. 6, which allowed the determination of the required shaft diameter. This ensured that the shaft could safely transmit torque without risk of mechanical failure.

Assuming Factor of Safety of 2:

$$FoS = \frac{\sigma_{yield}}{\sigma_{von\ misses}} \qquad (5)$$

$$2 = \frac{250MPa}{\sigma_{von\ misses}}$$

$\sigma_{von\ misses} = 125MPa$

Torsional Shear Stress for solid shaft

$$\tau = \frac{16T}{\pi d^3} \qquad (6)$$

$\tau = \sigma_{von\ misses} = 125MPa$

Where: T = torque
d = diameter
τ = torsional stress

$$125Mpa = \frac{16(299781\,N - mm)}{\pi d^3}$$

$d = 23.03mm \approx 25.4mm$

Power Requirement. The mechanical power needed to rotate the rollers was calculated using eq. 7 where N is the rotational speed. This confirmed that a 1 hp gasoline engine was sufficient to drive the system under load.

$$P = 2\pi TN \qquad (7)$$

$$P = 2\pi(299.781N - m)\left(15\frac{rev}{\text{min}}\right)\left(\frac{min}{60s}\right)$$

$$P = 470.726\,Watts\left(\frac{1hp}{746watts}\right) = 0.631\,hp$$

Chain Drive Selection. The chain drive was sized using eq. 8, with adjustments for design life. Although theoretical calculations suggested Chain No. 25, Chain No. 40 was selected due to its superior strength, higher load capacity, and better performance under shock loads, making it more suitable for agricultural applications [6,7].

$$Power\ rating = power\ to\ be\ transmitted\ \times service\ factor \quad (8)$$
$$Power\ rating = 746Watts\ \times 1.4$$
$$Power\ rating = 1044.4Watts$$

Required Power Rating:

$$\text{PR} = \frac{\text{Designed power} \times Designed\ Life}{15000hrs} \quad (9)$$
$$= \frac{1044.4Watts \times 5000h}{15000h} = 348.13\ Watts$$

Gear Ratio. The gear ratio was determined by using eq. 10, yielding a 100:1 reduction. This was achieved through staged gear and chain reductions, ensuring that the rollers operated at low speed but high torque, which is optimal for dehusking.

Engine RPM=1500
Roller RPM=15

$$\frac{N_{engine}}{N_{roller}} \quad (10)$$

$$= \frac{1500}{15} = 100:1$$

Stage gear reduction

$$3 \times 3.2 \times 3.2 \times 3.2 = 100$$

Belt drive:

$$R1 = \frac{driven\ pulley}{driver\ pulley} = \frac{0.24}{0.08} = 3$$

Chain drive stage 1:

$$R2 = \frac{driven\ sprocket}{driver\ sprocket} = \frac{70}{20} = 3.2$$

Chain drive Stage 2:

$$R3 = \frac{driven\ sprocket}{driver\ sprocket} = \frac{70}{20} = 3.2$$

Chain drive Stage 3:

$$R4 = \frac{driven\ sprocket}{driver\ sprocket} = \frac{70}{20} = 3.2$$

Output RPM:

$$Output = \frac{1500}{100} = 15RPM$$

The belt length was calculated using Eq. 11, where C is the center distance and $D_L + D_S$ are pulley diameters.

$$L = 2C + \frac{\pi}{2}(D_L + D_S) + \frac{(D_L - D_S)^2}{4C} \quad (11)$$
$$L = 2(0.5) + \frac{\pi}{2}(0.24 + 0.08) + \frac{(0.24 - 0.08)^2}{4(0.5)}$$
$$L = 1.52m$$

Belt speed was determined by using eq. 12. The design power rating confirmed that a Type A V-belt was appropriate, offering reliability and ease of replacement in rural settings.

$$V = \frac{\pi \times D_S \times N_S}{1000} \quad (12)$$

Where: V= belt speed
Ds=diameter of small pulley
Ns= RPM of small pulley

$$V = \frac{\pi \times 80mm \times 1500RPM}{1000\frac{mm}{m}} = 376.99\frac{m}{min}$$

Bearings. Bearing loads were evaluated using eq. 14, where R is the radial load and T is thrust load. The required dynamic capacity was calculated with Eq. 15, ensuring that the selected bearings could sustain operational loads over the intended service life.

$$Radial\ Load = \frac{L_S}{2} \quad (13)$$

Where: Ls= Load on Shaft

$$Radial\ Load = \frac{2997.81}{2} = 1498.91N$$

Calculate equivalent load P for various bearing types. Use X and Y values from Table 4 select the value from single-row ball bearing (assume zero thrust load).

$$P = XR \times YT \quad (14)$$

$$P = (1.0 \times 1498.91) + 0 = 1498.91N$$

Calculate the required capacity, Cr. Use a rated life value of 3,000 h (Table 6) and a k and Z value of 3 and 25.6, respectively.

$$C_r = \frac{P(L_{10}N)^{\frac{1}{k}}}{Z} \quad (15)$$

Cr = required capacity, N N = rotational speed, rpm
L10 = rated life, h (Table 6) K = constant K = 3 for ball bearings
P = equivalent radial load, N Z = constant Z = 25.6 for ball bearings

$$C_r = \frac{(1498.91N)(3000hr \times 1500rpm)^{\frac{1}{3}}}{25.6}$$

$$C_r = 9666.58\ N$$

These calculations collectively ensured that each component—shaft, motor, chain, belt, and bearings—was appropriately sized for the operational demands of coconut dehusking. By systematically applying mechanical design principles, the prototype achieved both structural integrity and functional efficiency, while remaining practical for small-scale farmers.

Results and Discussion

Results of Tests

Table 3. Evaluation test for processing time

Trial	No. of Coconut Input	Ave. Coconut Size (in)	Total Dehusking Time (s)	Average Time per Coconut (s)	Std. Dev. (s)	Error Margin (±s)
1	10	8.10	88.64	11.08	0.9	±1.2
2	10	7.20	71.61	10.23	1.0	±1.3
3	10	7.50	99.76	12.47	1.1	±1.4
4	10	7.80	121.86	13.54	1.2	±1.5
5	10	8.05	125.64	13.96	1.3	±1.6
Average	10	7.730	101.50	12.26	1.10	±1.4

Across the five trials, the average dehusking time per coconut was 12.26 seconds, with a standard deviation of 1.1 seconds and an error margin of ±1.4 seconds. These values indicate consistent performance with limited variability, supporting the reliability of the prototype's time efficiency results.

Table 4. *Evaluation test for efficiency*

Trial	No. of Coconut Input	Successfully Dehusked	Damaged Nut	Efficiency (%)	Loss (%)	Std. Dev. (Efficiency)	Error Margin (±%)
1	10	8	2	80	20	2.5	±3
2	10	7	3	70	30	2.8	±3.5
3	10	8	2	80	20	2.5	±3
4	10	9	1	90	10	2.2	±2.8
5	10	9	1	90	10	2.2	±2.8
Average	**10**	8.2	1.8	82	18	3.1	±4

Table 4 presents the efficiency results across five trials, showing an average efficiency of 82% with a standard deviation of 3.1% and an error margin of ±4%. These statistical indicators demonstrate consistent performance with limited variability, thereby supporting the reliability of the prototype's efficiency outcomes.

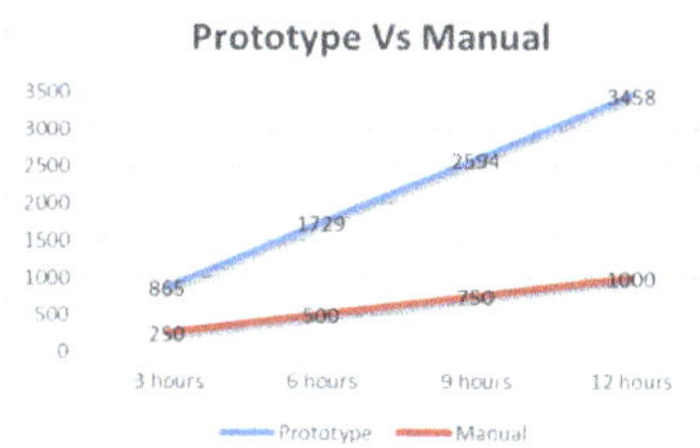

Figure 1. Prototype vs manual dehusking

Coconut dehusking efficiency by the proposed prototype machine versus manual techniques of dehusking over the time has been given in graphical form. The prototype is better than manual intervention in every respect, dehusking a far greater number of coconuts at every time interval. For example, the manual method requires about 12 hours to dehusks 1000 coconuts, while the prototype takes only 3.47 hours for the same number, about 3.46 times faster.

For the entire period of 12 hours, the prototype was able to dehusks 3458 coconuts in contrast to only 1000 by manual effort, which implies a productivity increase of 245.8%. This clearly demonstrates that the prototype is a very efficient and scalable solution for coconut dehusking.

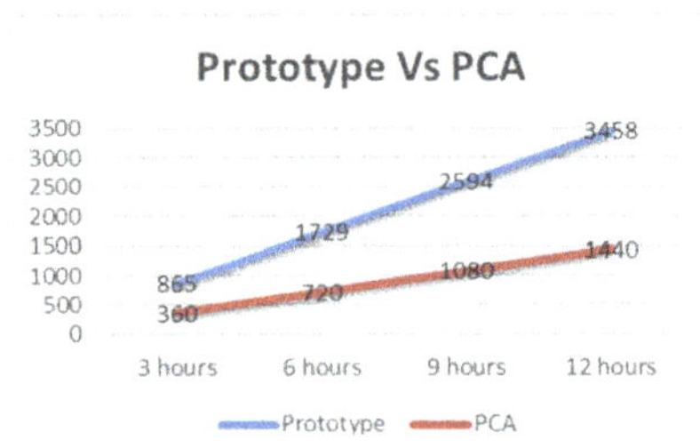

Figure 2. Prototype vs PCA

The performance of the Philippine Coconut Authority's standard dehusking system and the proposed prototype was compared. While a PCA design was able to dehusk 1,440 coconuts in a 12-hour operation, the prototype design accomplished dehusking of 3,458 coconuts in the same time frame.

In fact, almost 140.14% improvement in productivity is achieved, which clearly indicates increased efficiency of the prototype. In terms of speed, the prototype has approximately 2.4 times the speed of the PCA design. This improvement assures the prototype's capability and great potential to now be competent in saving labor hours, increasing productivity, and thus providing operational efficiency in coconut operation.

Alongside the technical performance, the economic viability of the prototype is an important consideration. The fabrication cost of the machine was Php 42,820, while the annual cost of manual dehusking is approximately Php 12,600. Using these figures, the estimated payback period is about 3.4 years, meaning that the investment in the prototype could be recovered within this timeframe through savings in labor costs. Over a five-year horizon, the return on investment (ROI) is projected at roughly 47%, indicating that the prototype not only improves efficiency and safety but also offers long-term financial benefits for small-scale farmers. These preliminary calculations highlight the potential economic advantage of mechanized dehusking, though future studies should refine this analysis by incorporating variability in labor rates, maintenance costs, and farm sizes.

Prototype model

Figure 3. Conceptual Model (isometric view)

Figure 4. Conceptual Model (top view)

Figure 5. Conceptual Model (side view)

Conclusion

This study successfully designed, fabricated, and evaluated a gas-powered coconut dehusking machine tailored for small-scale farmers in Barangay Tumbaga II, Sariaya, Quezon Province. The prototype demonstrated substantial improvements in operational efficiency, reducing dehusking time from 12 hours manually to only 3.47 hours for 1,000 coconuts. Over a 12-hour period, the machine processed 3,458 coconuts compared to 1,000 coconuts manually, representing a 245.8% increase in productivity. When benchmarked against the Philippine Coconut Authority's (PCA) machine, the prototype achieved a 140.14% improvement in output and operated at approximately 2.4 times the speed. The design incorporated durable materials, safety features, and user-friendly operation, making it adaptable for rural farming conditions. Although the efficiency rate of 82% was high, the 18% damage rate highlights the need for refinements in blade alignment, feed systems, and pressure control.

Overall, the results confirm that the prototype is a workable and scalable solution that lessens physical strain, increases safety, and greatly boosts coconut growers' production. It contributes to the sustainable future of coconut farming communities by offering a practical substitute for both manual dehusking and current mechanized technologies. Based on the results, it is recommended that future studies expand the sample size of experimental trials to strengthen statistical validity, refine blade and feed mechanisms to minimize coconut damage, and conduct durability testing under diverse farm environments to ensure robustness. In addition, exploring cost optimization strategies will make the machine more affordable and accessible to small-scale farmers, thereby maximizing its impact as a community asset. In addition to these technical and economic considerations, future work should also include a formal safety assessment and risk analysis to quantitatively validate the claimed safety improvements and further enhance farmer protection.

References

[1] Varghese, A., & Jippu, J. (2014). Proceedings of the International Conference on Emerging Trends in Engineering and Management (ICETEM14), December 30-31, 2014, Ernakulam, India. International Journal of Design and Manufacturing Technology.

[2] Varghese, F. K., & Jippu, J. (2016). A study of physical and mechanical properties of the Indian coconut for efficient dehusking. *Journal of Natural Fibers*. https://doi.org/10.1080/15440478.2016.1212760

[3] Philippine Coconut Authority. (2018). *PH Coconut Industry Statistics*. Retrieved from http://www.pca.da.gov.ph/index.php/2015-10-26-03-15-57/2015-10-26-03-22-41

[4] Punchihewa, P. G., & Arancon, R. N. (1999). *Coconut: Post harvest operations*. Asian and Pacific Coconut Community (APCC). Retrieved from www.apcc.org.sg

[5] Ghosal, M. K., & Mohanty, S. K. (2011). Coconut de-husking machine (US patent No. 4,708,056).Dinanath, Chaguanas, Trinidad and Tobago.

[6] Khurmi, R. S., & Gupta, J. K. (2005). A textbook of machine design (14th ed.). Eurasia Publishing House.

[7] Oberg, E., Jones, F. D., Horton, H. L., Ryffel, H. H., & McCauley, C. J. (Eds.). (2016). Machinery's handbook: A reference book for the mechanical engineer, designer, manufacturing engineer, draftsman, toolmaker and machinist (30th ed.). Industrial Press

Design of Local Exhaust Ventilation with Addition of Electro Cyclone Filter Using Computational Fluid Dynamics

Kyle Ysabel Baquir[1,a], Randell John Andulan[1,b], James Kenneth Bollena[1,c], Raighnailt Canja[1,d], Christian Aldwin Canlapan[1,e], Luvisminda Marcelo[1,f*]

[1]Mechanical Engineering Department, National University, 551 M.F. Jhocson St. Sampaloc, Manila, Philippines

[a]baquirkt@students.national-u.edu.ph, [b]andulanrg@students.national-u.edu.ph, [c]bollenajs@students.national-u.edu.ph, [d]canjari@students.national-u.edu.ph, [e]cadcanlapan@national-u.edu.ph, [f]ldmarcelo@national-u.edu.ph

Keywords: Local Exhaust Ventilation (LEV), Electro-Cyclone, Filtration, Capture Efficiency, Separation Efficiency

Abstract. This study features a dust collection system consisting of local exhaust ventilation (LEV) and electro cyclone filtration. The researchers designed a dust collection system to achieve the desired air quality in compliance with air and safety regulations. Computational Fluid Dynamics was performed to calculate the system's capture and separation efficiency. The simulation consisted of five tests with different time dependencies--1 second, 1 minute, 1 hour, 8 hours daily, and 40 hours weekly. The boundary conditions set were -- inlet and outlet velocities, ambient temperature, atmospheric pressure, and wood dust particle diameter (1 micron, 5 microns, and 10 microns). The calculated overall capture efficiency of the LEV was 78.89507%., while the electro-cyclone's separation efficiency was 74.18424% for the mass flow rate and 74.1727% for the volume flow rate. Achieving the objective efficiency above 70% makes the design project practical and feasible.

Introduction

Wood manufacturing industries are among the leading industries in the world and the Philippines. Furniture, modular, and fit-out productions are the most common products from the wood industry. The latest data from 2018 shows that the world demands 188 million tons of wood for its production and over 66 million tons for trading. The country's wood production in Philippines alone consumes six million tons of wood. Dust is generated during the production of woodwork, and when these particles are not agglomerated and disposed of properly, it becomes not only pollution but also a health hazard. The amount of wood dust generated from manufacturing amounts to 6% of the total wood used, roughly around 11.28 million tons annually. This dust poses a significant threat to the environment and human health.

The Philippines enacted Republic Act No. 8749, known as the "Philippine Clean Air Act of 1999," which defines an air pollutant as any particulate matter in the environment other than the naturally existing elements and compounds like oxygen, nitrogen, carbon dioxide, etc. This law solidifies that dust produced from wood processes like edging, trimming, sawing, and sanding is an air pollutant once it dissipates in the atmosphere without proper filtration and mitigation.
Aside from being an air pollutant, wood dust can also cause contamination and health-related effects. Dust particles smaller than 10 micrometers are non-respirable, which causes allergic reactions and is detrimental to human health, especially those directly exposed to wood dust.

For a wood manufacturing plant like Cornersteel Systems Corporation, a plant base in Brgy Pulo, Cabuyao, Laguna, Philippines it is essential to have a safe and clean space to produce good quality products and maintain the occupational safety and health of their workers. In the sanding area of their wood processing, the facility is not equipped with a sound ventilation system that

Materials Research Proceedings 66 (2026) 213-223 https://doi.org/10.21741/9781644904152-19

captures the dust particles and a filter before releasing the air to the surroundings. To regulate the circulation of wood dust in the area, the administrators decided to open the facility and let natural airflow, which made the wood dust spread.

In different studies conducted about the relevance of good ventilation and filtration systems in the workplace, it is believed that a well-designed dust collection system can increase the workers' productivity, improve employees' morale, and comply with safety regulations. Dust collection involves the process of dirt and dust removal from a flowing air stream to abide by health and safety regulations. This control system uses the principles of ventilation in capturing dust-filled airstream and removing the particles away from exposure through the ductwork and a filtering device to remove the particulates from the airstream before its release into the atmosphere. [1]

To help provide better air quality, the designers aim to design a dust collection system composed of local exhaust ventilation (LEV) to capture the dust particles and an electro-cyclone device, a filter less and centrifugal operated system, in separating dust particles from the airstream.

Literature Review

Overview of the Review Process. In this review process, the research focused on modeling and simulating ventilation and filtration system designs suitable for a woodwork sanding strategy. Particular solutions to be considered in this design project are discussed, namely hybrid ventilation, local exhaust ventilation (LEV), and natural ventilation on wood manufacturing set-up, as well as an electro-cyclone, cyclone with cartridge filter, and baghouse filter for micro and nanoparticle filtration are discussed on this chapter. The use and importance of Computational Fluid Dynamics (CFD) analysis for ventilation and filtration were also presented through literary studies.

Woodwork. Woodworking pertains to creating cabinetry, instruments, fixtures, and other objects out of wood. It involves jointing, planning, routing, sawing, drilling, gluing, sanding, and finishing [2]. Sanding is a machining operation used to produce a smooth and good surface quality of metal and wood.

Wood Dust. Wood dust is an accumulation of any wood particulate generated during the processing or handling of wood. When this dust becomes airborne, workers can inhale it, leading to mucosal irritation, allergies, and respiratory system cancer [3].

Wood Dust Pollution. Woodwork manufacturing industries are among the leading contributors to dust emissions [3]. A research study conducted in Nigeria recorded that the most significant air pollution problem in the country comes from atmospheric wood and sawdust produced by local sawmills [3].

Ventilation. Ventilation is the introduction of external air into a place. The primary purpose of ventilation is to manage indoor air quality by diluting and dispersing indoor contaminants. There are three types of ventilation systems: natural, mechanical, and hybrid ventilation [4].

Local Exhaust Ventilation. The primary purpose of the LEV system is to remove contaminants from our activities in the building to achieve healthy air in the building. LEV has many functions, such as controlling highly toxic pollutants, handling dust clouds, and a small amount of makeup air due to another amount of air being exhausted. Usage and maintenance of the LEV system have been published as standard safety. The LEV system needs to be functional. For instance, if grinding is involved in the process, the system must likely be created to collect and absorb any dust that is discharged from the elusive source. A thorough knowledge of pollutants is necessary before designing or installing an LEV system, and the needs of the procedure are also required.

Filtration. Filtration pertains to the process of separating other particulate matter from the fluid. Filters are components or systems that eliminate impurities, such as dust, dirt, electronic signals, etc., as they filter material or devices. Filters are available for filtering electrical and optical phenomena, air, gases, and fluids. Filters are used to clean the air. Fluid filters clean fluids of particles. While hydraulic filters clean up pollutants from hydraulic fluids, fuel filters clean up contaminants from fuels. Water treatment facilities use filters to clean the water before it is used for swimming and drinking. These many filter types are used in industrial filtration to filter fluids or gases [5].

Electro-Cyclone Filtration System. To collect particles smaller and finer than 1 μm, it was studied to employ additional means. In a Ph.D. dissertation conducted by Wadenpohl, the author studied the addition of an electrostatic effect on the cyclone filtration system to collect nanosized particles. In this study, a modified cyclone filter was used by adding a central electrode; diesel soot particulates were separated from an automotive diesel engine. "Wadenpohl showed that primary particles of 10–30 nm diameter are transported to the cyclone wall through electric field forces, forming fairly large aggregates of few microns in diameter. Collected particles may be recharged to the potential of the collecting electrode. Thus, the field lines of the electric field are focused so that the collected particles act as nuclei for the dendritic growth of aggregates. Two mechanisms can explain the separation of the aggregates from the collecting surface. Both image forces induced in the collected structures, which are directed towards the spray electrode, and fluid forces in the boundary layer can lead to the removal of the aggregates from the collecting surface. The particles are then transported in the boundary layer flow to a small outlet where 5–10% of the total flow rate is extracted. This principle was upscaled to flow rates of several 1000 m/hr. This cyclone was tested after a stationary diesel engine. The mass-specific separation efficiency was found to range from 50% to 80%, depending on the flow rate and the applied voltage." [6]

Result of the Review of Literature. Through the conduct of a literature review, the designers were able to assess what type of ventilation and filtration system best suited the project. Considering the location of the sanding area, it is nearly impossible to use purely natural ventilation due to its proximity to other facilities inside the plant. Thus, more natural ventilation is needed to consider. This goes with hybrid ventilation as well, since hybrid ventilation partly relies on the natural air supply. The designers concluded that designing a local exhaust ventilation system is the best option for the sanding area. Supporting this design claim was a literature study that defines to capture inhalable wood dust, the use of local exhaust ventilation (LEV) is recommended because this design type is suitable to control dust with higher kinetic energy [7].

Different considerations based on the literature review were also made for filtration. Baghouse filters can capture microparticles but cannot capture nanosized dust particles, which is not applicable for a wood sanding process that produces less than one micrometer of dust diameter. Both cyclones with cartridge and electro-cyclone filtration can capture micro- to nanoparticles, so the designers studied a cost-benefit analysis. It has been concluded that a cyclone with a cartridge is considered low-cost but high-maintenance, while an electro-cyclone is high-cost but low-maintenance [8].

After several considerations agreed upon by the researchers with past studies and literature to support their design option, local exhaust ventilation and electro-cyclone will be utilized. The ducting system of the area will be centralized and connected to a blower with the required ventilating capacity. Then, the system will be connected to an electro-cyclone filter separating bigger wood dust particles from the air through centrifugal force and an electrode system that will capture smaller dust particles.

Methodology

Design Requirement and Specification

Location. The researchers designed local exhaust ventilation for the Sanding Area of Cornersteel Systems Corporation, a manufacturing plant in Brgy. Pulo, City of Cabuyao, Laguna. The plant is comprised of different facilities for metalworks and woodworks. The woodwork area is comprised of three *stations, namely, sanding, pre-varnishing, and varnishing* stations. The total area of the three working areas is 174.42 sqm.

Exhaust Rate. Based on the American Society of Heating, Refrigerating, and Air-Conditioning Engineers (ASHRAE) Handbook, Table 6-2, Minimum Exhaust Rates, the recommended exhaust rate for a woodwork shop is 0.5 cfm/ft2 [9,10].

Particle Size. Particle size plays a role in determining the overall efficiency of the electro-cyclone, especially in determining the particle trajectory. For woodwork manufacturing, specifically in the sanding process, particle sizes vary from 1-10 micrometers.

Volumetric Flow Rate. For calculating the volumetric flow rate, see Eq. 1:

$$Q = AV \tag{1}$$

Wherein: Q = volumetric flow rate, cfm
A = Area, ft^2
V = velocity/ exhaust rate, cfm/ft^2

Air Change per Hour. For an adequate ventilation and dust-collecting system design, it is essential to consider the air change per hour needed for a space. For a wood workshop, it is recommended to use eight (8) ACH [11].

Test Plan

Design Modeling. Modeling is creating a three-dimensional object to make a design come to life without the process of prototyping. In software engineering, design modeling represents the software's architecture, user interface, and component-level details—all essential for engineers to create software efficiently. Like an architectural blueprint for a house or building, design modeling offers a range of various viewpoints of the system. The design model uses various techniques, including data-driven, pattern-driven, and object-oriented methodologies. For creating a model, each of these techniques follows a set of design principles. [12]

Computational Fluid Dynamics. Computational Fluid Dynamics (CFD) is a type of simulation testing performed to determine if a design system works well without the cost of prototyping. The equations regulating fluid motion can be applied to various complex circumstances using computational fluid dynamics (CFD), which offers insight and quantitative forecasts. Discrete approximations at grid points are used in place of the fluid equations, and they must be near enough for the solution to be independent of the grid point spacing [13].

The CFD simulation was performed using FreeCAD with an integrated CFD solver to analyze the airflow behavior within the local exhaust ventilation and electro-cyclone system. The simulation solves the Reynolds-Averaged Navier–Stokes (RANS) equations to model turbulent airflow conditions present in the ventilation system. The boundary conditions applied in the simulation include an inlet velocity of 20.32 m/s, outlet velocity of 17.78 m/s, ambient temperature of 30°C, atmospheric pressure of 101325 Pa, and wood dust particle diameters of 1 μm, 5 μm, and 10 μm.

Efficiency. One measure of exhaust hood performance is capturing efficiency. This is defined as the ratio between the rate of wood dust flow velocity captured by the exhaust system to the duct velocity. For capture efficiency, refer to Eq. 2

$$CE = \frac{average\ velocity\ captured}{air\ duct\ velocity}(100\%) \tag{2}$$

The ratio of the amount of material already separated in the system (such as a filter or centrifuge) to the amount of material that must be separated when it enters the system is known as separation efficiency. For efficiency of separation, refer to Eq. 3:

$$SE = \left[\frac{Cv\ out}{Cv\ in}\right]\ x\ 100\% \tag{3}$$

The designers followed the fundamentals of adequately designing the area's ventilation system. Identifying the components to be filtered is to attain the proper values to use in computing the flow rate and other design requirements. In this case, the fumes to be screened are wood dust. The total flow rate required is Q_{total}≈5820 cfm.

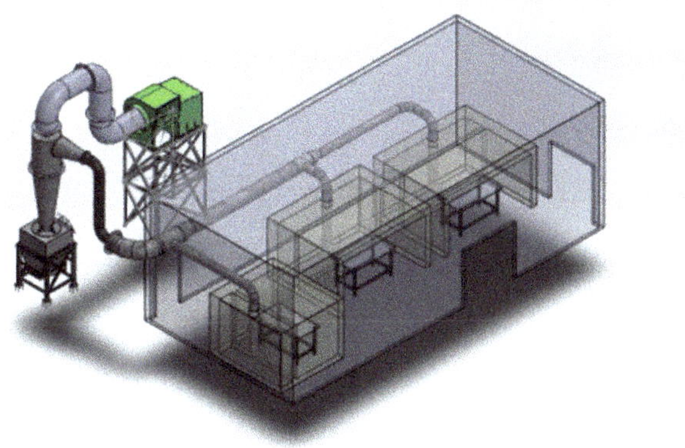

Fig. 1. Ventilation Layout on Solidworks 2019, Isometric View

The schematic diagram in Fig. 2 shows how the system operates from the LEV to the electro-cyclone. Generated wood dust from using the wood sander will be enclosed in the booth to prevent the dust particles from dissipating around the sanding area. The distance of the person using the wood sander from the grilles of the booth was 0.8 meters, based on the preferred measures of the stakeholder. Through the exhaust duct of the LEV, it will suction the wood dust that will flow through the branches up to the central vent. The primary duct is linked to the electro-cyclone, where the wood dust particles are separated from the airstream. Through centrifugal forces acting in the electro-cyclone and the rod, wood dust particles with diameters of 1, 5, and 10 micrometers will sit on the outlet underflow of the system. Filtered air, with little to no wood dust particles, will exit the electro-cyclone to the outlet overflow and go directly to the centrifugal fan to eject the filtered air into the atmosphere.

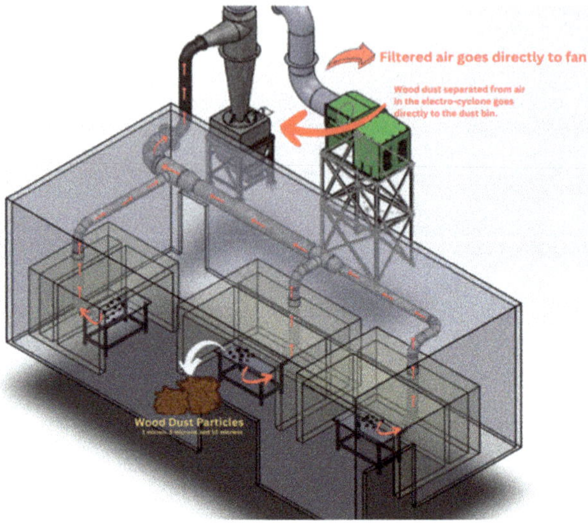

Fig. 2. Schematic Diagram of the System Operation

Results

Displayed in Table 1 is the summary of results of the efficiency generated for the mass and volume flow rate in each of the test simulations conducted. Based on the mass and volume flow rate in the first second of the electro-cyclone's operation, the system generated a separation efficiency of 76.75~9% and significantly dropped to 73.54% as the operation time went by. As the operation time of the electro-cyclone increases, the efficiency obtained decreases.

Table 1. Result Summary of Electro-cyclone Efficiency v Time

Mass Flow Rate (kg/s)			Volume Flow Rate (m^3/s)		
Test	**Time**	**Efficiency (%)**	**Test**	**Time**	**Efficiency (%)**
1	1 s	76.7879%	1	1 s	76.74885%
2	1 min	73.5275%	2	1 min	73.5241%
3	1 hr.	73.5275%	3	1 hr.	73.546%
4	1 day	73.5146%	4	1 day	73.5113%
5	1 week	73.5367%	5	1 week	73.5332%

The separation efficiency of the electro-cyclone over the time set per test is depicted in Figure 67. The graph shows that the system's efficiency slightly declined as the trial progressed for the mass and volume flow rates entering and leaving the system.

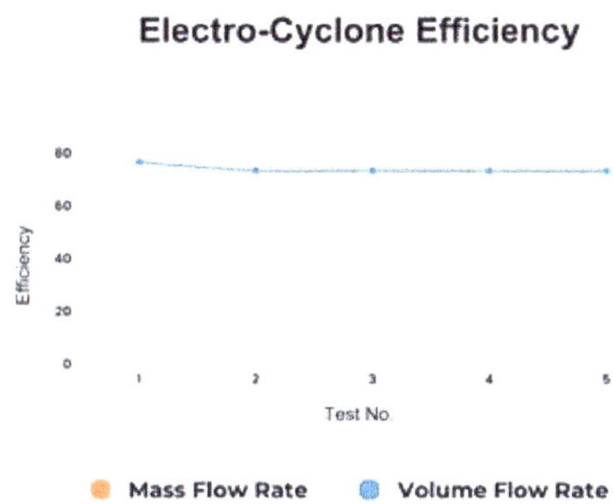

Fig. 3 Electro-cyclone efficiency v test

Lastly, based on the mass accumulation rate generated from the test 4 of the LEV, with 8 hours of working time, equivalent to a day of operation, the accumulated rate is 81.27498 kg/ m^2-day with a surface area of 0.128345654 m^2.

Table 2. Accumulation Rate of Wood Dust a Day

Mass Accumulation Rate (kg/ m^2-day)	Surface Area (m^2)
81.27498	0.128345654

Comparatively, the average amount of wood dust that can be absorbed by the LEV within a day of operation, based on the data gathered from Solidworks 2019 Flow Simulation is 10.4312905 kg. This result is 347.7096833% higher than the current amount of wood dust that Cornersteel Systems Corporation capture in the sanding area using the existing ventilation design, which is approximately 3 kg per day.

Additionally, the wood dust bin has a storing capacity of 1.3125 m^3, given that the wood dust accumulation is 10.4312905 kg and the separation efficiency of the electro-cyclone is 73.511334%, the wood dust that can be filled in the bin in a day is 7.668108 kg. With a density of 545 kg/ m^3 , the bin can be filled within 93 days.

Discussion

Based on the data the designers gathered through the simulation for each of the five time-dependent tests with input parameters of inlet velocity, 20.32 m/s, outlet velocity of 17.78 m/s, ambient temperature of 30°C, standard environmental pressure at 101325 Pa, and wood dust particle diameters of 1, 5, and 10 micrometers. Test 1 for LEV showed that the mass and volume flow rate ejected by the system was the same as the rate entering the system. With a mass flow ratio of 1.00000032 and a volume flow rate ratio of 1.004823856, the LEV suctioned most of the air dust in the sanding area. Whilst the average capture airflow velocity in the LEV runs at 14.02608592 m/s, garnering a capture efficiency of 78.8868714%. For test 2 of the LEV with a running operation time of 1 minute, the recorded mass and volume flow rate ratio were 0.999999901 and 0.995177834, respectively, and the average velocity rated at 14.02736702 m/s, having an efficiency of 78.89407773%. Test 3 obtained a capture efficiency of 78.89389904% with an average speed of 14.02733525 m/s within an hour of operating time. On the recorded data from Test 4 of LEV's capture efficiency, the recorded ratio for the mass and volume flow rate were 0.9999997144 and 0.99517662011, respectively. The rated velocity was at 14.02817191 m/s, leading to an efficiency of 78.89860467%. Lastly, for Test 5, with an operation time of 40 hours a week, the calculated capture efficiency was 78.9018843% based on a mean velocity of 14.02875503 m/s, having a mass and volume flow ratio of 0.999999676 and 1.00482367 on the inlet and outlet surfaces of the LEV. According to these results, the capture efficiency of the system slightly increased as the operation time went by.

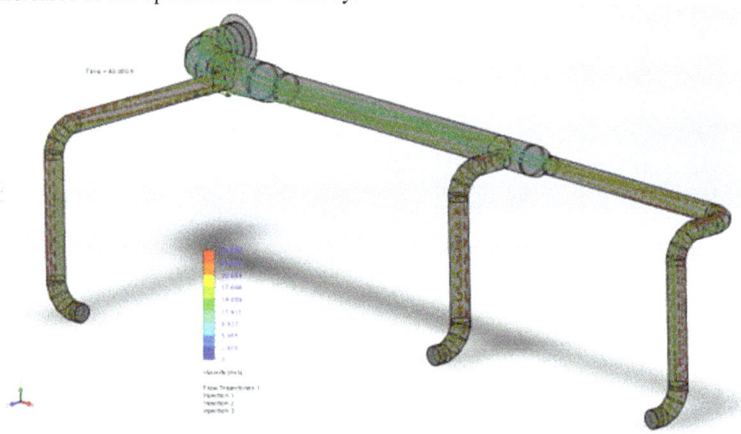

Fig 4. Particle Study on LEV

On the other hand, the recorded data from the electro-cyclone displays how efficient is the separation rate of the wood dust particles from the flowing air. On test 1, under an operation time of 1 second, the system showed a high separation rate of 76.78793419% in terms of mass flow rate and 76.74884669% in terms of volume flow rate. Thus, more than three-quarters of the wood dust particles flowing inside the sanding area were separated from the airstream before its release into the atmosphere. For test 2, with a minute of operating time, the separation efficiency recorded was 73.52749238% and 73.52410351%, based on the mass and volume flow rate recorded on the inlet and outlet underflow, respectively. Under an hour of running time on test 3, the separation efficiency of the electro-cyclone was calculated to be 73.52752393% based on the mass flow, and as for volume flow rate, the efficiency is 73.54597377%. For test 4, with an operation time of eight hours within a day, the recorded separation rate was 73.51462994% from the mass flow rate, and as for the volume flow rate, the efficiency was 73.511334%. Lastly, under a 40-hour operation time in a week, the efficiency of the mass flow rate resulted in 73.53667719%; for the volume flow rate, the efficiency is 73.53322512%. Based on these data of separation efficiencies, the designed electro-cyclone can effectively separate the wood dust particles flowing from the air; also, the values range from 73% to ~79%, which is already a high separation rate in terms of the particle diameter it separates. Lastly, based on the figures gathered from the simulation tests, it can be observed that the air velocity within the electro-cyclone flowed from a high-velocity rate and then exited the device at an average speed.

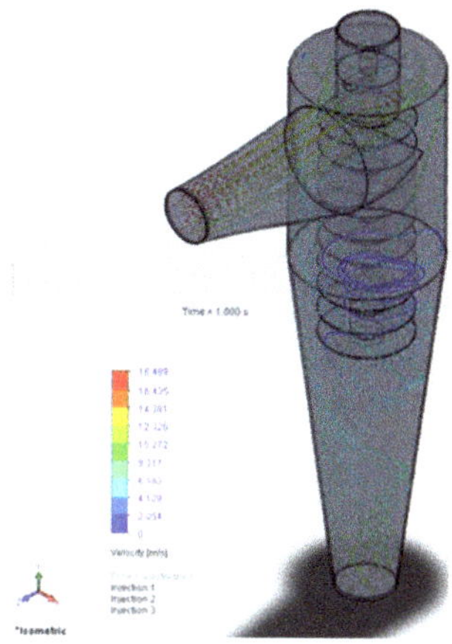

Fig 5. Particle Study on Electro-cyclone

Moreover, based on the mass accumulation rate, 10.4312905 kg of wood dust are captured by the LEV within a day of operation and the separated amount of wood dust by the electro-cyclone is 7.668108 kg.

Importantly, the health relevance of the proposed system can be contextualized against occupational exposure standards. The Occupational Safety and Health Administration (OSHA) sets a permissible exposure limit (PEL) of 5 mg/m^3 for respirable wood dust and 15 mg/m^3 for total wood dust (8-hour TWA), while the American Conference of Governmental Industrial Hygienists (ACGIH) recommends a threshold limit value (TLV) of 1 mg/m^3 for certain hardwood dusts due to their carcinogenic potential. In the Philippines, occupational health regulations under the Labor Code adopt threshold limit values consistent with international practice. By maintaining capture and separation efficiencies above 70%, the designed LEV and electro-cyclone system demonstrates its potential to keep airborne wood dust concentrations below these limits, thereby ensuring compliance with both national and international occupational safety standards [14,15,16].

Conclusion

The study aims to provide a dust collection system consisting of local exhaust ventilation and electro-cyclone filtration for the sanding area of the woodwork section of Cornersteel Systems Corporation. The designers redesigned the sanding site to achieve the desired air quality in compliance with air and safety regulations. In conclusion,

1. The researchers calculated the appropriate ventilation requirements of the sanding room with an overall rate of 8,500 CFM. The electro-cyclone required a volumetric flow rate of 2,300 CFM, while the booths with sander equipment required 5,820 CFM.
2. The study analyzed local exhaust ventilation and electro-cyclone flow, revealing a velocity gradient as air velocity decreases from the central duct to the outlet. The highest particle velocity was recorded in the duct inlet — more particles sit on the outlet underflow, indicating more wood dust exits and goes to the dust bin.
3. The pressure loss calculation for exhaust and fresh air ducts is directly proportional to the velocity. The total pressure drop from the blower to the central duct is 75.6175 Pascal, while 152.33083 Pascal is for the branch duct—72.05939 Pascal for the fresh air ducts, and 69.72308 Pascal for the branches.

4. The researchers achieved a target capture efficiency of over 70% for the LEV, with an average of 78.89507% for the local exhaust ventilation system. The capture efficiency increased with time input, while the electro-cyclone separation efficiency was above 70%. The separation efficiency of the electro-cyclone was 74.18424% in terms of mass flow rate and 74.1727% for volume flow rate, with an inverse relationship with simulation time.

Although the results demonstrate that the designed local exhaust ventilation and electro-cyclone system can effectively capture and separate wood dust particles based on CFD simulations, the present study relies solely on computational analysis. Experimental validation was not conducted in this work. Therefore, future studies are recommended to perform experimental testing and field validation of the proposed system to further verify the numerical findings and enhance the real-world applicability of the design.

Additionally, mesh independence testing is recommended to confirm that the CFD results are not sensitive to grid resolution. Conducting simulations with progressively refined meshes and comparing velocity fields, pressure drops, and efficiency values will strengthen the reliability of the reported capture and separation efficiencies and provide greater confidence in the robustness of the proposed design

References

[1] Bhuiyan, Z. (n.d.). Design analysis of Dust collection system QA & QC of a Denim Fabrics and Quality Manual View project Silver-Incorporated Pressure Garment View project. https://www.researchgate.net/publication/338541015

[2] Arzt, K. (n.d.). Woodworking 101: A Beginners Guide to Getting Started. Retrieved from The Crucible: Guides to Industrial Art: https://www.thecrucible.org/guides/woodworking/

[3] Proto, A. R., Zimbalatti, G., & Negri, M. (2010). THE MEASUREMENT AND DISTRIBUTION OF WOOD DUST. J. of Ag. Eng. - Riv. Di Ing. Agr., 1, 25–31.

[4] Ventilation at work. (n.d.). Retrieved from Safework: https://www.safework.nsw.gov.au/hazards-a-z/ventilation-at-work

[5] Industrial Filtration Process: Steps, Importance, and Uses. (n.d.). Retrieved from Thomas: A Xometry Company: https://www.thomasnet.com/articles/machinery-tools-supplies/industrial-filtration-process/

[6] Peukert, W., & Wadenpohl, C. (2001). Industrial separation of fine particles with difficult dust properties. Powder Technology, 118(1–2), 136–148. https://doi.org/10.1016/S0032-5910(01)00304-7

[7] Ojima, J. (2016). Generation rate and particle size distribution of wood dust by handheld sanding operation. Journal of Occupational Health, 58(6), 640–643. https://doi.org/10.1539/joh.16-0136-BR

[8] Chen, C. J., & Wang, L. F. S. (2001). Cost-benefit analysis of electro-cyclone and cyclone. Resources, Conservation and Recycling, 31(4), 285–292. https://doi.org/10.1016/S0921-3449(00)00086-0

[9] American Society of Heating, Refrigerating, and Air-Conditioning Engineers. (2019). ASHRAE Handbook—HVAC Applications. Atlanta, GA: ASHRAE

[10] Falih, M. (2020). Minimum exhaust rates for woodwork shops: Application of ASHRAE standards. International Journal of Engineering Research and Technology, 9(5), 1123–1130.

[11] Air Change Rates in Typical Rooms and Buildings. (2005). ASHRAE Handbook—Fundamentals. Atlanta, GA: American Society of Heating, Refrigerating, and Air-Conditioning Engineers (ASHRAE)

[12] Tawde, S. (2022). Design Modeling in Software Engineering. Retrieved from EducBA: Tawde, S. (2022). Design Modeling in Software Engineering. Retrieved from EducBA:

[13] Tryggvason, G. (2016). Computational fluid dynamics. In Fluid Mechanics (pp. 153–190). Elsevier. https://doi.org/10.1016/B978-0-12-405935-1.00006-X

[14] Occupational Safety and Health Administration (OSHA), Wood Dust – Permissible Exposure Limits (PEL). U.S. Department of Labor, 2023. [Online]. Available: https://www.osha.gov/wood-dust

[15] American Conference of Governmental Industrial Hygienists (ACGIH), Threshold Limit Values (TLVs) and Biological Exposure Indices (BEIs). Cincinnati, OH: ACGIH, 2016.

[16] Department of Labor and Employment (DOLE), Occupational Safety and Health Standards, Rule 1070 – Threshold Limit Values for Airborne Contaminants. Bureau of Working Conditions, Manila, Philippines, 1990.

Harnessing AI and Machine Learning for Sustainable Development: Applications, Challenges, and Future Trends

K. ARORA[1*], K. JOSHI[1], ROHINI[2], UMANG[3], R. P. JAYASWAL[4], S. Y. BUKHARI[4]

[1]Department of Computer Science & Engineering, Uttaranchal Institute of Technology, Uttaranchal University, Dehradun-248007, Uttarakhand, India

[2]Chandigarh School of Business, Department of Sciences, Chandigarh Group of Colleges, Jhanjeri, Mohali-140307, Punjab, India

[3]Department of Computer Applications, DSB Campus, Kumaun University, Nainital

[4]Department of Medical Lab Technology (UIAHS), Chandigarh University, Mohali-140413, Punjab, India

*porwalkanupriya00@gmail.com

Keywords: Artificial Intelligence, Machine Learning, Smart Energy Systems, Sustainable Development, Environmental Monitoring

Abstract: We are just beginning to understand how transformative technologies such as artificially intelligent (AI) and machine learning (ML) can potentially transform the world of sustainability in a world where environmental issues are common and sustainable development is strongly prioritized. The efficiency, flexibility, and responsiveness of existing decay detection and air quality or energy monitoring strategies are undoubtedly rudimentary. They are especially skilled in predictive analytics, real-time robotics, and intelligent-centric decision-making; when utilized in the right manner, they totally revolutionize any sustainability plan towards the accomplishment of those goals on a global scale. Therefore, the article presents a brief overview of all the applications of AI and ML in sustainable practice, from precision farming, waste reduction, climate simulation, smart energy grids, and renewable energy forecasts. Every one of these uses takes into account how these newer technologies can enhance their operations, conserve resources, and reduce their environmental footprint. These technologies also create some thought-provoking questions about their own detriments, such as bad data quality, expensive processing, ethical dilemmas, and strict regulations. Among the outcomes of the more recent inventive creations, pragmatic uses, and academic research on the topic are expressions of the critical significance, utmost care, and responsible behavior of artificial intelligence and machine learning toward a democratic and sustainable future. The action graciously defies the technological challenges; the technologies must find new ways of advancing justice and environmental awareness.

1. Introduction

That is, the application of AI and machine learning in sustainability is slowly changing our ability to solve and ultimately end problems with the environment of the planet. With the imperative of climate change, the depletion of resources, and environmental deterioration, and in addition to the benefits of the hitherto unknown capacity that they unlock for better assessment of complex systems, improved optimization of the use of resources, and facilitation of application-based data in all areas, AI and ML have matured from being obstacles to achieving sustainability objectives to requiring introspection as tools for taking stock of one's ethical and environmental imprints. "AI for sustainability" thus implies the use of smart technology to advance across all the Sustainable Development Goals (SDGs), such as clean energy, sustainable consumption, and climate action. The energy needs, environmental effects, and social consequences of creating and deploying AI systems are the opposite of sustainability in AI. Thus, the two conflicting positions will effectively deprive innovation of the necessary balance between leveraging AI's optimal potential and restraining its potential for unwarranted harm [1].

The most recent body of literature shows a growing collection of studies being published on the intersection of AI and sustainability. For example, a systematic review by Kar, Choudhary, and Singh (2022) highlights AI applications within environmental monitoring, smart agriculture, energy systems, and sustainable manufacturing. In Figure 1, the areas of AI application with the most potential for sustainability are shown across sectors.

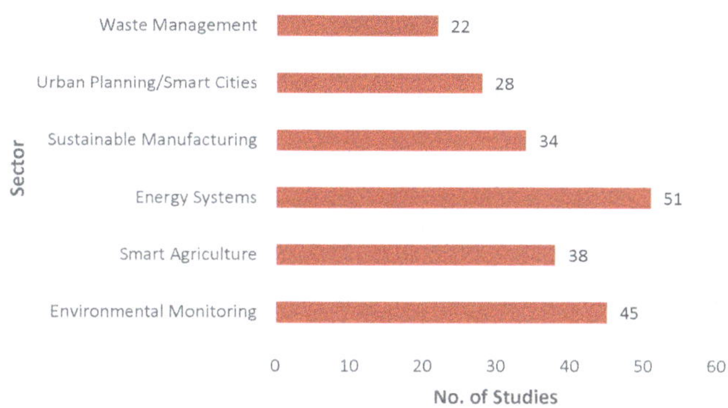

Figure 1. *AI Applications Across Sustainability Sectors [2]*

Nonetheless, this resulting interdisciplinary method, a lower type of sustainable AI systems making, requires technical, social, and policy-oriented views around ethical, data-hiding, algorithmic bias, and, finally, the digital divide in renewable energy integration towards countries threatened by flooding as they are forecasted [3-4]. In energy and environmental contexts, AI and deep learning techniques most efficiently optimize renewable energy to predict pollution and climate behavior models; however, these could also enable smarter urban systems and circular economy practices for the conservation of biodiversity and therefore take sustainability into a digitally enabled and dynamically managed goal [5]. To overwhelmed constraints, it talks about edge computing powered by AI and advanced communication protocols. Case studies are used to evaluation the contribution of embedded systems to energy efficiency and reliability. Future expansions, such as hardware evolution, machine learning for predictive management, and IoT-based smart ecosystems, are also predicted to increase efficiency by 15–25% over the next ten years [6]. Similar to Begum et al. (2024), who now state that AI for a future has not come without limits. Computing infrastructures employ a cutting-edge model when it is trained in a cost-demanding environment, with potential environmental cost implications, particularly in the wake of artificial intelligence advancements. Figure 2 presents the sustainability challenge in terms of training extremely large AI models.

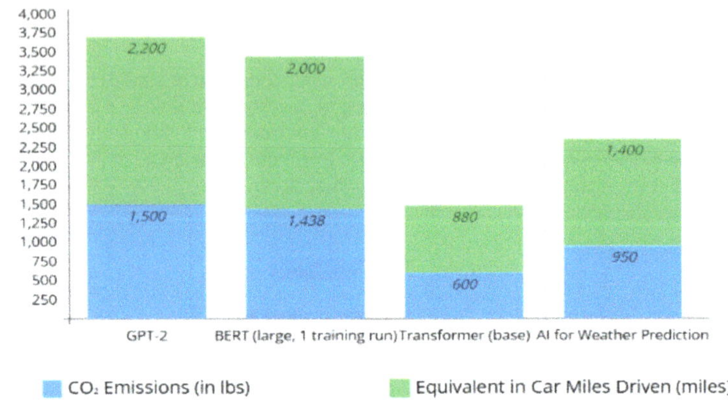

Figure 2. *Carbon Footprint of Selected AI Models (Training Phase) [7]*

Thus, sustainable AI would be deeply rooted in the energy effectiveness, openness, and equity principles [8]. Briefly, artificial intelligence (AI) and machine learning (ML) contribute dually to sustainability by enabling green transformation and covering areas of sustainability research fields Consequently, this duality continues to facilitate a long-term process that is based on the full integration of research or action from each of these resources, which would facilitate the implementation of principles of sustainability.

The authors of this paper present an analysis of artificial intelligence and machine learning technology which they use to study sustainable solutions across multiple sectors. The novelty of this review lies in three key contributions. The paper uses a dual-perspective framework to analyze AI for sustainability and the sustainability of AI which the researchers apply to energy and waste management and urban systems and decarbonisation to produce a unified analytical framework that previous studies failed to achieve. The manuscript presents a comparison of AI and ML methods which researchers evaluated across different sustainability fields to demonstrate successful applications while showing the algorithmic limitations and data-dependency problems and energy-cost trade-offs which occur when using large-scale AI systems. The review employs a systems perspective which studies emerging research gaps for model energy efficiency and data governance and ethical deployment and cross-sector integration which organizations need to develop sustainable solutions that are both scalable and fair. The manuscript develops a detailed understanding of responsible AI and ML usage for long-term sustainable development goals through its integrated critical approach which goes beyond basic descriptive content.

ML and AI have become the main drivers of environmental sustainability due to more insightful, adaptive, and data-driven approaches. These technologies make tracking, simulation, and regulation of environments that are typically too intricate to be handled through traditional means possible. Machine learning and AI models are providing insightful data for sustainable decision-making across numerous sectors while processing huge amounts of data such as satellite pictures, weather stations, and the Internet of Things. Some of the uses include predictive maintenance of power infrastructure, maximizing the efficiency of renewable energy, and optimizing energy consumption through the use of machine learning in smart grids. Trash management can employ AI in identifying the recyclables or sorting more efficiently. Other possibilities include employing machine learning with a focus on agriculture to detect pests, predict crop yields, and allocate resources for sustainable and safe food. Some examples of how

AI will assist green technology, carbon emissions reduction, and resource preservation are as follows [9].

There are a number of and well-established benefits of machine learning and artificial intelligence to sustainability. They can enhance the accuracy in sustainability modeling and forecasting, mechanize tedious, time-consuming procedures, and allow for real-time tracking of environmental indicators. Aside from doing away with much of human error and biases in environmental research, AI systems can be utilized to aid in the amplification of sustainable practices. But many will be the hurdles to be cleared when implementing AI in sustainability projects. The environmental impacts of developing and operating AI systems, especially big models usually resource and energy costly, pose a significant issue. Social-technical challenges (unequal accessibility in regions, algorithm transparency, data secrecy), too, can all contribute to operational and ethical challenges [10]. The twofold structure in significant issues of sustainability is presented in Table 1.

Table 1. *The Dual Perspective: AI for Sustainability vs. Sustainability of AI [1-10]*

Perspective	Focus Area	Examples	Key Challenges
AI for Sustainability	Using AI to promote environmental goals	Renewable energy forecasting, smart agriculture, waste sorting	Data availability, algorithm bias
Sustainability of AI	Reducing AI's ecological and ethical footprint	Energy-efficient AI models, ethical frameworks	High energy use, transparency, inequality

The current research continues to concentrate on how AI sustainability technologies boost organizational performance while there exists limited research that examines the operational impacts which arise when organizations implement AI technology throughout their operations. The deep learning systems used in smart grids and renewable forecasting applications achieve high accuracy in their predictions however their training process consumes excessive energy which creates doubts about their environmental sustainability. The computer vision-based waste sorting systems enhance recycling efficiency but they require expensive sensing equipment and high-quality labeled data for proper operation. The sustainability results depend on how algorithms function but they also depend on model complexity and data availability and energy consumption and governance systems which require research to develop comprehensive evaluation standards.

2. AI in Energy Management

Energy systems are radically transformed due to the implementation of artificial intelligence (AI), transforming into more flexible, efficient, and sustainable methods. Through the application of complex data analysis, neural network algorithms, and system monitoring in real time, artificial intelligence (AI) can potentially transform energy production and consumption. Smart grids and prediction of demand in energy consumption are two areas of application where the impact is highly visible.

2.1 Smart grids and predictive analytics

To establish a clean and doubly resilient energy infrastructure, smart grids utilizing artificial intelligence will play an important role. Opportunities for real-time data collection at IoT devices and sensors have enhanced power flow, anomaly detection, and self-supply/demand capabilities. One can learn to rely on AI algorithms to predict load-flow fluctuations, maximize energy-distributed facilities, and interconnect renewable power sources such as solar and wind energy into the grid successfully from this information. Power grids will real-time rearrange dynamically according to conditions for fault detection, load balancing, and predictive maintenance under AI and ML, says Saravanan et al. (2024). Thus, operational effectiveness will increase and energy losses will decline. Smart grids powered by AI can go even further in the future by promoting distributed energy systems and consumer involvement [11].

2.2 Energy demand forecasting

Accurate forecasting of energy demand and generation is essential for the planning and optimization of energy systems. Artificial intelligence (AI) models using deep learning and time series analysis are particularly well positioned to identify complex consumption patterns at the level of buildings, cities, or specialised industrial areas. These forecasts allow energy to be allocated effectively, waste to be prevented, and demand-side management strategies to be introduced. As stated by Khan et al. (2023), AI modalities incorporated into intelligent building energy management systems can forecast energy demand profiles with accuracy based upon historical usage, occupancy dynamics, various weather inputs, and data at the appliance level. The systems provide demand-side energy optimization but are also a part of the scheduling process needed for using renewable energy and battery storage systems for a more sustainable energy ecosystem [12].

3. ML for Renewable Energy Optimization

Machine Learning (ML) is crucial in optimizing renewable energy systems, as it enhances efficiency, reliability, and predictability of energy generation and storage. Two of the most important areas that ML facilitates the advancement of renewable energy is in solar and wind energy forecasts and battery energy management systems.

3.1 Solar and wind energy prediction

Because renewable energy is stochastic, such as solar and wind energy, forecasting is central to the stable integration and reliability of grids. Hybrid solar power and wind systems are beginning to generate extremely precise power outputs through the use of adaptive artificial intelligence (ML) and ensemble learning methods. For hybrid PV-wind farms, Zafar et al. (2022) have developed an adaptive machine learning-based forecasting technique that mimics the complex and nonlinear behavior of weather-sensitive power systems. They are founded on the application of short- and long-term electricity generation from historical energy output and current meteorological information, enhancing scheduling and minimizing reliance on fossil fuel-based backup systems. Aside from ensuring greater overall efficiency and financial viability of renewable power projects, this would enhance grid stability [13].

3.2 Battery management systems

This means that all in the area of machine learning will be of less importance after October 2023. To counter the problem of intermittency, there is a need for efficient storage facilities as decentralized renewable energy generation goes up. Smart BMSs that can monitor, predict, and optimize the lifespan and performance of energy storage units are already employing machine learning. To forecast battery degradation and supply optimal energy dispatch methods, ML models can potentially work with vast amounts of data containing battery health, charging and discharging cycle information, temperature changes, and consumption levels (Yang et al., 2022). The safe incorporation of sources of clean energy into power systems would be enabled through the optimization of energy availability and operational life using the safe and efficient utilization of storage [14].

4. AI in Waste Management

Increasingly, it is realized that incorporating artificial intelligence (AI) into waste management systems has the potential to enhance output, reduce the impact on the environment, and enable the creation of a circular economy.AI is disrupting the waste management cycle by employing smart automation, enabling real-time data processing, and making predictions throughout the phases of waste collection, sorting, processing and conversion into valuable resources. Broadly, there are two distinct applications of AI in the sector, namely waste sorting and optimization for recycling, and predictive maintenance.

4.1 Waste sorting and recycling optimization

Effective waste sorting is crucial to achieving higher recycling rates and reducing the amount of waste entering landfills. Artificial intelligence comes up with a better solution through automatic sorting, taking advantage of its accuracy and speed abilities. Traditional methods of sorting using manpower have developed their own inefficiency through massive labor inputs and considerable opportunities for error. The integration of robotic automation, machine learning, and image recognition will be able to carry out much more accurate separation and sorting of waste materials like organics, metal, glass, and plastics. The development of computational intelligence (AI) and algorithmic learning (ML) for smart waste management platforms that facilitate waste sorting with minimal human intervention and can, in turn, increase recycling efficiency is addressed by Gupta et al. (2019). Computer vision and deep learning models are employed to further enhance AI processes. These models analyze the material, form, and composition of the wastes and result in faster and more accurate sorting. The recycling process thus turns out to be highly efficient from both ecological and economic viewpoints [15].

4.2 Predictive maintenance

Sortation through AI and predictive maintenance for equipment by AI would constitute a large percentage of the work of engineers at waste processing plants. This requires using equipment usage data to generate models that will forecast potential failures even before they occur. That reduces time waste caused by production stoppage, promotes a longer machinery lifespan, and guarantees continuous operation of waste-to-energy conversion systems, as well as other waste-reducing technologies. According to Manoharan (n.d.), AI predictive maintenance systems are created to facilitate circular economic practices in their settings through maximizing performances by optimizing waste management infrastructure. These systems collect and analyze data from sensors that are installed on that equipment to detect wear patterns, predict maintenance needs, and schedule repair in advance. Consequently, its operating expenses are reduced, and the environmental threat entailed by equipment breakdowns and unplanned shutdowns is also minimized [16].

5. Applications and Case Studies

AI has been a tangible tool with power and magnitude in several industries since its very beginning, not merely to speculate about how it might aid in solving sustainability issues. Various social and environmental challenges, such as prevention of pollution, wildlife conservation, climate change adaptation, and sustainable urban planning, are being tackled by AI. Various case studies from all over the world show how artificial intelligence (AI) technologies are being used to develop more intelligent, sustainable, and resilient systems.

Taghikhah et al. (2022) provide a number of examples in which AI has been applied in the real world, showing positive outcomes for sustainability. They say, for example, that AI-assisted environmental sensing systems have monitored air and water quality in real time with relevant interventions quickly made by government authorities and other stakeholders for pollution incidents. Usage of artificial intelligence in agriculture, such as yield forecasting and computerized crop monitoring, optimizes agricultural production for food security using reduced water and land. Artificial intelligence (AI) has greatly enhanced urban sustainability in most environments, including smart cities. For instance, AI traffic prediction and control systems have reduced traffic levels in urban neighbourhoods and, in turn, carbon emissions. In addition to risk assessment models that allow for faster response and preparation for mitigation methods, AI detection systems have also been utilized in disaster-risk zones as a response to droughts, wildfires, and flooding [17]. These examples demonstrate how AI can contribute to the UN's sustainable development goals (SDGs). AI brings data-driven decision making, system-performance optimization, and timely intervention into direct solutions to some of the most glaring sustainability issues in the contemporary context.

The reviewed literature study shows that AI-based sustainability solutions from multiple areas are starting to become similar to each other. The smart energy grid predictive models have been modified to support urban traffic control and emissions management, while waste sorting computer vision methods now serve environmental monitoring and biodiversity evaluation. Existing studies continue to exist as separate research efforts because they restrict successful AI implementation to particular industry fields. The review establishes that AI-powered sustainable development will achieve its maximum potential through two requirements: cross-sector algorithm reuse and interoperable data systems and shared sustainability measurement standards.

6. Future Prospects (AI and ML in decarbonisation)

AI and ML are coming to the forefront in accelerating the transition to a low-carbon economy as the global community collectively works to address climate change. Their ability to analyse complex data characteristics, identify options for emissions reductions, and optimize decision-making across systems enable the decarbonization of multiple new and existing sectors.

According to Vorozheykina (2022), AI and ML can be a major source of innovation and enable decarbonization through real-time emissions monitoring, support for low carbon development related to energy planning and improving sustainability of industrial processes. AI modelling can help analyse energy use and propose alternative options that have a lower carbon intensity, while ML algorithms can be applied to optimize coordination, and improve fuel efficiency and other environmental impacts of production chains. AI can support carbon footprint measurement and predictive modelling to achieve emissions reductions in heavy emitting sectors such as manufacturing, transportation, and energy production. AI will support the confidence and verification of carbon markets and trading by forecasting carbon prices based on market dynamics and type of policy, as well as transparency and verification of offsets [18]. Looking into the future, with the introduction of emerging technologies - quantum computing, edge AI, and federated learning - the use of AI in decarbonisation is now permanently extended. These new innovations will provide even more accurate and privacy-preserving sustainability solutions. However, for the promises of AI-enabled decarbonisation to have future impact, data governance, energy efficiency of these AI systems, and access to digital infrastructure will have to be in place. The block diagram depicted in figure 3 illustrates that AI and ML contribute to decarbonisation by translating raw environmental and operational data into insights that can be applied to mitigate emissions. Data sourced from IoT sensors, satellite imagery, and other sources will all be processed by AI and ML models to develop models to forecast emissions and detect operational inefficiencies before employing targeted solutions across priority sectors, from energy to transportation to manufacturing, by optimizing smart grid, fuel-efficient routes, and low-carbon industrial processes. The insights generated by this will be used for decision making and operating carbon markets, while the continuous feedback loop provides the capacity to improve the model, and provide updates to the policies over time. The framework will capture the potential of AI and ML in a catalyst role to enable a low-carbon future.

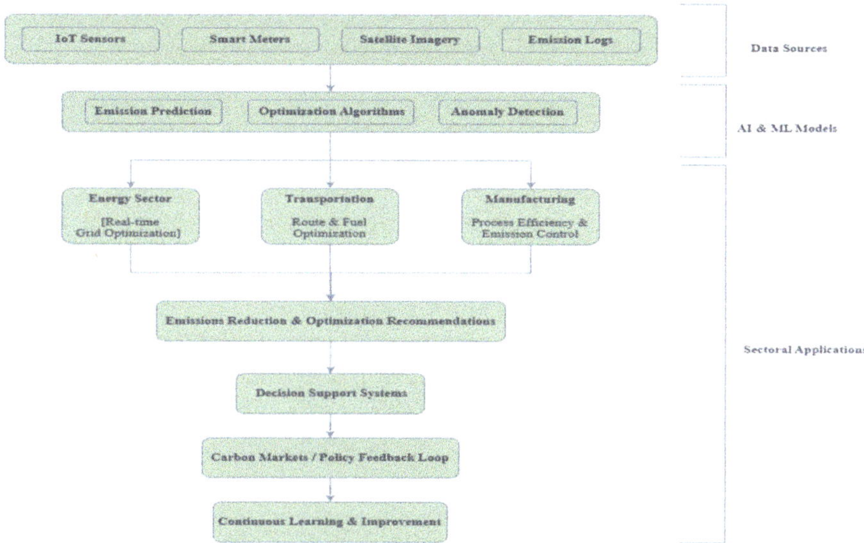

Figure 3. *AI & ML-Driven Decarbonisation Framework*

7. Conclusion

Artificial intelligence together with machine learning technologies shows major potential to improve sustainability practices through their ability to power data-based system optimization and instant system monitoring and their ability to forecast future system performance across various domains including energy operations and waste processing and city development and methods for reducing carbon emissions. The review demonstrates that AI solutions which companies use in their operations can achieve three goals: they enhance operational efficiency and decrease environmental damage and help organizations meet their United Nations Sustainable Development Goals. The research results show that two opposing forces exist because AI technology supports sustainable practices while its creation and usage result in problems which include energy use and data management and biased algorithms and problems with fair access. The solution to these problems requires two things: first we need new technical solutions and second we require different groups including policymakers and industry experts and university researchers to work together. The review establishes a research foundation by analyzing how AI technology applies to different sustainability areas and showing their common limitations and research needs. The future development process requires organizations to create algorithms which use less energy together with systems that enable clear and responsible AI usage and methods which connect different sectors to achieve positive environmental and social results through AI implementation.

References

[1] Van Wynsberghe, A. (2021). Sustainable AI: AI for sustainability and the sustainability of AI. AI and Ethics, 1(3), 213-218. https://doi.org/10.1007/s43681-021-00043-6

[2] Kar, A. K., Choudhary, S. K., & Singh, V. K. (2022). How can artificial intelligence impact sustainability: A systematic literature review. Journal of Cleaner Production, 376, 134120. https://doi.org/10.1016/j.jclepro.2022.134120

[3] Nishant, R., Kennedy, M., & Corbett, J. (2020). Artificial intelligence for sustainability: Challenges, opportunities, and a research agenda. International journal of information management, 53, 102104. https://doi.org/10.1016/j.ijinfomgt.2020.102104

[4] Kar, A. K., Choudhary, S. K., & Singh, V. K. (2022). How can artificial intelligence impact sustainability: A systematic literature review. Journal of Cleaner Production, 376, 134120. https://doi.org/10.1016/j.jclepro.2022.134120

[5] Fan, Z., Yan, Z., & Wen, S. (2023). Deep learning and artificial intelligence in sustainability: a review of SDGs, renewable energy, and environmental health. Sustainability, 15(18), 13493. https://doi.org/10.3390/su151813493

[6] Srivastava, S.S., Satyanarayana, G.S., Dhasmana, A., Rawat, V., Rana, A.S. and Bisht, Y.S., 2025. Role of Embedded Systems in Smart Energy Management: Challenges, Innovations, and Future Trends. Solar Energy and Sustainable Development Journal, 14(STR2E), pp.27-50. https://doi.org/10.51646/jsesd.v14iSTR2E.797

[7] Strubell, E., Ganesh, A., & McCallum, A. (2020, April). Energy and policy considerations for modern deep learning research. In Proceedings of the AAAI conference on artificial intelligence (Vol. 34, No. 09, pp. 13693-13696) https://doi.org/10.1609/aaai.v34i09.7123

[8] Begum, A., Naim, A., & Sabahath, A. (2024). The Impact of AI on Sustainability. In Harnessing High-Performance Computing and AI for Environmental Sustainability (pp. 99-113). IGI Global. https://doi.org/10.4018/979-8-3693-1794-5.ch005

[9] Asha, P., Mannepalli, K., Khilar, R., Subbulakshmi, N., Dhanalakshmi, R., Tripathi, V., ... & Sudhakar, M. (2022). Role of machine learning in attaining environmental sustainability. Energy Reports, 8, 863-871. https://doi.org/10.1016/j.egyr.2022.09.206

[10] Gundeti, R., Vuppala, K., & Kasireddy, V. (2024). The future of AI and environmental sustainability: Challenges and opportunities. Exploring Ethical Dimensions of Environmental Sustainability and Use of AI, 346-371. https://doi.org/10.4018/979-8-3693-0892-9.ch017

[11] Saravanan, S., Khare, R., Umamaheswari, K., Khare, S., Gowda, B. K., & Boopathi, S. (2024). AI and ML Adaptive Smart-Grid Energy Management Systems: Exploring Advanced Innovations. In Principles and Applications in Speed Sensing and Energy Harvesting for Smart Roads (pp. 166-196). IGI Global. https://doi.org/10.4018/978-1-6684-9214-7.ch006

[12] Khan, S. U., Khan, N., Ullah, F. U. M., Kim, M. J., Lee, M. Y., & Baik, S. W. (2023). Towards intelligent building energy management: AI-based framework for power consumption and generation forecasting. Energy and buildings, 279, 112705. https://doi.org/10.1016/j.enbuild.2022.112705

[13] Zafar, M. H., Khan, N. M., Mansoor, M., Mirza, A. F., Moosavi, S. K. R., & Sanfilippo, F. (2022). Adaptive ML-based technique for renewable energy system power forecasting in hybrid PV-Wind farms power conversion systems. Energy Conversion and Management, 258, 115564. https://doi.org/10.1016/j.enconman.2022.115564

[14] Yang, Y., Bremner, S., Menictas, C., & Kay, M. (2022). Modelling and optimal energy management for battery energy storage systems in renewable energy systems: A review. Renewable and Sustainable Energy Reviews, 167, 112671. https://doi.org/10.1016/j.rser.2022.112671

[15] Gupta, P. K., Shree, V., Hiremath, L., & Rajendran, S. (2019). The use of modern technology in smart waste management and recycling: artificial intelligence and machine

learning. Recent advances in computational intelligence, 173-188. https://doi.org/10.1007/978-3-030-12500-4_11

[16] Manoharan, R. Waste to Energy Conversion and Predictive maintenance in Circular Economics Through AI. IJSAT-International Journal on Science and Technology, 16(1).

[17] Taghikhah, F., Erfani, E., Bakhshayeshi, I., Tayari, S., Karatopouzis, A., & Hanna, B. (2022). Artificial intelligence and sustainability: solutions to social and environmental challenges. In Artificial intelligence and data science in environmental sensing (pp. 93-108). Academic Press. https://doi.org/10.1016/B978-0-323-90508-4.00006-X

[18] Vorozheykina, T. M. (2022). Challenges and prospects of decarbonization of the economy in the age of AI. Frontiers in Environmental Science, 10, 952821. https://doi.org/10.3389/fenvs.2022.952821

Design and Prototyping of Sand Sieving Machine

Katrine Anne A. GARCIA[1,a*], Mhicxon B. SARMIENTO[1,b], Jake D. ERATO[1,c], Pocholo R. ACUEZA[1,d], Wilfred M. TENORIO[1,e]

[1]Mechanical Engineering, National University, Manila, Philippines

[a]wmtenorio@national-u.edu.ph

Keywords: Sand Sieving Machine, Mesh Size, Sifter, Mechanized Filtration, Productivity

Abstract: One of the problems encountered by suppliers of sand here in the Philippines is low productivity of workers and high operational costs due to manual sifting done by laborers. This paper details the design and prototyping of a sand sieving machine to reduce labor, boost efficiency and productivity, and lower costs. The study explains the chosen optimal design selected after evaluating and trading-off various alternatives in terms of material cost and specifications. The machine's performance was validated through multiple tests. Results showed it reduced sieving time from an average of 31.63 seconds to 14.52 seconds and increased throughput, yielding consistent sand quality. This project offers a practical solution for sand sieving, improving working conditions and productivity, lowering operational costs and supporting future innovations in mechanized filtration.

Introduction

Sand is a naturally occurring material of granular materials and a mixture of finely divided rock particles. The act of passing sand through a sieve is an integral unit operation required to determine the particle size distribution of sand [1]. The mechanical process that separates particles based on their size is done with the help of a sieve of the required mesh size. It is a method of classification often used in agriculture to identify the size distribution of particles that may compromise the quality, as in food processing and mining.

In the Philippines, especially for sand suppliers of construction projects, manual hand sifters or sieves filled with sand then shaken, stirred over a screen, or blown off to remove larger particles, are mostly used. Manually sifting sand under the sun for extended period of time can pose challenges in terms of health such as fatigue and heat stroke. Also, this poses a larger operational cost since the workers are taking a long time to manually sieve all the sand to be supplied.

With this, the researchers suggested that a sand sieving machine could be used in the hardware to prevent workers from working under the sun for long and to increase productivity thus reduce operational costs in the long-term.

Materials and Methods

The sand sieving machine was prototyped by constructing a main frame from welded galvanized square tubes, ensuring stability with flat metal sheet footings. A funnel-shaped outlet was welded to the frame using thin metal sheet. The sieve mesh consists of a 1/4" screen mesh supported by a perforated metal sheet. Ball bearings were integrated into the sieve mesh frame for smoother sliding. A motor was installed at the rear, linked to the sieve frame by a mild steel plate and rod. A Variable Frequency Drive (VFD) was connected to control motor speed. A removable housing was added for safety, covering the motor and moving parts. Finally, the machine was coated with green and black paint for durability and corrosion protection. The list of components are given in Table 1 below.

Table 1 List of components

Name	*Quantity [pc]*
Frame	1
Siever	1
Sand catcher	1
Motor	1
VFD (Variable Frequency Drive)	1
Stopper	1
Plate	1
Rod	1

For area of screen mesh:

Area of screen mesh = (Length of screen mesh frame)(Width of screen frame) (1)

Area of screen mesh = (0.644525 m)(0.473075) = 0.304909 $m^2 \approx 0.3\ m^2$ (2)

Getting the total weight of the screen mesh m_{sm}:

$$m_{sm} = A\rho = (area)(density) = 0.74\ kg \quad (3)$$

Mass of sieve frame m_{sf}:

$$m_{sf} = (Length\ of\ frame)(density) + (Width\ of\ frame)(density) \quad (4)$$

$$m_{sf} = \left[(0.88741\ m)\left(3.92\ \tfrac{kg}{m}\right)\right] + \left[(0.752475\ m)\left(3.92\ \tfrac{kg}{m}\right)\right] = 6.42833\ kg \quad (5)$$

Total mass m_{total}:

$$m_{total} = m_{sm} + m_{sf} + m_{sand} = 0.74\ kg + 6.4283\ kg + 13\ kg = 20.1683\ kg \quad (6)$$

For force of sieve F_s:

$$F_s = (m_{total}\ of\ sieve\ frame)(acceleration) = 197.85132\ N \quad (7)$$

For load torque T:

$$T = (F_s\ on\ sieve)(length\ of\ sieve) = 175.57524\ N - m \quad (8)$$

For force of rod F_{rod} and plate F_{plate}:

$$F_{net}\ at\ plate = F_{rod} \quad (9)$$

$$F_{rod} = \sqrt{(rod\ force\ x\ component^2 + rod\ force\ y\ component^2} \quad (10)$$

$$F_{rod} = F_{plate} = \sqrt{(F_{rodx})^2 + (F_{rody})^2} = 494.6382285\ N \quad (11)$$

For plate torque τ:

$$W_{plate} = (mass)(gravity) = (0.67399\ kg)\left(9.81\ {}^{m}/_{s^2}\right) = 6.6118\ N \quad (12)$$

$$\tau = (Force)(radius) = (494.64\ N + 6.61\ N)(0.05\ m) = 18.49\ ft - lb \quad (13)$$

For motor HP selection P :

$$P = \frac{2\pi TN}{33000\frac{ft-lb}{HP-min}} = \frac{2\pi(Torque)(rotational\ speed)}{33000\frac{ft-lb}{HP-min}} = 1.056\ HP \approx 1\ HP \quad (14)$$

Motor HP selection proofing P [2]:

$$P = T\omega = (Material\ factor\ of\ safety)(Torque)\ (angular\ speed) \quad (15)$$

$$Factor\ of\ safety = \frac{Yield\ stress}{Working\ stress} = \frac{300\ MPa}{200\ MPa} = 1.5 \quad (16)$$

$$Torque = \frac{Power\ (5252)}{RPM} = \frac{1\ HP\ (5252)}{200\ rpm} = 26.26\ N - m \quad (17)$$

$$P = (26.26\ N - m)\ (1.5)\left(\frac{200\ rpm\ x\ 2\pi}{60\ min}\right) = \frac{824.9822\ Watts}{\frac{746\ Watts}{1\ HP}} = 1.1059\ HP \approx 1\ HP \quad (18)$$

As per the results of the computations above, the required power for the plate to maintain its motion is 1 HP. A 1 HP motor is sufficient to handle the sand sieving machine load of 494.63823 N and torque of 25.06250 N-m or 18.48515 ft-lb.

Results and Discussion

After designing the sand sieving machine using the trade-offs and considerations in the Methodology part, the machine is evaluated by running a rate test compared with the manual set-up and its efficiency of filtering the debris on the sand. The manual method has an average sieving rate of around 31.63 seconds while the sieving machine achieves an average rate of around 14.52 seconds. The sieving rate was reduced by almost 50% with the use of the sand sieving machine. The sieving machine's efficiency is also tested to be an average of 98.2 %, meaning an average of 0.982 kg is filtered by the machine for a 1 kg debris on the sand. The tables and figure below show the actual results gathered. The study proves that the sand sieving machine is more efficient than manual sand sieving, which can significantly reduce the time required for the sieving process and the quality of sand results. .

Table 2 Sieving Rate Test Result between Sieving Machine and Manual Method

	Motorized [s]	Manual [s]
1	14.46	28.26
2	13.46	35.38
3	15.11	33.54
4	16.04	31.65
5	13.53	29.33

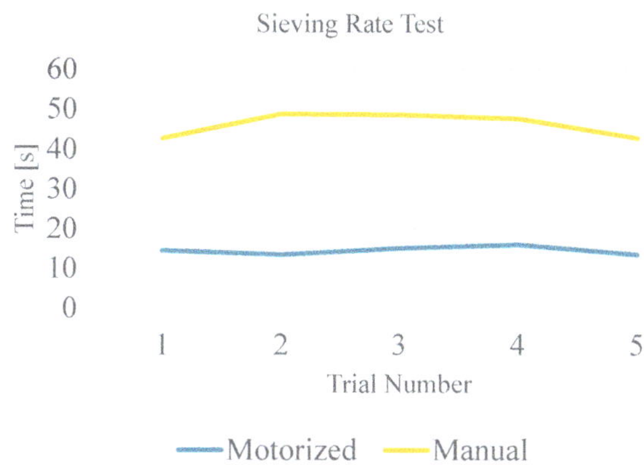

Fig. 1 Sieving rate test results

Figure 1 and Table 2 shows that actual sieving rate test results. The sieving machine achieves an average sieving rate of 14.52 seconds while the manual process has an average sieving rate of 31.63 seconds. This means that it takes less time for the sieving machine to sift the sand compared to the manual method.

Table 3 Sieving Effectiveness Test Result

	Input [kg]	Debris [kg]	Debris Collected [kg]	Efficiency [%]
1	5	1	0.97	97
2	5	1	0.98	98
3	5	1	0.99	99
4	5	1	0.98	98
5	5	1	0.99	99
		Average	0.982	98.2

Table 3 presents the average efficiency of 98.2 % debris collection of the sieving machine. This means that the fabricated sand sieving machine is effective in separating the debris from the sand.

Conclusion

According to the study, a motorized sand-sieving machine was designed and developed, taking into consideration the pressing requirement for productivity improvement and reduction of operational costs associated with sand processing. The machine was evaluated for structure, adequacy of materials for sustained operations, and the separation of debris with effectiveness in the range of 98.2%. Experimental evidence demonstrates the undoubted superiority of the machine over any manual procedures in terms of sieving, thus decreasing processing time from an average of 31.63 seconds in manual procedures to an average of 14.52 seconds. User and expert assessments have also rated the machine consistently high in terms of ease of use, effectiveness, and practicality, attaining over 88% of the design objectives. On an economic note, it is strong on the bankroll due to its monthly returns and savings. Thus, the project provides a strong, effective, economical solution for sand sieving, enhancing working conditions, increasing productivity while lowering operational costs, and advances in future mechanization of small-scale filtration.

Acknowledgement
Our most profound appreciation to our advisors for their precious guidance during this work. The suggestions from our associates raised the quality of the work considerably, while the funding from the department was indispensable to complete it. We gratefully acknowledge the technical staff for their unwavering support and funding agencies for their vital funds. We also appreciate the collaboration and moral support from my colleagues. Finally, we are forever thankful to our families for their unconditional love and infinite patience.

References

[1] P. Poullet, J. J. Muñoz-Perez, G. Poortvliet, P. Lopez-Garcia, Influence of Different Sieving Methods on Estimation of Sand Size Parameters, MDPI Water (2019) 1. https://doi.org/10.3390/w11050879

[2] R. C. Hibbeler, Engineering Mechanics: Statics and Dynamics, fifteenth edition, Pearson, United States of America, 2023.

[3] W. D. Callister, Jr., D. G. Rethwisch, Materials Science and Engineering, tenth edition, John Wiley & Sons, United States of America, 2018.

[4] J. K. Nisbeth, Shigley's Mechanical Engineering Design, McGraw-Hill LLC, New York, 2024.

Advanced Materials and Sustainable Energy Technologies – AMSET2025 Materials Research Forum LLC
Materials Research Proceedings 66 (2026) 239-247 https://doi.org/10.21741/9781644904152-22

Sustainable Concrete Solutions for Green Buildings: A Study on Nano-Silica and Dolomite Powder Integration

Gaurav THAKUR[1,a*], Ajay SINGH[2,b], Arun Kumar AGARWAL[1,c], Jitender GAIROLLA[3,d], Shanoo SHARMA[4,e], Nighat SHAMEEM[5,f], Vineeta SINGH[6,g]

[1]Uttaranchal Institute of Technology, Uttaranchal University, Dehradun-248007, Uttarakhand, India

[2]School of Applied & Life Sciences, Uttaranchal University, Dehradun-248007, Uttarakhand, India

[3]Department of Telemedicine, AIIMS, Rishikesh, Uttarakhand, India-249203

[4]Department of Medical Lab Technology, University Institute of Allied Health Sciences Chandigarh University Mohali, Punjab 140413

[5]Chandigarh Group of Colleges, Jhanjeri, Mohali, Punjab, 140307, Chandigarh School of Business, Department of Radiology

[6]Department of Computer Science & Engineering, School of Engineering and Technology (Formerly UIET), Chhatrapati Shahu Ji Maharaj University Kanpur

[a]gauravthakur649@gmail.com, [b]ajay21singh@yahoo.com, [c]akagarwal_ongc@yahoo.co.in, [d]jeetugairolla1985@yahoo.com, [e]shanoo.e11565@cumail.in, [f]nighatshameem103@gmail.com, [g]vineeta.singh.cs@gmail.com

Keywords: L9 Orthogonal Array, Larger-the-Better Approach, Optimization, Signal-to-Noise Ratio (S/N), Taguchi's Method

Abstract. To meet the standards of sustainability and efficiency, the cement and concrete sector need positive development of concrete mix strength. The current study is about the interaction of a couple of dolomite powder and nano-silica cementitious additives on workability and compressive strength of concrete M30. The application of the L9 orthogonal array of the Taguchi method was done to obtain the optimal mix, where the nano-silica was tested at the levels of 2%, 3%, and 4% and the dolomite powder was used in the amounts of 15%, 20%, and 25% of cement's weight with a standard water-cement ratio of 0.40. The findings revealed that there were different behaviors in terms of workability and mechanical strength. The greater the amount of nano-silica was used, the lower the workability was, but on the other hand, 25% of dolomite powder addition made the slump value 200 mm which was considered as an efficient filler for the flowability enhancement. The mechanical performance was evaluated based on the analysis of the Taguchi Signal-to-Noise (S/N) ratio wherein the compressive strength demonstrated a non-linear pattern of increase and reached its maximum at the 3% nano-silica level. The most effective combination was determined to be 3% nano-silica plus 20% dolomite powder, resulting in a mean compressive strength of 39.5 MPa at 28 days, which surpassed the M30-grade concrete target strength successfully. When the quantity of nano-silica was increased to more than 3%, the strength dropped a little bit and this was due to agglomeration of the nanoparticles. The results indicate that the harmonious combination of 3% nano-silica + 20% dolomite not only produces but also the sustainable concrete with the highest performance which is perfect for green building applications.

Introduction

Unprecedented difficulties are faced by the construction industry in the 21st century as it has to meet the dual demands of economic growth and environmental protection. Concrete, the most extensively used construction material, is at the centre of this issue since its production not only releases a huge amount of CO_2 but also consumes huge amounts of natural resources and creates a lot of waste [1]. The traditional way of making Ordinary Portland Cement (OPC) is under heavy criticism for using a lot of energy and releasing CO_2 in large amounts [2]. Thus, the focus of the development of sustainable concrete technologies has been shifting toward the utilization of industrial by-products, the reduction of clinker content, and the addition of new materials that are environmentally friendly without compromising performance [3,4].

Nanotechnology is a path that holds great potential for the future of eco-friendly concrete. Among the various nanomaterials, nano-silica (SiO_2) is a crucial one, which has a large specific surface area and very high pozzolanic reactivity that are able to convert the cementitious mixtures' microstructure [5]. Recent microstructural studies validate that optimized integration of nano silica at nano scale accelerates the hydration reaction thus resulting in increased early-age strength and better matrix density for sustainable cement composites [6–8]. If nano-silica is added in the right amount, it can fill in the micro-voids, thus, promoting hydration and producing a denser matrix, which usually results in enhancement of mechanical properties and factor [9]. Yet, the addition of nano-silica usually makes the mixture less workable because of its water demand that is high, thus, requiring a mixture design that is perfectly balanced [10].

In order to manage workability and sustainability at the same time, naturally occurring minerals like dolomite powder [$CaMg(CO_3)_2$] are great alternative materials for doing so. Recent investigations have shown that dolomite microparticles not only work as efficient physical filler in improving the workability but also provides nucleation sites which enhances early-stage strength development [11,12]. Apart from this, advanced fractal pore analyses reveal that blending dolomite powder refines the gel pore structure of the cement-based materials [13,14]. One of the major benefits of using dolomite powder either as a partial replacement for cement or as fine aggregate is producing low-cost buildings, which are less material consuming and taking advantage of its good particle arrangement that results in better workability [15]. Moreover, the use of nano-silica along with dolomite powder is a step towards green building because it helps to produce less harmful construction materials [16].

Even though each of these materials has been recognized for their specific benefits, the overall impact of the mixture of nano-silica and dolomite powder is still a matter that needs to be looked into more deeply. On one hand, nano-silica gives a concrete mixture strength and durability; on the other hand, dolomite powder does not change so much but increases flowability and may even modify the dynamics of hydration [17]. Furthermore, the introduction of these additives in practice will necessitate comprehensive experimental research to reveal their synergistic or antagonistic effects on the properties of fresh concrete and compressive strength.

In order to come up with the most efficient and stable concrete mixes, we are going to apply the Taguchi method of optimization in this paper. This main technique, which is based on statistics, not only allows the detection of the most important parameters but also practices the use of orthogonal arrays [18] to obtain the least resource consumption and experimental variation. The robustness of the Taguchi design of experiments- DOE in optimizing the complex and multi-variate concrete matrices is increasingly being validated, especially when it comes to balancing the conflicting mechanical and physical parameters of the pozzolanic and nanomaterial additives [19–21]. The traditional methods were completely different since they needed a complete testing of all possible combinations; however, on the contrary, the Taguchi method provided a performance statistic in the form of the (S/N) ratio to produce robust results [22,23]. In the case of concrete strength in particular, this paper is going to practice the "larger-the-better" S/N criterion to measure performance in different conditions and also to select the strongest and best performing mix. According to Eq. 1, the S/N ratio is calculated as follows:

Larger-the-better: $S/N = -10\log_{10}(\frac{1}{n}\sum_{i=1}^{n}\frac{1}{Y_i^2})$ (1)

where Y_i denotes the observed value (compressive strength) and n signifies the total number of observations. The S/N ratio assists in measuring the performance in different conditions, thus making it possible to choose a mix design that not only reaches the highest strength but also has the least variability.

The paper structure is as follows: The section titled Experimental Investigation describes the materials and the experimental program that has been developed using Taguchi L9 orthogonal array. The section entitled Results and Discussion offers the experimental results about workability and compressive strength changes, including the Signal-to-Noise ratios analysis for determining the most efficient mix design through identifying the optimized parameter levels. Lastly, the Summary gives the conclusions drawn from the study and also points out the directions for future research.

Experimental Investigation

Materials Used. The materials used in the test program were carefully selected, not only for practicality but also for their scientific value. The primary binder was Ordinary Portland Cement (OPC), with a specific gravity of 3.15 and complying with the IS 1489 (Part 1): 1999 standard. Fine aggregate was sourced from a local river that is classified as Zone III according to IS 383: 1970. The sand had a specific gravity of 2.16 and a bulk density of 1457.3 [kg/m^3], which made it suitable for concrete production. The maximum size of crushed angular coarse aggregates was 20 [mm], and they were selected with a specific density of 2.89 and bulk density of 1108.3 [kg/m^3] to guarantee the strength required. The application of high-purity nano-silica powder, with particle sizes of 1 to 100 [nm] used as a partial cement replacement, was regarded as the key factor to the enhancement of microstructural properties due to the high specific surface area of nano-silica powder. Further, dolomite powder, which consists mainly of calcium magnesium carbonate [$CaMg(CO_3)_2$], was first obtained from vendors near the project site, then milled into a fine powder to be used as a sustainable cementitious additive. Potable tap water, free from impurities and conforming to IS 456:2000, was used for all mixing and curing processes [24].

Experimental Program. The manufacturing process was an orderly one that focused on the uniformity of product/output. Subsequently, the dry components such as cement, aggregates, nano-silica, and dolomite were first combined to form a homogeneous blend and then water was added. The resulting concrete mixture was poured into 150 × 150 × 150 [mm] steel moulds that had been cleaned and oiled. The process of compaction was done layer-wise with uniform tamping applied to each layer to remove air voids and obtain proper density. The moulds were then covered with wet cloths and kept undisturbed for 24 hours. After this period, the samples were taken out from the moulds and subjected to water curing for the designated test durations of 7 and 28 days under controlled conditions. The testing was done according to IS 516:1959 to determine compressive strength and the fresh property was evaluated by means of slump tests.

Taguchi Design of Experiment. To obtain the most efficient proportions of the mixture, the researchers used a two-factor, three-level experimental design based on Taguchi method. To evaluate the impact of the additives with the least number of tests done, an L9 (3^4) orthogonal array was applied. The independent factors were established as Nano-silica (2%, 3%, and 4%) and Dolomite powder (15%, 20%, and 25%) by weight of cement, according to Table 1. The nine different mixes had a set water-binder ratio of 0.40, which is consistent with the best practices and IS 10262:2009 standards.

Table 1. Factors and their percentage at different levels

Levels	Nano Silica (%)	Dolomite (%)	Water Binder Ratio
Level 1	2	15	0.4
Level 2	3	20	0.4
Level 3	4	25	0.4

Results and Discussion

Effect on Workability. The slump cone test was conducted to determine the workability of the freshly mixed concrete right away after mixing. The results show a notable change in the flow characteristics amongst the mixes depending on the additive amounts. The change in slump values due to dolomite percentage is depicted in Fig. 1. A distinct upward slope can be seen in the trend where it is shown that the addition of dolomite greatly improves the flow of the concrete. For example, the mixes containing 25% dolomite recorded a maximum slump of 200 [mm] (Mix M9). This is a strong indication that dolomite powder functions as a very good filler and lubricant, thereby lowering inter-particle friction and enhancing the mix's uniformity [25].

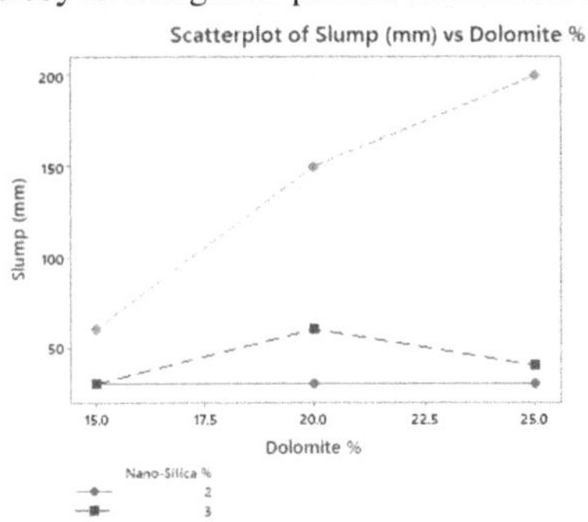

Fig. 1. Slump value variation as a function of dolomite powder percentage

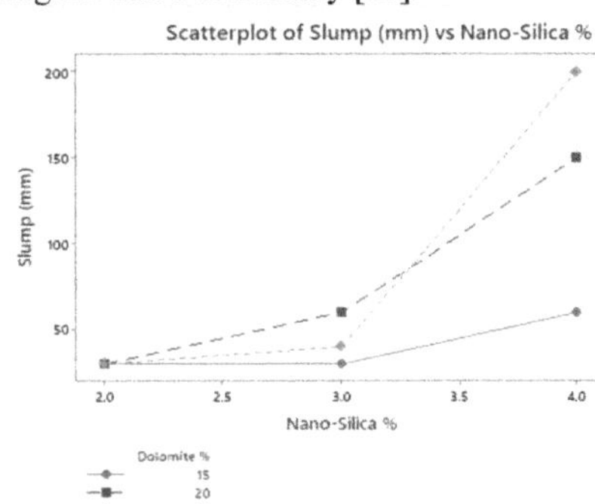

Fig. 2. Slump value variation as a function of nano silica percentage

On the contrary, the variation of slump with nano-silica percentage is shown in Fig. 2. Generally, the addition of nano-silica resulted in higher water usage thanks to its very high specific surface area [26]. At lower dolomite content (15%), the high-water absorption of nano-silica forced the production of stiff mixes with lower slump values (30 [mm]), thereby indicating dolomite's role in the mix to overcome the cohesiveness coming from the nanoparticles.

Effect on Compressive Strength. The compressive strength was the main performance indicator and was determined at 7 and 28 days. The values, given in Table 2, prove that the mixes developed not only achieved but also surpassed the target strength for M30 concrete. At 7 days, the compressive strength was the same throughout but showed slight variations among the specimens, ranging from 29.54 [MPa] to 29.89 [MPa]. Fig. 3 presents the 7-day strength in terms of dolomite powder, which shows a stable trend where the filler effect leads to early density without significantly hindering the hydration process. Likewise, Fig. 4 shows the 7-day strength versus the nano-silica percentage, suggesting that the stimulating effect of nano-silica on hydration is similar for all replacement levels [26] .The point at which the distinct non-linear optimization trend was observed is 28 days. The strength variation at 28 days with dolomite powder is given in Fig. 5. The results show that 20% replacement levels (Level 2) were the most powerful producing results, letting the mixing of cement with packing density be the main factor [27]. The critical bond between 28-day strength and nano-silica percentage is depicted in Fig. 6. The newly obtained data indicates a "bell curve" behavior where strength is increased as the amount of nano-silica rises from 2% to 3%, reaching the maximum at the latter level, contrary to the linear reduction noted in some studies [28]. The combinations with 3% nano-silica (M4, M5, M6) acquired the highest strength range, Mix M5 (3% nano-silica, 20% dolomite) being the one with the highest mean strength of 39.5 [MPa]. This boost is attributed to the perfect pozzolanic reaction and the packing of the interfacial transition zone (ITZ) [29]. On the other hand, the figure also shows a small decrease in strength when nano-silica was increased to 4%, which is likely the result of weak areas forming within the cement matrix due to the agglomeration of nanoparticles.

Table 2. Results of different tests and S/N values for all mixes

Mix	Nano-Silica (%)	Dolomite Powder (%)	Slump Value (mm)	7 Days Compressive Strength (MPa)			Mean 7 Days Compressive Strength (MPa)	S/N Ratio 7 Days	28 Days Compressive Strength (MPa)			Mean 28 Days Compressive Strength (MPa)	S/N Ratio 28 Days
				1st Cube	2nd Cube	3rd Cube			1st Cube	2nd Cube	3rd Cube		
M1	2	15	0	29.5	30.2	30.3	30	29.54	37.5	38.2	38.3	38	31.59
M2	2	20	0	30.1	30.6	30.8	30.5	29.68	38	38.6	38.9	38.5	31.71
M3	2	25	0	29.8	30.3	30.4	30.2	29.59	37.8	38.3	38.5	38.2	31.64
M4	3	15	0	30.5	31	31	30.8	29.78	38.7	39.2	39.1	39	31.82
M5	3	20	60	31	31.3	31.4	31.2	29.89	39.2	39.7	39.6	39.5	31.93
M6	3	25	40	30.4	31	31	30.8	29.77	38.6	39.2	39.1	39	31.81
M7	4	15	60	29.4	30.2	30.4	30	29.54	37.8	38.4	38.4	38.2	31.64
M8	4	20	150	29.7	30.4	30.6	30.2	29.61	38	38.6	38.8	38.5	31.70
M9	4	25	200	29.3	30	30.2	29.8	29.49	37.6	38.3	38.1	38	31.59

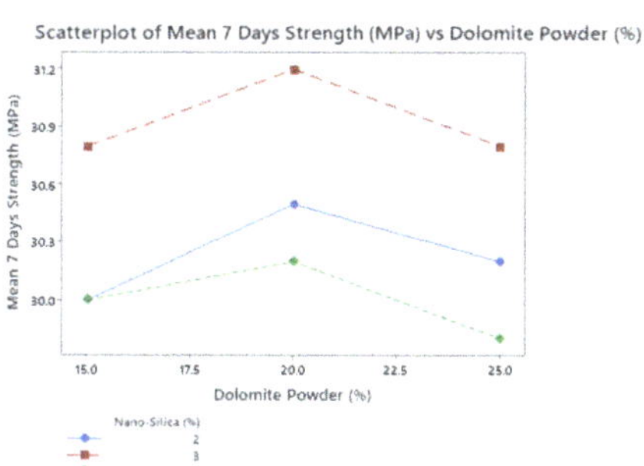

Fig. 3. 7 days compressive strength as a function of dolomite powder percentage

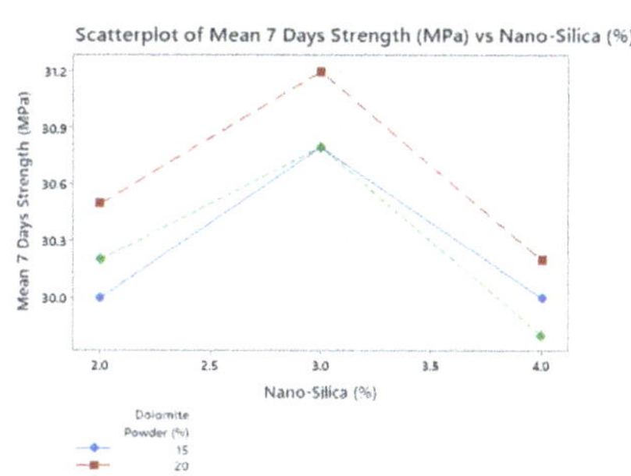

Fig. 4. 7 days compressive strength as a function of nano-silica percentage

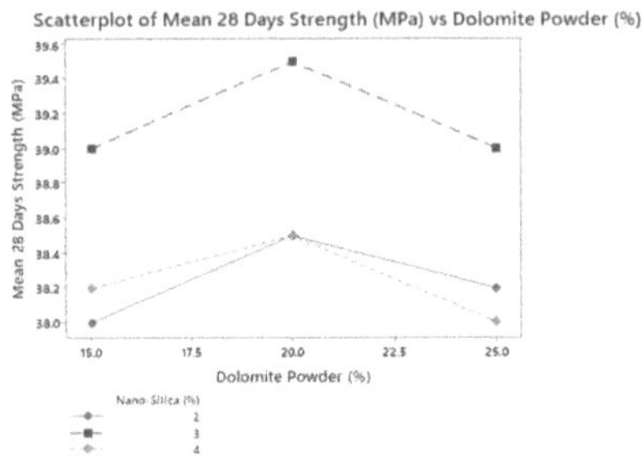

Fig. 5. 28 days compressive strength as a function of dolomite powder percentage

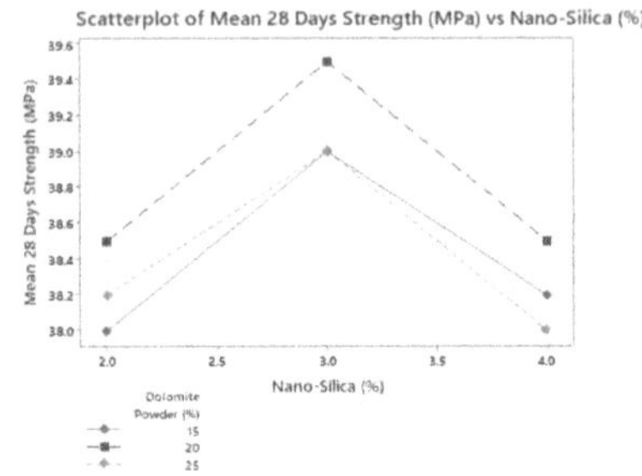

Fig. 6. 28 days compressive strength as a function of nano-silica percentage

Optimization of Parameters. In the Taguchi approach, the Signal-to-Noise (S/N) ratio was used to identify the best parameter levels, applying the "larger-the-better" principle. The analysis of the average response for the compressive strength after 28 days is shown in Table 3. The analysis indicates that Level 2 (3%) is the best dosage for nano-silica because it produces the highest average S/N ratio of 31.85. In the same way, Level 2 (20%) is the best dosage for dolomite powder with an S/N ratio of 31.78. Moving to Level 3 (4% nano-silica) leads to a lower S/N ratio (31.64), which proves the negative effect of overdose/agglomeration. Thus, the statistical analysis is in agreement with the experimental data—the mixture of 3% nano-silica and 20% dolomite powder gives the best mechanical performance for sustainable M30 concrete.

Table 3. Average S/N values of 28-days compressive strength for different parameter levels

Levels	% of Nano-silica	Average S/N Value for 28 Days Compressive Strength	% of Dolomite Powder	Average S/N Value for 28 Days Compressive Strength
1	2	31.65	15	31.68
2	3	31.85	20	31.78
3	4	31.64	25	31.68

Summary

The combination of nano-silica and dolomite powder examined using the Taguchi method enabled the creation of a sustainable concrete mix that was both highly durable and eco-friendly. One of the main targets of the investigation was to improve the mechanical characteristics of green concrete without sacrificing its workability. From the experimental data and the analysis of Signal-to-Noise (S/N) ratios, the authors concluded the following:

Workability. The addition of nano-silica brought about a marked increase in the demand for water, which in turn affected the slump values of the mixes containing low dolomite amounts. Nonetheless, the addition of dolomite powder turned out to be a very effective means of enhancing flowability. The 25% increase of dolomite as a substitute brought about a tremendous improvement in workability, which resulted in a maximum slump of 200 [mm] (Mix M9). This confirmed the use of dolomite as a sustainable filler that enhances the flow of concrete.

Mechanical Strength. Compressive strength tests showed that the mixtures developed satisfied the design strengths of M30 grade concrete. The relationship between the amount of nano-silica and strength was not linear. The compressive strength reached its maximum at an addition of 3%

nano-silica on account of the optimal interplay between micro-pore filling and pozzolanic reactivity. The strength decreased slightly at 4% because of the agglomeration of nanoparticles. For dolomite, the 20% replacement level gave the best mechanical performance.

Optimization. The Taguchi method of analysis revealed the best parameter levels as Level 2 for Nano-silica (3%) and Level 2 for Dolomite powder (20%). The combination (Mix M5) reached the highest mean compressive strength of 39.5 [MPa] at 28 days.

Future Scope. This study has determined the best mix in terms of strength and workability, but future research should focus on the long-term durability of the concrete [30], covering resistance to carbonation, freeze-thaw cycles, and microstructure analysis by SEM and XRD to confirm the binding mechanisms.

References

[1] Andrew RM. Global CO_2 emissions from cement production. Earth Syst Sci Data 2018;10:195–217. https://doi.org/10.5194/essd-10-195-2018.

[2] Habert G, Miller SA, John VM, Provis JL, Favier A, Horvath A, et al. Environmental impacts and decarbonization strategies in the cement and concrete industries. Nat Rev Earth Environ 2020;1:559–73. https://doi.org/10.1038/s43017-020-0093-3.

[3] Griffiths S, Sovacool BK, Furszyfer Del Rio DD, Foley AM, Bazilian MD, Kim J, et al. Decarbonizing the cement and concrete industry: A systematic review of socio-technical systems, technological innovations, and policy options. Renewable and Sustainable Energy Reviews 2023;180:113291. https://doi.org/10.1016/j.rser.2023.113291.

[4] Thakur G, Agarwal AK, Singh A, Yamsani N. Enhancing Compressive Strength Prediction in Fiber Reinforced Geopolymer Concrete Through Machine Learning Algorithms. RAICS - IEEE Recent Advances in Intelligent Computational Systems, Institute of Electrical and Electronics Engineers Inc.; 2024. https://doi.org/10.1109/RAICS61201.2024.10690118.

[5] Singh LP, Karade SR, Bhattacharyya SK, Yousuf MM, Ahalawat S. Beneficial role of nanosilica in cement based materials – A review. Constr Build Mater 2013;47:1069–77. https://doi.org/10.1016/j.conbuildmat.2013.05.052.

[6] Erdem Y, Kocak Y. Microstructural characterization, hydration reactions and mechanical properties of cement-based composites produced with silica fume used to develop green, clean and sustainable cement. Powder Technol 2026;467:121545. https://doi.org/10.1016/j.powtec.2025.121545.

[7] Niu W, Dou T, Li M, Xia S. Microscopic Deterioration Mechanism and Different Reinforcement Methods of Concrete Under Freeze–Thaw Environment: A Review. Processes 2025;13:4064. https://doi.org/10.3390/pr13124064.

[8] Joudah ZH, Khalid NHA, Baghban MH, Faridmehr I, Talip ARA, Huseien GF. Development sustainable concrete with high-volume wastes tile ceramic: Role of silica nanoparticles amalgamation. Case Studies in Construction Materials 2024;21:e03733. https://doi.org/10.1016/j.cscm.2024.e03733.

[9] Althoey F, Zaid O, Martínez-García R, Alsharari F, Ahmed M, Arbili MM. Impact of Nano-silica on the hydration, strength, durability, and microstructural properties of concrete: A state-of-the-art review. Case Studies in Construction Materials 2023;18:e01997. https://doi.org/10.1016/j.cscm.2023.e01997.

[10] Berra M, Carassiti F, Mangialardi T, Paolini AE, Sebastiani M. Effects of nanosilica addition on workability and compressive strength of Portland cement pastes. Constr Build Mater 2012;35:666–75. https://doi.org/10.1016/j.conbuildmat.2012.04.132.

[11] Samchenko S V., Larsen OA, Kozlova I V., Alpackiy DG, Alobaidi DAN. Concrete Modification for Hot Weather Using Crushed Dolomite Stone. Buildings 2023;13:2462. https://doi.org/10.3390/buildings13102462.

[12] Woldesenbet DT, Mohammed JJ, Negedu SD, Henriques J, Bekele EA. Experimental investigation on partial cement replacement with binary blended bagasse ash and calcined dolomite for enhanced C-25 grade concrete performance. Sci Rep 2025;15:22844. https://doi.org/10.1038/s41598-025-98019-7.

[13] Zhang X, Wei Y, Zuo J, Luo Y, Wang B, Yao W. Segmented Fractal Pore Structure in Cement-Based Materials Blended with Dolomite Powder. Journal of Materials in Civil Engineering 2021;33. https://doi.org/10.1061/(ASCE)MT.1943-5533.0003944.

[14] Hu H, Ma X, Zuo J, Wei Y, She A, Yao W, et al. Improvement of concrete performance through a ternary aggregate system: Microstructural insights into pore structure and ITZ characteristics. Cem Concr Compos 2026;166:106388. https://doi.org/10.1016/j.cemconcomp.2025.106388.

[15] Nguyen H-A, Chang T-P, Shih J-Y, Suryadi Djayaprabha H. Enhancement of low-cement self-compacting concrete with dolomite powder. Constr Build Mater 2018;161:539–46. https://doi.org/10.1016/j.conbuildmat.2017.11.148.

[16] Zeyad AM. Sustainable concrete Production: Incorporating recycled wastewater as a green building material. Constr Build Mater 2023;407:133522. https://doi.org/10.1016/j.conbuildmat.2023.133522.

[17] Xu J, Chen J, Lu D, Xu Z, Hooton RD. Effect of dolomite powder on the hydration and properties of calcium sulfoaluminate cements with different gypsum contents. Constr Build Mater 2019;225:302–10. https://doi.org/10.1016/j.conbuildmat.2019.07.050.

[18] Dave S V., Bhogayata A. The strength oriented mix design for geopolymer concrete using Taguchi method and Indian concrete mix design code. Constr Build Mater 2020;262:120853. https://doi.org/10.1016/j.conbuildmat.2020.120853.

[19] Siddiq MU, Anwar MK, Almansour FH, Qurashi MA, Adeel M. AI-Driven Optimization of Fly Ash-Based Geopolymer Concrete for Sustainable High Strength and CO2 Reduction: An Application of Hybrid Taguchi–Grey–ANN Approach. Buildings 2025;15:2081. https://doi.org/10.3390/buildings15122081.

[20] Bakhbergen U, Shon C-S, Zhang D, Ryeol Kim J, Liu J. Optimization of mixture parameter for physical and mechanical properties of reactive powder concrete under external sulfate attack using Taguchi method. Constr Build Mater 2022;352:129023. https://doi.org/10.1016/j.conbuildmat.2022.129023.

[21] Ul Haq MZ, Singh S, Vora T, Ak D, Tomar A, S V, et al. Sustainable concrete mix optimization with recycled glass powder and Nano-Silica: a Taguchi method analysis. Journal of Structural Integrity and Maintenance 2026;11. https://doi.org/10.1080/24705314.2025.2546210.

[22] Thakur G, Singh Y, Singh R, Prakash C, Saxena KK, Pramanik A, et al. Development of GGBS-Based Geopolymer Concrete Incorporated with Polypropylene Fibers as Sustainable Materials. Sustainability (Switzerland) 2022;14. https://doi.org/10.3390/su141710639.

[23] Zarehparvar-Shoja M, Eskandari-Naddaf H. Optimizing Compressive Strength of Micro- and Nano-silica Concrete by Statistical Method. Civil Engineering Journal 2017;3:1084–96. https://doi.org/10.28991/cej-030939.

[24] Divekar RS, Sawant RM. Dolomite Powder in Concrete: A Review of Mechanical Properties and Microstructural Characterization. Civil Engineering and Architecture 2023;11:3314–21. https://doi.org/10.13189/cea.2023.110607.

[25] Barbhuiya S. Effects of fly ash and dolomite powder on the properties of self-compacting concrete. Constr Build Mater 2011;25:3301–5. https://doi.org/10.1016/j.conbuildmat.2011.03.018.

[26] Kong D, Su Y, Du X, Yang Y, Wei S, Shah SP. Influence of nano-silica agglomeration on fresh properties of cement pastes. Constr Build Mater 2013;43:557–62. https://doi.org/10.1016/j.conbuildmat.2013.02.066.

[27] Kishore K, Pandey A, Wagri NK, Saxena A, Patel J, Al-Fakih A. Technological challenges in nanoparticle-modified geopolymer concrete: A comprehensive review on nanomaterial dispersion, characterization techniques and its mechanical properties. Case Studies in Construction Materials 2023;19. https://doi.org/10.1016/j.cscm.2023.e02265.

[28] Chithra S, Senthil Kumar SRR, Chinnaraju K. The effect of Colloidal Nano-silica on workability, mechanical and durability properties of High Performance Concrete with Copper slag as partial fine aggregate. Constr Build Mater 2016;113:794–804. https://doi.org/10.1016/j.conbuildmat.2016.03.119.

[29] Nili M, Ehsani A. Investigating the effect of the cement paste and transition zone on strength development of concrete containing nanosilica and silica fume. Mater Des 2015;75:174–83. https://doi.org/10.1016/j.matdes.2015.03.024.

[30] Du H, Pang SD. High-performance concrete incorporating calcined kaolin clay and limestone as cement substitute. Constr Build Mater 2020;264:120152. https://doi.org/10.1016/j.conbuildmat.2020.120152.

Advancing Energy Storage and Conversion Technologies: Innovations, Challenges, and Future Prospects

S. P. GAIROLA[1*], R. RASTOGI[2], S. A. SHEIKH[3], S. G. ADUSUMALLI[4], R. K. CHOUDHARY[5], K. SARASWATI[6]

[1]Department of Aerospace Engineering, Uttaranchal Institute of Technology, Uttaranchal University, Dehradun-248007, Uttarakhand, India

[2]Department of Information Technology, NIET, Greater Noida, UP, India

[3]Dialysis Therapy Technology, USAHS, Rayat-Bahra University, Mohali-140104, Punjab, India

[4]Biochemist, Micropath Labs, New Railway Road, Gurugram, Haryana

[5]School of Health and Allied Sciences, Sanskaram University, Jhajjar, Haryana

[6]School of Business, Department of Sciences, Chandigarh Group of Colleges, Jhanjeri, Mohali-140307, Punjab, India

suryapg07@gmail.com

Keywords: Energy Storage Systems, Renewable Energy Integration, Energy Conversion Technologies, Fuel Cells and Hydrogen Storage, Grid Stability

Abstract: The solar and wind resources being variable and intermittent need to ensure the security of power supply with some large energy storage and energy conversion systems. The huge issue of installing a large storage system, e.g., limited lifespan, low energy-storage capacity versus money available, and high cost, has been taken into account in this study. It examines storage techniques with research on lithium-ion and solid-state batteries, supercapacitors, chemical storage through hydrogen, and emerging energy conversion techniques. It also elaborates on the role of fuel cells, thermoelectric converters, and piezoelectric devices in optimal utilization of energy and balancing grid stability. Application of new nanomaterials and solid electrolytes is regarded as crucial to propel safety, performance, and cost virtues. The paper also discusses the opportunities within hybrid storage technology systems, AI-driven energy management, and policy framework as real drivers of sustainable energy technologies and international decarbonization.

1. Introduction

In today's energy systems, advanced energy storage and conversion are essential to ensure power grid, transport, and industrial stability and sustainability. With the world's increasing energy demand and emerging at breakneck speed towards alternative sources of energy, energy storage and conversion must be highly efficient in terms of being able to fleece the problem of intermittency and build energy reliability. Energy storage systems are intended for in-situ excess energy harvesting and bridging the gap between demand and production by storing energy to use later on in a bid to maximize the use of resources as well as stabilize the system [1]. Energy storage technologies have come leaps and bounds from when they were first started. There are several storage technologies available, ranging from mechanical, electrochemical, thermal, to chemical. Each technology is accompanied by an advantage and disadvantage; therefore, the choice relies on elements like efficiency, capacity, cost, and their environmental effect. Consequently, the dominant system under consideration for engineering analysis was batteries, lithium-ion batteries, and flow batteries in general, because of the high energy density and therefore scalability. Other options like hydrogen storage and SMES could be suitable for long duration and efficiency [2]. Figure 1 shows the average efficiencies of all systems and per unit time energy costs for

Materials Research Proceedings 66 (2026) 248-261 https://doi.org/10.21741/9781644904152-23

mechanical storage systems, batteries, hydrogen, and SMES as a function of the different investment-performance compromises. Embedded systems are the key driver of real-time monitoring, control, and optimization as the energy industry is transformed by smart grids, renewable energy, and operative energy management. However, there are matters with cost, cybersecurity, scalability, dependability, and energy efficiency. Optimized embedded controllers minimize power consumption by 30–50%, while battery management systems increase EV life by 20–40%. With smart grids able to manage 100,000 nodes, scalability is crucial [3].

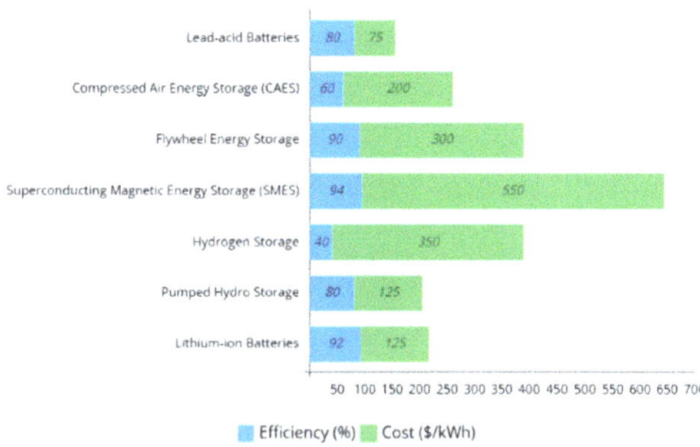

Figure 1. *Comparison of Energy Storage Technologies: Efficiency vs. Cost [4]*

They range from their capacity to simulate system flexibility, i.e., energy security, and, thereby, decentralized integration of renewable energy. Key issues in such energy-storage technologies involve cost, technology, and regulation [4]. In today's times, energy can be stored and converted till the time of their effective use and administration. Power storage systems should be allocated as per requirement: to store energy when excess is generated, say, to balance supply and demand fluctuations; to provide energy security, and to introduce an added assurance level to the grid. We have labeled this technology as those potentialities that should be utilized as green optimizers for the best use of resources, and in the process, minimize the consumption of fossil fuels [1]. The best mode of energy conversion, therefore, should be emphasized wherein the stored energy in various forms is converted into worthful energetic forces for various applications, like transport, industrial, and domestic electricity supply, and so on.

Energy storage is faced with many opportunities for expanding due to the scope for interfacing with renewable energy and grid support requirements. Several storage techniques are considered for various purposes like energy efficiency, sustainable development, etc., such as from electrochemical, mechanical, thermal to chemical storage. Lithium-ion energy batteries in demand in the market comprise several small units that make up very clean and rechargeable units that may either give their output in the required timeframe, in other words, compatible with high energy density and easy variations. Emerging technologies include hydrogen storage and superconducting magnetic energy storage (SMES) to provide a long-term storage solution. Despite some of the benefits of adopting an alternative, there still exist major challenges to widespread upsizing of energy storage technologies. The major setbacks are a huge initial cost, availability-of-materials-for-adsorption, loss of efficiency, ecological chain-reaction, blockage due to policy, and regulatory outlook. To arrest these bottlenecks, the policy and the legislation ought to be complimented by research and innovation in the improvement of energy storage technologies, the reduction in costs, and greener deployment thereof [5]. Figure 2 presents a pictorial overview of how different energy storage technologies were expected to develop over the years, emphasizing the growing role of lithium-ion, hydrogen, and SMES storage.

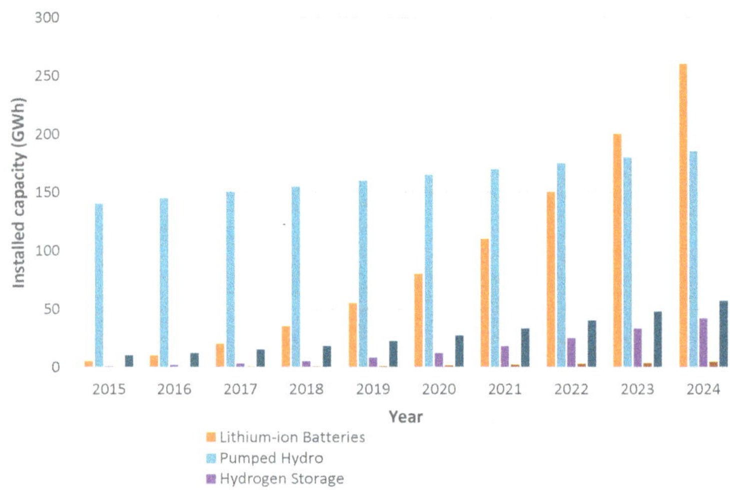

Figure 2. *Global Energy Storage Market Growth (2015–2035) [6]*

Energy storage is among the key drivers of grid integration into the growing portion of renewable energies. Solar and wind are self-intermittent, so to support the grid, one needs storage facilities to provide power assurance. Furthermore, in hours of peak demand or during peak generation hours, the system, that is to say, charged with excess power so that in low generation hours assured power could be supplied. From the decentralization perspective, storage also gives support to power systems so that communities can fulfil their energy needs with minimal assistance from restoration central fossil-fuel-fuelled power plants. With such characteristics from an increasing number of technologies with respect to renewable energy integration, the energy conversion and storage devices will take another whirl on the strong pillar of the economy in making it sustainable, strong, and carbon-free. The chapter will discuss the basic principles, technological achievement, and utilization of energy conversion and energy storage and an understanding of transformational capacity in harmony with present-day energy systems [6].

The research has developed separate energy storage and conversion technologies until 2023. The existing research base assesses technology-specific performance metrics but fails to evaluate system-level integration and scalability and long-term sustainability trade-offs. The majority of studies focus on achieving laboratory-based efficiency gains while only a small number of studies evaluate the actual deployment obstacles which include material supply limitations and environmental effects throughout the product lifecycle and challenges with grid connection. The existing comparative studies of mechanical and electrochemical and thermal and chemical storage technologies create obstacles for researchers who need to find suitable hybrid storage solutions for integrating renewable energy sources at large scales. The existing performance evaluation

Materials Research Proceedings 66 (2026) 248-261 https://doi.org/10.21741/9781644904152-23

framework requires improvements through a complete critical assessment which will measure technological performance and financial viability and safety standards and policy compliance.

2. Types of Energy Storage Systems

2.1 Mechanical storage: Pumped hydro, compressed air, flywheels

Mechanical energy storage systems are systems that convert movement or potential energy into energy and vice versa. The largest implemented mechanical-energy storage method is pumped hydroelectric storage (PHS), which makes use of meanwhile available surplus electricity for pumping water to a higher elevation. Under increased demand for electricity, the water is released to produce power by passing through turbines. PHS, therefore, with its highly efficient storage capacity for long durations, stands as very crucial to the grid stabilization technology. CAES, however, stores energy by compressing the air either into underground caverns or into high-pressure tanks. The generation of power is done by expanding the compressed air through a turbine, backfilling energy during peak demand. CAES offers masses of capacity and several suitable geological formations for effective storage. Another design is Flywheel Energy Storage, which uses kinetic energy stored in a spinning mass to back convert into electricity. Basic principles characterize a fast response from charge to discharge with a power output suitable for short-period applications [7].

2.2 Electrochemical storage: Batteries (lithium-ion, solid-state, flow batteries)

The batteries stored energy in chemical forms and later released them through electrochemical means. In fact, batteries have been essential for current energy storage applications. Lithium-ion batteries are currently the preferred technology for portable electronics and electric vehicles due to their energy density, cycle life, and efficiency. Despite this, problems with availability of raw materials and, especially, safety risks due to thermal runaway have made this area of research subject to unprecedented attention looking for alternatives. Solid-state batteries, which replace liquid electrolyte with a solid electrolyte, are expected to substantially improve safety, energy density, and longevity. The promise of such technology seems bright for the electric vehicles of the near future as well as grid storage systems. Further, flow batteries, such as vanadium redox flow batteries, are storage systems that generally use liquid electrolytes in external tanks to hold the energy. Due to their scale and long cycle life, flow batteries are well suited for large-scale renewable energy storage. Their services in power grid stabilization and grid support between renewable sources further make them an attractive option [8].

2.3 Thermal storage: Molten salt, phase change materials

The TES concept hopes to hold waste heat for application at a later time. TES takes center stage in concentrated solar power generation and the industry. When molten salts are heated by the solar radiations obtained from the solar collector during the daytime, this is considered one of the most salient methods of thermal energy storage in a CSP plant. This heat is discharged to generate steam at night hours to produce electricity, an energy supply from the plant even in night hours. Another important TES concept is embodied by phase change materials, which bear heat during phase transition, e.g., freezing, and release heat during the reverse transition, e.g., melting. In buildings, PCMs work to heat and cool, while in industry they may provide more stable or controlled heat output by rejecting temperature variations. Such technology would further enhance operational efficiencies and overall sustainability of renewable energy and industry [9].

2.4 Chemical storage: Hydrogen, ammonia

Various chemical energy technologies are storing energy in the form of chemical bonds, which will be released at future times by controlled reactions. Thus, chemical storage appears to be one of the suitable alternatives that can feasibly function with the management of energy in large quantities. Hydrogen storage is such a technology. Hydrogen is among the most universal energy

Materials Research Proceedings 66 (2026) 248-261 https://doi.org/10.21741/9781644904152-23

carriers; hydrogen can be generated using excess renewable energy through electrolysis. Compressed gas, liquid, or solid-state hydrogen can also be used as hydrogen storage. Hydrogen stored could be converted into electricity once needed by means of fuel cells, thereby making it the core promoter of a hydrogen economy. Hydrogen storage is, in a way, the reverse; ammonia storage, however, is getting attention for its high energy density and much easier handling through storage and transport. One could think of using ammonia formed from any combination of hydrogen either in the internal combustion engine or for fuel cell power, making this a renewable means of decarbonization and long-term energy storage dependent on its own production. All these kinds of energy storage make renewable energy systems more reliable and stable [10].

The deployment of mechanical storage systems which include pumped hydro and CAES systems faces two main challenges because these systems require extensive funding and they can only be installed in specific locations. The electrochemical storage technologies deliver better operational flexibility because they have higher energy storage capacity but face problems with material availability and safety hazards and performance deterioration. The industrial and solar applications of thermal storage systems achieve effective performance but the technology does not work in all situations whereas chemical storage solutions including hydrogen and ammonia storage systems offer effective long-duration storage solutions which require complex infrastructure and result in efficiency losses. Existing studies often evaluate these technologies in isolation, whereas comparative, application-driven assessment is essential to determine context-specific suitability.

3. Energy Conversion Technologies

Energy conversion processes form the nucleus around which energy systems are developed today. In the broadest sense, these processes ensure the sustainable use of energy in its generation, storage, and utilization. Various conversion methods for electrical energy, be it the chemical energy route, the thermodynamic one, or mechanical energy, are being practiced for a host of applications-from renewable energy resources to conventional industrial applications under green processes-electrochemical, heat, and kinetic conversion systems. They exhibit certain merits; for instance, better fuel usage in fuel cells attracts much attention, whereas piezo or triboelectric means are interesting for their harvesting advantages. The energy conversion technology lines can hence be considered and have been shown in Figure 3. Environmental energy sources (chemical, thermal, and mechanical) offer their own kinds of conversion. From electrochemical, thermal, and kinetic conversion, we have electrochemical conversion (fuel cells, electrolysers), thermal conversion (thermoelectric generators and thermophotovoltaic systems), and kinetic or mechanical conversion (piezoelectric and triboelectric generators). So they are capable of electrical power supplies, hydrogen fuels, and power wastes and heat for the application, storage, and grid integration. Hence such a scheme covers the variety and importance of these conversion technologies in provision of viable means toward attaining sustainable energy solutions.

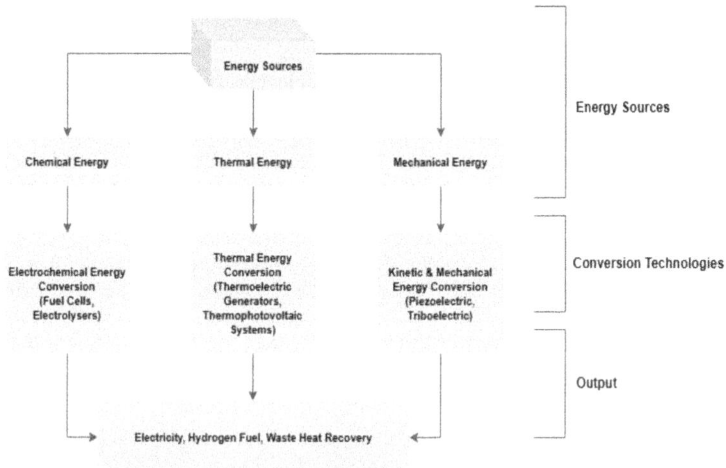

Figure 3. *Energy Conversion Technologies and Their Outputs*

3.1 Electrochemical energy conversion (fuel cells, electrolysers)

Electrochemical energy transformation technologies are responsible for energy generation via or chemical reactions and energy storage in the form of electricity. They form an important pillar of renewable energy systems-interconnected when using fuel cells as fuel. These fuel cells directly convert chemical energy into electricity through electrochemical reactions, typically involving hydrogen and oxygen. Fuel cells, as part of these systems, provide an efficient, low-emission, and continuous electricity supply as long as fuel is available. PEMFCs and SOFCs are examples of fuel cell technologies that are widely investigated for use in transportation, power generation in stationary applications, and powering small electronic devices. Electrolysis is the opposite process where electricity is applied to split water into hydrogen and oxygen through electrolysis. Hydrogen is produced in this way for use in energy storage and clean fuels. Some of the barriers to the application of electrochemical energy conversion have been associated with thermal management and efficiency losses and with material degradation. Despite such challenges, electrochemical conversion technologies hold great potentials in improving energy efficiencies and enhancing the transition to low-carbon energy [11].

3.2 Thermoelectric and thermophotovoltaic systems

Thermal energy can be transformed into electrical energy in a direct manner by thermoelectric or thermophotovoltaic (TPV) systems. These kinds of installations actively generate electrical power from thermodynamic gradients or radiation in an effective way. Thermoelectric generators (sometimes known as TEGs) utilize the Seabeck effect-under which an electric current is induced in different materials maintained at a temperature gradient. They find applications widely in waste heat recovery, space missions, and wearable electronics. Yet their efficiencies are still limited by the shortage of high-performance thermoelectric materials. TPV systems, on the other hand, work by converting infrared radiation emitted by a heat source into electricity, using specially designed photovoltaic cells. TPV technology is therefore versatile for compact power generation in industrial heat recovery and off-grid applications. More recently, advances in selective thermal emitters and high-efficiency photovoltaic materials have further boosted the production efficiency of TPV systems. However, beyond that, the technology still suffers from other challenges, which may include high costs and limitations on materials [12].

3.3 Kinetic and mechanical energy conversion (piezoelectric, triboelectric)

Creating electric energy out of kinetic energy caused by motion and vibration is the attracting concept of piezoelectric and triboelectric effects as one of the innovative solutions toward self-powered systems. Piezoelectric energy harvesting is mainly based on the phenomenon of electricity generation in materials converting mechanical deformation into polarization, making them great candidates for applications in energy harvesting floors, biomedical sensors, and smart textiles. These recent advancements in nanostructured piezoelectric materials have considerably improved efficiency and flexibility, making them acceptable to integrate into various devices. Even triboelectric nanogenerators (TENGs) are constructed on the principle of charge transfer during contact separation or friction between two materials. These systems exhibit great potential for ambient energy harvesting using wind, human movement, and ocean waves as their energy sources. Additionally, integrating triboelectric, electromagnetic, and piezoelectric effects leads to hybrid systems that have proven to be more efficient energy harvester candidates than any of the effects integrated with each other [13].

4. Emerging Materials and Technologies

Energy storage technologies have continuously evolved with the participation of new materials and hybrid systems that exhibit advantageous performance in various areas such as efficiency, lifetime, and sustainability. New technologies are really changing the energy landscape, creating ever more efficient and scalable solutions for modern energy needs, and these include advanced nanomaterials, solid electrolytes, and hybrid energy storage systems.

4.1 Advanced nanomaterials for storage

Nanomaterials are becoming key components of future energy storage systems because of their existing high electrical conductivity, large accessible surface areas and tunable property. Carbonized nanomaterials like graphene, carbon nanotubes and carbon nanofibers have excellent capability in improving the performance of supercapacitors and batteries. Charge transfer is improved, energy density is increased, and life cycles are extended for energy storage devices using these materials. Besides them, many more metal oxide nanomaterials and 2D transition metal dichalcogenides (TMDs) have been extensively explored for enhanced performance and energy conversion efficiency in lithium-ion battery applications. Specific structural aspects of these materials enable very rapid ion diffusion and testimony is owed by high-capacity energy storage too, making them eligible for next-generation battery technologies [14].

4.2 Solid-state electrolytes and supercapacitors

Being safer and more effective alternatives to liquid electrodes in supercapacitors or batteries, the exiting SSEs are the present investigation targets. Since SSEs discard any leakage, such-type of storage devices comply with safer working in association with long-lived hardware. Quasi-solid-state electrolytes that instantiate the desired properties of liquid and solid electrolytes need to offer higher ionic conductivity while maintaining structural stability for supercapacitors. Building on the recent developments in polymer-based gel electrolytes, more advancements have been made concerning charge storage capacity and electrochemical stability, making them a very promising option in flexible and wearable energy storage systems [15].

4.3 Hybrid energy storage systems

In the Hybrid Energy Storage Systems (HESS), because various storage technologies-GESS Power Generation is made to include batteries, supercapacitors, and flywheels-are characterized by their performance and reliability, a mix is made accordingly. Batteries stand for storage of energy, but their charge-discharge rate stands comparatively sluggish against the supercapacitor. Looking from the perspective of efficiency the combined operation of power source-storage, involving batteries or supercapacitors, would be a few layers away from the best choice. Relevant has been an instance herein at the HESS system with renewables where relevant coordinated storage systems are expected to counteract any time variations which intervene with power generations from solar and wind. In addition, controlling energy flows amongst the storage subsystems in an adaptively reactive manner yields a hybrid system with potential to maximize grid stability, enhance power quality, and cut down on fossil-fuel-based backup power generation [16].

5. Efficiency, Performance, and Challenges

There are several performance metrics of energy storage systems including energy density, power density, and efficiency, but cost-effectiveness other than scalability, safety, and lifespan is the largest hindrances to mass-market adoption. An answer to these issues is necessary to lead energy storage solutions into the future as well as expand their uses across different industries.

5.1 Energy density, power density, and efficiency factors

Three chief performance attributes of energy storage systems are: energy density, power density and efficiency. Energy density, with units in Wh/kg or Wh/L, is the energy that a storage system can hold per mass or per unity of volume. It is preferable when applications have extended durations, such as electric vehicles and grid-scale applications. PEFC and lithium-ion batteries are the perfect examples of an energy-dense system, ideal for portable and transportation usage. Power density (W/kg) is a measure of how energy can be supplied by a system mass in unit time; where high-power density provides consumption levels at an extremely rapid discharge, such as in supercapacitors and flywheels. Efficiency is the ratio of useful energy output to energy input. Internal resistance, heat loss, and charging-discharging losses introduce some inefficiency. Fuel cells have shown extremely high efficiencies in energy using optimized electrode kinetics and mass transport processes and are thus good devices for clean energy storage systems [17].

5.2 Challenges in scalability and cost-effectiveness

While energy storage has developed rapidly, its large-scale implementation still contends with several hurdles. One of the main challenges is material and manufacturing costs, as the very high cost of raw materials such as lithium, cobalt, and rare earth metals restricts the affordability of advanced batteries. Accordingly, researchers are developing alternative materials, such as sodium-ion and solid-state batteries, which will lower costs and provide sustainable storage solutions. The next major hurdle is scale, with large-scale energy storage needing high-capacity systems that can effectively store and then distribute energy. Pumped hydro storage and compressed air energy storage (CAES) are examples of technologies that can be scaled; however, their successful implementation is largely thwarted by their geographical and infrastructural constraints. Finally, the integration with renewable sources presents more challenges due to the intermittency of solar and wind power. With these challenges in mind, hybrid energy storage systems are currently being developed, uniting batteries and supercapacitors with complementary technologies for enhanced flexibility and responsiveness to optimize the performance of renewable energy applications [18].

5.3 Safety concerns and lifespan issues

To the maximum extent, safety and endurance have always been the prime concerns of energy storage systems, particularly lithium-ion batteries and fuel cells. The risks with the most impact are thermal runaway and fire danger they are susceptible to from over-heating. For these purposes, much emphasis is being laid on downstream advancements in battery management systems (BMS) and ongoing solid-state electrolyte advancements for improved stability and safety. One of the problems with involved degradation and cycle life of such systems is the issue of capacity fade and attendant loss of efficiency over the course of time from charge-discharge cycles. Electrode material degradation, electrolyte instability and the formation of dendrites in batteries result in progressive loss of lifespan. Out of all applications, battery reliability and safety under varying temperature and load conditions are particularly challenging to meet for electric vehicle and grid application systems. Moreover, grid-scale storage systems need rigorous safety measures that can be taken against failures resulting in instability and unreliability of power supply [19].

6. Applications and Case Studies

Energy storage technologies have dramatically changed the modern energy system. They can make the operation of these systems more efficient, reliable, and sustainable. These technologies find applications not only in wide-scale grid integration but also in portable electronics and electric vehicles. Some recent advances in energy storage technologies have led to innovative solutions that improve performance for all applications.

6.1 Grid-level energy storage

Grid-scale energy storage is integral to maintaining the stability of a power supply, adding renewable sources to a grid, and improving overall grid efficiency. Lithium-ion batteries have become the solution of choice for large-scale energy storage due to their high energy density, fast response times, and scalability. As the adoption of solar and wind energy rises, grid-level storage plays a vital role in mitigating renewable generation fluctuations. A battery energy storage system (BESS) helps to balance supply and demand by absorbing energy during peak generation and releasing it whenever needed. Energy storage contributes to modifying grid stability by damping frequency fluctuations and shifting load demands to off-peak hours whenever possible, increasing energy efficiency and reducing reliance on fossil-fueled backup power systems. Various countries such as the United States, China, and Germany have come up with large-scale lithium-ion battery farms to assist their power grids. For instance, the Tesla Megapack installations in California stay firm in stabilizing energy supplies and mitigating blackouts, demonstrating the efficacy of grid-sized battery storage in providing a reliable and resilient energy framework [20]. Figure 4. visually illustrates the rapid growth of grid-scale energy storage and the growing dependence on battery energy storage systems (BESS).

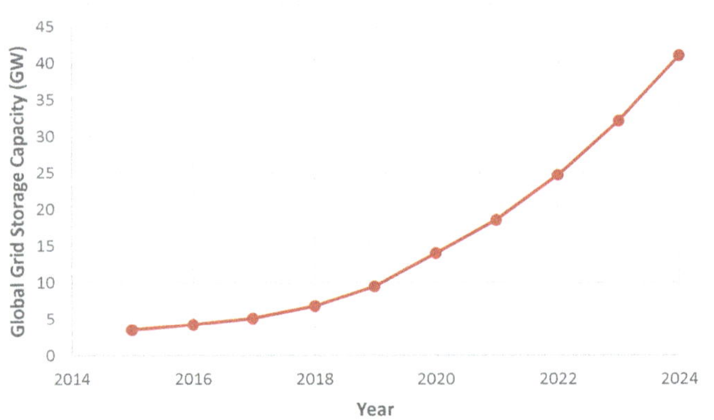

Figure 4. *Growth of Grid-Level Energy Storage Capacity (2015–2024)*

6.2 Portable electronics and electric vehicles

Energy storage systems prove most important in portable electronic devices as well as electric vehicles (EVs), where weight, dimensions as well as, most importantly, efficiency play key roles. For portable applications, from rechargeable lithium-ion batteries to supercapacitors, the development of a wide range of consumer devices from smartphones to laptops and finally wearables is now evident. All of the above applications must have storage options that are compact, lightweight, and high-capacity to provide long battery life and fast charging capabilities. Transitioning from internal combustion engines to electric drivelines in the EV sector therefore turns heavily on advanced battery systems. Currently, lithium-ion batteries are the best choice to perform in EVs because of the elevated energy density and fast charging potential. Nevertheless, a number of challenges regarding the battery life, restricted charging infrastructure, and battery costs still persist as the key obstacles to the vehicle electric drive penetration. Enhancing the performance and even charging time of any vehicle while adding to safety promises a bright future for energy storage market innovations with the advent of futuristic technologies such as solid-state batteries or supercapacitors that can be electroplated with their charges in a snap [21]. Table 1 gives a comparative overview of the respective strengths and trade-offs in real-world applications of the various energy storage technologies.

Table 1. *Comparison of Energy Storage Technologies for Portable Electronics and EVs*

Technology	Energy Density (Wh/kg)	Charge Time	Lifespan (Charge Cycles)	Main Applications
Lithium-Ion Batteries	150–250	30 min – 2 hrs	1,000–3,000	Smartphones, Laptops, EVs
Solid-State Batteries	250–400	15 min – 1 hr	5,000+	Next-gen EVs, Wearables
Supercapacitors	5–10	Seconds	1,000,000+	Wearables, Fast-charging devices
Hydrogen Fuel Cells	1,000+	Minutes	10,000+	Heavy-duty transport, Backup power

6.3 Case studies on recent breakthroughs

This innovation in energy storage comes with greater advancement in itself to very many other sectors, making them more efficient, safe, and sustainable. Among the battery technologies, emerging so far, research on the solid-state version of the technology shows much promise in improving energy density while eliminating safety risks associated with liquid electrolytes, involving industry leaders such as Quantum Scape or Toyota in successfully developing commercial applications of such technologies. Hydrogen is emerging as a clean form of energy storage and mobility solutions. Hydrogen fuel cells provide long-duration storage and clean energy output; thus, they are growing rapidly in heavy-duty transport and backup power systems. Hybrid Energy Storage Systems (HESS) combine batteries with supercapacitors to optimize energy and power density. It synthesizes energy density for long-term storage and power density for fast discharge, recently being adopted in a large number of applications such as Electric Vehicles (EVs) and renewable energy storage [22].

7. Future Trends and Research Directions

Three things of interest with respect to the evolutions of the energy storage technology are material innovations; smart systems integration; and government policy frameworks that facilitate them.

Given that there is an ever-growing demand required for highly efficient, sustainable, scalable energy application, research is underway on next-generation battery chemistries, smart grid integration with AI, and the regulations of energy storage of the future.

7.1 Next-generation battery chemistries

The advances on next-generation battery chemistries target not just future gains in energy density and safety, but also resource constraints, all of which stalled the deployment of current lithium-ion technology. The highest rated of the new contenders are the lithium-sulphur (Li-S) batteries, which have theoretical energy densities several times higher than the lithium-ion chemistry. The major issues in solving the unwanted dissolution of sulphur and rapid fading of capacity need hugely novel material-engineered solutions for stabilizing and enhancing the lifespan. Solid-state batteries do have a very promising technology future in so far as they replace liquid electrolytes with solid ones, with significantly higher energy densities and safety to deliver. Much work is pushing a new frontier to the development of solid electrolytes that will achieve high ionic transport for long cycles. Moreover, sodium-ion batteries and multivalent-ion batteries, such as those based on magnesium and zinc, are becoming more prominent in the area of renewable and cheap substitutes for lithium-based batteries. All of these chemistries offer potential for mass market energy storage applications, as they have the ability to mitigate issues in lithium availability and cost [23].

7.2 Integration with smart grids and AI

In order to develop further the main argument that the energy grids are inherently dynamic, one must implement storage energy within the AI system and the smart grid to guarantee maximum efficiency and reliability. AI-based solutions incorporate predictive management of energy. Energy generation and consumption patterns are studied using AI algorithms so they may be stored and supplied in a way that avoids grid congestion and attains the highest global energy efficiency. In this, the early warning signs of battery degradation or faults are detected through machine learning techniques in fault avoidance, which helps to enhance the safety of the storage system as well as extend its lifetime. The microgrid uses AI technology to organize decentralized energy generation via real-time trades and supplies. So this facilitates peer-to-peer trading of electricity, which diminishes reliance on central generation stations for electricity transmission and strengthens the grid [24].

7.3 Policy and regulation perspectives

Government policy and regulatory frameworks are still at the center of determining the fate of energy storage deployment. One such trend is the creation of incentives for the use of energy storage so that the majority of countries are providing subsidies or tax credits to promote investment in grid-scale energy and battery storage solutions. Furthermore, as the energy storage system is deployed more and more, standardization along with regulation pertaining to safety will become more important. Reliability and risk avoidance associated with physical failure and exposure to the environment can be improved by ensuring uniform safety and performance standards. Regulative systems would also incorporate provisions for proper integration of storage system within renewable energy policy and enhancing stability of the grid in facilitating decarbonization efforts [24]. Government policy and regulatory structures are still at the core of deciding the destiny of energy storage deployment. One of these trends is the development of incentives for deploying energy storage so that most nations are offering subsidies or tax credits to encourage investment in grid-scale energy solutions and battery storage. In addition, as the energy storage system is increasingly deployed, standardization as well as regulation related to safety will become increasingly significant. Reliability and risk prevention linked with physical failure and environmental exposure can be increased through the implementation of consistent safety and performance standards. Regulative frameworks would also include mechanisms for effective integration of storage system into renewable energy policies and boosting grid stability in enabling decarbonization work [25].

Conclusion

Energy storage technologies together with conversion technologies determine the ability of energy systems to deliver energy in a reliable way while maintaining high efficiency and producing low carbon emissions. The development of solid-state batteries and hybrid battery systems has brought major advancements to energy storage because these technologies deliver higher energy capacity and improved safety features and longer product lifetime which enables their use in both grid power systems and transportation vehicles. The storage technologies of mechanical energy and thermal energy and chemical energy continue to play essential roles in providing support for power systems which require extended operational periods to manage their energy demands. The combination of energy storage systems with smart grid technology and artificial intelligence systems has created a new technology which allows for facilities to conduct maintenance works through predictive analytics while their operations run at optimal levels and energy resources get managed from multiple locations. The new system improvements create better operational performance which enables higher use of energy that comes from renewable sources which produce electricity in an unpredictable manner. The industry faces many obstacles because materials are hard to find and expensive to obtain and cannot be produced on a large scale and products are unsafe and do not comply with rules of government authorities. The solution to these problems depends on scientists working together to develop new materials and engineers creating complete systems and policymakers establishing necessary legislation. The energy storage industry requires ongoing innovation which must combine with safety standards that have reached official recognition to achieve effective energy storage deployment which will create a sustainable energy base for future energy systems that need to run without carbon emissions.

References

[1] McLarnon, F. R., & Cairns, E. J. (1989). Energy storage. https://doi.org/10.1146/annurev.eg.14.110189.001325

[2] Elalfy, D. A., Gouda, E., Kotb, M. F., Bureš, V., & Sedhom, B. E. (2024). Comprehensive review of energy storage systems technologies, objectives, challenges, and future trends. Energy Strategy Reviews, 54, 101482. https://doi.org/10.1016/j.esr.2024.101482

[3] Srivastava, S.S., Satyanarayana, G.S., Dhasmana, A., Rawat, V., Rana, A.S. and Bisht, Y.S., 2025. Role of Embedded Systems in Smart Energy Management: Challenges, Innovations, and Future Trends. Solar Energy and Sustainable Development Journal, 14(STR2E), pp.27-50. https://doi.org/10.51646/jsesd.v14iSTR2E.797

[4] Faisal, M., Hannan, M. A., Ker, P. J., Hussain, A., Mansor, M. B., & Blaabjerg, F. (2018). Review of energy storage system technologies in microgrid applications: Issues and challenges. Ieee Access, 6, 35143-35164. https://doi.org/10.1109/ACCESS.2018.2841407

[5] Vaghela, P., Pandey, V., Sircar, A., Yadav, K., Bist, N., & Kumari, R. (2023). Energy storage techniques, applications, and recent trends: A sustainable solution for power storage. MRS Energy & Sustainability, 10(2), 261-276. https://doi.org/10.1557/s43581-023-00069-9

[6] Thellufsen, J. Z., & Lund, H. (2016). Roles of local and national energy systems in the integration of renewable energy. Applied Energy, 183, 419-429. https://doi.org/10.1016/j.apenergy.2016.09.005

[7] Jones, C. R., Hilpert, P., Gaede, J., & Rowlands, I. H. (2021). Batteries, compressed air, flywheels, or pumped hydro? Exploring public attitudes towards grid-scale energy storage

technologies in Canada and the United Kingdom. Energy research & social science, 80, 102228. https://doi.org/10.1016/j.erss.2021.102228

[8] Zhao, Y., Ding, Y., Li, Y., Peng, L., Byon, H. R., Goodenough, J. B., & Yu, G. (2015). A chemistry and material perspective on lithium redox flow batteries towards high-density electrical energy storage. Chemical Society Reviews, 44(22), 7968-7996. https://doi.org/10.1039/C5CS00289C

[9] Li, M. J., Jin, B., Ma, Z., & Yuan, F. (2018). Experimental and numerical study on the performance of a new high-temperature packed-bed thermal energy storage system with macroencapsulation of molten salt phase change material. Applied energy, 221, 1-15. https://doi.org/10.1016/j.apenergy.2018.03.156

[10] Tawalbeh, M., Murtaza, S. Z., Al-Othman, A., Alami, A. H., Singh, K., & Olabi, A. G. (2022). Ammonia: A versatile candidate for the use in energy storage systems. Renewable Energy, 194, 955-977. https://doi.org/10.1016/j.renene.2022.06.015

[11] Rashidi, S., Karimi, N., Sunden, B., Kim, K. C., Olabi, A. G., & Mahian, O. (2022). Progress and challenges on the thermal management of electrochemical energy conversion and storage technologies: Fuel cells, electrolysers, and supercapacitors. Progress in Energy and Combustion Science, 88, 100966. https://doi.org/10.1016/j.pecs.2021.100966

[12] Mustafa, K. F., Abdullah, S., Abdullah, M. Z., & Sopian, K. (2017). A review of combustion-driven thermoelectric (TE) and thermophotovoltaic (TPV) power systems. Renewable and Sustainable Energy Reviews, 71, 572-584. https://doi.org/10.1016/j.rser.2016.12.085

[13] Huang, M., Liu, C., Zhang, Z., Wang, J., Zu, Q., Zhou, L., ... & Guo, S. (2024). Dual-mode electromagnetic-triboelectric-piezoelectric multifunctional self-charging energy system for efficient capture of kinetic energy. Nano Energy, 128, 109819. https://doi.org/10.1016/j.nanoen.2024.109819

[14] Dai, L., Chang, D. W., Baek, J. B., & Lu, W. (2012). Carbon nanomaterials for advanced energy conversion and storage. small, 8(8), 1130-1166. https://doi.org/10.1002/smll.201101594

[15] Amaral, M. M., Venancio, R., Peterlevitz, A. C., & Zanin, H. (2022). Recent advances on quasi-solid-state electrolytes for supercapacitors. Journal of Energy Chemistry, 67, 697-717. https://doi.org/10.1016/j.jechem.2021.11.010

[16] Bocklisch, T. (2015). Hybrid energy storage systems for renewable energy applications. Energy Procedia, 73, 103-111. https://doi.org/10.1016/j.egypro.2015.07.582

[17] Srinivasan, S., Velev, O. A., Parthasarathy, A., Manko, D. J., & Appleby, A. J. (1991). High energy efficiency and high power density proton exchange membrane fuel cells-electrode kinetics and mass transport. Journal of Power Sources, 36(3), 299-320. https://doi.org/10.1016/0378-7753(91)87009-Z

[18] Gür, T. M. (2018). Review of electrical energy storage technologies, materials and systems: challenges and prospects for large-scale grid storage. Energy & environmental science, 11(10), 2696-2767. https://doi.org/10.1039/C8EE01419A

[19] Hannan, M. A., Hoque, M. M., Mohamed, A., & Ayob, A. (2017). Review of energy storage systems for electric vehicle applications: Issues and challenges. Renewable and Sustainable Energy Reviews, 69, 771-789. https://doi.org/10.1016/j.rser.2016.11.171

[20] Chen, T., Jin, Y., Lv, H., Yang, A., Liu, M., Chen, B., ... & Chen, Q. (2020). Applications of lithium-ion batteries in grid-scale energy storage systems. Transactions of Tianjin University, 26(3), 208-217. https://doi.org/10.1007/s12209-020-00236-w

[21] Kularatna, N. (2014). Energy storage devices for electronic systems: rechargeable batteries and supercapacitors. Academic Press.

[22] Mitali, J., Dhinakaran, S., & Mohamad, A. A. (2022). Energy storage systems: A review. Energy Storage and Saving, 1(3), 166-216. https://doi.org/10.1016/j.enss.2022.07.002

[23] Zhang, X. Q., Zhao, C. Z., Huang, J. Q., & Zhang, Q. (2018). Recent advances in energy chemical engineering of next-generation lithium batteries. Engineering, 4(6), 831-847. https://doi.org/10.1016/j.eng.2018.10.008

[24] Ali, S. S., & Choi, B. J. (2020). State-of-the-art artificial intelligence techniques for distributed smart grids: A review. Electronics, 9(6), 1030. https://doi.org/10.3390/electronics9061030

[25] Kyriakopoulos, G. L., & Arabatzis, G. (2016). Electrical energy storage systems in electricity generation: Energy policies, innovative technologies, and regulatory regimes. Renewable and Sustainable Energy Reviews, 56, 1044-1067. https://doi.org/10.1016/j.rser.2015.12.046

Advanced Materials and Sustainable Energy Technologies – AMSET2025 Materials Research Forum LLC
Materials Research Proceedings 66 (2026) 262-276 https://doi.org/10.21741/9781644904152-24

Enhancing the Mechanical Properties of 3D Printed Clay Ceramics Artificial Coral Reefs Using Rice Husk, Corn Kernels, and Eggshell Additives

Angela May D. FLORES[1,a *], Arnaldo D. VALINO[1,b], Eisley John S. TIONGSON[1,c], Rugi Vicente RUBI[1,d]

[1]Adamson University, Ermita Manila, Philippines

[a]angela.may.flores@adamson.edu.ph, [b]arnaldo.valino@adamson.edu.ph, [c]eisley.john.tiongson@adamson.edu.ph, [d]rugi.vecente.rubi@adamson.edu.ph

Keywords: Mechanical Properties, 3D-Printed, Clay Ceramics, Artificial Coral Reefs, Sustainable

Abstract. This study aimed to improve the mechanical properties and environmental alignment of 3D-printed clay ceramics by using agricultural additives such as rice husk ash, cornstarch, and powdered eggshells. Each additive served a specific function to improve the structural integrity of the artificial coral reefs. Multiple combinations of the said materials were tested through 3D printing, compression test, and saltwater test, with the best-performing mix identified as 85% clay, 5% rice husk ash, 5% cornstarch, and 5% powdered eggshells. This ratio showed a significant improvement in the durability to seawater by the resistance to mass loss and water penetration. The study recommended further testing, trying other waste materials, and more time for underwater testing to assess long-term performance. The results support the Sustainable Development Goal (SDG) 14, preserving life under the ocean, seas, and marine.

Introduction

The enhancement of alternative coral reefs has been greatly improved since the start of the new generation of Philippine technology. From natural to synthetic forms of artificial reefs, 3D-printed artificial reefs are one of the many forms of technology used by marine life conservationists wherein their main goal is to provide solutions to the rapidly increasing problem of coral reef deterioration. Based on the research conducted by Palminteri [1], the materials and methods of installation for an artificial reef must be carefully chosen and prepared, or the structure may do more harm than good for the marine environment. With so many factors to consider, such as white band disease in corals, ocean acidification, and coral bleaching, the materials used for the enhancement of 3D-printed reefs must be chosen carefully to achieve the main purpose of this study with fewer complications along the way.

Because of its abundant material, clay has been used in various applications since prehistoric times. Clay's versatility as a raw material is what makes it more ideal for types of real-life applications wherein it is used to be molded and formed into different shapes that, when cured, reach final hardness and are difficult to break. Ceramics are generally made by taking mixtures of clay, earthen elements, powders, and water and shaping them into desired forms. Once the ceramic has been shaped, it is fired in a high-temperature oven known as a kiln. Often, ceramics are covered in decorative, waterproof, paint-like substances known as glazes [2]. In line with this, Clay Ceramic 3D Printing of Coral Reef was proven in July 2018 through a project conducted on Summer Island in the Maldives—named the Mars Coral Reef Restoration Program (MARS). The project reimagines farming structures as a three-dimensional lattice that can be deployed from small boats and implemented by divers, like an underwater Lego set. Ceramic is one of the best transplanting materials as it is completely inert and can be molded with rough surfaces. Once implemented, the system provides a permanent rigid structure for transplanting coral but also acts as habitat protection for other species in the area [3].

The agriculture sector in the Philippines constitutes a major source of employment (around 25 percent) in the country [4]. According to Zafar and Zafar [5], the Philippines is mainly an agricultural country with a land area of 30 million hectares, 47 percent of which is agricultural. The most common agricultural wastes in the Philippines are rice husks, rice straw, corn husks, coconut husks, coconut shells, and bagasse. The country has good potential for biomass power plants, as one-third of the country's agricultural land produces rice and consequently large volumes of rice straw and hulls. As a result, waste from the manufacturing of rice can be an ideal source of 3D printing material, as well as a less expensive and more sustainable solution to protect the environment while adapting to expanding technological innovations.

Moreover, one often overlooked but abundant waste source in the Philippines' poultry industry is eggshells. The continuous generation of this waste product contributes to environmental concerns if disposed of improperly. However, research proves that one of the primary components that make up an eggshell is calcium carbonate, which is not only useful to humans but also beneficial to the environment. It is also proven that the structure of coral reefs is composed of many layers of calcium carbonate—the same material eggshells are made of—which keeps the corals safe from ocean acidification, a chemical occurrence in the ocean that increases acidity and breaks down coral over time. Like the study conducted in India, where eggshell by-products were utilized in an environmentally friendly and sustainable way by building artificial coral reef structures [6].

Artificial reef restoration has increasingly relied on innovative technologies to overcome ecological, environmental, and structural challenges. Based on the study of Dotson Kirk, a system and method were introduced to enhance coral growth while simultaneously protecting coral against predatory species such as parrotfish. The concept revolves around attaching coral to a non-biodegradable base using biodegradable pins that offer initial protection while gradually disintegrating, allowing coral to acclimate and merge with the natural reef system [7].

By the integration of these biomasses, commonly abundant in Philippine farms, the researchers determined that the continuous deterioration of coral reefs—whether due to human impact or natural occurrences—can be addressed sustainably

Materials and Methods

Terracotta clay was the main material for this ratio mixed with natural additives. The Rice Hus (RHA) and powdered eggshells have passed through a fine mesh. For the cornstarch, corn kernels were soaked, blended, filtered and let the starch to separate to the liquid and then let it dry. The initial ratio shown in Table 1,that was tested had 85% clay, 5% RHA, 5% cornstarch, and 5% powdered eggshell. To perform compression testing, Universal Testing Machines was used on the blocks, and saltwater testing to see how the material responded to maritime environments. Every ratio has a significant outcome for its development.

Table 1. Percentage of a pre-determined Ratio for Additives

Ratio	Rice husk	Corn	Eggshells	Clay
RCE_{555}	5	5	5	85
RCE_{752}	7.5	5	2.5	85
RCE_{725}	5	2.5	5	85
RCE_{572}	5	7.5	2.5	85
RCE_{527}	5	2.5	7.5	85
RCE_{257}	2.5	5	7.5	85
RCE_{275}	2.5	7.5	5	85

It is needed some tools and supplies to do this study well. To make clay-ceramic mixtures that will work well in water, it's very important to get the materials ready properly. First, RHA needs to go through a furnace that for a temperature of 700 °C in 30 minutes. After that, it needs to be ground up with a pulverizer and then passed through a 600-micron mesh. To make cornstarch, soak corn kernels in water, mix them into a slurry, strain them through cheesecloth several times, let the starch settle out of the liquid, and then dry it in the oven before sieving. Powdered eggshells that is available in the market.

The experiment examines the application of natural additives—rice husk ash, corn kernel starch, and eggshells—in ceramic-based artificial reefs, owing to their accessibility, eco-friendliness, and capacity to enhance porosity, thereby facilitating coral larvae attachment. First, a simple 3D model is made in SolidWorks. It is then scaled down to 1×1 inch for testing in seawater and 2×2 inch for testing in compression.

To achieve the balance of porosity and structural strength, adjustments will be made based on trial and error. The clay mixture is then prepared for 3D printing, with printing parameters calibrated and modified for quality and consistency.

Modular Block System for Artificial Reef Formation:

With the Modular Block System, concrete blocks of standardized shapes of the size of 1x1 and 2x2 were designed as approved structural techniques to build artificial reefs. The system promotes simplicity and ease of fabrication with repeated results, which are ideal for testing. Therefore, it does not concern itself with mimicking the natural corals in their elaborate shapes.

The smaller version of the block, the 1x1, is to be used in seawater exposure tests to determine how well the material resists corrosion by seasaltwater. On the other hand, the larger 2x2 will be used for mechanical testing, where the blocks will be evaluated for parameters such as compressive strength and yield strength in order to determine their behavior under pressure and the depth under the sea to withold the pressure.

Artificial Reef Sample Block Procedure:

The researchers depend on SolidWorks application for a design of solid blocks of 1x1 inch and 2x2 inch by keeping them in the dimensions and shapes as constant as possible, such that they could be used for reliable testing. The 3D model, as shown in Fig. 1, enables them to view the blocks in detail without having to fabricate it. Such an arrangement could help to trace and rectify possible design problems in advance, while fine-tuning the blocks to the specifications for seawater corrosion testing in particular.

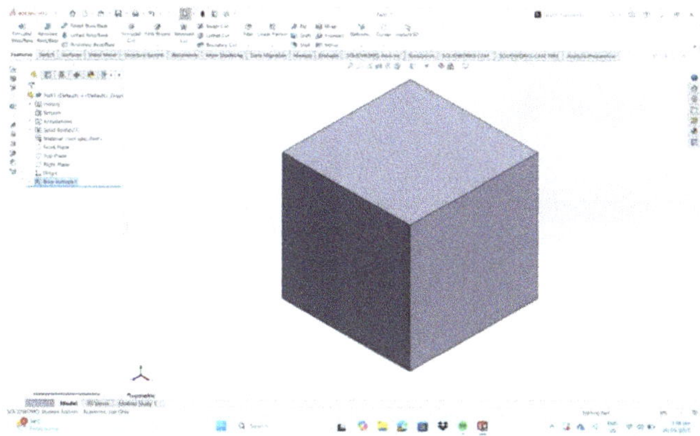

Fig. 1 2x2 inch AR Sample Block model in SolidWorks.

The researchers used Prusa Slicer to prepare the solid block models 1x1-inch and 2x2-inch for 3D printing, it is fine-tuned most of the significant parameters such as layer height, infill density, and print orientation so that it generally had similar geometry and good quality for each block-preferences that matter to reliable testing. Prusa Slicer also provided clear model preview, as shown in Fig. 2, thus making it easy to identify problems with the model before the real printing starts.

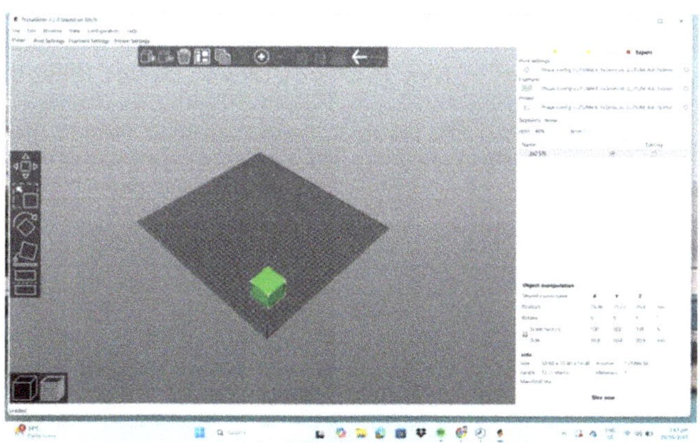

Fig 2. Sample 2x2 inch Block calibrated in Pura Slicer.

To prepare the solid block models for 3D printing, the researchers used Repetier-Host software. The software allowed careful control of slicing settings and ensured consistent geometry, which is critical for accurate and repeatable test results as shown in Fig. 3. With precise control over printing parameters, they were able to fine-tune the models, reduce potential errors, and ensure the final printed blocks met the standards needed for seawater corrosion and mechanical strength testing. Additionally, adjustments to the height between the ceramic slab and extruder nozzle, along with calibration of the slab's exact position, were performed to optimize the extrusion process.

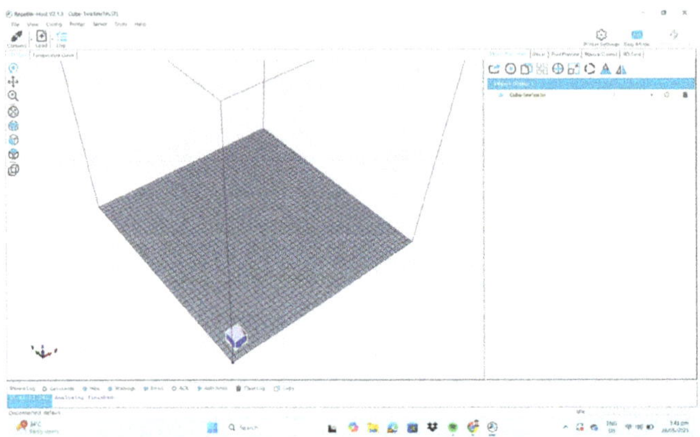

Fig. 3 Sample 2x2 inch Block prepared and ready for printing in Repetier-Host.

3D-Printing Trials and Drying Process:

Two types of 3D-printed solid block models sized 1 by 1-inch (1x1) and 2 by 2-inch (2x2) for 3D printing were created using a LUTUM v4.6 PRO printer for simulation and evaluation purposes only. The trial involved seven different material ratios in Table 1.The use of the LUTUM v4.6 PRO printer resulted in a machine time of approximately 1 minute and 30 seconds for the 1x1 block samples and 9 minutes and 45 seconds for the 2x2 samples. Several important observations were made during and after the printing and drying processes. For initial 3D printing preparation, the clay must be thoroughly mixed with the additives in the selected ratio. While hand-mixing is an option as a mixing method, inconsistent mixing is observed during the process especially when working with powdered additives. The use of a mechanical mixer ensures uniformly mixed material which prevents 3D printing irregularities and defects in the structure of the sample blocks.

Fig. 4 2x2 3D-Printed Sample Block Print Irregularities due to Excessively Wet Mixture.

Fig. 4, illustrates an excessively wet mixture resulting in noticeable shrinkage and overall inconsistency in form. It is observed that if the clay is too dry, the clay extrusion will be difficult due to increased resistance in the clay extruder, which may result in an incomplete layer in the creation of block during printing process. If the mixing is too wet, the printing process tends to lose its structural integrity, appearing watery and unstable. This condition makes the printed block shrink a lot during or after printing, which makes it hard to keep the intended geometric shape and size.

Fig.5 2x2 3D-Printed Sample Showing Non-Uniform Extrusion Due to Air Entrapment and Pressure Backflow.

As shown in Fig.5, non-uniform extrusion was caused due to air bubbles inside the clay cartridge and pressure backflow during printing. To ensure that no air bubbles are present, prevent air leakage, and backflow during this process, it is important to tightly seal the clay cartridge before printing process. The samples undergo an air-drying phase for 24 hours to gradually reduce ambient moisture and prevent print defects. The samples are kept in a closed storage with silica gel to keep them from getting wet from the air, which can cause mold to grow. The samples were then put in a kiln for more curing. This controlled the drying process, which made them stronger and removed more moisture.

Fig.6 2x2 3D-Printed Sample Showing consistent extrusion

The better performance in these ratios is due to the addition of corn kernels and egg shells, which made the paste flow more smoothly and made the structure stick together better. However, the higher amount of rice husk ash made the printed sample's surface look grainy. Figure 6 shows that it had relatively good printability, making forms that were structurally stable and had consistent extrusions.

The precise quantification and analysis of water ratios were considered beyond the scope of this study. However, the amount of water has a big effect to each ratio on how it will hold and print the mixture together. Starting with 100ml for each ratio and water was added in 10ml increments for the right consistency. This is to narrow the study's focus on the effects of the additive compositions.

After printing, the models are air-dried for 24–48 hours to minimize cracking, then fired at 1150°C for 7 hours to burn out any organic components. The fired blocks undergo testing to determine their strength and suitability. For testing the compressive strength, the researcher used A Universal Testing Machine (UTM). 1x1 blocks are put in seasaltwater to test durability and possible biological interactions in under marine environments. For analysis, all observations, data, and measurements are carefully written down.

Results and Discussion

To determine which material are more effective in enhancing the mechanical properties as well as the structural integrity of the clay ceramic artificial reef, by evaluating in different testing. The additives were systematically analyzed and investigated by the researchers to determine the optimal ratio composed of 85% terracotta clay and the remaining 15% were allocated among three additives.

Compressive Strength:

For the Compressive Strength Testing was conducted as one of the optimal ratio determinations to evaluate the structural analysis and mechanical performance of the clay ceramic artificial reef. Artificial reefs block samples measuring 2 x 2 inch were prepared after several drying processes. For this testing, 3 samples tested for each composition ratio which is a common practice in experimental research and material testing to ensure basic repeatability and also to allow easy identification through median comparison. The results of the mechanical testing were analyzed using the minimum, average, and maximum values for each sample composition

Hydrostatic Pressure:

In an underwater marine environment, reef structures are subjected primarily to hydrostatic pressure wherein it is a type of pressure exerted uniformly by the surrounding seawater due to the fluid depth. Therefore, in order to simulate how the long-term behavior of the samples under sea water pressure, the researchers used the load (kN) values that were achieved using Ultimate Testing Machine and converted it into its equivalent Hydrostatic Pressure using the fundamental fluid mechanics relationship formula;

$$P = \rho g h \qquad (1)$$

Where P (pressure) in Megapascal, ρ density of sea water (1,022 $\frac{kg}{m^3}$), gravitational acceleration (9.807 $\frac{m}{s}$), and h (depth). Through this conversion, it was able to gauge the theoretical pressure and sea depth that the samples can withstand before failure assuming static fluid conditions.

Materials Research Proceedings 66 (2026) 262-276 https://doi.org/10.21741/9781644904152-24

Table 2 Ratio to Stress, Force, and Computed Depth

Ratio	Stress (MPa)	Load (kN)	Computed Depth (m)
RCE_{555}	15.04	33.768	1490.50
RCE_{752}	11.738	26.353	1171.14
RCE_{725}	14.728	33.066	1469.46
RCE_{572}	16.491	37.024	1574.12
RCE_{527}	17.294	38.828	1725.47
RCE_{257}	13.634	30.611	1360.30
RCE_{275}	15.442	34.669	1540.69

The researchers utilized the values from the Ultimate Testing Machine shown in table 2 , specifically the force/load and corresponding compressive strength, and used it in order to calculate the equivalent hydrostatic pressure and determine the optimal depth for potential deployment of the sample for further continuation of this study. It is also shown that RCE572 and RCE527 perform better than the other ratios. This is because they have wide interquartile ranges and high median values, which means they can handle some variability while still being able to handle large applied stresses. These traits are very important in the real world, where materials have to deal with sudden or cyclic loads. RCE752, on the other hand, has the lowest median load capacity and the widest spread, which supports earlier observations of inconsistency.RCE275 and RCE555 work okay, but their distributions are much narrower, which means they are stable and predictable but don't have much load. RCE752, on the other hand, has the lowest median load capacity and the widest spread, which supports earlier findings of inconsistency.RCE275 and RCE555 work okay, but their distributions are much narrower, which means they behave in a stable and predictable way, but with very little load. These traits might be useful when consistency is more important than strength.

This load-based test backs up the recommendations for RCE572 and RCE527 for structural use. However, RCE752 may need more stabilizations when used to bear weight. These observations indicate that corn kernel starch may serve as a binder; however, it must be equilibrated with the mineral component derived from eggshells to maintain mechanical integrity post-processing.

Using Eq. 1 to see how compressive strength and hydrostatic pressure are related, the data show that RCE527, which has more eggshell in it, had the highest stress values. This means that it worked to make the ceramic structure stronger and more stable. RCE572 also showed more stress, but with less variation. This is probably because it had a good mix of rice husk and eggshell, which made it strong and gave it shape. RCE752, on the other hand, had the lowest and most variable stress values, which means that the structure wasn't strong enough. This might be because the cured material relies too much on the organic part (the corn kernel), which breaks down and makes the material weaker. Even though RCE257 is weaker, it can still be useful in situations where mechanical consistency is more important than peak strength. This is because it had a pretty tight distribution, which meant that performance was easy to predict but limited.

Finally, the calculated hydrostatic pressure, the constant values for seawater density and gravitational acceleration, and the estimated failure depth for each sample, as shown in Fig. 7.

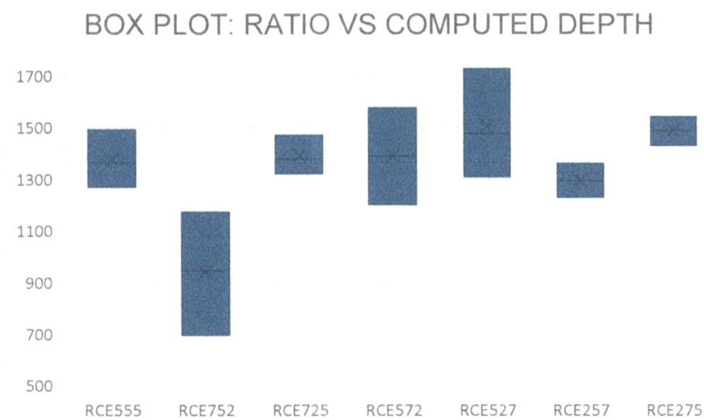

Fig. 7 Ratio to Computed Depth changed using the numbers from UTM.

The fibers in rice husk help make microchannels, which makes capillarity and absorption better. This helps coral grow or move nutrients because it makes small habitats inside the structure. RCE752 and RCE275 were not as extreme, but they were the same depth, which shows that the pore structure was controlled. This was perfect for finding the right balance between strength and permeability. But RCE752 had the widest spread and the least deep estimate. This means that it didn't mix well with water and its internal structure wasn't clear because it had a lot of organic binder and not much structural filler. This could be a problem for uses underwater where biological compatibility or a steady supply of nutrients are important.

Water Absorption:

To find out if the samples could handle the harsh underwater environment and the corrosion of sea water, the researchers did Sea Saltwater Testing. To test how well concrete holds up in harsh marine conditions, the 1x1 inch AR Sample solid blocks were put in seawater. The researchers observed the samples over the duration of one (1) month.

Water absorption measures the amount of water a porous material can absorb relative to its dry weight. This property is a key indicator of material porosity and long-term durability, particularly in applications exposed to harsh marine environments. The test is conducted by soaking the samples in sea water and is calculated by comparing the increase in weight of a sample after soaking with its original dry weight using the formula

$$\text{Mass Absorption\%} = \frac{\text{Weight of water after sea salt testing-Total dry weight}}{\text{Total dry weight}} x100 \qquad (2)$$

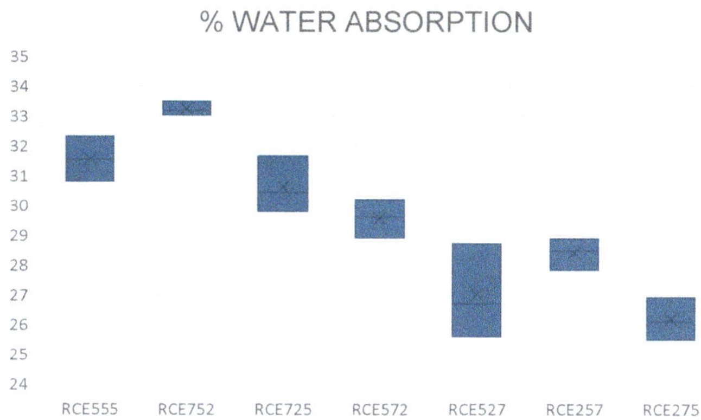

Fig. 8 %Water Absorption to Ratio Graph

The data presented by Fig.8 shows that the range of absorption range from a high of 33.482% to a low of 25.510, indicating significant variability due to the different additive combinations and also indicates moderate to high porosity that are generally typical for porous ceramic-based materials. High water absorption was observed in mixtures including RCE752 (33.482%), RCE555(33.182%), and RCE725 (32.308%), indicating the importance of naturally porous and fibrous component from rice husk and corn kernels. These components improve the structure's ability to retain moisture, which is crucial for promoting biological growth on artificial reefs. Lower absorption values, on the other hand, such those in RCE275(28.912%) and RCE257 (29.796%), indicate higher content of eggshells, which are less water-absorbent and denser because of their calcium carbonate content. Since sponges, algae, and coral larvae like somewhat damp and porous surfaces, water absorption is crucial during the early stages of colonization.

Mass Loss:

In sea water, wherein corrosion and abrasion from currents or sediments occur continuously, it is essential to assess the material's resistance to degradation in order to ensure long-term structural integrity. In line with this, in order to ensure that the AR structure can endure there harsh marine conditions, the researchers test the sample's % mass loss which serves as a critical indicator of how well the samples withstand the prolonged exposure to this type of environment. The formula for % mass loss is as follows

$$\text{Mass Loss (\%)} = \frac{\text{Initial Mass} - \text{Final Mass}}{\text{Initial Mass}} \text{x } 100\% \qquad (3)$$

Where the percent mass loss is determined by subtracting the Initial weight of the sample to its Final weight after dehydrating it and dividing it to its initial weight. This is to achieve the formulation for sustainable underwater applications.

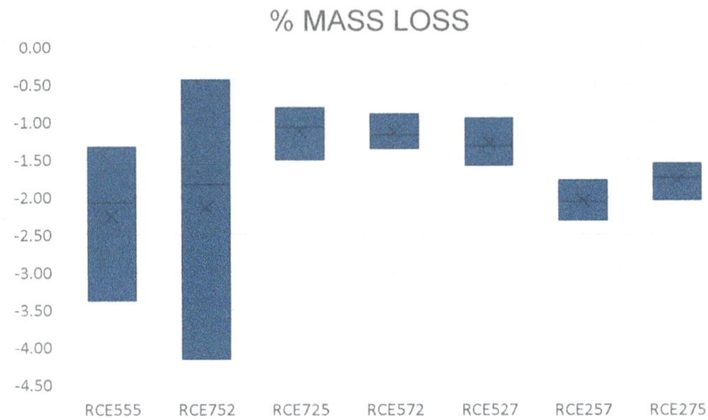

Fig. 9 %Mass Loss to Ratio Graph.

In Fig. 9 shows that all of the mass loss values are negative, which at first glance might seem bad. In this case, negative values are acceptable and show that after being submerged the material's mass decreased .People usually say that this loss is caused by natural processes like leaching of soluble parts, erosion of the surface, and slow degradation in marine conditions. These values are prevalent in durability testing and are generally acceptable for materials that are going to be used underwater, as long as the loss stays within acceptable limits. They don't mean that the material is bad.

To achieve the long life of the artificial reefs, it's important to look at how much mass they lose when exposed to simulated marine conditions. From worst (-4.16%) to the best (-0.45%) mass loss in this study . The mixtures RCE527 (-0.45%), RCE725 (-1.34%), and RCE257 (-1.51%) these showed impact resistance to degradation. This suggests that adding eggshells makes structures much stronger. The eggshells act as a binder and slowing down the breakdown of organic parts and making the material stronger in wet and salty conditions. The ratios of RCE752 and RCE572, decreased more mass, with losses of -4.16% and -3.38%, respectively. This is likely because they had many portions of corn kernels and rice husk, they make the material more porous. which can break down faster.

Porosity:

The ratio of empty or pore spaces in a material to its total volume is Percent porosity. To make the Artificial coral reefs more like real one, it should have high porosity because it makes them which helps water flow, nutrient exchange, and places for marine life to attach. It directly affects the reef material's ability to support ecological growth and biodiversity in underwater habitats. Percent porosity is good for the marine environment, but it weakens the structure. As porosity increases in value, the material gets lighter and more permeable, but there is less solid content to hold loads or stand up to outside forces. Generally, coral reefs structures are more fragile and easily to break if it is with many holes. Also, easy to erode, or crack, specifically in marine life with a lot of energy. So, even though high porosity makes it easier for marine life to live on an artificial reef, it could make the reef less stable and last longer if it is not measured equally with additive materials. To find the %porosity,the formula is as follows.

$$\text{Sample Density} = \frac{\text{Sample Mass}}{\text{Sample Volume}} \quad (4)$$

Given the calculated density of samples, the researchers obtained the %porosity of each samples given the constant density of seawater.

$$\text{Porosity\%}=\frac{\text{Sea water density}}{\text{Sample density}}\text{x}100 \quad (5)$$

High porosity was made possible in utilizing Eq. 5 particularly with the ratio having high percentage on cornstarch or rice husk.

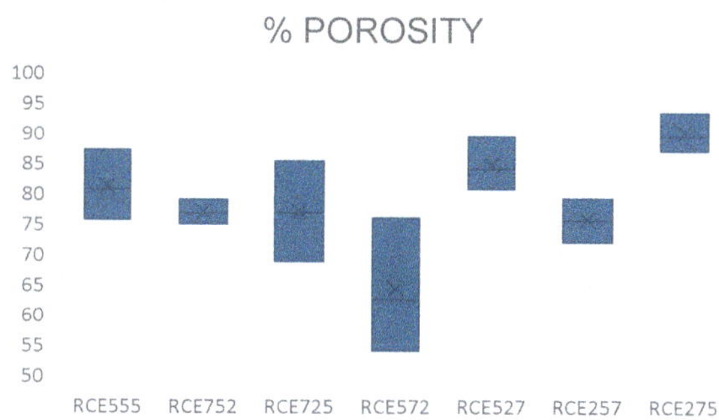

Figure 10 %Porosity to Ratio Graph.

The experimental assessment of natural additive combinations, namely rice husk, corn kernels, and eggshells, illustrates the potential for customizing 3D printed coral reef materials for ecological betterment or structural durability. As shown in Fig. 10, the porosity values increased from 54.199% to 93.562%. This large range has a direct effect on how well the artificial reef materials work in the environment. For reef surfaces that are meant to support marine environment, high porosity values (above 85%) are best. For example, 93.562% RCE275, 87.227% RCE555, and 89.559% RCE527 are all suitable for alternative for coral reefs. These samples probably have more corn kernels or rice husks in them, which help create interconnected pore networks and open up the inside of the structures. These traits are necessary for better nutrient exchange, water flow, and coral larval settling.

Table 3. Summary of Ratio to Sample Weight, %Water Absorption, %Mass Loss, and %Porosity

	Post Submersion	After 1 month of submersion					
Ratio	Initial Mass (grams)	Initial Mass	Pre-dry	Final dry	% Water Absorption	%Mass Loss	%Porosity
RCE_{555}	21.16	28.8	27.8	21.9	-2.27	81.24	80.72
RCE_{752}	22.1	29.3	28.2	22.4	-1.34	87.23	87.23
RCE_{725}	21.18	29.4	28.5	22.1	-4.16	75.1	75.1
RCE_{572}	21.9	29.4	28.7	22.1	-2.15	77.02	76.88
RCE_{527}	22.3	29.9	28.9	22.4	-0.45	79.07	79.07
RCE_{257}	21.9	28.7	27.9	22.1	-1.51	68.92	68.92
RCE_{275}	19.6	26.2	25.4	19.9	-1.13	77.1	76.95

In table 3, shows the summary of the mass and physical properties after being submerged in seasaltwater after a month.

Retention Ratio:

To find out how strong the clay-ceramic composite compositions were after being in seawater, the used of the %Retention Ratio to provide a quantitative test of strength retention. Eq.6 compares the average compressive stress of samples that were exposed to saltwater (1x1 inch samples) to the average compressive stress of samples that were not exposed (2x2 inch samples). This will give us an idea of how resistant the samples are to marine degradation. The formula is this:

$$\%\text{Retention Ratio}=\frac{\text{Mean Stress}_{1x1}}{\text{Mean Stress}_{2x2}}\text{x}100 \qquad (6)$$

Retention testing was done to see how strong the materials were before and after being in seawater. RCE527 had the highest strength and came in first place, followed by RCE752 and RCE572 in a study that looked at sustainability, cost, durability, and manufacturability.

Table 4 Computed Retention Ratio to Stress

Ratio	Retention Ratio	Stress 2x2	Stress 1x1
RCE_{555}	55.33	13.71	7.59
RCE_{752}	77.62	9.46	7.34
RCE_{725}	53.06	13.81	7.33
RCE_{572}	61.02	14.77	9.01
RCE_{527}	77.66	14.78	11.48
RCE_{257}	46.29	12.96	6.00
RCE_{275}	54.25	14.89	8.08

Retention testing was used to measure strength before and after exposure to seawater. The retention ratios of the tested samples showed that RCE527 had the highest ratio at 77.66%, which means it was better at low chances of being damage under sea water. In the same way, the ratios of RCE752and RCE572kept their retention value at 7, which is still high at 77.62% and 61.02% for each. On the other hand, RCE257 had the lowest value for retention that is 46.29%, which means that the saltwater exposure made it much weaker. This could be because it had less binder or biocompatibility additive, which could have made it break down faster when it was in the harsh sea water. These retention ratios show how important it is to make materials that can last in the environment and how well the materials keep their structural integrity in harsh marine environments.

Conclusion

Using 3D printed solid blocks model of 1×1-inch and 2×2-inch 3D-printed , it was possible to find the best additive ratios that make 3D-printed clay ceramics stronger. A multi-criteria framework that looked at printability, compressive strength, salinity resistance, and structural integrity while also considering sustainability, environmental, economic, and manufacturability limits made it possible to choose strategic ratios. Material preparation and ambient moisture had a big effect on printability and integrity, but RCE527 showed steady extrusion. It was shown in a thorough way that it had strong mechanical properties, was resistant to saltwater damage, was easy to print, and was easy to make. Its high compressive strength and resistance to the environment make it more reliable as a structure, and its low-cost production methods make it easy to use on a large scale.

Researchers can use 3D-printed solid block models that are 1×1 inches and 2×2 inches to find the best ratio that makes 3D-printed clay ceramics stronger.

The application of multiple evaluation criteria such as printability, compressive strength, salt water resistance, structural analysis, and trade-offs—guided by a multi-criteria design constraints encompassing sustainability, environmental impact, economic feasibility, and manufacturability, provides a comprehensive approach and strategic selection of an optimal ratios that is best suitable for manufacturing of high-performance clay-based ceramic components.

Material Preparation during the fabrication of 3D-printed Artificial Reef solid block models identified as a crucial factor affecting printing process and selection in determining the optimal ratio that enhances the mechanical properties of clay for artificial reef production. Moreover, exposure to ambient moisture was determined to affect the structural integrity of 3D-printed clay samples by promoting mold growth.

Among the tested ratios composed of 85% clay and 15% additives that were distributed for rice husk, corn kernel starch, and eggshells, RCE275 and RCE527both exhibited relatively good printability producing structurally stable forms with consistent extrusions.

The use of the Universal Testing Machine provides clear evidence for determining the optimal material ratio by evaluating the mechanical properties of each formulation. RCE572, and RCE527, demonstrated superior mechanical performance, confirming their effectiveness for load- bearing applications. These characteristics are vital in real-world applications where materials face cyclic or sudden loads. Their ability to withstand higher hydrostatic pressure and greater depth before failure highlights their suitability as optimal formulations for future structural design and material selection for 3D printed clay ceramic that will be used for artificial coral reef production. Researchers determined that corn kernel starch and eggshells need a well balance mixture with each other to sustain post processing mechanical integrity. Researchers determined that corn kernel starch and eggshells need a well-balanced mixture with each other to sustain post-processing mechanical integrity. RCE572, RCE527, and RCE275 was determined to be an optimal ratio in terms of load bearing Applications. the top-performing ratios, proved that even though Rice Husk Ash (RHA) was an additive mainly for strength purpose, compressive strength can also be optimized by balancing the intended functions of each additive

The evaluation using Sea Salt Testing effectively assesses samples and determine optimal ratio that can withstand harsh underwater environments and sea water corrosion. RCE527 (33.482%), RCE555 (33.182%) and RCE725 (32.308%) were determined to have high water absorption. Researchers concluded that fibrous and porous component from rice husk and corn kernels improve the structure's ability to retain moisture, which is crucial for promoting biological growth on artificial reefs.

To test durability and determine general suitability for underwater applications, the mass loss of each additive ratio was evaluated. Excellent resistance to degradation was shown by mixtures RCE527 (-0.45%), RCE725(-1.34%), and RCE257 (-1.51%), suggesting that the addition of eggshells greatly enhances structural durability. By reducing the breakdown of organic components and enhancing the material's resilience in situations that are saline and damp, eggshells serve as a stabilizing agent.

Researchers effectively evaluated the Clay-Ceramic Additive Composition based on Compressive Strength after Sea Water Exposure using the Universal Testing Machine. RCE527, RCE752, and RCE555 were identified as the most suitable clay-ceramic composite formulations for seawater-exposed environments, based on their computed retention ratios. These ratios, shown in, were selected due to their superior ability to maintain mechanical strength after one month of immersion in seawater.

Evaluation of percent porosity was compared to each additive ratio to determine desirable characteristics for artificial coral reefs. RCE275(93.562%), RCE555 (87.227%), and RCE527

Materials Research Proceedings 66 (2026) 262-276 https://doi.org/10.21741/9781644904152-24

(89.559%) were observed to be optimal as it produces high porosity values over 85% which is desirable characteristic for artificial coral reefs because it replicates the natural complexity of coral structures, improving water flow, nutrient exchange, and marine life attachment sites.

RCE527 that has additive proportions of 5%, 7.5%, and 2.5%—was determined to be the most effective and optimal formulation overall. It exhibited a well-rounded combination of strong mechanical performance, resistance to saltwater degradation, high printability, and ease of manufacturability. Its excellent compressive strength and environmental durability enhance its structural reliability, while its efficient production characteristics make it suitable for scalable deployment.

Acknowledgement
The researchers are deeply grateful to Engr. Eisley John Tiongson, Engr. Lester Olasiman, and Engr. Arnold Valino, Engr. Mark John Obsid and Mr. Jules Christian Ramos for his technical know-how, helpful criticism, and invaluable advice that helped to enhance the research. Also, to the students, Lyra Albino, Pouline Arcallana, Alyza Nicole Carlos who extend their help for collaborative effort to finish this research.

References

[1] "MARS - Alex Goad," Mars Coral Reef Restoration Program, 2018. [Online]. Available: [source].

[2]. D. Palminteri, "Materials and methods for artificial reefs must be carefully chosen," Marine Conservation Research, 2019.

[3]"What is a Ceramic?", Ceramic Materials Resource, Available: [source]. (n.d.).

[4] M. Zafar and N. Zafar, "Agricultural land use and waste in the Philippines," Agricultural Studies Journal, 2023.

[5] Pillai, "Use of eggshell by-product for sustainable artificial coral reefs," Environmental Sustainability Research, 2020

[6] Benavides-Reyes, C., Folegatti, E., Dominguez-Gasca, N., Litta, G., Sanchez-Rodriguez, E., Rodriguez-Navarro, A. B., & Faruk, M. U. (2021). Research note: Changes in eggshell quality and microstructure related to hen age during a production cycle. Poultry Science, 100(9), 101287. https://doi.org/10.1016/j.psj.2021.101287

[7] Dotson, K. (2024). 3D printing porous coral and preparation method (CN113831110A). China National Intellectual Property Administration.

Advanced Materials and Sustainable Energy Technologies – AMSET2025 Materials Research Forum LLC
Materials Research Proceedings 66 (2026) 277-285 https://doi.org/10.21741/9781644904152-25

Graphene-Based Nanofluids Effectiveness for Thermal Applications

Naser ALI[1,a], Ali ALSAYEGH[2,b], Ayas AL-ZANKI[1,c], Mohammad BANYAN[1,d], Mumayaz ALENEZI[1,e], and Ammar M. BAHMAN[2,f *]

[1]Nanotechnology Applications Program, Energy and Building Research Center, Kuwait Institute for Scientific Research, Safat 13109, Kuwait

[2]Department of Design Manufacturing and Engineering Management, University of Strathclyde, 75 Montrose Street, Glasgow, Scotland G1 1XJ, United Kingdom

[3]Mechanical Engineering Department, College of Engineering and Petroleum, Kuwait University, P.O. Box 5969, Safat 13060, Kuwait

[a]nmali@kisr.edu.kw, [b]A.alsayegh@strath.ac.uk, [c]azanki@kisr.edu.kw, [d]mbanyan@kisr.edu.kw, [e]maenezi@kisr.edu.kw, [f]a.bahman@ku.edu.kw

Keywords: Colloidal, Suspension, Nanomaterial, Thermophysical Properties, Figures of Merit

Abstract. The surge in global energy demand necessitates efficiency improvements in thermal systems. Nanofluids have emerged as promising working fluids because of their improved thermal performance. This investigation assesses the thermal and pumping behavior of graphene–water nanofluids with 0.01–0.10 vol.% graphene nanoplatelets and varied sodium dodecyl sulphate concentrations over 10–70°C. It also explores the benefit of these suspensions for applications operating under laminar and turbulent flow conditions. The thermophysical data were used to assess laminar effectiveness via the thermal conductivity-to-viscosity ratio ($C\lambda/C\mu > 0.25$) and turbulent performance via the Mouromtseff number ($Mo > 1$), alongside heat-transfer ($FoM_{heat\ transfer}$) and pumping (FoM_{pump}) figures of merit. Results show that higher graphene loadings (0.1 vol.%) with lower surfactant levels maximize heat-transfer performance, especially at elevated temperatures, while the lowest pumping performance occurs at 10 °C with 0.10 vol.% graphene and higher surfactant. These findings reveal that both graphene concentration and temperature critically influence nanofluid effectiveness for enhanced thermal applications.

Introduction

The continued rise in energy needs, including both water and electricity, possess a growing global challenge that is driven primarily by the expansion in population, progress in technology, and ongoing industrialization [1, 2]. Consequently, policymakers have been actively pursuing new strategies to mitigate for such pressure in energy demands. One of the common tactics that can be adopted to overcome such obstacle is to build additional power-generation facilities to increase the energy supply over the demand. However, this approach is not feasible, construed by land availability, time consuming, and would increase the greenhouse emissions. A more convenient option is to improve the efficiency of existing power plants by integrating advanced technologies into their operational cycles.

In the past years, nanotechnology has emerged as a transformative force in the energy sector, as it offers novel avenues to improving thermal systems performance, and thus can help in addressing the increasing demands for sustainable energy [3-5]. For instance, within thermal energy systems, such as solar-driven desalination setups, replacing the conventional working fluid with nanofluids can significantly improve the process overall heat transfer efficiency [6, 7]. The concept of forming suspensions by itself is well known; however, dispersing nano-scaled materials in a basefluid was first introduced by Masuda et al. [8], in 1993, after which it was formally termed 'nanofluid' by Choi and Eastman [9] two years later. In principle, nanofluids are suspensions made of nanoparticles of size less than 100 nm. These particles need to be uniformly distributed within a non-dissolving hosting liquid.

To fabricate these types of suspensions, scholars have established three principal routes: the single-step (or one-step), two-step, and the effervescent-tablet approach [10]. The one-step method relies on synthesizing and dispersing the nanoparticles the directly within the base liquid, and therefore minimizing the variety of achievable suspensions [11]. On the other hand, the two-step approach allows for a broader array of nanofluid compositions, as dry nanopowders is added to any non-dissolving liquid then dispersed to form the suspension [12]. In contrast, the effervescent tablet route utilizes the bubbles generated from the reaction of the effervescent agents with the liquid as the particles mixing mechanism for producing the nanofluid [13]. Comparing the three production methods, the effervescent tablet approach is the fastest in terms of suspension production duration, user-friendly, and much feasible [14]. However, it can only be applied to aqua liquids.

In solar thermal applications, carbon-based nanomaterials outperform their counterparts due to their relatively low cost, high thermal conductivity, and outstanding solar absorption capability [11]. As such, scientists prefer using them in their nanofluid recipe. For example, Ali [15] has shown that adding graphene to water can increase its thermal conductivity up to 125.72%, thereby demonstrating the significant thermal role of such type of material. Ahmadi et al. [16] showed an 18.87% enhancement in the efficiency of their solar collector test-rig over the basefluid due to the solar absorption characteristics of the dispersed graphene. Li et al. [17] demonstrated an increase in the photothermal conversion efficiency by 172% in their graphene-based suspension. Sainz-Manas et al. [18] has shown that dispersing 0.2 g/L and 0.3 g/L of graphene in a direct absorption solar collector (DASC) can increase the overall system photo-thermal conversion efficiency by 62.3% and 74.3%, respectively. Sathish et al. [19] investigated the thermal performance of heat pipe-based solar tube collectors using 0.1-0.3 vol.% of graphene-based suspensions, where a maximum increase of 41.3% was achieved. In the literature review conducted by Pradeep et al. [20] on cooling towers, the authors have mentioned high cooling rate enhancement from using graphene-based nanofluids, such as 67% (0.1 wt.%) and 36.2% (0.02 wt.%). The previous studies pinpoint the effectiveness of carbon-based nanomaterials in improving the suspensions performance.

In an economic point of view, the use of theoretical means to determine the effectiveness of a suspension before experimentally testing it is more feasible for real-life applications. One way of doing so is via calculating the different figure of merits (FoM) of the fabricated nanofluid that are related to the thermal behavior of the working fluid, its effect on the pumping power, and its effectiveness under laminar and turbulent conditions [21-23].

Due to the previous, this research investigates the thermal and pumping performance of graphene nanoplatelets (GN)-based nanofluids under turbulent and laminar conditions. According to the authors awareness, such research work has not been explored previously in the literature, and hence demonstrates the novelty of this work. The thermophysical properties of GN-water suspensions, of 0.01-0.10 vol.%, were obtained at 10°C-70°C from the study of Ali [15]. Afterwards, the effectiveness of the nanofluids under both turbulent and laminar situations was determined. Next, the effectiveness of the suspensions in terms of their heat transfer capability and influence on the pumping power were determined. Lastly, the performance of the nanofluids was compared to reveal the optimum production recipe.

Materials and Methods

Nanofluid Thermophysical Properties. Thermophysical properties of GN-water suspensions were gathered from the previous work of Ali [15]. Furthermore, the water-based nanofluids fabricated by the scholar used GN of > 99.5 wt. %, 2 to 8 nm (thickness), and 4 to 12 μm (diameter), along with an average of 3 to 6 graphene layers. The production process consisted of initially mixing 1.0:0.5 to 1.5 weight ratio of GN (0.01-0.10 vol.%) and sodium dodecyl sulphate (SDS) within deionized water (DIW), for 10 min, through a magnetic stirrer. Then, the mixture was intensively segregated for 90 min, at 10°C-70°C, via an ultrasonic microprocessor to reach the suspensions final phase. Afterwards, the density, specific heat capacity, thermal conductivity, and viscosity were measured, and their values were reported. Table 1 tabulates each suspension content. The thermophysical properties of the basefluid and different nanofluids (i.e., NF 1 to NF 9) can be found at Ali's work [15].

Table 1. Suspensions mixed content in 75 mL of basefluid.

Denotation	**vol.%**	**GN:SDS weight ratio**
NF 1	0.01	1.0:0.5
NF 2	0.01	1.0:1.0
NF 3	0.01	1.0:1.5
NF 4	0.05	1.0:0.5
NF 5	0.05	1.0:1.0
NF 6	0.05	1.0:1.5
NF 7	0.10	1.0:0.5
NF 8	0.10	1.0:1.0
NF 9	0.10	1.0:1.5

Suspension Performance. Thermophysical properties data were utilized to determine the effectiveness of the nanofluid under both laminar and turbulent flow status. In general, a suspension can be considered beneficial under laminar condition if the ratio between the viscosity enhancement coefficient and the thermal conductivity improvement coefficient ($C\lambda/C\mu$) is > 0.25. However, for a nanofluid to be beneficial under a turbulent flow status, its Mouromtseff number (Mo) should be > 1. The following equations were used to obtain $C\lambda/C\mu$ and Mo [24]:

$$\frac{C_\lambda}{C_\mu} = \frac{(\lambda_{NF}-\lambda_{BF})/\lambda_{BF}}{(\mu_{NF}-\mu_{BF})/\mu_{BF}} \tag{1}$$

$$Mo = \left(\frac{\lambda_{NF}}{\lambda_{BF}}\right)^{0.67} \times \left(\frac{\rho_{NF}}{\rho_{BF}}\right)^{0.8} \times \left(\frac{C_{NF}}{C_{BF}}\right)^{0.33} \times \left(\frac{\mu_{NF}}{\mu_{BF}}\right)^{-0.47} \tag{2}$$

where λ, μ, ρ, and C are the thermal conductivity, viscosity, density, and specific heat capacity respectively. The subscript BF and NF in Eqs. (1) and (2) refer to basefluid and nanofluid, respectively.

Furthermore, the performance of both pumping power and heat transfer of the nanofluids were determined through their pumping power figure of merit (FoM_{Pump}) and heat transfer figure of merit ($FoM_{Heat\ transfer}$), consecutively. The equations used in determining both figure of merits are shown below [21]:

$$FoM_{\text{Pump}} = \left(\frac{\rho_{NF}}{\rho_{BF}}\right)^{-2} \times \left(\frac{C_{NF}}{C_{BF}}\right)^{-2.8} \times \left(\frac{\mu_{NF}}{\mu_{BF}}\right)^{0.2} \tag{3}$$

$$FoM_{\text{Heat}\ transfer} = \left(\frac{\lambda_{NF}}{\lambda_{BF}}\right)^{0.6} \times \left(\frac{\rho_{NF}}{\rho_{BF}}\right)^{0.34} \times \left(\frac{C_{NF}}{C_{BF}}\right)^{0.06} \times \left(\frac{\mu_{NF}}{\mu_{BF}}\right)^{-0.44} \tag{4}$$

Results and Discussion

Suspension Effectiveness Under Different Flow Condition. The effectiveness of the nanofluids denoted as NF 1 to NF 9 under laminar condition is shown in Fig. 1. It can be seen from the plotted data that all of the different nanofluids are beneficial for usage under such flow condition; apart from the NF 2 type of suspension at 10°C, which had a $C\lambda/C\mu$ value of 0.08. This is mainly because the $C\lambda/C\mu$ for the different dispersions at various temperatures had a value higher than 0.25, which is the threshold that distinguish between the effectiveness and non-effectiveness of the working fluid. Nevertheless, at a low temperature range (i.e., 10°C to 30°C), the higher vol.% suspensions (NF 7 – NF 9) showed to outperform their counterpart. The previous can be linked to the effectiveness of the SDS surfactant on the dispersion stability at such temperature range and the higher thermal conductivity gains caused by increasing the nanoparticles concentration. On the other hand, raising the temperature reduces the influence of the SDS surfactant on the physical stability of the dispersed particles. This causes higher particles concentrations to cluster faster into forming sedimentation. Hence, the suspension becomes less physical stable, and therefore its thermal conductivity reduces. The previous can be noted from the 70°C scenario, where the NF 1 – NF 3 (i.e., 0.01 vol.% suspensions) showed to be more beneficial under laminar condition compared to the other variants. In general, increasing the temperature was found to increase the effectiveness of the working fluid under laminar condition, simply because the thermal conductivity of the basefluid itself (i.e., water) increases with the rise of its temperature. The highest obtained $C\lambda/C\mu$ was 132.8, which corresponds to the NF 1 suspension at 70°C. This $C\lambda/C\mu$ value is 72.8% higher than lowest $C\lambda/C\mu$ case (i.e., NF 9) at the same temperature.

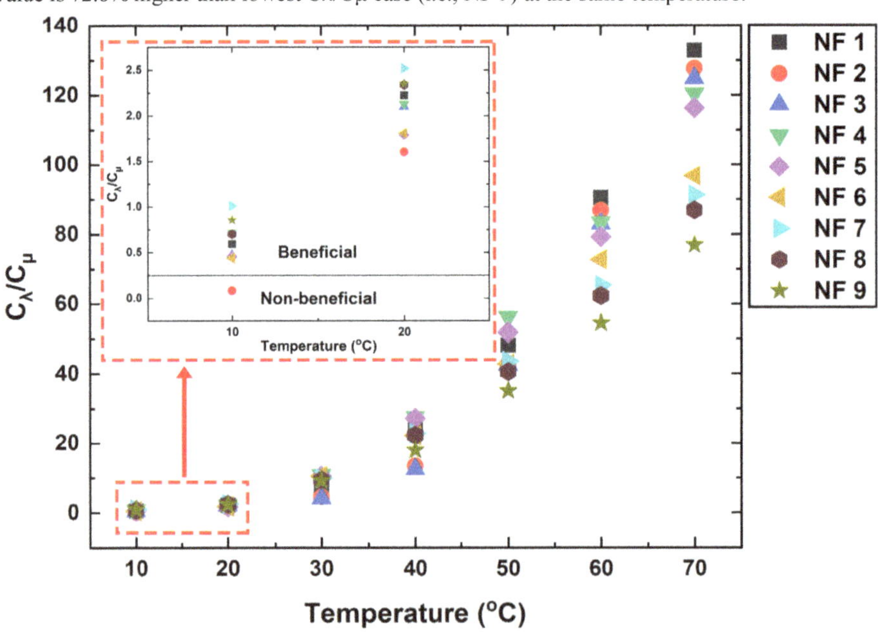

Fig. 1. Ratio between the viscosity enhancement coefficient and the thermal conductivity improvement coefficient of different nanofluids at various temperatures.

As for the effectiveness of the suspensions under turbulent condition (Fig. 2), it was found that all of the different nanofluids were not beneficial for usage in applications operating at 10°C because they had a Mo value less than 1. Also, at 20°C, only NF 1 – NF 4 were able to overcome the Mo > 1 threshold, thus demonstrating their effectiveness under turbulent flow condition. Beyond the 20°C point, all types of nanofluids were found to be effective under turbulent status, with the highest being NF 7 at 70°C with a Mo of 1.61. It is to be noted that, unlike the laminar from cause, increasing the concentration of GN causes an improvement in the Mo value, and therefore enhancing its effectiveness as a working fluid under turbulent flow condition. The previous is reasonable when considering that turbulent flow improves the mixing of the suspended particles. Hence, the nanofluid can be seen as physically stable and relies considerably on the vol.% of the suspension and the effect of basefluid temperature, which corresponds both to the overall thermal conductivity of the suspension.

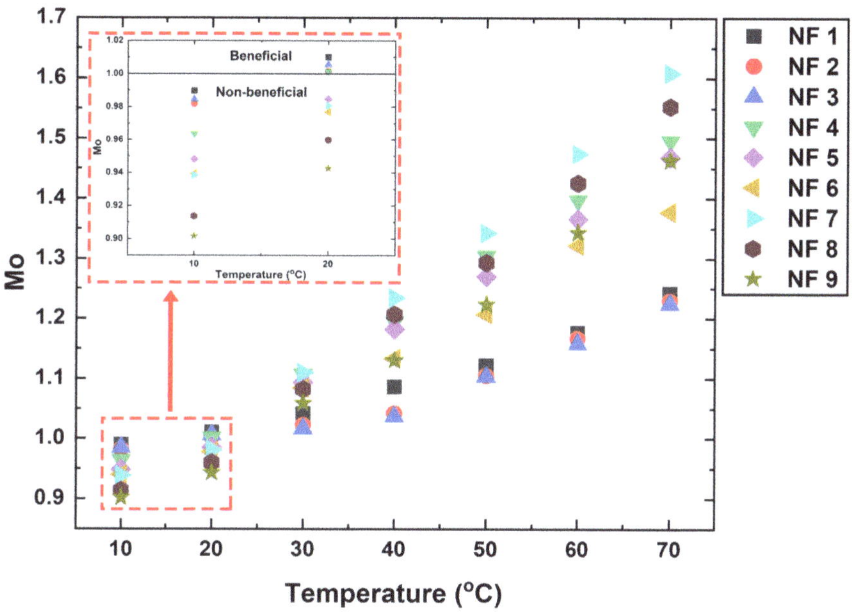

Fig. 2. Mouromtseff number of different nanofluids at various temperatures.

Nanofluid Performance. FoM_{Pump} and $FoM_{Heat\ transfer}$ results are shown in Fig. 3 and Fig. 4, respectively. From Fig. 3, it can be seen that NF 1 – NF 3 had the highest FoM_{Pump}, which is due to these nanofluids having the lowest dispersed GN concentration (i.e., 0.01 vol.%). Increasing the vol.% showed to decrease the FoM_{Pump}, as in the causes of NF 4 – NF 9. Notably, increasing the SDS surfactant concentration was also shown to cause a decrease in the FoM_{Pump}. For example, the level of reduction in the FoM_{Pump} between the NF 7 (GN:SDS weight ratio of 1.0:0.5) and NF 9 (GN:SDS weight ratio of 1.0:1.5) was found to be ~25%. The previous is likely due to the increase in suspension density, which plays a crucial role in calculating the FoM_{Pump}. Furthermore, the temperature of the suspension, as a parameter, had a negligible influence on the previous figure of merit.

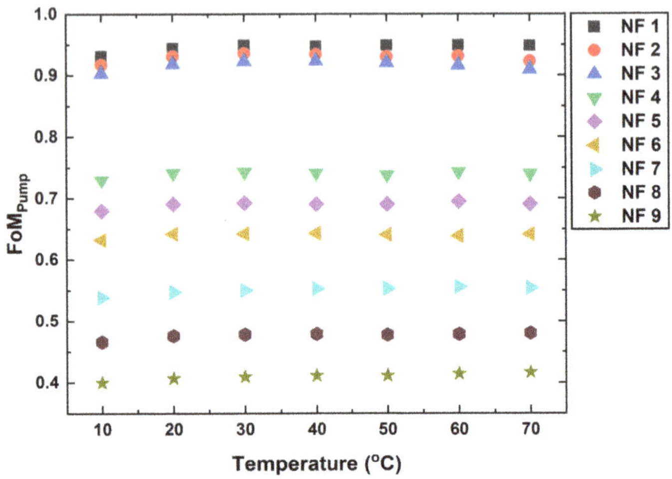

Fig. 3. Pumping power figure of merit of different nanofluids at various temperatures.

In contrast, analysis of the outcomes in Fig. 4 demonstrates that the liquid temperature and GN concentration are both crucial parameters for obtaining a higher $FoM_{Heat\ transfer}$ in the working fluid. In addition, surfactants are well-known to reduce the effective thermal conductivity of the suspension [11]. This is why the nanofluids of high nanoparticles vol.% with lower SDS concentration is seen to overperform thus with similar vol.% but higher SDS concentrations (e.g., NF 7 and NF 9). The highest $FoM_{Heat\ transfer}$ was achieved with the NF 7 suspension at 70°C (i.e., 1.62), whereas the lowest was found to be 1.01 with the NF 2 dispersed at 10°C.

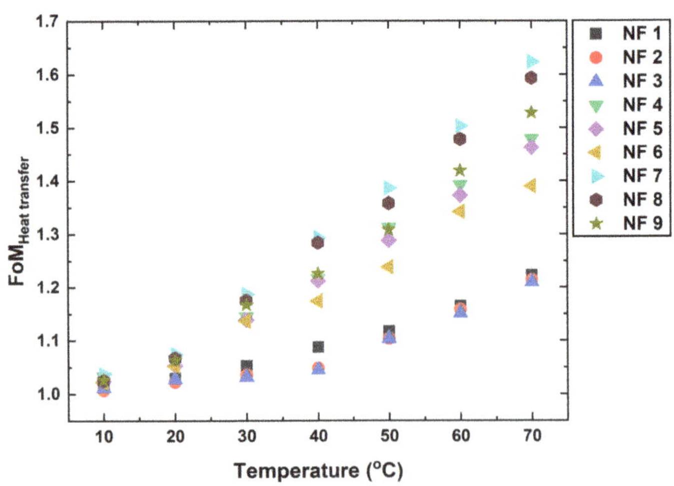

Fig. 4. Heat transfer figure of merit of different nanofluids at various temperatures.

Conclusions

An evaluation of the thermal-hydraulic performance of dispersed graphene in water nanofluids. The explored concentrations of GN were 0.01 vol.% to 0.10 vol.%, across a temperature range of 10°C to 70°C. Sodium dodecyl sulphate surfactant of 0.5 to 1.5:1 (SDS to GN weight ratio) was also encountered in the study due to its importance in maintaining the suspension physical stability. Afterwards, the effectiveness of the suspensions under laminar and turbulent flow conditions were encountered along with the pumping power figure of merit and the heat transfer figure of merit.

The results have shown that all types of nanofluids were effective under laminar flow condition except for the NF 2 case at 10°C, which had a $C\lambda/C\mu < 0.25$. Furthermore, the highest $C\lambda/C\mu$ was found to be 132.8, corresponding to the NF 1 case at 70°C. As for turbulent flow conditions, all of the suspensions at 10°C and most of them at 20°C demonstrated to be non-beneficial due to their Mo value being lower than 1. Beyond these temperature points, all types of nanofluids showed > 1 Mo values, with the highest achieved with the NF 7 dispersion at 70°C (Mo = 1.61). Thus, reflecting the crucial role of dispersed particles concentration and liquid temperature on the effectiveness of the fluid. A clear trade-off was found between the FoM_{Pump} and $FoM_{Heat\ transfer}$, where the first favored the lower particles loading as a pose to the second. Accordingly, the highest FoM_{Pump} was achieved with the 0.01 vol.% suspensions, whereas the 0.10 vol.% nanofluids were more effective in terms of the $FoM_{Heat\ transfer}$.

The results herein can be used by researchers and decision makers as guidelines to examine the effectiveness of similar (or other) suspensions before employing them in their thermal systems. However, other factors should also be considered, such as the techno- and thermo-economic of the dispersion, to fully ensure the success of the operation.

Acknowledgement

The authors would like to express their appreciation to the Kuwait Foundation for the Advancement of Sciences (KFAS) for financially supporting this investigation under project no. (CN22-35EM-1547), and the Kuwait Institute for Scientific Research (KISR) for their financial support under project no. EA117C.

References

[1] G.S. Mutumba, T. Odongo, N.F. Okurut, V. Bagire. A survey of literature on energy consumption and economic growth. Energy Reports. 7 (2021) 9150-239. https://doi.org/10.1016/j.egyr.2021.10.107

[2] A. Zaharia, M.C. Diaconeasa, L. Brad, G.-R. Lădaru, C. Ioanăș. Factors Influencing Energy Consumption in the Context of Sustainable Development. Sustainability. 11 (2019) 4147. https://doi.org/10.3390/su11154147

[3] E. Serrano, G. Rus, J. Garcia-Martinez. Nanotechnology for sustainable energy. Renewable Sustainable Energy Rev. 13 (2009) 2373-84. https://doi.org/10.1016/j.rser.2009.06.003

[4] N. Ali, J.A. Teixeira, A. Addali. Effect of Water Temperature, pH Value, and Film Thickness on the Wettability Behaviour of Copper Surfaces Coated with Copper Using EB-PVD Technique. J Nano Res. 60 (2019) 124-41. https://doi.org/10.4028/www.scientific.net/JNanoR.60.124

[5] A.K. Hussein. Applications of nanotechnology in renewable energies-A comprehensive overview and understanding. Renewable Sustainable Energy Rev. 42 (2015) 460-76. https://doi.org/10.1016/j.rser.2014.10.027

[6] A.E. Kabeel, Z.M. Omara, F.A. Essa. Improving the performance of solar still by using nanofluids and providing vacuum. Energy Conversion and Management. 86 (2014) 268-74. https://doi.org/10.1016/j.enconman.2014.05.050

[7] T. Elango, A. Kannan, K.K. Murugavel. Performance study on single basin single slope solar still with different water nanofluids. Desalination. 360 (2015) 45-51. https://doi.org/10.1016/j.desal.2015.01.004

[8] H. Masuda, A. Ebata, K. Teramae. Alteration of thermal conductivity and viscosity of liquid by dispersing ultra-fine particles. Netsu Bussei. 7 (1993) 227-33. https://doi.org/10.2963/jjtp.7.227

[9] S.U.S. Choi, J.A. Eastman. Enhancing thermal conductivity of fluids with nanoparticles. Conference: 1995 International mechanical engineering congress and exhibition, San Francisco, CA (United States), 12-17 Nov 1995; Other Information: PBD: Oct 1995. Argonne National Lab., IL (United States)1995. p. Medium: ED; Size: 8 p.

[10] N. Ali, H. Bahzad, N.F. Aljuwayhel, S.A. Ebrahim, A.T. Hamoud. Effervescent tablets for carbon-based nanofluids production. Journal of Molecular Liquids. 390 (2023) 123083. https://doi.org/10.1016/j.molliq.2023.123083

[11] N. Ali, A.M. Bahman, N.F. Aljuwayhel, S.A. Ebrahim, S. Mukherjee, A. Alsayegh. Carbon-Based Nanofluids and Their Advances towards Heat Transfer Applications-A Review. Nanomaterials. 11 (2021) 1628. https://doi.org/10.3390/nano11061628

[12] N. Ali. Crystal layer growth with embedded carbon-based particles from effervescent tablet-based nanofluids. Scientific Reports. 14 (2024) 18554. https://doi.org/10.1038/s41598-024-69605-y

[13] N. Ali, H. Bahzad, N.F. Aljuwayhel, S.A. Ebrahim, A.T. Hamoud, H. Al-Mazidi, et al. A Trihybrid Approach for Enhancing Crude Oil Recovery Using Effervescent-Tablet-Based Nanofluids. Global Challenges. 9 (2025) 2400202. https://doi.org/10.1002/gch2.202400202

[14] N. Ali, H. Bahzad, N.F. Aljuwayhel, S.A. Ebrahim. Techno-economic assessment of effervescent tablet-based nanofluids. PloS one. 20 (2025) e0319265. https://doi.org/10.1371/journal.pone.0319265

[15] N. Ali. Graphene-Based Nanofluids: Production Parameter Effects on Thermophysical Properties and Dispersion Stability. Nanomaterials. 12 (2022) 357. https://doi.org/10.3390/nano12030357

[16] A. Ahmadi, D.D. Ganji, F. Jafarkazemi. Analysis of utilizing Graphene nanoplatelets to enhance thermal performance of flat plate solar collectors. Energy Conversion and Management. 126 (2016) 1-11. https://doi.org/10.1016/j.enconman.2016.07.061

[17] Z. Li, A. Kan, K. Wang, Y. He, N. Zheng, W. Yu. Optical properties and photothermal conversion performances of graphene based nanofluids. Appl Therm Eng. 203 (2022) 117948. https://doi.org/10.1016/j.applthermaleng.2021.117948

[18] M. Sainz-Manas, A. Vossier, R. Garcia, C. Caliot, F. Bataille, G. Flamant. On-sun performance and stability of graphene nanofluids in concentrating direct absorption solar collectors. Sustainable Energy Technologies and Assessments. 83 (2025) 104605. https://doi.org/10.1016/j.seta.2025.104605

[19] T. Sathish, J. Giri, M. Al-lehaibi, A.O. Hourani, A. Anderson. Performance enhancement of heat pipe-based solar tube collectors using graphene nanofluids: Experimental analysis and thermal optimization. Case Studies in Thermal Engineering. 73 (2025) 106460. https://doi.org/10.1016/j.csite.2025.106460

[20] E. Pradeep, N. Ali, A.M. Bahman. Nanomaterial-enhanced cooling towers: Advances, applications, and performance optimization strategies. Appl Therm Eng. 292 (2026) 130451. https://doi.org/10.1016/j.applthermaleng.2026.130451

[21] H. Althalab, N. Ali, S.A. Ebrahim, A.M. Bahman. Air conditioning compressor lubricant enhancement using carbon-based nanomaterials for hot climates. Renewable Sustainable Energy Rev. 217 (2025) 115672. https://doi.org/10.1016/j.rser.2025.115672

[22] Z. Xuan, Y. Zhai, M. Ma, Y. Li, H. Wang. Thermo-economic performance and sensitivity analysis of ternary hybrid nanofluids. Journal of Molecular Liquids. 323 (2021) 114889. https://doi.org/10.1016/j.molliq.2020.114889

[23] A. Alirezaie, M.H. Hajmohammad, M.R. Hassani Ahangar, M. Hemmat Esfe. Price-performance evaluation of thermal conductivity enhancement of nanofluids with different particle sizes. Appl Therm Eng. 128 (2018) 373-80. https://doi.org/10.1016/j.applthermaleng.2017.08.143

[24] M. Ma, Y. Zhai, P. Yao, Y. Li, H. Wang. Synergistic mechanism of thermal conductivity enhancement and economic analysis of hybrid nanofluids. Powder Technol. 373 (2020) 702-15. https://doi.org/10.1016/j.powtec.2020.07.020

Statistical Normalization of RAP-Based Mixtures: A Decision Framework for Sustainable Road Engineering

Justine John M. ALAS[1,a*], Alejandrino III P. ABRINA[1,b], Akhira Jaem A. ALI[1,c], Raphael B. AZARCON[1,d], Ervin Jayro B. BALBUENA[1,e], Keziah Lovelydel M. BALDIVIA[1,f], Roberto D. ROSARIO[1,g]

[1]Civil Engineering Department, FEU - Institute of Technology, P. Paredes St., Sampaloc, Manila, 1015, Philippines

[a]jmalas@fit.edu.ph, [b]apabrina@fit.edu.ph, [c]aaali@fit.edu.ph, [d]rbazarcon@fit.edu.ph, [e]ebbalbuena@fit.edu.ph, [f]kmbaldivia@fit.edu.ph, [g]rdrosario@feutech.edu.ph

Keywords: Reclaimed Asphalt Pavement, Sustainable Road Construction, Bio-Based Rejuvenators, Performance-Based Design

Abstract. Reclaimed Asphalt Pavement (RAP) is becoming significant for the sustainable road constructions. Despite its global success, the standardization of RAP as blending materials continues to be hampered by substantial differences and lack of field validation in tropical areas. This study addresses the challenges by utilizing a statistical normalization method to evaluate recent utilization of RAP-based mixtures in tropical countries, evaluating them across six essential factors: mechanical strength, durability, constructability, environmental impact, economic feasibility, and regulatory compliance. The result highlights that Warm Mix Asphalt with RAP (WMARAP) as the top-performing solution among twelve (12) different mixtures with a composite score of 4.44 out of 5.00, achieving up to 30% reduction in carbon emissions, 18% life-cycle cost savings, and full compliance with AASHTO. RAP offers notable benefits for those areas with limited availability of virgin aggregates and reducing construction expenses. In the Philippines, RAP-based mixtures present a promising solution for developing durable, low-emission, and cost-effective pavements. The study presents a practical decision framework for policymakers, engineers, and contractors aiming for sustainable road engineering.

Introduction

Reclaimed asphalt pavement (RAP) has recently become an increasingly important material due to its environmental and economic benefits. While RAP is used since the 1970s, its economic and mechanical viability has only gained recent recognition through testing and mix design advancements. Over 98 million tons of RAP were used in the United States in 2022, saving nearly 4.9 million tons of asphalt binder and 93 million tons of virgin aggregate. RAP also saw an increase in the use of that same year, from 15.6% in 2009 to 22.2% in 2022, further minimizing the reliance on virgin aggregates in asphalt pavements [1]. Research also shows that asphalt mixtures containing at least 30% RAP perform comparably to virgin asphalt in terms of durability and structural integrity [2]. Moreover, other innovations such as fractioned RAP and softer binders further enhanced the feasibility of using higher RAP contents without compromising performance [3].

Globally, reports from Australian Road Research Board in June 2022 highlighted RAP as the most cost-effective recycled material when used in surface and base layers, being capable of reducing greenhouse gas emissions up to 98% and its overall environmental impact by 99% [4]. Additionally, using RAP with other sustainable technologies such as warm-mix asphalt (WMA) can reduce carbon emissions and energy consumption in road construction [5]. The Department of Public Works and Highways (DPWH), Philippines shows that RAP can replace up to 40% of coarse aggregates in specific mixes while maintaining quality standards if it follows strict technical specifications and quality control. RAP mixtures can perform similarly to conventional mixes under local climate and traffic conditions based on field trials in Metro Manila, with 25% and 40% mixtures producing comparable results in rut depth, skid resistance, and sand patch tests. The RAP Mixtures also meet DPWH specifications and British standards, supporting the materials viability. The mixture containing 25% RAP is generally recommended for better surface layers, while 40% RAP is preferable for base or lower pavement layers. The blending percentages allow greater use of the material without compromising overall pavement integrity [6]. However, challenges would remain despite of RAP's advantages. The increased stiffness and inconsistencies of aged binders can cause the mixture to be susceptible to moisture and cracking, which contribute to the degradation of materials [7]. Moreso, the long-term performance of RAP is still not fully established under the local climate and traffic variability. Further, the limited field trials restrict a broader application of the material. Performance disparities in aged binder also complicates mix design and quality control [8], emphasizing the need for thorough research to optimize RAP use and ensure pavement sustainability.

This study assesses the mechanical soundness, applicability and nature-related ability, and legal conformity of different RAP mixes. The study sorts to find the best mix designs that balance durability, ease of use, and cost viability to fill gaps in the standardization of RAP and identifies the adaptability issues of RAP in tropical areas.

Methodology

This study uses a systematic secondary data analysis approach to evaluate the performance and feasibility of implementing RAP technologies for Philippine Road infrastructure. The methodology involves reviewing and analyzing existing publications, technical reports, and standardized testing results in various academic and technical databases such as ASCE Library, Elsevier, Scopus, and Web of Science from 2000 - 2025. This methodology facilitates integrating varied experimental conditions, material compositions, and testing protocols from different research. The method produces a comprehensive performance matrix reflecting the current applications on sustainable pavements and road construction standards (see Fig. 1).

Materials Research Proceedings 66 (2026) 286-299 https://doi.org/10.21741/9781644904152-26

Figure 1. Input-Process-Out Diagram

Job-Mix Designs of Various Experiments

The interest in using Reclaimed Asphalt Pavement in road construction has started in the 1970s [9]. Since then, knowledge regarding this matter has been better understood because many researchers have focused on it, reducing the construction industry's dependence on virgin asphalt binders.

The content of a Reclaimed Asphalt Pavement in an asphalt mix is associated with its mechanical properties. Resistance to cracking and rutting is especially crucial as it dramatically impacts the pavement's long-term performance. Most notably, depressions along wheel paths where water tends to pool can cause rutting, posing significant safety issues such as aquaplaning [10]. That is why it is important to ensure that when designing an asphalt mix, it is both resistant to cracking and rutting.

Similarly, RAP content is limited to around 15% in most applications. This is because the higher the RAP content in an asphalt mixture, the more noticeable its properties change [9]. In line with this, [Refer with: Tab 1] showing the findings of various studies the effects on asphalt mixture with varying RAP contents.

Table 1. *Findings on stiffness/cracking and deformations/rutting based on varying Reclaimed Asphalt Pavement (RAP) contents*

The author(s)	Location	RAP content	Findings on Stiffness/Cracking	Findings on Permanent Deformations/Rutting
In 2022, Abdel-Jabjer et al. [11]	Amman University, Amman, Jordan	0%, 25%, 50%, 75%	As RAP content increases, stability also increases—indicating increased stiffness.	Flow values decreased slightly with higher RAP, while Voids Filled with Asphalt (VFA) and compaction increased. These trends suggest that higher RAP content improves resistance to permanent deformation or rutting.
In 2007, Al-Qadi et al. [12]	University of Illinois Urbana-Champaign, Illinois, USA	0% — 60%	Increased RAP content results in a decrease in fatigue resistance	Rutting resistance is directly proportional to RAP content; with increased RAP, rutting resistance also increases
In 2020, Tarsi et al. [13]	University of Bologna, Bologna, Italy	Not specified, but implicit in RAP mixes	RAP performed equal to or better than virgin aggregates for various cracking types (71-97% of cases)	RAP performed equal or better than virgin aggregates for rutting (67%) and raveling (93%)
In 2011, Copeland [14]	McLean, Virginia, USA	0%, 15%, 30%, 60%	Higher RAP content results in lower fatigue. Also, using soft virgin binders is effective in counteracting the stiffening effect of the RAP.	Higher RAP increases rutting resistance; However, soft binders may compromise this resistance.
In 2023, Abd Ali et al. [15]	University of Technology, Baghdad, Iraq	0%, 20%, 30%, 40%, 50%	Marshall Stability generally increases from 0% to 40%, except for 50%, which indicates increased stiffness as RAP content also increases.	Since Marshall Flow has an inverse relationship with RAP contents, which indicates increased resistance to permanent deformation.
In 2017, Arshad et al. [16]	Universiti Teknologi MARA, Shah Alam, Selangor, Malaysia	15%, 25%, 35%	Higher RAP content creates stiffer asphalt pavements compared to virgin asphalt	Higher RAP content has lower rutting resistance than virgin aggregates.
In 2022, Plati et al. [17]	National Technical University of Athens, Greece	0%, 10%, 20%, 30%, 40%	Higher RAP content can handle overall higher axial loads compared to virgin aggregates, indicating a stiffer material.	Higher RAP shows increased deformation resistance compared to virgin aggregates, with the 20%-80% ratio of RAP-VA aggregates having the highest Modulus of Resilience (464MPa).
In 2024, Adriati et al. [18]	Bangkinang-Petapahan crossroad in Kampar Regency, Indonesia	0%, 30%	30% RAP has a 16% increase in stability (2088 kg in RAP mixture compared to 1800 kg in virgin aggregates), implying a stiffer material.	30% RAP shows higher rutting resistance compared to virgin aggregates.
In 2021, Antunes et al. [19]	Instituto Superior Técnico Universidade de Lisboa, Portugal	0%, 75%	75% Unaged RAP mixture has a higher stiffness modulus than virgin aggregates. Aged RAP also shows a higher stiffness modulus compared to unaged RAP.	70% aged and unaged RAP shows higher rutting resistance compared to aged and unaged virgin aggregates.
In 2024, Putra et al. [20]	Lampung University, Indonesia	25%, 50%, 75%	Increased RAP content shows a decrease in stability, showing potential risks of cracking.	Higher RAP content shows a direct relationship with the level of degradation, indicating lower long-term durability.

Studies on various RAP content imply that higher RAP commonly increases overall stiffness and deformation resistance, although with some notable exceptions. Studies by Abdel-Jabber et al. and Abd Ali et al. demonstrate that increasing RAP content enhances Marshall Stability while decreasing Flow values, displaying improved mix stiffness and rutting resistance [11,15]. Abd Ali et al. specifically found an increase of the Indirect Tensile Strength (ITS) and Punching Strength of the material up to 40% RAP, indicating better cracking resistance, while reduced rutting depths confirmed superior rutting performance compared to virgin mixes [21]. Al-Qadi et al. and Tarsi et al (2007). confirmed that RAP-modified mixes frequently show enhancement of rutting resistance. However, Copeland (2011) notes that using soft virgin binders to counteract initial stiffening could compromise this resistance [12,13,14]. Antunes et al. (2021) observed higher stiffness and rutting resistance in high RAP mixtures with rejuvenators. Putra et al. (2024) found that increasing RAP content without rejuvenators decreases material stability over time [19,20]. Certain studies highlight deformation concerns, with Plati et al. (2022) reporting that RAP specimens show greater permanent deformation than virgin aggregates under certain conditions [9]. Arshad et al. (2017) also found that varying RAP-Virgin Aggregate ratios slightly underperformed compared to virgin aggregates by 2%, though differences remained within Malaysian Public Works Department specifications [16].

The content percentages and stress levels significantly influenced the deformation behavior of RAP. The 100% RAP shows more distortion at low stress levels, while virgin aggregates deform more at higher stress levels. Triaxial repeated load testing showed that 50-50% RAP-virgin blends exhibited less permanent deformation than pure virgin mixes, with only minor increases in residual stresses up to 50% RAP content. Studies found that using more than 75% RAP content indicated a critical point when higher levels of deformation became an issue; this implies that while adding RAP typically increases stiffness and rut resistance, its application must carefully balance to ensure long-lasting pavement performance, especially when choosing the base binder, creating the mix, and modifying the quantity of recycling agents [12,13,14].

Logistic Sustainability. Logistical sustainability in asphalt pavement engineering focuses on cutting down energy use, material transport, and operational demands during the production, construction, and maintenance of roads. Using Reclaimed Asphalt Pavement (RAP) along with warm mix asphalt (WMA) and other recycled materials has proven to offer clear logistical benefits in many settings. Studies around the world show that high RAP-content mixes can support sustainability goals without sacrificing performance [19]. For example, Antunes et al. found that using 75% RAP in pavement rehabilitation significantly lowered energy use, greenhouse gas emissions, and material handling, while still meeting mechanical and functional standards. Moura et al. also reported a 23% drop in global warming potential and a 58% cut in water use when RAP was combined with steel slag. Likewise, Abdalla et al. showed that asphalt mixes with RAP, fly ash, and bio-oil rejuvenators greatly reduced the need for virgin materials and shortened transport distances [21,22].

Life-cycle assessments strongly support the logistical advantages of RAP-based technologies as it shows clear signs of improvement in supply chain efficiency. Liu et al. [22] determined that RAP and WMA mixtures lead to a 30.7% decrease in greenhouse gas emissions, a 25% decrease in energy use, and a 17.9% reduction in life-cycle costs. These results came mainly from a 45.8% decrease in aggregate transport and a 41.9% reduction in binder usage. Chen et al. also showed that cold-recycled RAP mixtures can reduce about 12,976 kg of CO_2 per kilometer of roadway by cutting down fresh aggregate transport and removing the need for high-temperature processing. Similar tactics have worked well in the Philippines. Viado's work, for instance, eliminated the need for long-distance virgin aggregate hauling by reusing RAP on rural roads. In order to process RAP on-site and eliminate transportation, the runway project at Lubang Airport used mobile HMA plants [23]. RAP-based construction is a workable, effective way to enhance logistics and increase infrastructure resilience because S. Chen, et. al, (2019) demonstrated that current facilities and supply chains can manage these techniques [24].

Material Comparison

The preceding chapters established the mechanical behavior and material interactions of Reclaimed Asphalt Pavement (RAP) across diverse binder formulations and environmental conditions. However, individual material characterizations provide limited insight for practical implementation decisions. This chapter bridges laboratory findings with real-world applications through a systematic comparative analysis using choice matrix methodology with statistically normalized scoring.

This analytical framework merges performance metrics, cost-efficiency, environmental impacts, constructability constraints, and regulatory compliance into a single decision-support tool. The scoring methodology (Refer with: Tab 2) evaluates of twelve distinct RAP-containing mix designs, providing evidence-based guidance for material selection, field deployment, and specification development.

Table 2. RAP Mix Designs

Abbreviation	Mix Design	Description	References
CM-RAPZW	Cement Mortar + RAP + Zinc Waste	Fine RAP aggregates with industrial waste valorization	[23][24]
WMARAP	Warm Mix Asphalt + RAP	Temperature-reduced production with high RAP content	[25]
WEOA-HMARAP	Waste Engine Oil + Hot Mix + RAP	Bio-rejuvenator integration with conventional HMA	[21]
PAC13-RAP	PAC-13 Asphalt + RAP	Porous asphalt concrete with recycled content	[26]
40RAP3SOY	40% RAP + 3% Soybean Oil	High RAP content with bio-based rejuvenation	[27][28]
50RAP-RA	50% RAP + Recycling Agents	Ultra-high RAP with chemical rejuvenation	[29]
RAPSPc	RAP + Superplasticizer (in Concrete)	Concrete application with superplasticizer as workability enhancement	[30]
RAPSPa	RAP + Superplasticizer (in Asphalt)	Asphalt application with superplasticizer as flow improvement	
RCRS	Concrete + Crumb Rubber + RAP + Slag	Multi-waste stream integration	[30]
RAP-EpoA	RAP + Epoxy Asphalt	High-performance binder modification	[31]
RAPRASRA	RAP + Recycled Asphalt Shingles + RA	Dual reclaimed material utilization	[32]
RAPCC	RAP + Castor Wax-Corn Oil	Bio-based rejuvenator combination	[33]

Evaluation Framework. Six evaluation criteria were established based on engineering standards and industry best practices:

Table 3. Score Calibration using Linear Interpolation

Category	Indicators	Min (Score = 60)	Max (Score = 90–95)	References
Mechanical	Modulus (GPa), rutting (mm), fatigue life	Modulus <6, rut >7mm	Modulus >10, rut <3.5 mm	[34]
Durability	TSR %, freeze–thaw %, stiffness loss	TSR <70%	TSR >90%	[35]
Constructability	Mixing temp, viscosity, compaction energy	>160°C, poor workability	≤130°C, excellent slump	[36][37][38]
Environmental	RAP %, CO_2 reduction %, bio-content	RAP <10%, no CO_2 ↓	RAP ≥40%, 30% CO_2 ↓	[35][39]
Economic	Cost savings (%), LCCA net present value	Negative or neutral LCCA	≥20% cost savings	
Regulatory	AASHTO/FHWA/ ASTM conformance	Not covered by specs	Fully compliant and used	[34][36][40][41][42][43]

Scoring Methodology. Performance evaluation employed a systematic five-step process:

1. Data Collection: Quantitative performance data extracted from peer-reviewed sources, including journals, FHWA reports, and standards (AASHTO M323, ASTM D6927)
2. Benchmark Establishment: Performance ranges defined from literature representing minimum acceptable, average, and best-in-class levels
3. Linear Interpolation Scoring: Raw scores (0-100) assigned using the formula:

$$Raw\ Score = \left(\frac{x_{min} - x}{x_{max} - x_{min}}\right) \times [(Score_{max} - Score_{min}) + Score_{min}]$$

   Where:
   $x = actual\ measured\ value\ from\ data$
   $x_{min}, x_{max} = benchmark\ minimum\ and\ maximum\ values\ for\ that\ indicator$
   $Score_{min}, Score_{max} = typically\ 60 - 95\ for\ realism\ instead\ of\ 0 - 100$

4. Category Aggregation: Multiple indicators within categories averaged with appropriate weighting
5. Scale Normalization: Raw scores converted to 5-point descriptive scale for interpretation

Where:
$x = actual\ measured\ value\ from\ data$
$x_{min}, x_{max} = benchmark\ minimum\ and\ maxumum\ values\ for\ that\ indicator$

Table 4. 5-point Evaluation Scale

Score	Description	Interpretation	Application to Pavement Material Properties
5	Excellent	Exceeds performance benchmarks; optimal for implementation	Exceeds AASHTO/ASTM specs, superior durability (TSR >90%), ideal workability, >40% RAP, cost-efficient [28,34,36,41,44,41,42,43,48]
4	Good	Meets or slightly exceeds benchmarks; minimal issues	Meets all specs, slight room for optimization (e.g., TSR ~85–90%), RAP >30%, moderate cost savings [28,34,36,41,42,43,44,49]
3	Acceptable	Meets minimum standards; performance dependent on conditions	Passes specs, average durability (TSR ~75–85%), workability fair, CO_2 reduction or RAP ~20–30% [20,21,28,34,36,41,43,44,50]
2	Marginal	Inconsistent performance; needs further testing or optimization	Near or below spec limits (e.g., TSR ~70–75%), RAP <20%, workability issues, low constructability [20,21,22,28,34,36,41,51]
1	Deficient	Fails to meet critical standards; major limitations	Below spec thresholds, major cracking/rutting, durability failures, excessive cost or emissions [20,21,22,28,34,36,41]
0	Inapplicable	Inapplicable or lacking essential performance	Not implementable under current standards or fails laboratory testing

Results: The comprehensive evaluation yielded clear performance differentiation among the twelve mix designs (Refer with: Tab 5)

Table 5. Comparative Performance Results based on Previous Experiments

Mix Design	MP	D	C	EI	EF	RC	Average	Normalized (5-point scale)
CM-RAPZW [23][24]	65	66	65	74	74	70	69.00	3.45
WMARAP [25]	88	90	85	92	88	90	88.83	4.44
WEOA-HMARAP [21]	82	84	81	89	87	85	84.67	4.23
PAC13-RAP [26]	63	64	63	73	72	68	67.17	3.36
40RAP3SOY [27][28]	72	70	70	80	79	78	74.83	3.74
50RAP-RA [29]	77	76	72	85	82	81	78.83	3.94
RAP-SP (Concrete) [44]	63	64	63	73	72	68	67.17	3.36
RAP-SP (Asphalt) [44]	79	78	79	86	83	83	81.33	4.07
RCRS [30]	61	62	60	72	70	65	65.00	3.25
RAP-EpoA [31]	70	68	66	75	76	76	71.83	3.59
RAP-RAS-RA [32]	75	74	72	82	81	79	77.17	3.86
RAP-CC [33]	85	86	83	90	86	87	86.17	4.31

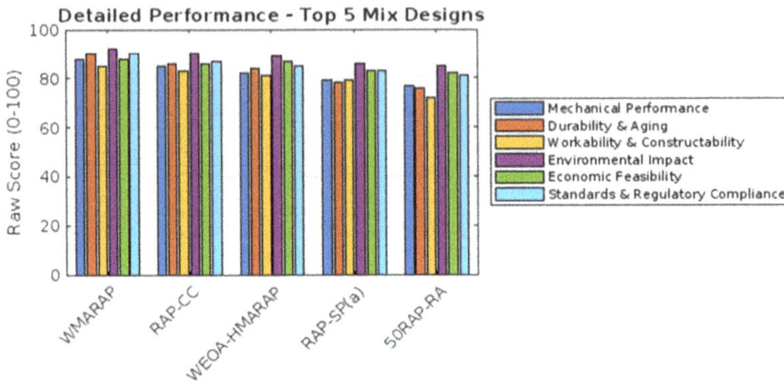

Figure 2. Performance of Top 5 Mix Designs on every category

The following (Refer with Fig. 2) present the twelve RAP-based pavement mix designs. These designs are quantitatively compared based on six important criteria: Mechanical Performance (MP), Durability & Aging (D), Workability & Constructability (C), Environmental Impact (EI), Economic Feasibility (EF), and Regulatory Compliance (RC). (Refer with: Tab 5 and Fig. 2) show the results based on the criteria given. The normalized 5-point scores obtained using statistically balanced scoring techniques described in the previous chapters represent the aggregate integrability of each mixture.

Based on the results (Refer with: Fig 2), WMARAP stands out with the highest normalized score of 4.44/5. This scoring is due to its excellent mechanical strength, being resistant to aging, having significant economic and environmental capabilities, and adhering to regulatory standards [25]. As such, Philippine Road infrastructure can significantly benefit from broadly implementing WMARAP, mainly in primary road networks where reliable mechanical properties and emission reduction are needed. RAP-CC and WEAO-HMARAP follow with scores of 4.31 and 4.23 [21,33]. RAP-CC, in turn, is perfect for mixed-traffic regional roads due to its balanced profile, high fatigue resistance, effective rejuvenation, and substantial compliance with regulatory standards. WEOA-HMARAP, such as WMARAP, provides excellent environmental performance and economic viability due to using engine oil as a rejuvenator; however, WEOA-HMARAP still falls slightly short in feasibility compared to WMARAP.

RAP-SP (asphalt) and 50RAP-RA mixtures finally complete the top five with normalized scores of 4.07 and 3.94, respectively [29,44]. The notably high performance of RAP-SP (asphalt) in durability and constructability suggests that it is an applicable material for urban rehabilitation projects. Even so, the relatively high economic and environmental scores of 50RAP-RA mean that it can handle more extensive use if its mechanical reliability is further locally validated. On the other hand, blends such as RCRS and CM-RAPZW, with respective scores of 3.25 and 3.45, display lower serviceability mainly due to their low durability and mechanical strength [23,31]. Compared to traditional pavement construction, these materials show better uses in experimental or specialized green construction. While the other blends display less consistent results, most notably in workability and durability, (Refer with: Fig. 2) shows the consistently high performance of WMARAP and RAP-CC.

The results validated that RAP-SP(a) and 50RAP-RA for secondary implementation following further pilot validation; and WMARAP, RAP-CC, and WEOA-HMARAP for direct deployment. In the Philippines pavement sector, the data-scoring approach highlights the worth of multi-criteria analysis in balancing practical constructability, regulatory compliance, and mechanical, environmental, and economic objectives.

Conclusion and Recommendations

The evaluation reveals a direct correlation between mechanical and environmental performance due to aged binder properties and high RAP content. At the same time, durability and workability mainly rely on proper rejuvenation strategies. This highlights the importance of performance-based design frameworks such as Balanced Mix Design (BMD) over prescriptive RAP content limits. RAP mixtures for the Philippines and countries with similar economic status offer a path to cost-efficient, low-emission, high-performance pavements. The recommended strategy involves rapid deployment of WMARAP in primary road networks, combined with controlled field trials for experimental compositions to develop future standards further.

The study shows WMARAP mixture containing 40% recycled materials, decreases greenhouse gas emissions by 30%, and provides nearly 18% savings in lifecycle costs due to considerable decreases in aggregate transportation and binder consumption. Following this, RAP-CC and WEOA-HMARAP should be introduced for specialized applications, with controlled field trials of high-RAP bio-rejuvenated mixtures conducted over 12 to 24 months to establish reliable long-term performance data.

. The inconsistencies in RAP and rejuvenator quality also require careful quality control measures and standardized protocols to maintain a consistent mixture performance. Moreover, operational challenges such as equipment availability, contractor experience, and costs for adapting Warm Mix Asphalt must be managed through capacity-building and resource allocation. Training and resource allocation must address operational challenges such as equipment availability, expertise, and cost considerations. Likewise, life-cycle assessments, including the entire process from RAP collection to road construction, will further strengthen RAP's economic and environmental justification.

Furthermore, RAP-CC and WEOA-HMARAP provide proven substitutes, particularly in urban and sub-urban settings where resource limitatios. It is advised that high-RAP formulations using bio-based rejuvenators (such as RAP-RAS-RA and 50RAP-RA) be implemented gradually through upscale implementation. These formulations have shown technical viability under controlled conditions. On the other hand, although mixtures like CM-RAPZW and RAP-CR-Slag provide distinctive waste valorization pathways, additional validation against local performance requirements is necessary before they can be adopted in the Philippine context, especially with regard to shrinkage, interfacial bonding, and early-life durability.

References

[1] B. Williams and J. Willis, Asphalt Pavement Industry Survey on Recycled Materials and Warm-Mix Asphalt Usage 2022 Information Series 138 13th Annual Survey. 2024. doi: 10.13140/RG.2.2.22994.16328.

[2] A. Milad, A. Mohd Taib, A. G. F Ahmeda, M. Solla, and N. I. Md Yusoff, "A REVIEW OF THE USE OF RECLAIMED ASPHALT PAVEMENT FOR ROAD PAVING APPLICATIONS," J Teknol, vol. 82, no. 3, Apr. 2020. https://doi.org/10.11113/jt.v82.14320.

[3] G. A. AL-Mosawe and H. M. Al-Mosawe, "Study the Effect of using RAP in Warm Mix Asphalt Pavement.," Journal of Engineering, vol. 29, no. 7, pp. 153–168, Jul. 2023. https://doi.org/10.31026/j.eng.2023.07.10.

[4] "Best Practice Expert Advice on the Use of Recycled Materials in Road and Rail Infrastructure: Part A Technical Review and Assessment Author: Australian Road Research

Board Prepared for: Commonwealth Sustainable Procurement Advocacy and Resource Centre (C-SPARC), Department of Agriculture, Water and the Environment," 2022.

[5] A. Hasan, U. Hasan, A. Whyte, and H. Al Jassmi, "Lifecycle Analysis of Recycled Asphalt Pavements: Case Study Scenario Analyses of an Urban Highway Section," CivilEng, vol. 3, no. 2, pp. 242–262, Mar. 2022. https://doi.org/10.3390/civileng3020015.

[6] R. Singson, "Use of Reclaimed Asphalt Pavement (RAP) in Hot Mix Asphalt," Manila, May 2015. Accessed: Jul. 04, 2025. [Online]. Available: https://www.dpwh.gov.ph/dpwh/sites/default/files/issuances/DO_071_S2015.pdf

[7] M. Alavi, D. Jones, Y. He, P. Chavez, and Y. Liang, "Investigation of the Effect of Reclaimed Asphalt Pavement and Reclaimed Asphalt Shingles on the Performance Properties of Asphalt Binders: Phase 1 Laboratory Testing."

[8] F. Zaremotekhases, H. Sadek, M. Hassan, C. Berryman, and M. I. Hossain, "The Use of Reclaimed Asphalt Pavement Materials and Warm-Mix Asphalt Mixtures in the South-Central States: Challenges and Limitations," in Tran-SET 2020, Reston, VA: American Society of Civil Engineers, Jan. 2021, pp. 185–194. doi: 10.1061/9780784483305.019.

[9] C. Plati, M. Tsakoumaki, and K. Gkyrtis, "Physical and Mechanical Properties of Reclaimed Asphalt Pavement (RAP) Incorporated into Unbound Pavement Layers," Applied Sciences, vol. 13, no. 1, p. 362, Dec. 2022. https://doi.org/10.3390/app13010362.

[10] S. Mirzapour Mounes, M. R. Karim, A. Khodaii, and M. H. Almasi, "Improving Rutting Resistance of Pavement Structures Using Geosynthetics: An Overview," The Scientific World Journal, vol. 2014, pp. 1–6, 2014. https://doi.org/10.1155/2014/764218.

[11] M. Abdel-Jaber, R. A. Al-shamayleh, R. Ibrahim, T. Alkhrissat, and A. Alqatamin, "Mechanical properties evaluation of asphalt mixtures with variable contents of reclaimed asphalt pavement (RAP)," Results in Engineering, vol. 14, p. 100463, Jun. 2022. https://doi.org/10.1016/j.rineng.2022.100463.

[12] I. L. Al-Qadi, M. Elseifi, and S. H. Carpenter, "RECLAIMED ASPHALT PAVEMENT-A LITERATURE REVIEW CIVIL ENGINEERING STUDIES Illinois Center for Transportation Series No. 07-001 Determination of Usable Residual Asphalt Binder in RAP," 2007.

[13] G. Tarsi, P. Tataranni, and C. Sangiorgi, "The Challenges of Using Reclaimed Asphalt Pavement for New Asphalt Mixtures: A Review," Materials, vol. 13, no. 18, p. 4052, Sep. 2020. https://doi.org/10.3390/ma13184052.

[14] A. Copeland, "Reclaimed Asphalt Pavement in Asphalt Mixtures: State of the Practice," Apr. 2011, Accessed: Jul. 04, 2025. [Online]. Available: https://rosap.ntl.bts.gov/view/dot/40918

[15] N. S. Abd Ali, H. H. Joni, and R. H. A. Al-Rubaee, "Effect of asphalt modified with waste engine oil on the durability properties of hot asphalt mixtures with reclaimed asphalt pavement," Open Engineering, vol. 14, no. 1, Feb. 2024. https://doi.org/10.1515/eng-2022-0529.

[16] A. K. Arshad, Z. Abd Karim, E. Shaffie, W. Hashim, and Z. Abd Rahman, "Marshall properties and rutting resistance of hot mix asphalt with variable reclaimed asphalt pavement (RAP) content," IOP Conf Ser Mater Sci Eng, vol. 271, p. 012078, Nov. 2017. https://doi.org/10.1088/1757-899X/271/1/012078.

[17] C. Plati, M. Tsakoumaki, and K. Gkyrtis, "Physical and Mechanical Properties of Reclaimed Asphalt Pavement (RAP) Incorporated into Unbound Pavement Layers," Applied Sciences, vol. 13, no. 1, p. 362, Dec. 2022. https://doi.org/10.3390/app13010362.

[18] Y. Adriati, Elizar, B. Kurniawan, and Novrianti, "Characteristics of Recycled Asphalt Mixtures Utilizing Reclaimed Asphalt Pavement (RAP) Materials," Journal of Sustainable Engineering: Proceedings Series, vol. 2, no. 1, pp. 50–54, Oct. 2024. https://doi.org/10.35793/joseps.v2i1.7137.

[19] V. Antunes, J. Neves, and A. C. Freire, "Performance Assessment of Reclaimed Asphalt Pavement (RAP) in Road Surface Mixtures," Recycling, vol. 6, no. 2, p. 32, May 2021. https://doi.org/10.3390/recycling6020032.

[20] S. Putra, D. A. Islamiyah, and D. Heryanto, "The Effect of Aggregate Degradation Level on Durability of Reclaimed Asphalt Pavement," 2024, pp. 429–446. doi: 10.2991/978-94-6463-475-4_36.

[21] N. S. Abd Ali, H. H. Joni, and R. H. A. Al-Rubaee, "Effect of asphalt modified with waste engine oil on the durability properties of hot asphalt mixtures with reclaimed asphalt pavement," Open Engineering, vol. 14, no. 1, Feb. 2024. https://doi.org/10.1515/eng-2022-0529.

[22] N. Liu, L. Liu, M. Li, and L. Sun, "A comprehensive review of warm-mix asphalt mixtures: mix design, construction temperatures determination, performance and life-cycle assessment," Road Materials and Pavement Design, vol. 25, no. 7, pp. 1381–1425, Jul. 2024. https://doi.org/10.1080/14680629.2023.2268194.

[23] S. K. Saini, S. K. Sahdeo, G. D. Ransinchung RN, and P. Kumar, "Performance of cement mortar mixes containing fine reclaimed asphalt pavement aggregates and zinc waste," Advances in Cement Research, vol. 36, no. 4, pp. 1–33, Apr. 2024. https://doi.org/10.1680/jadcr.23.00046.

[24] S. Chen, Z. You, N. P. Sharifi, H. Yao, and F. Gong, "Material selections in asphalt pavement for wet-freeze climate zones: A review," *Constr. Build. Mater.*, vol. 201, pp. 510–525, Mar. 2019. https://doi.org/10.1016/j.conbuildmat.2018.11.276.

[25] A. Saleh and L. Gáspár, "Modelling indirect tensile strength of warm mix asphalt with variable reclaimed asphalt pavement (RAP) content," Dorogi i mosti, vol. 2024, no. 30, pp. 157–173, Oct. 2024. https://doi.org/10.36100/dorogimosti2024.30.157.

[26] F. Tang, J. Fan, T. Ma, and Y. Sun, "Study on the Performances of PAC-13 Asphalt Mixture Containing Reclaimed Porous Asphalt Pavement," Buildings, vol. 15, no. 9, p. 1395, Apr. 2025. https://doi.org/10.3390/buildings15091395.

[27] N. Baldino, O. Mileti, Y. M. Marchesano, F. R. Lupi, D. Gabriele, and M. Paolini, "Rheological Performance and Differences between Laboratory-Aged and RAP Bitumen," Materials, vol. 17, no. 16, p. 3954, Aug. 2024. https://doi.org/10.3390/ma17163954.

[28] R. Babagoli and A. Ameli, "Influence of Soybean Oil on Binder and Warm Mixture Asphalt Properties," Advances in Materials Science and Engineering, vol. 2021, no. 1, Jan. 2021. https://doi.org/10.1155/2021/6860878.

[29] I. Elnaml, L. N. Mohammad, G. Baumgardner, S. Cooper, and S. Cooper, "Sustainability of Asphalt Mixtures Containing 50% RAP and Recycling Agents," Recycling, vol. 9, no. 5, p. 85, Sep. 2024. https://doi.org/10.3390/recycling9050085.

[30] M. Hajiebrahimi, S. A. Hosseini, and M. R. Fariman, "Experimental Evaluation of Mechanical Characteristics of Roller-Compacted Concrete Containing Crumb Rubber, RAP and Slag," Arab J Sci Eng, vol. 49, no. 10, pp. 14459–14471, Oct. 2024. https://doi.org/10.1007/s13369-024-09038-x.

[31] A. Elnihum, Q. Lu, M. Alharthai, M. Alamri, C. Chen, and A. Elmagarhe, "Evaluation of an Asphalt Mixture Containing a High Content of Reclaimed Asphalt Pavement (RAP) Materials with Epoxy Asphalt," Sustainability, vol. 16, no. 12, p. 4988, Jun. 2024. https://doi.org/10.3390/su16124988.

[32] A. Epps et al., Evaluating the Effects of Recycling Agents on Asphalt Mixtures with High RAS and RAP Binder Ratios. Washington, D.C.: Transportation Research Board, 2020. doi: 10.17226/25749.

[33] K. Kim and T. H. M. Le, "Feasibility and Sustainable Performance of RAP Mixtures with Low-Viscosity Binder and Castor Wax–Corn Oil Rejuvenators," Buildings, vol. 13, no. 7, p. 1578, Jun. 2023. https://doi.org/10.3390/buildings13071578.

[34] P. Konca et al., "Sustainable Infrastructure: Recycled Concrete Aggregates for Cycle Paths," Materials, vol. 18, no. 1, p. 131, Dec. 2024. https://doi.org/10.3390/ma18010131.

[35] W. A. Hamad and G. J. Khoshnaw, "Mechanical properties and durability of asphalt-concrete composites enhanced with recycled concrete and porcelain aggregates," MethodsX, vol. 14, p. 103296, Jun. 2025. https://doi.org/10.1016/j.mex.2025.103296.

[36] R. Bonaquist, "Critical Factors Affecting Asphalt Concrete Durability," 2016.

[37] C. Petit, A. Millien, F. Canestrari, V. Pannunzio, and A. Virgili, "Experimental Study on Shear Fatigue Behavior and Stiffness Performance of Warm Mix Asphalt by adding Synthetic Wax."

[38] H. Majidifard, N. Tabatabaee, and W. Buttlar, "Investigating Short-term and Long-term Binder Performance of High-RAP Mixtures Containing Waste Cooking Oil."

[39] M. K. D. Rout, S. K. Sahdeo, S. Biswas, K. Roy, and A. K. Sinha, "Feasibility Study of Reclaimed Asphalt Pavements (RAP) as Recycled Aggregates Used in Rigid Pavement Construction," Materials, vol. 16, no. 4, p. 1504, Feb. 2023. https://doi.org/10.3390/ma16041504.

[40] F. C. G. Martinho et al., "Mechanical and Environmental Performance of Asphalt Concrete with High Amounts of Recycled Concrete Aggregates (RCA) for Use in Surface Courses of Pavements," Sustainability, vol. 16, no. 1, p. 248, Dec. 2023. https://doi.org/10.3390/su16010248.

[41] S. M. Piryonesi and T. E. El-Diraby, "Data Analytics in Asset Management: Cost-Effective Prediction of the Pavement Condition Index," Journal of Infrastructure Systems, vol. 26, no. 1, Mar. 2020. https://doi.org/10.1061/(ASCE)IS.1943-555X.0000512.

[42] J. Bryce, R. Boadi, and J. Groeger, "Relating Pavement Condition Index and Present Serviceability Rating for Asphalt-Surfaced Pavements," Transportation Research Record: Journal of the Transportation Research Board, vol. 2673, no. 3, pp. 308–312, Mar. 2019. https://doi.org/10.1177/0361198119833671.

[43] S. S. Wadalkar, R. K. Lad, and R. K. Jain, "Study of Pavement Performance Indicators: RII Approach," Indian J Sci Technol, vol. 11, no. 22, pp. 1–4, Jun. 2018. https://doi.org/10.17485/ijst/2018/v11i22/121101.

[44] A. Alwathaf, M. A. Jaber, and Y. Hunaiti, "Experimental investigation on the Effect of Superplasticizer on Concrete Containing Recycled Asphalt Pavement (RAP)," E3S Web of Conferences, vol. 586, p. 04002, Nov. 2024. https://doi.org/10.1051/e3sconf/202458604002.

[45] A. K. Arshad, Z. Abd Karim, E. Shaffie, W. Hashim, and Z. Abd Rahman, "Marshall properties and rutting resistance of hot mix asphalt with variable reclaimed asphalt pavement (RAP) content," IOP Conf Ser Mater Sci Eng, vol. 271, p. 012078, Nov. 2017. https://doi.org/10.1088/1757-899X/271/1/012078.

[46] V. Antunes, A. C. Freire, and J. Neves, "Life Cycle Assessment of road pavements rehabilitation re-using high content of Reclaimed Asphalt Pavement," Transportation Research Procedia, vol. 72, pp. 3371–3378, 2023. https://doi.org/10.1016/j.trpro.2023.11.785.

[47] X. Liu, L. Deng, H. Zhang, and J. Yu, "Quantitative study on carbon emissions of modified recycled asphalt mixture based on life cycle assessment method," Developments in the Built Environment, vol. 20, p. 100584, Dec. 2024. https://doi.org/10.1016/j.dibe.2024.100584.

[48] R. D. Rosario, F. Poso, A. Victoria, and M. de Guzman, "Bio-Engineered Fiber-Reinforced Rigid Pavement: A Durability, Strength Recovery and Self-Healing Evaluation," pp. 625–637, 2026. https://doi.org/10.1007/978-981-95-4534-6_58.

[49] R. D. Rosario, A. Alvarez, R. C. Quinto, and M. de Guzman, "Intelligent Transportation Systems: Enabling Sustainable Transportation and Efficient Traffic Management—A Review," Communications in Computer and Information Science, vol. 2378 CCIS, pp. 311–323, 2025. https://doi.org/10.1007/978-3-031-84148-4_24.

[50] R. Rosario D and M. J. Viado, "Encapsulating immobilized ureolytic bacteria yields self-healing concrete apropos sustainable transportation materials: A review," E3S Web of Conferences, vol. 488, p. 03019, (Feb. 2024). https://doi.org/10.1051/e3sconf/202448803019.

[51] R. D. Rosario, J. A. Padilla, N. G. Bonto, R. C. De Mesa, and O. G. Dela Cruz, "Value engineering on car curbing ownership in metro Manila," in TRANSPORT, ECOLOGY - SUSTAINABLE DEVELOPMENT: EKOVarna2022, AIP Publishing, Aug. 2023, p. 020017. doi: 10.1063/5.0162414.

[52] R. D. Rosario, A. D. La Cruz, and M. P. De Guzman, "A Review of Biomineralization as Solution for Roads and Infrastructures Concrete Sustainability," Civil Engineering Journal, vol. 10, no. 8, pp. 2745–2760, Aug. 2024. https://doi.org/10.28991/CEJ-2024-010-08-020.

Experimental Evaluation of a Small Storage Cooking Pot using a 12 V Battery-Powered Solar DC Cooking Stove

Ashmore MAWIRE[1,a*], Patrick TSOPBOU NGUEAGNI[1,b], Katlego DIRATSAGAE[1,c]

[1]Material Science, Innovation and Modelling (MaSIM) Research Focus Area, Department of Physics, Mmabatho 2745, Mahikeng, South Africa

[a]ashmore.mawire@nwu.ac.za, [b]nguetsop.ricks@gmail.com, [c]kdiratsagae77@gmail.com

Keywords: DC Stove, Erythritol, Storage Cooking Pot, Water Heating

Abstract. A small storage cooking pot is experimentally evaluated using a 12 V DC cooking stove powered by AGM batteries. Two 12 V 100 AH AGM deep-cycle batteries supply power to the stove and are charged via a 450 W solar panel and a maximum power point tracking (MPPT) charge controller, enabling continuous 24 hr operation. The storage cooking pot consists of a black 1-litre stainless steel pot placed inside a larger 5-litre black stainless-steel pot. The bottom space between the two pots contains 0.5 kg of erythritol as the storage material. The cooking setup, including the storage pot and the 12 V DC stove, is placed inside a wonderbag insulating cooker to reduce heat loss and store thermal energy during experimental heating tests. The storage cooking pot is heated using three water-heating loads of 1.0, 0.50, and 0.25 kg in the inner pot for 3 hrs to investigate simultaneous cooking and storage of thermal energy. After 3 hrs of heating, the power is switched off, and fresh water is loaded into the inner pot to investigate the utilisation of stored heat using the same water heating loads of 1.0, 0.50, and 0.25 kg for an additional 3 hrs. During the storage cooking period with the power switched off, the maximum temperatures achieved by water are around 54, 74, and 86 °C for water loads of 1.0, 0.5, and 0.25 kg, respectively, showing the potential of the storage cooking pot integrated with the DC cooker and the wonderbag. After the 3-hour heat utilisation period, the final water temperatures are approximately 45, 62, and 65 °C for loads of 1.0, 0.5, and 0.25 kg, respectively.

Introduction

Solar cookers offer an environmentally friendly option that does not emit greenhouse gases [1-3]. There has been a lot of recent research on traditional solar cookers, such as parabolic dish, box, and panel solar cookers [4-8]. Traditional solar cookers cannot be used during cloudy periods or at night without thermal energy storage (TES); thus, recent work has also examined integrating TES with traditional solar cookers, especially the box and parabolic dish solar cookers [9-13]. Under continuous cloudy conditions, traditional solar cookers, even with storage, do not perform well since there is limited energy to charge the TES effectively.

The DC cooking option, powered by batteries and PV panels, is particularly attractive during cloudy periods. Only a few recent works have reported DC PV cookers without using expensive inverters [14-16]. A 12 V DC hotplate showed better performance than parabolic dish solar cookers during cloudy periods, as reported recently by Mawire et al. [16], without the need for expensive DC-DC power boosters. However, prolonged use of the battery for cooking during cloudy periods, even at low power, tends to degrade it. A feasible option is to include TES in the cooking pot to reduce reliance on the battery during extended cooking periods, thereby improving its shelf life. To the best of our knowledge, there has been no reported work on a storage cooking pot integrated with a 12 V DC cooking hotplate that simultaneously cooks and retains heat for later use without solely relying on the battery. To address this research gap, a small storage cooking pot is experimentally evaluated during charging/storage cycles using 12 V batteries. It is also evaluated during discharging using the stored heat without using the battery. The results presented will help to understand the thermal performance of battery-powered solar DC hotplates with storage materials integrated into the cooking pot.

Materials and methods

The experimental setup was developed at North-West University, Mahikeng Campus, South Africa, to conduct cooking tests, as shown in Fig. 1. Two 12 V 100 AH AGM deep-cycle batteries were connected in parallel via a maximum power point tracking (MPPT) controller. A 450 W solar panel was connected to the MPPT charge controller. The MPPT controller output was connected to a 12 V DC hotplate with a storage cooking pot on its cooking surface. The temperatures of water and storage materials were measured with K-type thermocouples with an accuracy of ±2.0 °C. The thermocouple measuring the water was placed centrally, around 1 cm from the base of a small cooking pot (1 litre).

In contrast, K-type thermocouple measuring the temperature of the storage material was placed centrally on the left-hand side of the larger cooking pot, around 1 cm from its base. The storage material was placed in the larger cooking pot (5 litres), with the smaller cooking pot wholly enclosed. The batteries ensured an adequate supply of heating power during cloudy periods.

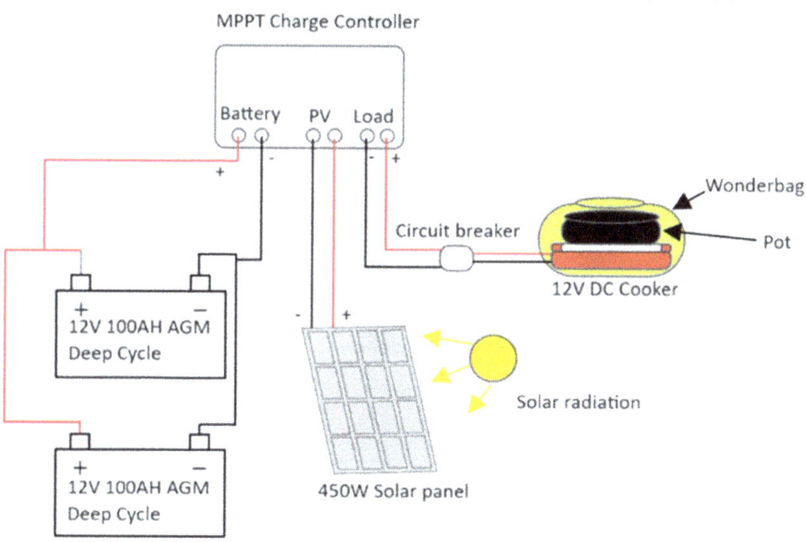

Fig. 1. A schematic diagram of the experimental setup of the solar PV DC cooker. The red colours indicate the positive terminal wires, and the black colours indicate the negative terminal wires.

A photographic view of the main components of the experimental setup is shown in Fig. 2. The storage cooking pot consisted of a black 1-litre stainless-steel pot placed in a larger 5-litre black stainless-steel pot. The space between the two pots contains 0.5 kg of erythritol as the storage material. The storage cooking pot and the 12 V DC stove were placed in a wonderbag insulator cooker to reduce heat losses as well as to store thermal energy during experimental heating tests. Heating experiments were conducted using three water loads of 1.0, 0.50, and 0.25 kg placed inside the inner pot, weighed with a digital balance (±0.001 kg). Each test lasted 3 hrs to evaluate the system's ability to perform cooking and thermal energy storage simultaneously. After 3 hrs of heating, the power was switched off, and fresh water was loaded into the inner pot to investigate the utilisation of stored heat using the same water heating loads of 1.0, 0.50, and 0.25 kg for 3 hrs. A multichannel data logger recorded temperatures at 1-minute intervals throughout the experiments. The voltage across the hotplate was also measured with a dedicated datalogger channel (accuracy ±0.1 %, resolution 0.01 V) set to voltage-measuring mode. During the water-heating period, the current was measured with a Testo clamp-on (accuracy ±0.2 %, resolution 0.1 A) with Bluetooth connectivity, and recorded on a smartphone every minute using the Testo smart app. The measured current and voltage were instantaneous values which were averaged at the end of the datalogging period to obtain the average power values. To avoid draining the battery and thereby affecting the cooking experiments, the battery was continuously charged during the experiments using a 450 W solar panel connected to an MPPT charge controller. The experiments were carried out when the battery was initially above 90 %. The experiments were carried out on sunny days, with average solar radiation of 677, 709, and 723 W/m^2 for test loads of 0.25, 0.50, and 1.0 kg, respectively. These solar radiation values are very comparable to each other.

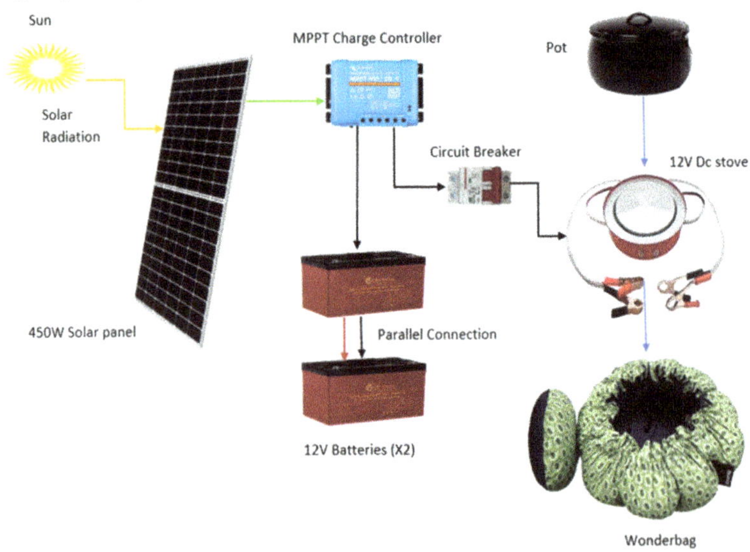

Fig. 2. A photographic view of the components of the PV cooker. The main components are the solar panel, MPPT charge controller, batteries, cooking pot, 12 V DC stove and the wonderbag.

The thermal properties of the erythritol storage material are shown in Table 1.

Table 1. Thermal properties of erythritol [17, 18]

Storage Material	Thermal conductivity k [W/mK]	Specific Heat Capacity Cp [J/kg K]	Density [kg/m^3]	Melting point [$^\circ$C]	Latent heat of fusion [kJ/kg]
Erythritol	0.733	1290	1.450	121	340

Results and discussion

Fig. 3 shows the water-heating (A) and storage-temperature (B) profiles during the 6-hr experimental test period. Water boils around 1, 2, and 2.5 hours for 0.25, 0.50, and 1.00 kg loads, respectively. The average electrical powers during the water-heating periods, calculated from measured voltages and currents, were 169, 137, and 152 W, respectively. As expected, the larger load takes a longer time to heat up. Using freshwater, with the hotplate switched off, after 3 hours, the maximum temperatures achieved by water are around 54, 74, and 86 °C for water loads of 1.0, 0.5, and 0.25 kg, respectively, showing the potential of the storage pot integrated with the DC cooker. After the 3 hr heat utilisation period, the final temperatures for the 1.0, 0.5, and 0.25 kg loads are around 45, 62, and 65 °C, respectively. The storage temperatures at the end of the storage period are around 109, 112, and 108 °C at the end for the 1.00, 0.50, and 0.25 kg loads, respectively. However, there are noticeable up-and-down fluctuations in the storage temperature profiles, especially with the 1.00 kg load, due to thermal switching of the DC hotplate to avoid overheating. With the power switched off, the final storage temperatures achieved after storage heating of water are around 63, 60, and 46 °C, respectively, for the 0.25, 0.50, and 1.00 kg loads.

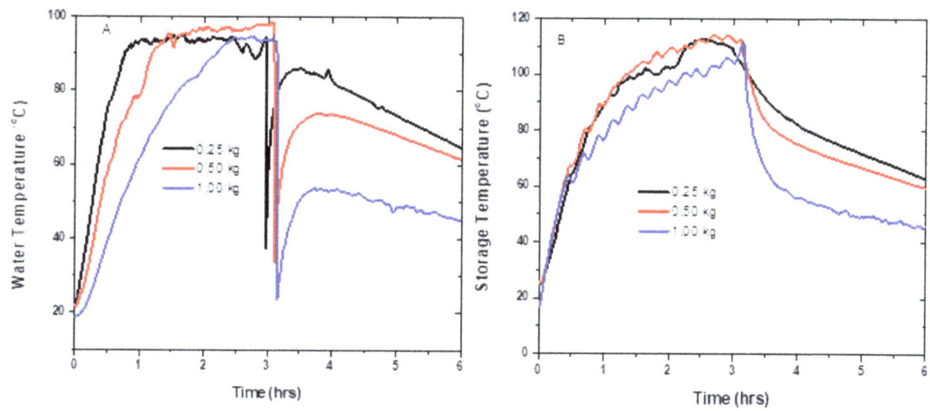

Fig. 3. Experimental (A) water heating and (B) storage temperatures (°C) for three loads (0.25, 0.50, and 1.00 kg) versus time (hrs). The trends show the variations in the water and heat storage temperatures with different loads. The smallest load shows the fastest rise in the water temperature.

The results show that the largest load is not well-suited for use with a small storage cooking pot. However, a larger DC hotplate and more PV panels could be added to improve the thermal performance, together with a larger storage cooking pot. An experimental test of cooking chicken using stored heat is shown in Fig. 4. The chicken cooked with stored heat was well-cooked, demonstrating that the proposed cooking method is viable. Erythritol, the phase change material (PCM), solidified after 60 minutes of cooking chicken.

Fig. 4. A photograph of a chicken after being cooked for 60 minutes using erythritol as TES. The chicken was well-cooked in 60 minutes.

Conclusion

The experimental results revealed the potential of using a storage cooking pot with a small storage mass to reduce the demand for battery energy for cooking. During the storage cooking period with the power switched off, the maximum temperatures achieved by water were around 54, 74, and 86 oC for water loads of 1.0, 0.5, and 0.25 kg, respectively, demonstrating the potential of a storage cooking pot with a DC hotplate. In real conditions, chicken was well-cooked in 60 minutes using stored energy with the battery switched off, showing that the concept of storage cooking is viable and can preserve the battery's lifetime. More tests are to be carried out with different foods and storage materials to investigate the effects of cooking different foods and of different storage materials. To reduce energy losses through better insulation, future studies should focus on optimising the interaction among the heating element, batteries, and the photovoltaic system.
Furthermore, examining sustainability, cost, life-cycle analysis, and assessing the socio-economic acceptance of people in rural areas will be crucial for the widespread implementation of battery-operated solar DC cooking stoves. The small size of the cooking pot is a major limitation for large-scale cooking. Future work will also look at upscaling the current reported system for family and community scale using more solar panels, larger panels and larger hot plates. However, the total cost will be more. The system's heat transfer is also not optimised. A CFD model is needed for parametric studies to optimise heat transfer.

References

[1] A. Saxena, A.A. Sagade, S.J Joshi, V.V. Tyagi, D.B Singh, M.A. Tawfik, A. Kumar, P. (2025). A thermodynamic review on concentrating type solar cookers, Solar Energy 286, 113159. https://doi.org/10.1016/j.solener.2024.113159.

[2] F. Odoi-Yorke, B, Baah, R, Opoku (2025). Scientific mapping of experimental research on solar cookers: Global trends, evolution, and future directions, Solar Energy Advances 5, 100093. https://doi.org/10.1016/j.seja.2025.100093.

[3] B. Paneru, B. Paneru, V. Alexander, S. Nova, N. Bhattarai, R. Poudyal, K.N. Poudyal, M.B. Dangi, J.J. Boland (2024). Solar energy for operating solar cookers as a clean cooking

technology in South Asia: A review. Solar Energy 283, 113004. https://doi.org/10.1016/j.solener.2024.113004.

[4] S. Tomassetti, A. Aquilanti, P.F. Muciaccia, M. Muccioli, G. Di Nicola (2025). Experimental characterisation of a panel solar cooker with adjustable geometry for sun tracking. Solar Energy, 287, 113178. https://doi.org/10.1016/j.solener.2024.113178.

[5] A.A. Sagade, X. Apaolaza-Pagoaga, C.R. Ruivo, A. Carrillo-Andrés (2024). Experimental determination of effective concentration ratio for two different designs of concentrating solar cookers. Solar Energy 275, 112608. https://doi.org/10.1016/j.solener.2024.112608.

[6] R. Jamal, S.K. Samdarshi, M. Singh, A.A. Sagade, P.S. Panja, A.U. Ansari (2024). Improvement in the opto-thermal performance of Indian standard solar box cooker by novel internal retrofit radiative control. Solar Energy 274, 112564. https://doi.org/10.1016/j.solener.2024.112564.

[7] R.R. Gil, B.C. Muñoz, H.E. González-Bravo, R.D. Rodríguez (2025). Development and experimental evaluation of a parabolic solar cooker for sustainable cooking in monsoon climates, Journal of Cleaner Production 523, 146416. https://doi.org/10.1016/j.jclepro.2025.146416.

[8] S. A. Kadhim, O. Al-Munaf Ibrahim, M.K.S. Al-Ghezi, A.M. Ashour, I. Omle, A Bouabidi (2025). Influence of the steel fibers inside the box on the solar cooker performance: An experimental approach. Results in Engineering 26, 105659. https://doi.org/10.1016/j.rineng.2025.105659.

[9] A. Kumar, M. Eswaramoorthy, R. Kumar, R.K. Goyal (2025). A comparative analysis on concrete and granite-based heat storage materials for evening cooking stability of box-type solar cooker. Solar Energy 296, 113577. https://doi.org/10.1016/j.solener.2025.113577.

[10] R. Khatri, R. Goyal, .R.K. Sharma (2023). Comparative experimental investigations on a low-cost solar cooker with energy storage materials for sustainable development. Results in Engineering 20, 101546. https://doi.org/10.1016/j.rineng.2023.101546.

[11] A. Mawire, K. Lentswe, A. Shobo, J. Darkwa, J. Calautit, M. Worall (2020). Performance comparison of two solar cooking storage pots combined with wonderbag slow cookers for off-sunshine cooking. Solar Energy 208, 1166-1180. https://doi.org/10.1016/j.solener.2020.08.053.

[12] M. Palaniappan, T.S. Keerthi, A. Muthu Manokar, S. Shanmugan (2026). Fe2O3-CeO2/PMMA nanocomposite-enhanced solar box cooker with efficient photothermal conversion and extended thermal storage. Results in Engineering 29, 109717. https://doi.org/10.1016/j.rineng.2026.109717.

[13] G. Coccia, A. Aquilanti, S. Tomassetti, G. Comodi, G. Di Nicola (2020). Design, realisation, and tests of a portable solar box cooker coupled with an erythritol-based PCM thermal energy storage. Solar Energy 201, 530-540. https://doi.org/10.1016/j.solener.2020.03.031.

[14] R. Opoku, B. Baah, C.K.K. Sekyere, E. A. Adjei, F. Uba, G. Y. Obeng, F. Davis (2022). Unlocking the potential of solar PV electric cooking in households in sub-Saharan Africa – The case of pressurised solar electric cooker (PSEC). Scientific African 17, e01328. https://doi.org/10.1016/j.sciaf.2022.e01328.

[15] J. Chaciga, D. Okello, K. Nyeinga, O.J. Nydal (2025). Experimental analysis on a solar photovoltaic indoor cooker integrated with an energy storage system: A positive step towards

clean cooking transition for Sub-Saharan Africa. Solar Compas 13, 100109. https://doi.org/10.1016/j.solcom.2025.100109.

[16] A. Mawire, O. P. Abedigamba, M. Worall (2024). Experimental comparison of a DC PV cooker and a parabolic dish solar cooker under variable solar radiation conditions. Case Studies in Thermal Engineering 54, 103976. https://doi.org/10.1016/j.csite.2024.103976.

[17] S. Höhlein, A. König-Haagen, D. Brüggemann (2017). Thermophysical characterisation of MgCl2·6H2O, xylitol and erythritol as phase change materials (PCM) for latent heat thermal energy Materials 10, 44. https://doi.org/10.3390/ma10040444.

[18] R. Jia, K. Sun, R. Li, Y. Zhang, W. Wang, H. Yin, D. Fang, Q. Shi, Z. Tan. (2017). Heat capacities of some sugar alcohols as phase change materials for thermal energy storage applications. The Journal of Chemical Thermodynamics, 115, 233-248. https://doi.org/10.1016/j.jct.2017.08.004,

Advanced Materials and Sustainable Energy Technologies – AMSET2025 Materials Research Forum LLC
Materials Research Proceedings 66 (2026) 307-314 https://doi.org/10.21741/9781644904152-28

Comparative Analysis of Hybrid Quantum-Classical and Artificial Neural Network for Sustainable Cyber-Physical Power System Anomaly Detection

Lloyd Aldrin T. PORNOBI[1,a *], Roman M. RICHARD[1,b]

[1]Computer Engineering Department, Technological Institute of the Philippines, Quezon City, Philippines

[a]lapornobi.cpe@tip.edu.ph, [b]rrichard.cpe@tip.edu.ph

Keywords: Hybrid Quantum-Classical Computing; Anomaly Detection; Sustainable Computing

Abstract. Smart grid infrastructures are becoming increasingly vulnerable to a range of physical and cyber-physical faults that can result in outages, equipment damage, and safety hazards. This paper presents a comparative evaluation of a hybrid quantum-classical neural network (HQNN) against a classical artificial neural network (ANN) for binary fault detection in a simulated smart-grid environment. Both models are trained and tested on a 4,965-sample dataset derived from a power-system simulator jointly developed by Mississippi State University and Oak Ridge National Laboratory. We encode input features into quantum states using Hadamard and RY rotations, employ entangling layers in PennyLane, and fuse the resulting quantum expectation values with classical fully connected layers. The ANN baseline, in contrast, utilizes two ReLU-activated hidden layers and dropout. Performance is measured in terms of precision, recall, F1-score, ROC AUC, and CPU-only inference latency, with 5-fold cross-validation implemented. Results show that the HQNN got a higher ROC AUC (0.7565 vs. 0.6763), higher precision (80% vs. 69%), recall (61% vs. 54%), and F1-Score (65% vs. 58%), while the ANN achieves a lower average inference time that runs almost 65x faster per sample. These findings demonstrate that hybrid quantum-classical models can enhance discriminative power for sustainable, low-power smart-grid fault detection, even in resource-limited settings, at the cost of slightly increased inference time.

Introduction

The use of advanced technologies in automation and digital communications has transformed traditional grid infrastructures into cyber-physical power systems [1]. However, despite technological advancements in these systems, they remain prone to numerous faults, which impact their performance compared to traditional systems [2]. Fourteen of these faults originate from power outages and eight from equipment damage, each capable of inflicting significant physical, economic, and safety impacts (e.g., unbalanced voltage/current, equipment fires) on the grid [3]. Machine learning has become a key technology for smart-grid applications, such as consumption forecasting, adaptive control, anomaly detection, and fault detection [4]. In this work, we introduce a hybrid quantum-classical machine learning approach to enhance defect detection in smart grid environments and compare its reliability and inference speed directly with those of a classical artificial neural network.

Related Work

Machine learning (ML) models are widely applied to detecting defects in smart grids; however, these also face several challenges. A primary challenge is scalability; related work [5] shows that integrating models can perform well in simulated environments, but does not perform as well in real-world grid environments. One primary reason for this scalability challenge is the inherent imbalance in the data resulting from infrequent fault occurrences. The related work further highlights the need for more efficient and reliable solutions. The use of machine learning models – sometimes referred to as classical machine learning – is further reinforced in [6], where the application is performed in the detection of defects in smart grid applications. This study is also another implementation of the technology, where disadvantages still occur. Due to a lack of computational efficiency, scalability issues still happen. Beyond the computational demands that cause these scalability issues; when data grows, there is also a need for better storage, and more time is also required for computational tasks. Studies, such as [7], have shown that hybrid quantum-classical machine learning approaches can surpass classical machine learning. Wherein, a one-class SVM is outperformed by a hybrid quantum-classical model, proving its ability to detect more complex patterns, especially when faced with imbalances and less observable anomalies. These related works show that the quantum approaches – recognized as an emerging technology approach – can enhance defect detection, being less prone to dependencies to features like classical machine learning solutions.

Methodology

In this section, we discuss the methodology for applying hybrid-quantum classical machine learning to the smart grid, in comparison to existing methods such as Artificial Neural Networks (ANNs).

A. Dataset and Pre-Processing

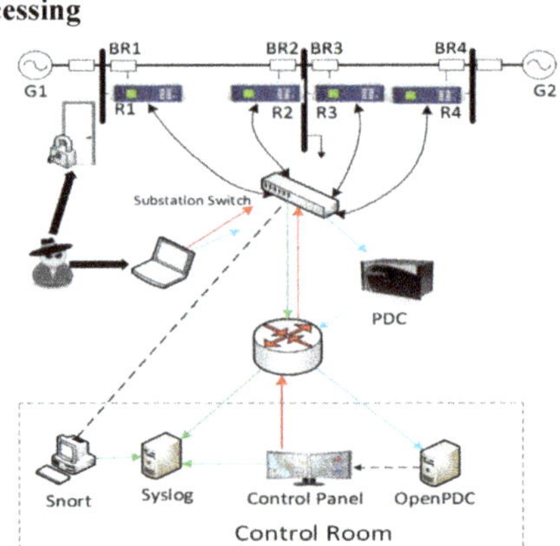

Fig 1. Power System Framework Configuration for Dataset Generation [8]

The Smart Grid Monitoring Power dataset, developed by Mississippi State University and Oak Ridge National Laboratory, and publicly available on Kaggle [8], is used in this study. The illustration in Fig. 1 shows that the dataset was gathered and generated from a simulated cyber-physical power system framework that consists of the following: Two generators (as G1 and G2), four Intelligent Electronic Devices (represented as the IEDs R1-R4), and lastly, the four breakers (represented as BR1-BR4), structured across two transmission lines. Each IED autonomously controls an associated breaker using distance protection schemes and can also be overridden for manual operation. With 4,965 data points, this selected dataset can support cyber-physical power system-related research, and is used as the training and testing dataset for the algorithmic solutions; aiming to simulate a real-world environment.

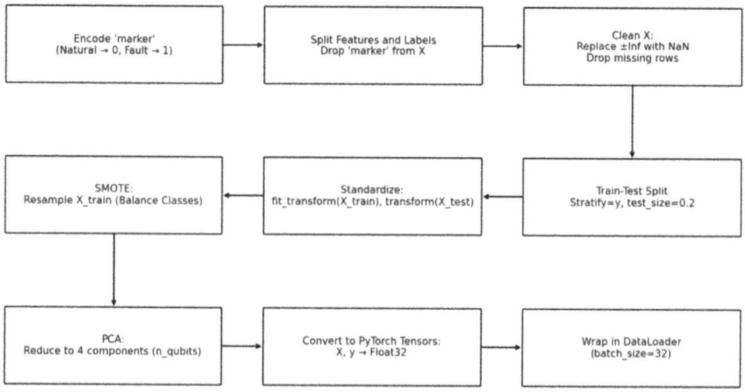

Fig. 2 Data Pre-processing Flowchart.

Fig. 2 shows the data pre-processing pipeline implemented in this study to ensure the usability of the data for the machine learning models. To transform the classification into a binary task, the labels were simplified into a "with fault" and "no fault" classes. This transformation involved removing all invalid values, such as NaN types. Once pre-processed, the data was split into train and test sets with stratified labels to ensure a consistent count of the two types in the dataset. The train dataset had 2,888 samples labeled as "with fault" and 806 samples labeled as "no fault." Since faults have a rarer occurrence than the normal, no fault class, the data had an inherent imbalance. On the training set, a Synthetic Minority Oversampling Technique was performed to balance the distribution of classes; changing the number of samples for both classes to 2,888 each. After this, standardization was applied using z-score normalization in order to ensure alignment of feature scales for the inputs. For the proper simulation of the hybrid quantum components, it was necessary to reduce the features so that they correspond to the four qubits used in the simulation – this was performed using a Principal Component Analysis (PCA) which retained the most significant elements in the data.

B. Computational Environment

The model development, training, and evaluation, were all conducted on a CPU runtime with the following specifications: Intel Xeon 2.20 GHz, ~13 GB RAM. In this environment, all code was written under Python version 3.11.13. The aim of utilizing a CPU-only runtime is to demonstrate that the solutions do not need specialized hardware and can run on many embedded platforms that have similar computational prowess to this environment.

C. Artificial Neural Network Architecture Design

A baseline ANN was built in PyTorch to simulate the classical ML approach to the smart-grid fault classification. The architecture is composed of an input layer, which will take in the PCA-reduced feature vectors, two fully connected hidden layers (16 and 8 neurons) with ReLU activations – which applies 30% dropout after each to reduce overfitting – and an output layer, which has a single sigmoid neuron. The model was trained with binary cross-entropy loss and the Adam optimizer.

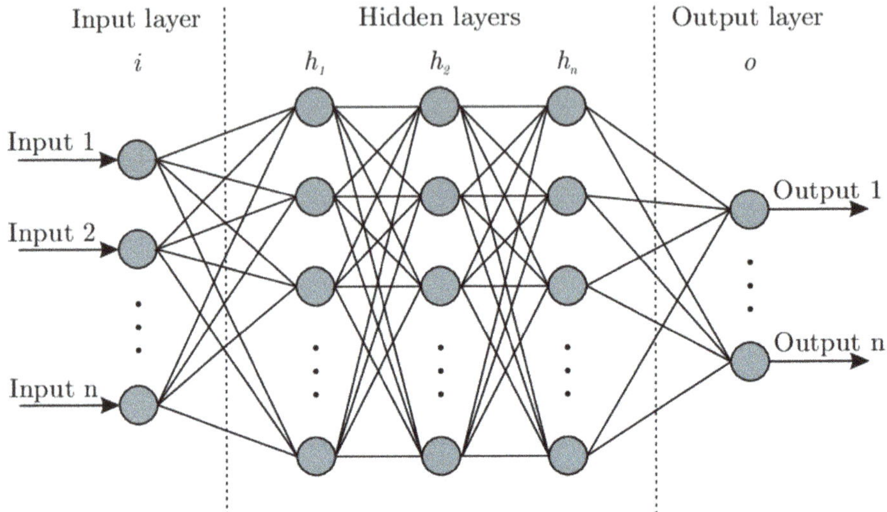

Fig. 3 Classic Artificial Neural Network Architecture [9]

D. Hybrid Quantum Neural Network Architecture Design

Inspired by research showing implementations of a similar architecture [10], the Hybrid Quantum Neural Network (HQNN) architecture integrates quantum computation with classical neural network layers. We implemented the HQNN using PennyLane's qubit simulator with the following steps:

- The inputs were normalized with a Tanh activation
- Then encoded into quantum states through Hadamard and RY gates.
- A sequence of BasicEntanglerLayers was applied
- And, lastly, extracted Pauli-Z expectation values as quantum features

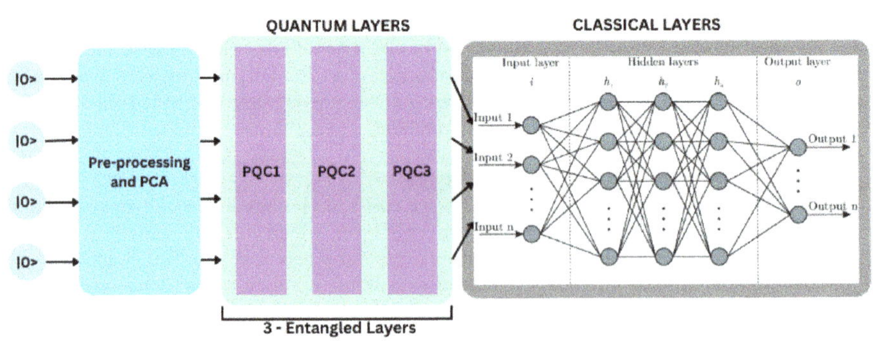

Fig. 4 Implemented Hybrid Quantum Neural Network Architecture.

Materials Research Proceedings 66 (2026) 307-314 https://doi.org/10.21741/9781644904152-28

In this architecture, a similar classical ANN model receives the features. The utilized quantum rotation angles and classical weights were trained together, combining the expressive power of quantum computing with the scalability of classical neural networks.

E. Evaluation Metrics

Evaluations of the enhanced smart grid monitoring are based on the two key performance aspects: the reliability of the result (confusion matrix metrics) and the speed of inference. In summary, Table 1 has the confusion matrix metrics.

Table 1. Summary of Confusion Matrix Metrics

Metric	Formula
Precision	$Precision = \frac{TP}{TP + FP}$
Recall	$Recall = \frac{TP}{TP + FN}$
Accuracy	$Accuracy = \frac{TP + TN}{TP + TN + FP + FN}$
F1-Score	$F1 = \frac{2 * Precision * Recall}{Precision + Recall}$

For the evaluation of HQNN and ANN models, we used the ROC AUC. The curve generated by comparing the predicted probabilities with the proper labels highlights the difference between the true positives and false positives. The AUC provides the evaluation of a single numerical score that will evaluate how well the models precisely distinguish between "with fault" and "no fault". This method gives a clear assessment of the model's performance, especially when dealing with an imbalanced class. To be considered as effective, the model must have achieved at least a "Fair" interpretation [11].

Table 2. Interpretation Table of AUC-ROC [11]

Area under the curve (AUC)	Interpretation
$0.9 \leq AUC$	Excellent
$0.8 \leq AUC < 0.9$	Good
$0.7 \leq AUC < 0.8$	Fair
$0.6 \leq AUC < 0.7$	Poor
$0.5 \leq AUC < 0.6$	Fail

In measuring the inference time of both the Hybrid Quantum Neural Network (HQNN) and Artificial Neural Network (ANN) using Python's time() function. For each test sample, the time was recorded before and after the model generated a prediction. The difference between these two timestamps represented the inference duration for that prediction.

Results & Discussion

The two comparatives of the Hybrid Quantum Neural Network (HQNN) and Classical Artificial Neural Network (CANN) were used as an evaluation for binary fault classification in a simulated environment of smart grid monitoring systems. The review is performed in terms of reliability and speed.

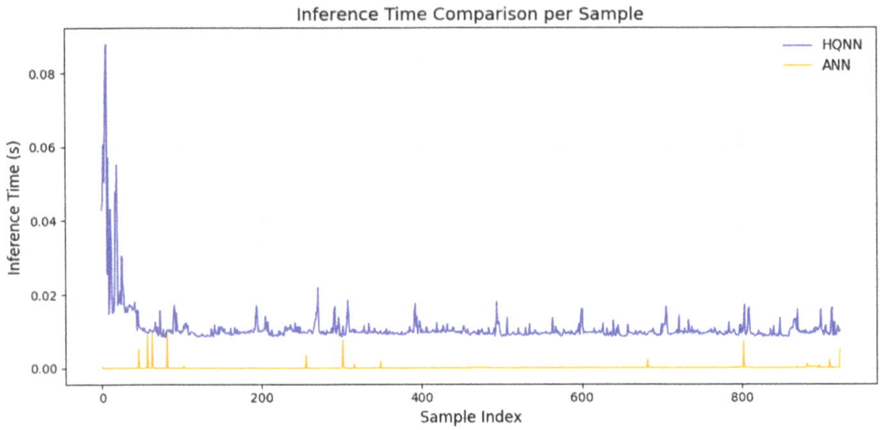

Fig. 5 Inference Time per Prediction for HQNN and CANN.

Fig. 5 shows that, from the 925 predictions, the ANN had a consistent inference time below 0.02s, while the HQNN also performed with sub-second inference time per prediction, but with a higher time to inference compared to the ANN model. The average inference time was 0.01090s for HQNN, while the average was 0.00017s for ANN. This indicates that the quantum feature maps (angle encoding + entangling layers) responsible for transforming inputs into a high-dimensional Hilbert space before being passed to the Neural Network layers incur computational overhead, making the HQNN model ~65 times slower.

Table 3. Performance Comparison of HQNN and ANN

Model	Precision	Recall	F1-Score	AUC-ROC	Avg. Inference Time
HQNN	80%	61%	65%	0.7533 (Fair)	0.01090 s
ANN	69%	54%	58%	0.6097 (Poor)	0.00017 s

Table 3 summarizes the comparison between the two models using performance metrics, including Precision, Recall, F1-score, and AUC-ROC. This clearly illustrates the gaps between the HQNN and ANN in specific performance areas. In this manner, the HQNN outperforms the ANN in most metrics, including precision, recall, F1-score, and AUC-ROC, which demonstrates the stronger detection capability of HQNN in both classes, "with fault" and "no fault". The AUC-ROC score was used in this research due to class imbalance in the dataset, which provides a more comprehensive assessment of the models. However, the ANN outperformed the HQNN in terms of inference time, with approximately 65 times higher inference time compared to the HQNN due to quantum circuit overhead. Still, the HQNN remains a possibility on CPU-only platforms. Future work, collaborating with an Electrical Engineer researcher for smart grid monitoring and quantum analytics, may be considered.

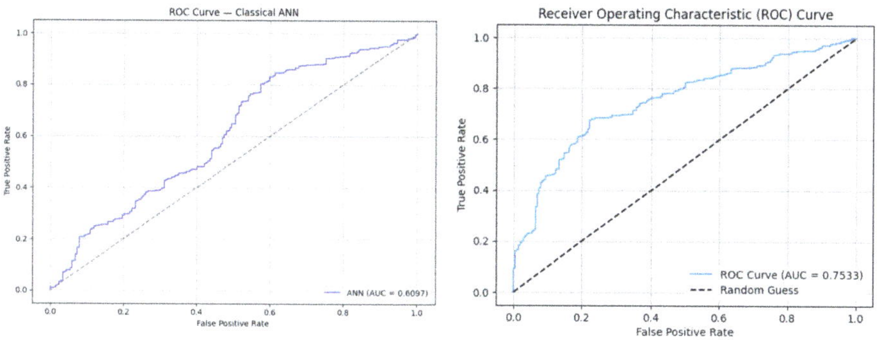

Fig. 6 Comparison of AUC-ROC for HQNN and ANN

Fig. 6 shows the comparison between the two models in terms of AUC-ROC. The HQNN achieved a score of 0.7533, outperforming the CANN's AUC-ROC with a score of 0.6097. This indicates that the HQNN has a higher ability to detect and distinguish between "with fault" and "no fault", showing that the HQNN has a "FAIR" ranking, whereas the ANN has a "POOR" ranking. Despite the high computational cost of HQNN, it still exhibits better classification ability and is more precise in distinguishing faults. It can run and do so effectively in a CPU-only, resource-limited environment, which supports the vision of environmentally friendly AI deployment, especially in low-cost or edge-based smart grid systems. The results demonstrate the practical value of quantum-classical hybrid models in actual grid fault prediction applications.

Conclusion

We demonstrate the practical viability of the Hybrid-Quantum Neural Network (HQNN) versus the classical ANN in application to anomaly detection in a simulated smart grid, all executed on low-cost, CPU-only hardware. Looking ahead, there is a need to monitor the computational costs if we decide to use much larger datasets or powerful GPUs. That said, our HQNN model performed significantly better across the chosen metrics, especially in precision (80% vs. 69%), recall (61% vs. 54%), and F1-score (65% vs. 58%). Its overall performance was rated as "Fair" while the ANN's was rated as "Poor." This performance was achieved even though it was about 60 times slower on each individual prediction. To put that in perspective, both models were still incredibly fast, handling over 950 samples in under a second. This shows that utilizing the quantum-enhanced approaches for monitoring smart grid faults can be considered as a realistic option, without the need for specialized hardware, showing its potential as a scalable solution. The results suggest that the HQNN has a balance between performance and practicality, making it a promising choice for building scalable and energy-efficient smart grid applications.

References

[1] M. Jeremiah and O. A. Olanrewaju, "Smart grid technologies and application in the sustainable energy transition: A review," International Journal of Sustainable Energy, vol. 42, no. 1, pp. 685–758, Sep. 2023. https://doi.org/https://doi.org/10.1080/14786451.2023.2222298

[2] A. E. Labrador Rivas and T. Abrão, "Faults in smart grid systems: Monitoring, detection and classification," Electric Power Systems Research, vol. 189, p. 106602, Dec. 2020. https://doi.org/https://doi.org/10.1016/j.epsr.2020.106602

[3] Zuzana Krivohlava, Stanislav Chren, and B. Rossi, "Failure and fault classification for smart grids," Energy Informatics, vol. 5, no. 1, Oct. 2022. https://doi.org/https://doi.org/10.1186/s42162-022-00218-3

[4] Krivohlava, Z., Chren, S., & Rossi, B. (2022). Failure and fault classification for smart grids. Energy Informatics, 5(1). https://doi.org/10.1186/s42162-022-00218-3

[5] Angel Esteban Labrador Rivas, Taufik Abrão, "Faults in smart grid systems: Monitoring, detection and classification" Electric Power Systems Research 189, 2020.

[6] H. A. Selmy, H. K. Mohamed, and W. Medhat, "Big data analytics deep learning techniques and applications: A survey," Information Systems, vol. 120, p. 102318, Feb. 2024. https://doi.org/https://doi.org/10.1016/j.is.2023.102318

[7] Dhruv Bhandarkar, "Hybrid Quantum-Classical Anomaly Detection: A Comparative Study(2025)," ResearchGate, Mar. 2025. https://doi.org/https://doi.org/10.13140/RG.2.2.19926.77122

[8] Power System. (2020, May 1). Kaggle. https://www.kaggle.com/datasets/bachirbarika/power-system

[9] F. Bre, J. M. Gimenez, and V. D. Fachinotti, "Prediction of wind pressure coefficients on building surfaces using artificial neural networks," Energy and Buildings, vol. 158, pp. 1429–1441, Jan. 2018. https://doi.org/https://doi.org/10.1016/j.enbuild.2017.11.045

[10] Deepak Ranga, Sunil Prajapat, Z. Akhtar, P. Kumar, and A. V. Vasilakos, "Hybrid Quantum–Classical Neural Networks for Efficient MNIST Binary Image Classification," Mathematics, vol. 12, no. 23, pp. 3684–3684, Nov. 2024. https://doi.org/https://doi.org/10.3390/math12233684

[11] Nahm, F. S. (2022). Receiver operating characteristic curve: overview and practical use for clinicians. Korean Journal of Anesthesiology, 75(1), 25–36. https://doi.org/10.4097/kja.21209

Effect of Electronically Driven Acoustic Disturbance to Lead-Acid Battery SOH Decay Rate

Alexis A. ARDIENTE[1,a], Napski Rudolph A. PACLEB[1,b], Cris Paulo G. HATE[2,c] and Gerard Francesco DG. APOLINARIO[1,3,d*]

[1]Advanced Batteries Center, Technological Institute of the Philippines, Philippines

[2]Computer Engineering Department, Technological Institute of the Philippines, Philippines

[3]Electrical Engineering Department, Technological Institute of the Philippines, Philippines

[a]aalexis.catalyst@tip.edu.ph, [b]nrpacleb.catalyst@tip.edu.ph, [c]cphate.cpe@tip.edu.ph, [d]gfrancesco.apolinario@tip.edu.ph

Keywords: Acoustic Desulfation, Lead-Acid Batteries, Micro CT, State of Health (SoH), Sulfation

Abstract. The study integrates the use of a potentiostat and X-ray Micro CT analysis in evaluating lead acid batteries (LABs) under various configurations and testing conditions. A charge/discharge approach using a potentiostat system was performed to show the significant reduction of the average decay rate for electronically driven acoustic disturbance. The experiment shows that the two transducer setup provides the most stable performance in reducing the decay rate. On the other hand, the use of X-ray Micro CT imaging highlights the differences in internal material distribution using voxel density data. LABs without transducers exhibited the highest voxel density of 24,146.77 and a standard deviation of 7,620.54, reflecting denser but less uniform material structures. The introduction of transducers reduces the voxel density to 11,108.69 and the standard deviation of 5,331.63 for two transducer setup. The results show that a decrease in voxel density happens when the transducer count increases.

1. Introduction

Battery performance, especially in lead–acid systems, declines over time due to cyclic usage and aging. A major contributor to this degradation is sulfation, which is the accumulation of lead sulfate ($PbSO_4$) on the electrodes during normal operation (see Fig. 1). While a natural process, persistent or irreversible sulfation increases internal impedance, reduces charge acceptance and capacity, and can render batteries difficult or impossible to recharge [1].

Fig. 1. Accumulation of Lead-Sulfates over Time (4 years)

Fig. 1 depicts the accumulation of sulfates: 0 cycle, 120 cycles/1 year, 240 cycles/2 years, 380 cycles/3 years, and 480 cycles/4 years (from left to right).

The key causes of the efficiency loss of lead-acid batteries are the sulfates; undercharging, deep discharge, or prolonged storage in a discharge state are the common triggers for these cases, which promote the formation of large crystal Lead(II) Sulfates ($PbSO_4$) [2]. While mild sulfation remains reversible, it is still manageable; severe cases increase the internal resistance. It impedes charge transfer and reduces lifespan [3]. It also affects the cyclability in the formation of manganese (II) sulfate ($MnSO_4$) on manganese oxide (MnOx) [5].

Carbon additives, such as carbon nanotubes (CNTs), overcome these effects by increasing conductivity and inhibiting crystal growth [4], and sulfation resistance can be enhanced using lead (IV) oxide (PbO_2) nanoparticles [5]. The reversal of sulfation without disassembly become possible using hemical treatments (e.g., sodium ethylenediaminetetraacetate, Na_2EDTA) [6] and inverse charging [7]. The use of carbon into the negative plate under partial state-of-charge (PsoC) [8] shows improvements in cycle durability and charge retention [9, 10].

Some measures like regular maintenance, electrolyte monitoring, and appropriate charging can slow down sulfation. Regrettably, reversing the sulfation process remains challenging without electrical or chemical intervention [11]. Several strategies in the reduction of sulfation such as graphene based additives in the negative active material (NAM) have enhanced cycle life under PSoC conditions [12]. Predictive insights into degradation and remaining useful life offered by the modeling approach provide support for better battery management [1]. Chemical methods like EDTA treatment effectively dissolve $PbSO_4$ without disassembly [11]. Other strategies includes: sensor-based systems using voltage and pressure data for localized sulfation control [13] and Al and K sulfate additives to enhance cycling and reduce sulfation through electrolyte modification [14].

This study contributes to the field of regeneration and maintenance of LABs and can be enumerated as follows.

a. A novel method that promote the breakdown of sulfate crystals, enhance mass transport, and electrolyte penetration into porous structures [15]. The method works in two ways:
   i. the electrolyte forms microbubbles and voids
   ii. the ultrasonic pressure fields cause mechanical agitation and localized turbulence on the electrode surfaces
b. The use of acoustic desulfation and its effects on charge retention, cycle life, and battery efficiency.
c. The use of ultrasonic transducer and its correlation to the charge recovery and SoH of LABs. The transducers are mounted on battery plates to apply ultrasonic vibrations as part of a novel mechanicl desulfation technique.
d. The study provides an observation of internal insights into sulfation by non-invasive structural changes in battery electrodes before and after desulfation using the micro-computed tomography (micro-CT).
e. In order to extend high-precision imaging to battery diagnostics, the study investigates the use of metal-based magnetic nanoparticles (MNPs) to improve micro-CT resolution in electrochemical systems, drawing on biomedical imaging techniques (such as PET, CT, SPECT, and MRI) [16, 17].

This study creates a thorough framework for identifying, tracking, and regenerating highly sulfated lead-acid batteries by fusing acoustic desulfation with high-resolution imaging.

This paper is organized as follows: Section 2 lists the study's setup and materials. The methodology is described in Section 3. The results and discussion are described in Section 4. Finally, Section 5 wraps up the conclusion and the suggested additional research.

2. MATERIALS
This section provides a brief overview of the tools and supplies used in the experimental setup. describing the equipment, chemicals, and tools used in the study.

2.1 Lead Acid Battery Cell
Typically, each cell produces approximately 2 volts—a basic unit of a lead–acid battery (Fig. 2). It comprises an anode—lead dioxide (PbO_2), a cathode—lead (Pb), an electrolyte—sulfuric acid (H_2SO_4), a separator (plastic or fiberglass), and durable housing components as a container. The battery is placed in a horizontal position within the CT scanner to optimize the visualization and reduce CT artifacts. This orientation keeps the planar surfaces from being perpendicular to the CT rotation axis and reduces the variance of the X-ray traversal-path lengths through the object [18]. Additionally, to enhance the accuracy of the imaging process, incorporate a cylindrical shape that can assist in improving X-ray detection. This method has been examined in earlier simulation-based studies concentrating on specimen orientation and is frequently used by skilled CT operators [19].

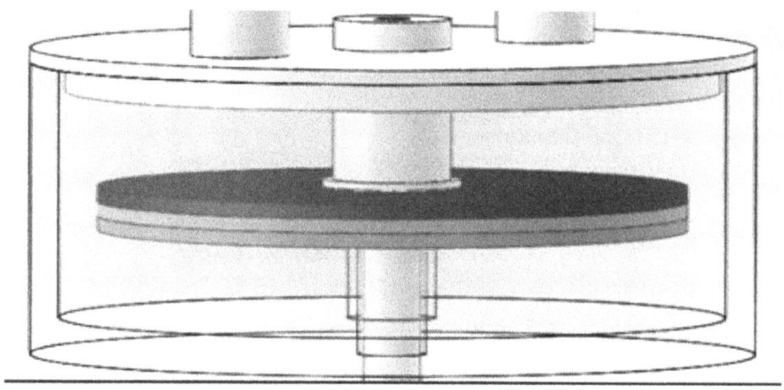

Fig. 2. 3D modeled Lead Acid Battery for X-ray Scanning

2.2 Potentiostat System
To control and measure the electrical potential (voltage) and current in a cell, a potentiostat system was used. A potentiostat's schematic diagram is shown in Fig. 3.

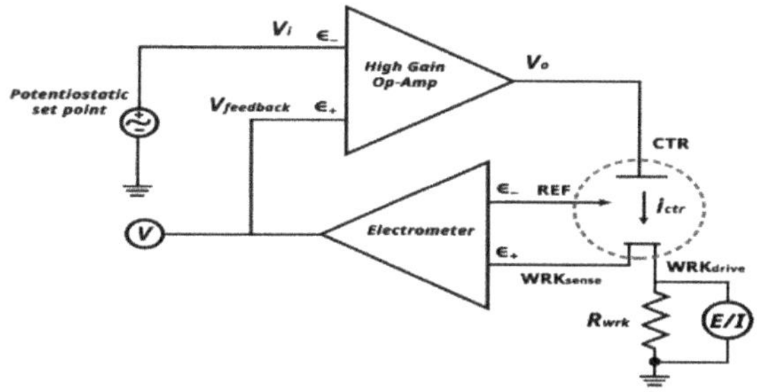

Fig. 3. A three-electrode potentiostat circuit schematic diagram.

The electrometer measures the potential difference between the reference electrode (REF) and the working electrode sense line (WRsense), and the operational amplifier maintains this potential steady with a feedback voltage (Vfeedback). The current flowing through the working electrode is determined by the voltage drop across a precision resistor (Rwrk). In voltammetric measurements, the current at the working electrode is monitored while the applied voltage is changed. This arrangement enables precise control of the potential between the working and reference electrodes regardless of current fluctuations, enabling accurate observation of electrochemical reactions and material behavior.

2.3 Micro-computed tomography (Micro-CT) scanner

The micro-CT scanner (Fig. 4) is a high-resolution imaging instrument that uses X-rays to non-destructively examine and visualize the internal structure of the scanned objects. It captures multiple cross-sectional images in 2D of a sample as it rotates with micron-scale resolution. A detailed 3D image is formed from the slices which enables internal assessment without physical sectioning. The technique is widely use in analyzation of biological tissues to inspection of aerospace components. It also provide precise structural insights without damaging the specimen. In this study, the internal structure of LABs is observed using these techiniques [17]. Table 1 shows the scanning parameters of the SkyScan 2214.

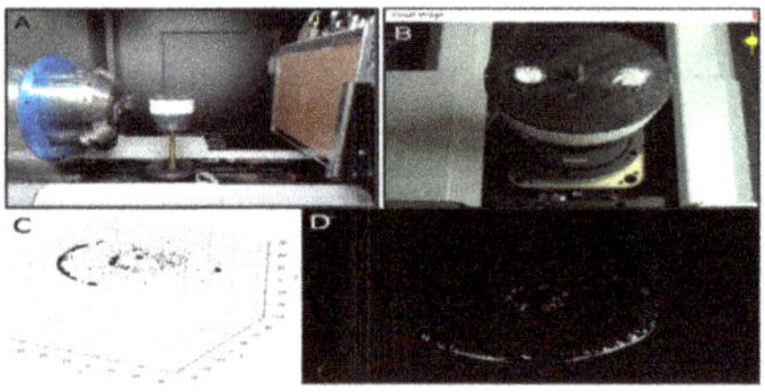

Fig. 4. Imaging and 3D reconstruction process: A composite specimen using X-ray micro-computed tomography (micro-CT).

Figure 4 shows (A) a mounted specimen of the experimental setup inside Skyscan 2214 Micro-CT, (B) before scanning, a visual image of the specimen, (C) initial 3D point cloud data showing defects or features in the specimen, and (D) reconstructed 3D volumetric image of the specimen, highlighting internal structures.

Table 1. Skyscan 2214 Scanning Parameters

Power Settings	Scan Settings
150kV	*21 microns*
100uA	*36.5 mm height*
Limiter Enable (can only reach 99au)	*Rotation step = 0.15*
2500ms-3000ms	*Frame Average =3*

2.4 Acoustic Desulfation Device

The acoustic desulfation system consists of acoustic transducers, a driver circuit, a signal generator, and a power supply. The signal generator produces sinusoidal waveforms, which are amplified and supplied to one or two transducers, depending on the configuration. Ultrasonic vibrations are transmitted via transducers to the battery cell. The generated acoustic excitation helps in breaking down the sulfate buildups on electron plates using ultrasonic vibrations. Table 2 summarizes the specification of the acoustic desulfation device.

Table 2. Desulfation Device Specification

Parameters	Values
Frequency	*1.700Mhz*
Input Voltage	*18Vpp*
Output Voltage	*60Vpp°*
No. of transducer/s	*0-2*

3. Methodology
The main objective of the study is to examine how ultrasonic vibrations effect the performance and internal structure of LABs. Figure 5 illustrates the experimental workflow.

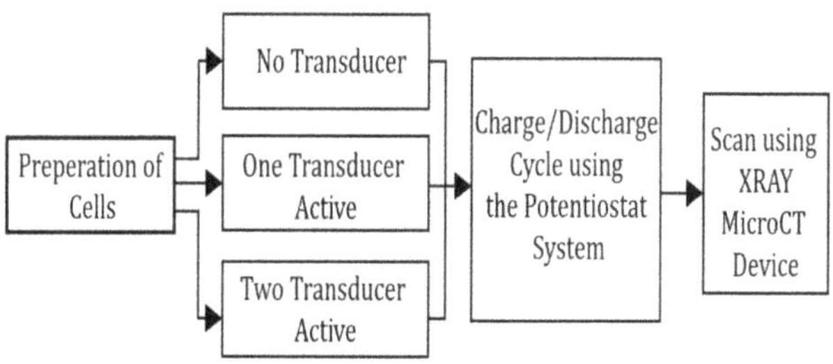

Fig. 5. Experimental workflow

3.1 Experiment Procedure
Series of test were conducted in this experiment. Each test can be classified into three configurations based on the number of active transducers (0, 1 and 2). The following steps are considered in doing the experiment:

a) Each cell configuration should complete the five charge/discharge cycles using the potentiostat system.
b) The state of health (SoH) of each sample was measured at the end of every cycle to track performance degradation.
c) All cells were scanned using an X-ray micro-CT system to examine the internal changes produced by the acoustic treatment.
d) The scans provide a visual of the sulfate crystals and material distribution inside each cell.

This allowed comparison between the three configurations.

3.2 Data Analysis
Linear regression was used by inputting the statistical slope of SoH over five cycles to calculate the rate of SoH decay. The relationship between transducer induced acoustic vibration and the SoH decay rate in LABs is evaluated using correlation analysis.

Additionally, to compute the voxel ratio, X-ray micro-CT scan data were exported in STL format, which will be processed using an image analysis algorithm. A volume of information within the scanned object is represented by a voxel, the 3D counterpart of a 2D pixel. Whereas low values suggest its absence, voxels with high binary values indicate the presence of sulfation. A reduced extent of sulfation within the battery cell is then interpreted by a lower voxel count.

4. RESULTS AND DISCUSSION
A total of 29 tests were performed on different LAB samples. Each of the test consists of five charge/discharge cycles, which are conducted under varying number of active transducers. SoH

measurements were gathered at the end of each cycle. The SoH decay rate for each sample was calculated using the Least Squares Method (LSM) after each test.

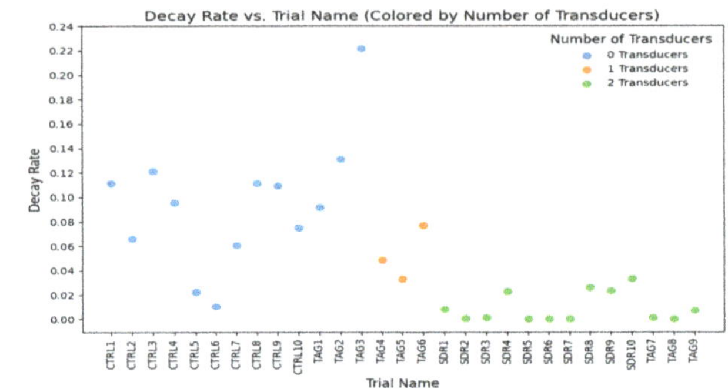

Fig. 6. Decay rate per trial grouped by transducer count.

As seen in Figure 6, LAB samples without transducer activation (CTRL and TAG with zero transducers) exhibit significantly poorer performance, lower final SoH percentages, and higher deration rates, some of which exceed 0.1. SDR trials using two transducers setup maintained SoH above 90% and shows minimal decay rates, typically around 0.01. Trials under TAG4 with one or two transducers also shows improvement in SoH stability, but with some variability in decay rates as seen in figs. 7 to 9.

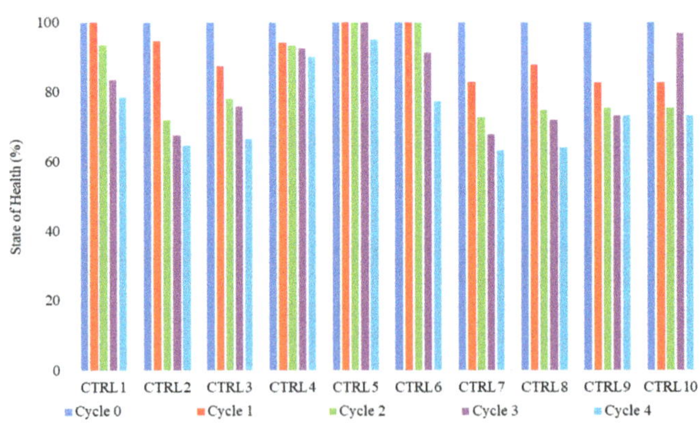

Fig. 7. State of Health (SoH) of CTRL Setup

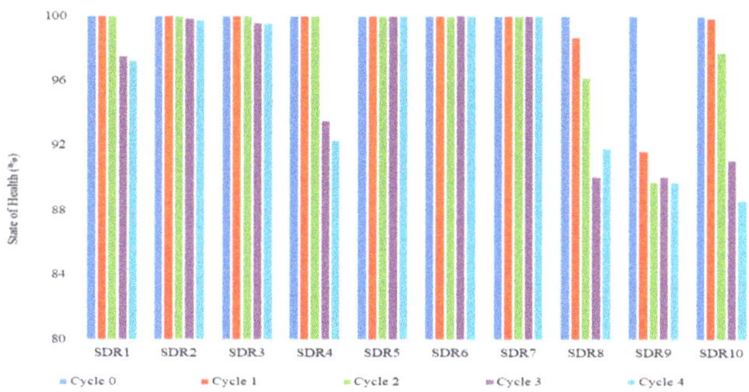

Fig. 8. State of Health (SoH) of SDR Setup

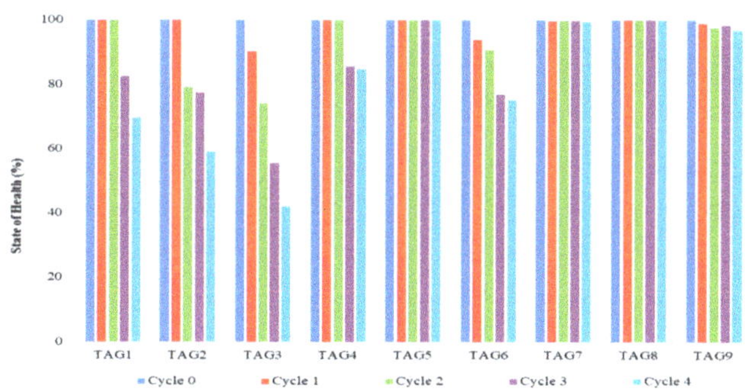

Fig. 9. State of Health (SoH) of TAG Setup

The findings on figs. 7 to 9 can be summarized as follows.

a) Trials with no transducer shows large fluctuation (exceeding 0.1) in SoH decay rates and higher overall degradation.
b) Increased in internal resistance and sulfation can be seen in trials with no transducer compared to trials with transducer/s.
c) A higher susceptibility to electrochemical wear and a reduction in LABs' performance and longevity can be seen on trials without transducers.
d) Trials with transducer/s present demonstrate relatively stable SoH readings throughout the test duration.
e) The stability of SoH readings for trials with transducer/s shows a positive effect of acoustic vibration in the mitigation of the degradation mechanism for LABs.
f) The constant SoH retention signifies that using ultrasonic energy helps in preserving internal electrochemical integrity.

Table 3 presents the summary of findings for each configuration.

Materials Research Proceedings 66 (2026) 315-325 https://doi.org/10.21741/9781644904152-29

Table 3. Average Decay Rates and Standard Deviations at Varying Transducer Counts

TRANSDUCER COUNT	Average Decay Rate	Standard Deviation
0	0.09442561	0.05292346
1	0.05309391	0.02221369
2	0.00957607	0.01229401

As seen in Table 3, the following findings can be seen.

a) Increasing the number of active transducers reduces the performance degradation rate of LAB.
b) Setup without transducers result in a high average decay rate of 0.0944 and standard deviation of 0.0529 indicating rapid deterioration and high variability.
c) The use of a single transducer improved performance by lowering the decay rate to 0.0531 and the standard deviation to 0.0222.
d) Lastly, using two transducer setup yields the lowest decay rate of 0.0096 and standard deviation of 0.0123.

The use of micro-CT scan analyses (see figs. 9 and 10) further validates the results that higher concentration of sulfation can be seen on trials with no transducer compared to that with transducer/s.

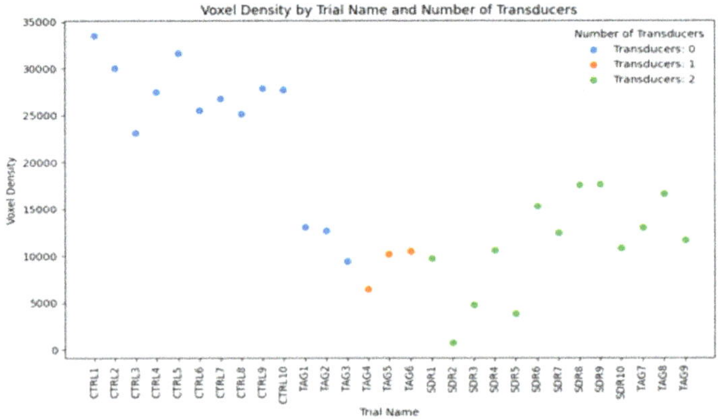

Figure 10. Voxel Density by Trial Name and Number of Transducers.

The voxel density of LAB electrodes across multiple trials can be seen in fig. 10. Each trial can be categorized based on the number of transducers used (0, 1, or 2). The findings can be summarized as follows.

a) Trials with no transducer involved show higher voxel densities indicating that these trials experienced more sulfation over time.
b) Trials that have 1 or 2 transducers exhibit lower voxel densities, suggesting that the application of acoustic excitation is effective in reducing sulfation and enhances the performance and longevity of LABs.

Table 4. Average Voxel Density and Standard Deviations at Varying Transducer Counts

TRANSDUCER COUNT	Average Voxel Density	Standard Deviation
0	24146.77	7620.537
1	9031.667	2249.829
2	11108.69	5331.628

Table 4 presents the average voxel density rate of the different test setups. The findings can be summarized as follows.

a) The setup with no transducers resulted to high voxel density, with an average of 24,146.77 and standard deviation of 7,620.54, indicating a dense but irregular internal distribution.
b) The use of a single transducer shows a decrease in the average voxel density to 9,031.67 and standard deviation to 2,249.83, which signifies an improvement in the structural uniformity of the material matrix.
c) The use of two transducers shows a slight increase in average voxel density to 11,108.69, while the standard deviation also rises to 5,331.63, indicating a denser but more variable distribution.
d) The observed reduction in both the mean voxel density and standard deviation shows that adding of transducer weakens sulfation process and enhanced internal homogeneity.

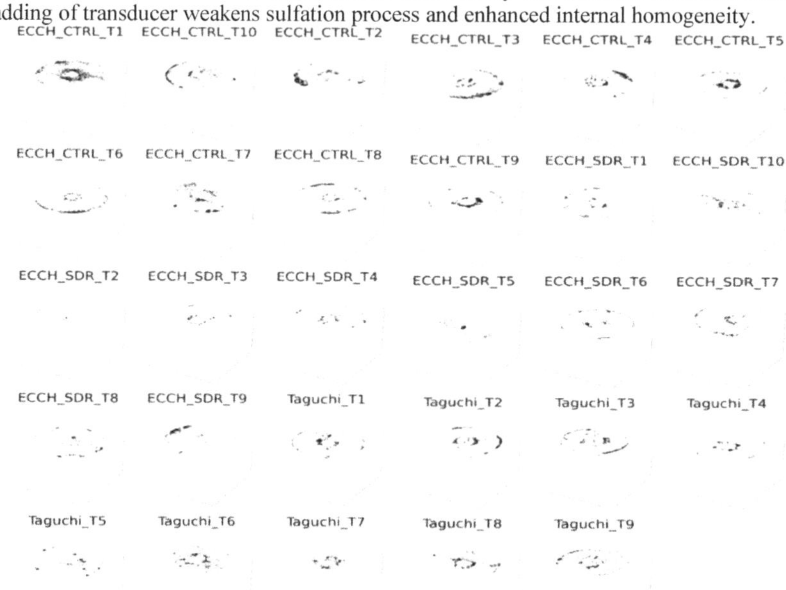

Fig. 11. 3D Image of Accumulated Sulfates

Fig. 11 further validates the findings in table 4 using the 3D micro-CT scans. As seen in fig. 11, the accumulated sulfates decreases as the number of active transducers increases. This indicates less sulfation and a more uniform internal structure for LAB.

5. CONCLUSION AND FUTURE WORKS

The study explores how acoustic desulfation using ultrasonic transducers affects the structure and performance of LABs with the use of both potentiostat electrochemical testing and micro-CT imaging. An increase in the number of ultrasonic transducers leads to the exposed cells retaining more charge, slower degradation over several cycles and having structural uniformity.

Potentiostat measurements also shows a reduction in average SoH decay rates and increasing the transducer count results to enhanced consistency across chagre/discharge cycles. The result validates the improvement in stability due to the acoustic excitation.

The use of micro-CT imaging in visualizing the internal structural changes of the pre- and post-desulfation is a novel aspect in this study. Results revealed that batteries with no transducers has higher sulfate accumalation and structural heterogeneity exhibited by higher mean voxel densities (24,146.77; SD = 7,620.54). In contrast, batteries with two transducers shows more uniform material distrubution and reduced compaction due to lower voxel densities (11,108.69; SD = 5,331.63).

The study also considered the application of post-analysis processing method often used in medical research.This approach provided a better view of internal changes without opening or damaging the cells, offering a useful direction for future non-invasive and non-destructive analysis.

Overall, the results suggest that the integration of the acoustic desulfation can extend the cycle life of lead–acid batteries and help maintain their efficiency. Testing of different vibration techniques and transducer placements maybe included as future works to improve the effect on desulfation. Application to large-scale energy storage system can also be considered for future works.

ACKNOWLEDGEMENT

The study received funding support from the Department of Science nad Technogology (DOST). The funding is form the DOST Grants in Aid (DOST-GIA), grant number 10047 under the Accelerated R&D Program for Capacity Building of Research and Developmen Institutions and Industrial Competitivenss: Niche Centers in the Region for R&D (NICER). This NICER Program aim to develop a R&D Center for Advanced Batteries. The authors also extend appreciation to the Technological Institute of the Philippines (T.I.P.) and TechnoCoRe team for providing the facilities and guidance necessary to complete this study.

REFERENCES

[1] K. S. Gandhi, "Modeling of sulfation in a flooded lead-acid battery and prediction of its cycle life," Journal of The Electrochemical Society, vol. 167, no. 1, p. 013538, Jan. 2020. https://doi.org/10.1149/1945-7111/ab679b

[2] C. H. A. Catherino, F. F. Feres, and F. Trinidad, "Sulfation in lead–acid batteries," J. Power Sources, vol. 129, pp. 113, 2004. https://doi.org/10.1016/j.jpowsour.2003.11.003.

[3] D. Pavlov, "Fundamentals of lead–acid batteries," in Lead-Acid Batteries: Science and Technology, Amsterdam, Netherlands: Elsevier, 2011, pp. 29-114. https://doi.org/10.1016/B978-0-444-52882-7.10002-3

[4] R. Shapira, G. D. Nessim, T. Zimrin, and D. Aurbach, "Towards promising electrochemical technology for load leveling applications: extending cycle life of lead acid batteries by the use of carbon nano-tubes (CNTs)," Energy Environ. Sci., vol. 6, pp. 587, 2013. https://doi.org/10.1039/C2EE22970F

[5] J.-e. Jin, D. Jin, J. Shim, and W. Shim, "Enhancing reversible sulfation of PbO2 nanoparticles for extended lifetime in lead-acid batteries," Journal of The Electrochemical Society, vol. 164, no. 7, p. A1628, 2017. https://doi.org/10.1149/2.1291707jes

[6] N. Vangapally, S. A. Gaffoor, and S. K. Martha, "Na2EDTA chelating agent as an electrolyte additive for high performance lead-acid batteries," Electrochimica Acta, vol. 258, pp. 1493-1501, 2017. https://doi.org/10.1016/j.electacta.2017.12.028

[7] C. Spanos, S. A. Berlinger, A. Jayan, and A. C. West, "Inverse charging techniques for sulfation reversal in flooded lead-acid batteries," Journal of The Electrochemical Society, vol. 163, no. 8, pp. A1612-A1618, Aug. 2016. https://doi.org/10.1149/2.0761608jes

[8] P. T. Moseley, "Consequences of including carbon in the negative plates of valve-regulated lead–acid batteries exposed to high-rate partial-state-of-charge operation," Journal of Power Sources, vol. 191, no. 1, pp. 134-138, Jun. 2009. https://doi.org/10.1016/j.jpowsour.2008.08.084

[9] I. K. Gibson and K. Peters, "Sulfation in discharged lead–acid batteries," J. Power Sources, vol. 22, no. 3, pp. 227–234, 1988. https://doi.org/10.1016/0378-7753(82)80052-9

[10] J. Yang, C. Hu, H. Wang, K. Yang, J. B. Liu, H. Yan, "Review on the research of failure modes and mechanism for lead–acid batteries," Int. J. Energy Res., vol. 40, pp. 2234–2248, 2016, vol. 41, no. 14, pp. 2014–2030, 2017. https://doi.org/10.1002/er.3613

[11] H. A. Catherino, F. F. Feres, and F. Trinidad, "Sulfation in lead–acid batteries," J. Power Sources, vol. 129, no. 1, pp. 113–120, 2004. https://doi.org/10.1016/j.jpowsour.2003.11.003

[12] K. K. Yeung, X. Zhang, S. C. T. Kwok, F. Ciucci, and M. M. F. Yuen, "Enhanced cycle life of lead–acid battery using graphene as a sulfation suppression additive in negative active material," RSC Adv., vol. 5, pp. 71314–71321, 2015. https://doi.org/10.1039/C5RA11114E

[13] Y. Shi, C. A. Ferone, and C. D. Rahn, "Identification and remediation of sulfation in lead–acid batteries using cell voltage and pressure sensing," J. Power Sources, vol. 221, pp. 177–185, 2013. https://doi.org/10.1016/j.jpowsour.2012.08.013

[14] C. E. Onu, N. M. Nwabunwane, and P. K. Igbokwe, "Additive effects of aluminium and potassium sulfates in enhancing the charge cycle of lead–acid batteries," Asian J. Chem. Sci., vol. 10, no. 3, pp. 11–19, 2021. https://doi.org/10.9734/AJOCS/2021/v10i319092

[15] M. Alvarado, A. Bulfa Jr., A. Aquino, "Ultrasonic influence on germination and early growth of bell pepper seeds," in Proceedings of International Exchange and Innovation Conference on Engineering & Sciences (IEICES), vol. 10, pp. 19-25, 2024. https://doi.org/10.5109/7323237

[16]N. Neogi, K. P. Choudhury, "Utilization of metal-based nanoparticles in biomedical imaging technologies," in Proceedings of International Exchange and Innovation Conference on Engineering & Sciences (IEICES), vol. 9, pp. 65-71, 2023. https://doi.org/10.5109/7157949

[17] J.M. Campillo-Robles, D. Goonetilleke, D. Soler, N. Sharma, D.M. Rodríguez, T. Bücherl, M. Makowska, P. Türkilmaz and V. Karahan, "Monitoring lead-acid battery function using operando neutron radiography," Journal of Power Sources, vol. 438, 226976, October 2019. https://doi.org/10.1016/j.jpowsour.2019.226976

[18] H. Villarraga-Gómez, A. Amirkhanov, C. Heinzl, S.T. Smith, "Assessing the effect of sample orientation on dimensional X-ray computed tomography through experimental and simulated data," Measurement, vol. 174, p. 109343, 2021. https://doi.org/10.1016/j.measurement.2021.109343

[19] R.A. Ketcham, W.D. Carlson, "Acquisition, optimization and interpretation of X-ray computed tomographic imagery: applications to the geosciences," Computers & Geosciences, vol. 27, no. 4, pp. 381-400, May 2001. https://doi.org/10.1016/S0098-3004(00)00116-3

Smart Inventory Management for Wastage and Losses in a Construction Company

Jose A. PEREZ JR.[1,a], Joshua Martin A. PERALTA[1,b], Arvin F. EUGENIO[1,c], Armand Sebastian E. BUENO[1,d], Risty M. ACERADO[1,e*]

[1]College of Computer Studies, Technological Institute of the Philippines, Quezon City, Philippines

[a]qjperezjr.@tip.edu.ph, [b]qjmperalta@tip.edu.ph, [c]qjaeugenio@tip.edu.ph, [d]qasbueno@tip.edu.ph, [e]racerado.is@tip.edu.ph

Keywords: PyTorch, Flutter, Construction Industry, Object Detection, Mobile Application

Abstract. Small and medium-sized construction enterprises often rely on manual inventory counting, which can lead to errors, material wastage, and operational inefficiencies. To address these challenges, this study developed a YOLOv5 object detection model for mobile-based smart inventory management system. A localized dataset of 671 images representing six construction material classes was collected from local hardware stores and construction sites in the Philippines and augmented to 6,000 images. The trained model achieved an average precision of 85.54%, a recall of 76.53%, an F1-score of 80.37%, an mAP@0.5 of 0.88, and an mAP@0.5:0.95 of 0.64. Thus, the system is accurate and efficient in managing the inventory of construction materials, thereby reducing manual errors, improving operational efficiency, and sustainable inventory management in the construction industry.

Introduction

According to the Department of Trade and Industry (DTI), only 10% of MSMEs in the Philippines are currently fully digitalized [1]. As of 2024, there are approximately 12,284 hardware stores and 10,034 construction companies. The number shows that competition is tight within the industry indicating operational efficiency with the use of technology is not an option but very significant and necessary for sustainability. Businesses relying on manual methods to manage their inventory would be critical in survival. Manual processes such as log and spreadsheets often result in human error and operational delays [3]. Thus, to sustain the business and be competitive, companies should consider innovating the management of inventory to reduce costs and maintain business operations. Mismanagement leads to resource waste that conflicts with global digital economy trends among modern enterprises [2].

Machine learning models address various challenges in inventory management to automate identification and counting of objects. Single-stage detection architectures have the potential in real-time applications due to their ability to process images quickly [4], [5]. The integration of such models into mobile systems is practical even in the area of limited internet connection [6]. There are still challenges in balancing the accuracy and hardware limitations in the implementation of artificial intelligence implementations in SMS. Thus, this study aimed to conduct a comparative sensitivity analysis of detection frameworks in determining the feasible algorithm. In addition, this study aimed to develop an object detection model for smart inventory management systems. The performance of the developed model would be evaluated using standard quantitative metrics to ensure its reliability in reducing manual errors and supporting sustainable resource management goals [7].

Research Methodology
This section presents the systematic approach adopted in the development of the object detection model from the data collection, model development and training, model evaluation, and mobile integration and deployment.

Data Collection and Annotation. The dataset is composed of 671 images of hollow blocks, rebar, skim coat, sand, cement, and gravel collected from local hardware stores and active construction sites across the Philippines. Several augmentation techniques were applied in the collected images considering the recommendations from the previous studies related to object detection [8, 9, 10]. Table 1, shows the composition of the dataset in developing the object detection model.

Table 1. Dataset Composition by Class

Material Class	Number of Images	Number of Annotations
Bistay Sand (in sacks)	114	6,006
Rebar	211	15,213
Gravel (in sacks)	63	2,175
Hollow Blocks	106	6,504
Skim Coat	88	4,567
Cement (in sacks)	89	1,949
Total	**671**	**36,414**

Fig. 1 shows an example of applying several augmentation techniques using Roboflow to reduce the model's overfitting [10].

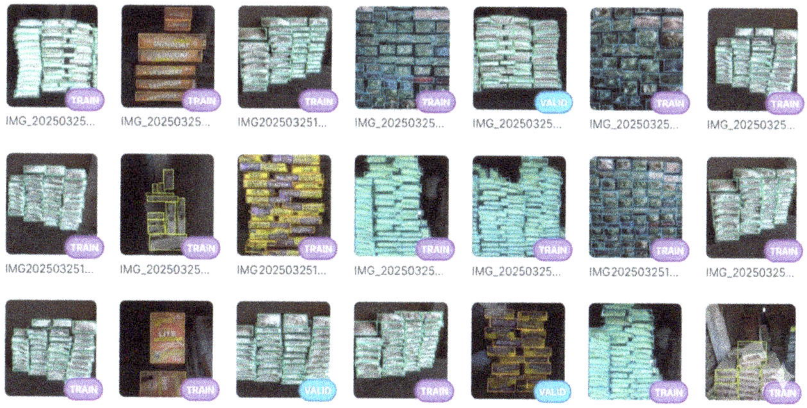

Fig. 1 Image Annotations

The specifications of the labeling is presented in table 2.

Table 2. Technical Details of the Dataset Annotation

Parameter	Specification
Annotation Platform	Roboflow
Labeling Method	AI-Assisted (Label Assist)
Annotation Type	2D Bounding Boxes (Rectangular)
Total Annotations	36,414
Output Format	YOLOv5 PyTorch

A comprehensive augmentation and balancing strategy was implemented as presented in Table 3 to address the class imbalances observed in Table 1. The training set was normalized to a target of 1,000 images per class, resulting in a balanced dataset of 6,000 total augmented images. The eliminated model bias toward majority classes while augmenting the representation of minority classes. These strategies were used to help the model adapt to different orientations found on local construction sites. Strengthening the model's accuracy on new data and preventing it from simply memorizing the training set [9].

Table 3. Data Augmentation and Class Balancing Strategy

Category	Technique/Parameter	Strategy Details & Purpose
Framework	Augmentation Library	Albumentations: Integrated via Python script for high-performance transformations.
Balancing	Target Dataset Size	Normalized to 1,000 images per class: Resulting in a balanced 6,000-image training set to eliminate bias toward majority classes.
Geometry	Rotation & Scaling	±15° to ±30° Rotation and Resizing: Improves robustness to varied camera angles and object distances.
	Flipping	Horizontal and Vertical Flips: Introduces spatial variation to prevent the model from memorizing fixed orientations.
Environment	Contrast & Lighting	Contrast Adjustment & Random Shadows: Simulates varying site lighting and uncontrolled shadows typical of construction areas.
	Weather Effects	Gaussian Blur & Rain Effects: Simulates varied and uncontrolled construction site conditions to ensure detection stability.

Model Development and Training. The most effective architecture was identified after analyzing and comparing several frameworks and detection algorithms. Single-stage detectors were investigated and compared against traditional computer vision libraries [5, 12, 15]. Two-stage detectors [6] were considered. Each architecture was evaluated based on its ability to handle the hardware constraints. To avoid the inefficiency of training from scratch, we developed our model using the YOLO framework [13, 14], as shown in the comparisons in Table 4. This allowed the system to inherit a foundation of visual recognition, which proved essential for accurately detecting construction materials in local environments.

Materials Research Proceedings 66 (2026) 326-333
https://doi.org/10.21741/9781644904152-30

Table 4. Framework Sensitivity Analysis Summary

Framework	Best Score	Worst Score	Key Advantage	Primary Limitation
PyTorch	255	249	Superior memory & size balance	Requires TorchScript conversion
TensorFlow Lite	236	232	Widespread documentation	Version/Dependency conflicts
OpenCV	270	261	Fast for simple tasks	Poor handling of lighting/shadows
YOLO (Native)	230.5	228.5	High accuracy	High coordinate scaling errors

Specific settings were chosen to balance training speed with accuracy, as shown in Table 5. A learning rate of 0.001 kept the training stable and prevented the model from missing the best results. A batch size of 4 was used to save memory on the device while keeping the calculations precise.

The model was trained for 100 epochs, with the validation loss monitored to ensure it learned real material textures rather than just memorizing the data. During testing, a confidence threshold of 0.25 was set to avoid missing any materials. This value also prevents the system from double-counting overlapping items, like rebar bundles. Together, these choices ensure accurate counting in busy, real-world construction sites.

Table 5. Hyperparameters Used in Training

Hyperparameter	Value
Learning Rate	0.001
Batch Size	4
Number of Epochs	100
Confidence Threshold	0.25

The dataset was split into 80% for training, 10% for validation, and 10% for testing. Once the model reached peak performance, the weights were exported in the TorchScript format. This conversion was essential for cross-platform compatibility, allowing the model to run efficiently within the Android app environment.

Table 6 provides the technical specifications for deploying the developed object detection model on a mobile device. Optimizing the YOLOv5 architecture and converting it to TorchScript enabled real-time detection on an **OPPO A5s**. With a compact model size of **42.8 MB**, memory usage remains low, while **PyTorch Mobile** facilitates full on-device edge computing. An average inference time of **245 ms** (approx. **4.08 FPS**) demonstrates that material counting can be performed locally. This localized processing effectively removes any dependency on cloud servers or stable internet access.

Table 6. Mobile Deployment and Inference Performance Specifications

Metric	Details
Model Architecture	YOLO
Model Format	TorchScript (.torchscript)
Model Framework	PyTorch Mobile
Model Size	42.8 MB
Input Image Size	640 x 640 pixels
Inference Time	~245 ms
Inference Speed (FPS)	4.08 FPS
Testing Hardware	OPPO A5s (CPH1909)
Device Processor	MediaTek Helio P35 (MT6765) Octa-core)
GPU	PowerVR GE8320
Development Tools	Android Studio, Google Colab, Roboflow
Processing Type	On-device Edge Computing (Local Inference)

Model Evaluation and Testing. Evaluation of the model's ability to identify construction materials relied on several standard computer vision metrics [12]. These results were benchmarked against previous object detection studies to establish a comparative baseline for accuracy. Mean Average Precision (mAP) served as the primary performance indicator, with higher values reflecting superior accuracy in localizing and classifying specific materials. Table 7 provides a detailed breakdown of all metrics utilized throughout this assessment.

Table 7. Model Evaluation Metrics

Metric	Description	Purpose
Precision	Measures the proportion of true positives among predicted positives	Evaluates the model's ability to avoid false positives.
Recall	Measures the proportion of true positives among actual positives	Evaluate how effectively the model identifies all relevant instances.
Intersection over Union (IoU)	The ratio between the overlapping area and the union area of predicted and actual bounding boxes	Ensures the accuracy of spatial localization for detected objects.
mAP@0.5	Mean Average Precision at IoU threshold of 0.5	A widely used indicator for evaluating overall object detection accuracy.
mAP@0.5:0.95	Averaged mAP across IoU thresholds from 0.5 to 0.95	Provides a more rigorous evaluation of the model across stricter localization criteria.

Mobile Integration and Deployment. The model was integrated into the mobile platform through the developed Android application. This application can be used offline and displays information in real-time. Using mid-range Android devices, the application was able to detect objects in real-time, displaying the number of detected construction materials within an image. Thus, it reduces human error in manual counting and improves the accuracy and overall efficiency of the inventory management. The study emphasized a more sustainable inventory management system and improved use of the resources to avoid waste and loss [6, 13].

Scope and Limitations. The study does not cover the other inventory management tasks, such as database management, user management, and data analytics.

Results and Discussions

The trained YOLOv5 model was tested using a set of unseen images to assess its generalization capability and object counting accuracy. The evaluation was conducted based on standard object detection metrics: precision, recall, and F1-score. Table 8 shows the evaluation results of the developed model.

Table 8. Object Detection Model Evaluation Results by Class

Material Class	Precision (%)	Recall (%)	F1-Score (%)	mAP@0.5	mAP@0.5:0.95
Bistay Sand	92.31	88.00	90.10	0.95	0.64
Skim Coat	88.00	85.00	86.47	0.94	0.78
Rebar	80.00	76.00	77.97	0.91	0.68
Gravel	85.71	81.25	83.43	0.93	0.70
Hollow Blocks	75.00	70.00	72.41	0.90	0.57
Cement	92.22	58.9	71.84	0.68	0.49
Average	**85.54**	**76.53**	**80.37**	**0.88**	**0.64**

The model showed the highest accuracy in detecting bistay sand with a score of 90.10% in F1. While, hollow blocks, rebar, and cement have a lower performance that could be attributed to visual blending and varying image quality. The results reflect common challenges in object detection.

Fig. 2 illustrates the confidence thresholding and bounding box drawing logic that visualizes detected objects on the mobile screen.

***Fig. 2** Snapshot of Confidence Thresholding*

Conclusion and Recommendations

This study developed and implemented an object detection machine learning model to automate the detection and counting of construction materials for a more efficient inventory management process. Several object detection frameworks were analyzed and compared YOLOv5 was selected to be the most feasible for this study integrated to Pythorch framework Based on a comparative evaluation of several detection frameworks, YOLOv5 integrated with PyTorch Mobile was identified as the most suitable model, effectively balancing detection accuracy and computational efficiency on mid-range Android devices. Thus, the system improves the business operations of the construction industry, specifically in the small and medium-sized enterprises, in tracking inventory management, thus reducing waste and loss.

The study recommends that additional types of construction materials can be used to train the model to increase the scope of the detection in the construction materials. Different variations and environmental conditions can also be added to increase the generalization of the model.

References

[1] Department of Trade and Industry (DTI), "DTI Strategic Plan 2023–2028: Transforming the Philippines through Inclusive, Innovative, and Digitally-driven Industries," DTI, Manila, Philippines, 2023.

[2] Eurostat, "Digital economy and society statistics - enterprises," Statistics Explained, 2024. [Online]. Available: https://ec.europa.eu/eurostat/statistics-explained/index.php?title=Digital_economy_and_society_statistics_-_enterprises.

[3] Department of Trade and Industry (DTI), "Profile of the Philippine Construction Industry," DTI, Manila, Philippines, 2023.

[4] J. Redmon and A. Farhadi, "YOLOv3: An incremental improvement," 2018. [Online]. Available: https://pjreddie.com/media/files/papers/YOLOv3.pdf

[5] Ultralytics, "YOLOv8: State-of-the-art object detection," Ultralytics, 2023. [Online]. Available: https://github.com/ultralytics/ultralytics

[6] J. Wang, C. Liu, D. Chen, and Y. Zhang, "Real-time object counting with edge-AI for intelligent construction," Automation in Construction, vol. 134, p. 104020, 2022. https://doi.org/10.1016/j.autcon.2021.104020.

[7] United Nations, "Sustainable Development Goals," United Nations. [Online]. Available: https://sdgs.un.org/goals

[8] J. Qin, W. Yan, H. Lu, and J. Wang, "Enhancing object detection robustness through environment-diverse training datasets," Sensors, vol. 24, no. 2, p. 315, 2024. https://doi.org/10.3390/s24020315.

[9] Y. Zhang, H. Wang, and D. Chen, "Real-time object detection on mobile devices using lightweight deep learning models," IEEE Access, vol. 9, pp. 123456–123467, 2021.

[10] C. Shorten and T. M. Khoshgoftaar, "A survey on image data augmentation for deep learning," Journal of Big Data, vol. 6, no. 60, 2019. https://doi.org/10.1186/s40537-019-0197-0.

[11] Roboflow, "Roboflow: Dataset annotation and management for computer vision," Roboflow, 2023. [Online]. Available: https://roboflow.com

[12] G. Jocher et al., "YOLOv5," Ultralytics, 2021. [Online]. Available: https://github.com/ultralytics/yolov5

[13] R. Padilla, S. L. Netto, and E. A. B. da Silva, "A survey on performance metrics for object detection algorithms," International Journal of Computer Vision and Applications, vol. 11, no. 3, pp. 45–59, 2021. https://doi.org/10.1007/s41095-021-0231-4.

[14] A. Ajayi, L. Oyedele, O. Akinade, and M. Bilal, "Waste reduction through material management strategies in construction," Resources, Conservation and Recycling, vol. 120, pp. 15–28, 2017. https://doi.org/10.1016/j.resconrec.2017.01.012.

[15] S. Sharma, K. Gulati, and S. Kumar, "Real-time object identification and tracking using OpenCV," International Journal of Computer Science and Mobile Computing, vol. 10, no. 5, pp. 1-7, 2021.

[16] S. Matani, "PyTorch Mobile: Bringing machine learning to edge devices," Towards Data Science, 2022. [Online]. Available: https://towardsdatascience.com/pytorch-mobile-bringing-machine-learning-to-edge-devices

Mix Design Development and Mechanical Evaluation of Self-Consolidating Lightweight Concrete using Perlite Aggregates

Ian Ceymark P. PALMERO[1,a *], Razon C. DOMINGO[2,b], and Meriam LEOPOLDO[3,c]

[1]FEU Institute of Technology, Sampaloc, Manila, Philippines, 1008

[2]Polytechnic University of the Philippines - Manila, Sta Mesa, Manila, Philippines

[3]Mapuan Malayan Colleges Mindanao, Davao City Philippines

[a]ippalmero@feutech.edu.ph, [b]rcdomingo@pup.edu.ph, [c]mpleopoldo@mcm.edu.ph

Keywords: Perlite, Mix Design, Aggregate, Lightweight Concrete, Compressive Strength

Abstract. Structural lightweight concrete (SLC) provides substantial advantages in reducing structural dead load, improving thermal performance in construction, and contributing to sustainable construction (SDG 9 and 11). Perlite, a naturally occurring volcanic glass, is recognized for its low density and excellent insulation properties, making it a promising alternative aggregate for SLC. This study aims to develop a mix design for self-consolidating lightweight concrete (SCLC) incorporating perlite aggregates as replacement for conventional coarse aggregates. Experimental mixes were prepared with perlite replacement ratios of 0%, 10%, 20%, 30%, 40%, 50%, and 60%, using 0.75% superplasticizer dosage, to evaluate compressive strength, split tensile strength, and density following ASTM standards. Results indicate that increasing perlite content generally leads to reductions in compressive strength, density, and tensile strength; however, structural-grade strength can still be achieved. It also shows segregation as replacement increases. Based on the results, a refined mix design is developed incorporating 100% perlite replacement, 0.45% superplasticizer, and 0.1% viscosity modifying agent (VMA) was introduced to address segregation issues. This optimized mix achieved a 28-day compressive strength of 33.70 MPa (4887 psi), density of 1790 kg/m³, and split tensile strength ranging from 1.557 MPa to 2.549 MPa. The results show the feasibility of perlite aggregates for structural lightweight concrete applications. It could be used for precast concrete, structural concrete, architectural components and other structural lightweight applications.

Introduction

With the growing demand for construction and dealing with environmental challenges, integrating Green Building principles in line with Sustainable Development would provide efficient use of our natural resources [1][2]. Structural lightweight concrete (SLC) helps these goals by decreasing dead loads, reducing structural requirements, enhancing seismic performance, maximizing materials handling, and enhancing thermal performance [3]. One of the available lightweight aggregates is perlite – a naturally formed volcanic glass that enlarges when heated, has low density, better insulation properties, and locally available in the Philippines [4]. Using perlite into self-consolidating concrete (SCC) can lower density while having acceptable mechanical properties, though challenges such as tensile strength must be addressed through admixtures and enhanced mix design [5], [6]. This study aims to develop a mix design for self-consolidating lightweight concrete (SCLC) with perlite aggregates that meets ASTM C330 requirements for strength and density while considering sustainable construction practices.

Structural lightweight concrete is made up of typical concrete material such as cement, water and aggregates. But the aggregate that is being used is lightweight aggregates whether it is a substitute to fine or coarse aggregate. There are many kinds of lightweight aggregates such as pumice, expanded clay, lightweight expanded clay aggregate, expanded shale, artificial lightweight aggregates, Liapore, silt, lytag, industrial waste, fly ash and perlite [7]. In the Philippines, one of the available lightweight aggregates is the perlite due to the mining industry.

Perlite is a volcanic material that has no definite shape and can expand up to 20 times its original volume [4]. It also has a lightweight property and has high water content and can be expanded when subjected to high temperature. One of the countries that have perlite mining industry is the Philippines. According to 2024 USGS data, about 20,000 metric tons of perlite had been produced in the Philippines. It has many uses such as concrete aggregate, fillers, filter aid, High temperature insulation, horticultural aggregate, formed products, low temperature insulation, plaster aggregate, and others. This shows the potential of perlite as possible lightweight aggregate in the Philippines. Utilizing this material for structural purposes would be a great advantage, making it a great choice as lightweight aggregate [4], [7].

Lightweight Concrete

The aggregates that are used for lightweight concrete are usually weaker than the normal weight aggregates. To compensate for this higher cement content are used, while water cement ratio is kept minimum by introducing water reducing admixtures. However, using stronger mortar alone is not sufficient unless you can consider other properties of the lightweight aggregate. Some of the advantages are the bonding to the cement matrix and thermal expansion coefficients to the concrete matrix, which leads to consistent concrete with little micro cracking [8][9]. The advantages of the compatibility between the cement matrix and aggregate are the reduction of permeability which leads to better durability and lessen the tendency to cracking due to shrinkage probably because of lack of micro cracks acting as stress raisers [10].

Compressive Strength of concrete varies depending on its application, for residential buildings, it can range from 2500 psi (17 MPa) up to 4000 psi (28 MPa). Compressive Strengths exceeding 10,000 psi (70 MPa) are applied to special or specified applications. The strength of the concrete is proportional to the days after its casting, thus making it stronger as days go by, up until it reaches its peak strength [11].

One of the problems the manufacturers of lightweight aggregates experience is if the lightweight aggregate is from the natural source their applications are very limited. While if the lightweight aggregate is manufactured, one of the problems is the limitations of raw material, method of processing, requirements of the market and the sustainability of the aggregates. Because of these dilemma lightweight aggregates manufacturers have suffered failures than other aggregate industry. However, most of the customers have not yet appreciated the advantage of lightweight concrete especially in reinforced concrete construction which results in lightweight concrete not being optimized in the construction industry. Some of the causes of it are the inadequate codes, very conservative and traditional construction industry. To remove the uncertainty of the industry, it requires demonstrated proof of stability and long-term durability of the lightweight concrete [12].

In a study on the effect of expanded perlite aggregate (EPA) on the attributes of lightweight concrete such as thermal conductivity, compressive strength, water absorption, modulus elasticity and capillarity coefficient. In the experimental investigation, concrete mixes were prepared by partially replacing aggregate by expanded perlite and it produces a lightweight concrete that has a weight between 700 kg/m^3 and 2000 kg/m^3. Test result shows that increasing perlite content means reduction of compressive strength and modulus of elasticity. However, a perlite content increases, water absorption and sorptivity also increases. The test results also show that thermal conductivity significantly improves using expanded perlite aggregates [8], [11].

Materials and Methods

The methodology of this research study involves different phases in developing a structural lightweight concrete using perlite aggregates which can be used for precast concrete and structural concrete. The whole cycle of this methodology includes four phases, and the first phase is the gathering information regarding lightweight concrete and perlite aggregates (see Figure **1**). In this study, the perlite aggregates used are made from crushing perlite ores conforming size distribution of ASTM C330 for combined coarse and fine aggregates (9.5mm to 0). The second phase focused on preparing concrete samples with varying perlite replacement ratios (0%, 10%, 20%, 30%, 40%, 50%, and 60%) for coarse aggregates, designed according to volumetric mix parameters and using Type I Portland cement, w/c ratio of 0.381, and polycarboxylate-based superplasticizer [5]. Materials and procedures complied with ASTM C330 requirements for lightweight aggregates for structural concrete, while concrete production followed proper batching, mixing, and curing protocols. The third phase consisted of testing the fresh and hardened properties, including slump flow (ASTM C1611), density, compressive strength (ASTM C39), and split tensile strength (ASTM C496). For the fourth phase, the results were analyzed, and a design mix is formulated and tested again.

Figure 1. Perlite aggregates

In this study, the Vp/Vs (ratio of volume of cement to volume of fine aggregate) is 0.7. The Vw/Vp that is used in this research is 1.2. The cement is Portland cement type 1, and for the superplasticizer, the admixture that was used is a polycarboxylate based superplasticizer. The recommended dosage from the supplier is 0.5% to 1.2%. The dosage that would be used in this study is 0.75%. While for Vm/Vg (ratio of volume of mortar and volume of coarse aggregates), it would be 2.05 [5]. Table 1 contains the estimated mix design parameters of different concrete mixes. For each mix design parameters, three cylinder specimens (100mm x 200mm) were tested for compressive strength at every curing age of 7, 14 and 28 days. While three cylinder specimens (100mm x 200mm) each mix design parameters were also tested for Split Tensile strength and density at curing age of 28 days.

Table 1. Mix Design Parameters

Concrete Mix	Mix design parameter			Vp(m³)	Vs(m³)	Vw(m³)	W/C (in mass)	Vg(m³)	PA(m³)
	Vp/Vs	Vw/Vp	Vm/Vg						
0%	0.7	1.2	2.05	0.185	0.265	0.222	0.381	0.328	0
10%	0.7	1.2	2.05	0.185	0.265	0.222	0.381	0.295	0.033
20%	0.7	1.2	2.05	0.185	0.265	0.222	0.381	0.262	0.066
30%	0.7	1.2	2.05	0.185	0.265	0.222	0.381	0.230	0.098
40%	0.7	1.2	2.05	0.185	0.265	0.222	0.381	0.197	0.131
50%	0.7	1.2	2.05	0.185	0.265	0.222	0.381	0.164	0.164
60%	0.7	1.2	2.05	0.185	0.265	0.222	0.381	0.131	0.197

Results and Discussion

Compressive Strength

Figure 2 shows the compressive strength of the different mix design (0% to 60%) in 7, 14 and 28 days curing. Based on the compressive strength test performed, it has shown that as partial replacement increases compressive strength also decreases. Also based on the concrete cylinder samples that were tested, it has shown that concrete is segregated and was shown evidently by the cracked samples after compressive strength as seen in Figure 3. Most perlite aggregates were at the top of the cylinder which is due to its lightweight characteristics that it causes itself to float during concrete placing. The segregation could be the cause why the compressive strength test result is not uniformly distributed [13]. But it clearly shows that by using 50% and 60% partial replacement of perlite aggregates to coarse aggregates, it has decreased its compressive strength [14], [15], [16].

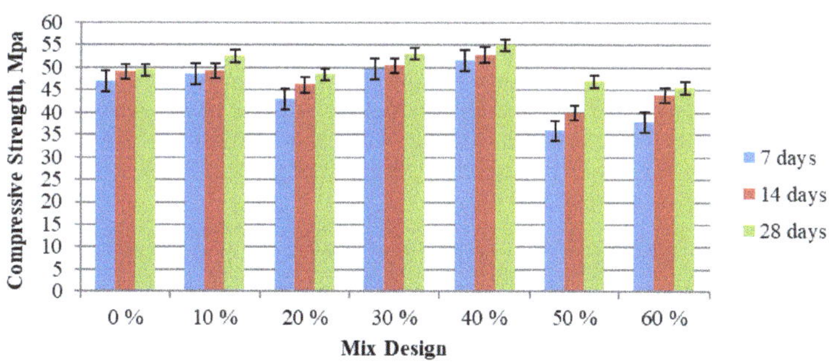

Figure 2. Compressive Strength Test Results

Table 2. Statistical Analysis of 28-day Compressive Strength Results

Concrete Mix	Compressive Strength, f'c (Mpa)			Mean f'c	Standard Deviation	COV (%)
	Sample #1	Sample #2	Sample #3			
0%	50.7	48.4	49.1	49.4	1.18	2.39%
10%	53.9	50.5	53.1	52.5	1.78	3.39%
20%	47.8	49.3	48.3	48.5	0.78	1.61%
30%	49.8	54.5	54.8	53.0	2.81	5.29%
40%	56.8	54.8	53.5	55.0	1.66	3.02%
50%	56.7	53.7	35.3	48.6	11.57	23.81%
60%	44.7	44.9	47.0	45.5	1.29	2.83%

Figure 3. Cracked concrete sample from Compressive test

Split Tensile Strength

Figure 4 shows the split tensile strength of the different mix design (0% to 60%) at 28 days curing. Based on the split tensile strength test at 28 days that was conducted, it shows that the results are not uniformly distributed. One of the causes could be the segregation of the concrete mixture [13], [17]. But based on the results of 50% and 60% partial replacement of perlite aggregates to coarse aggregates, it shows that the split tensile strength reduces from most of the design mixes. With this, it is safe to say that as partial replacement of perlite aggregates to coarse aggregates increases, the split tensile strength decreases[18], [19].

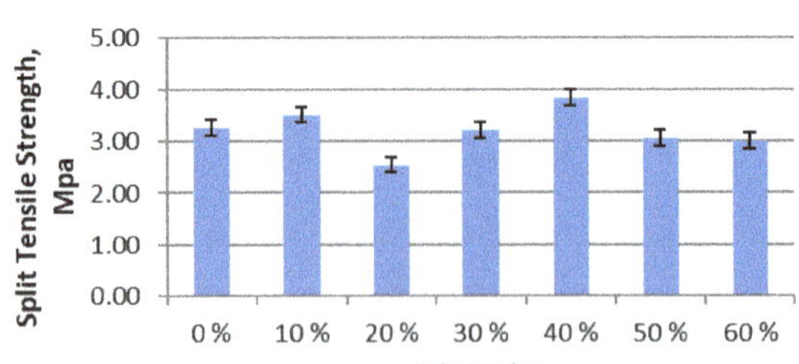

Figure 4 Split Tensile strength test results

Table 3 Statistical Analysis of 28-day Split Tensile Strength

Concrete Mix	Split Tensile Strength (Mpa)			Mean	Standard Deviation	COV (%)
	Sample #1	Sample #2	Sample #3			
0%	3.62	3.71	2.44	3.26	0.71	21.77%
10%	3.97	3.88	2.67	3.51	0.72	20.60%
20%	2.39	2.64	2.60	2.54	0.13	5.16%
30%	3.12	2.69	3.82	3.21	0.57	17.70%
40%	3.48	4.63	3.40	3.84	0.69	17.93%
50%	3.31	3.02	2.79	3.04	0.26	8.61%
60%	3.20	3.00	2.77	2.99	0.22	7.23%

Density

Figure 5 shows the density of different mix design (0% to 60%). Based on the results, it shows that as partial replacement of perlite aggregates to coarse aggregates increases, the density decreases. This is mainly because you replace the higher density aggregates with lightweight aggregates; it would decrease the density [20], [21], [22].

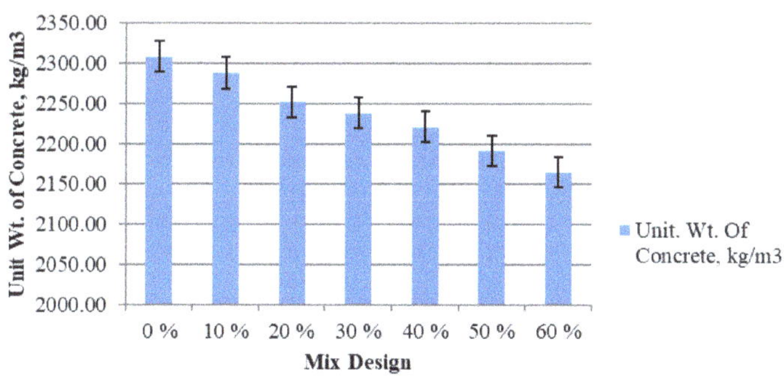

Figure 5. Density Test Results

Table 4 Statistical Analysis of Density Test Results

Concrete Mix	Density (kg/m^3)			Mean	Standard Deviation	COV (%)
	Sample #1	Sample #2	Sample #3			
0%	2317.4	2305.1	2303.2	2308.57	7.69	0.33%
10%	2288.5	2281.1	2294.6	2288.07	6.77	0.30%
20%	2249.2	2275.5	2232.2	2252.32	21.81	0.97%
30%	2246.8	2227.0	2241.4	2238.41	10.28	0.46%
40%	2201.6	2239.4	2223.8	2221.63	18.99	0.85%
50%	2185.8	2182.7	2207.3	2191.95	13.40	0.61%
60%	2164.3	2178.4	2152.6	2165.10	12.93	0.60%

Analysis of Mechanical properties

Based on the test results that have been conducted, the mechanical properties of lightweight concrete using perlite aggregates can now be analyze. The analysis would be also compared to standard specification of lightweight concrete. ASTM C330 specifies a structural lightweight concrete that has a combination of lightweight aggregates and normal weight aggregates with 4000 psi compressive strength at 28 days and must have a density of maximum 1840 kg/m^3.This is the target of the study. Analyzing the results that have been conducted, these were the findings.

1. As the replacement ratio of perlite aggregate to coarse aggregates increases, the compressive strength, split tensile and density decreases
2. The segregation of the concrete mix is also greater when increasing the replacement ratio
3. It shows that the replacement ratio of perlite aggregates can go up to 100% given the right mix design
4. With the use of superplasticizer, it helps the mix design attain high compressive strength[23].
5. The 0.75% dosage of superplasticizer produces high slump flow that could one of the causes of segregation
6. It is observed that using dry perlite aggregates float on the cylinder samples causing segregation [24]
7. 60% partial replacement of perlite aggregates to coarse aggregates does not pass the requirements of ASTM C330 due to higher density of 2165.10 Kg/m^3 compared to the required 1840 kg/m^3

Mechanism of Segregation and Mix Stabilization

One of the important issues need to be addressed is the segregation of lightweight aggregate. Which is due to the extremely low density and high porosity of the perlite aggregate[25]. Moreover, dry perlite has high water absorption capacity which leads to localized diminution on water-cement ratio effectivity as it quickly absorbs mixing water[26]. This affects the workability and the distribution of perlite aggregate resulting to floating.

Introducing pre-wetting of perlite aggregates minimizes these circumstances by saturating the interior pores before mixing[27]. It prevents the rapid absorption of mix water and maintains cement paste viscosity. Incorporating also with viscosity-modifying admixture (VMA) further improves the mixture stability by increasing the plastic viscosity and cohesion of the cement paste[28]. VMA composed of water-soluble polymers that dissolves in water and form an interconnected 3-dimensional network within the cement paste. This improves viscosity and cohesion of the concrete mixture. The network structure stabilizes the suspension of aggregates within the paste and limits the movement of particles that results to prevention of segregation.

Adjusted Mix Design

Based on the results of mechanical evaluation, it has been found that no mix design satisfies the requirements of ASTM specification. The 60% partial replacement mix design has only resulted to 2182.36 Kg/m3 density, which in terms of applicability with the construction industry it has no beneficial effect because it only lowers the density by 9%. To attain a conforming mix design, the researcher recommended various changes in the mix design.

1. Add 0.10% Viscosity modifying agent to mix design to minimize or avoid segregation [6], [29], [30]
2. Wet first the perlite aggregates to reduce floating of aggregates [24], [27]
3. Lower the dosage of superplasticizer to decrease the slump flow of the mix
4. Increase the replacement ratio of perlite aggregates to coarse aggregates by 100% to attain lesser density of concrete

Results of Adjusted Mix Design

The adjusted mix design using 100% perlite aggregates, 0.45% superplasticizer, w/c ratio of 0.381 and 0.10% viscosity-modifying agent (VMA), with pre-wetted perlite, successfully met ASTM C330 requirements. The mix achieved a 28-day compressive strength of 33.7 MPa and a density of 1790 kg/m^3, both within structural lightweight concrete limits of ASTM C330 (28 MPa and

1840 kg/m3). Perlite was uniformly distributed with no visible segregation as shown in Figure 6. While the split tensile strength ranges from 1.557 MPa to 2.549 MPa. Although no durability testing was conducted, the high compressive strength and the prevention of aggregate segregation promote long term durability of the concrete mix[31]. High compressive strength and uniform distribution of aggregates reduce permeability and porosity of concrete which increases the durability of concrete[32].

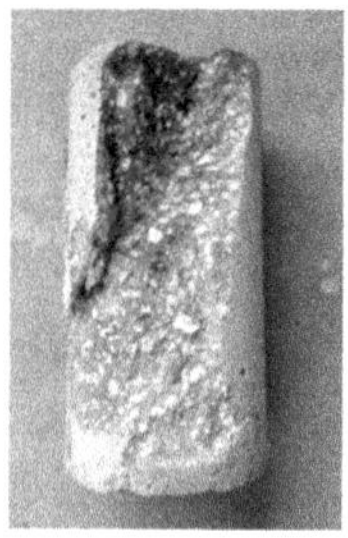

Figure 6. Cracked concrete sample of Adjusted mix design

Table 5 Statistical Analysis of Compressive Strength Results for Adjusted Design Mix

Concrete Mix	Curing age (days)	Compressive Strength, f'c (Mpa)			Mean f'c	Standard Deviation	COV (%)
		Sample #1	Sample #2	Sample #3			
100%	7	30.2	27.3	29.0	28.8	1.43	4.95%
100%	14	30.9	28.5	30.1	29.8	1.22	4.10%
100%	28	33.9	32.1	35.0	33.7	1.49	4.44%

Table 6 Statistical Analysis of Density Test Results of Adjusted Design Mix

Concrete Mix	Density (kg/m^3)			Mean	Standard Deviation	COV (%)
	Sample #1	Sample #2	Sample #3			
100%	1785.5	1784.9	1797.2	1789.22	6.93	0.39%

Table 7 Statistical Analysis of Split Tensile Strength of Adjusted Design Mix

Concrete Mix	Split Tensile Strength (Mpa)			Mean	Standard Deviation	COV (%)
	Sample #1	Sample #2	Sample #3			
100%	2.55	1.56	1.66	1.92	0.55	28.41%

Conclusion

The main objective of this study is to develop structural lightweight concrete which can be used in precast industry and small to medium rise buildings. Experimental research has been done to study the mechanical properties of lightweight concrete using perlite aggregates. The following conclusion has been made from the investigation results of this study.

1. As partial replacement of perlite aggregates to the normal coarse aggregates increases, the compressive strength, split tensile strength and density of concrete decreases.
2. The Perlite aggregates used in this study can be classified as combined coarse and fine aggregates (9.5mm to 0) based on grading requirements of ASTM C330.

3. The segregation occurs when using perlite aggregates because of its lightweight property but using Viscosity modifying agent helps prevent it.
4. Wetting the perlite lightweight aggregates first before concrete mixing helps prevent segregation
5. The percentage of replacement of perlite aggregates to normal coarse aggregates can go up to 100% with the right mix design. With the use of the right dosage of superplasticizer and Viscosity Modifying agent, it helps the concrete attain required compressive strength and workable concrete. Based on the testing results, it shows 32.1 MPa compressive strength at 28 days, 1790 kg/m3 density and 1.557 MPa to 2.549 MPa split tensile strength. The mix achieved a 28-day compressive strength of 33.7 MPa and a density of 1790 kg/m^3 satisfied the structural lightweight concrete requirements of ASTM C330 (28 MPa and 1840 kg/m3).

Acknowledgement
I would like to acknowledge and give my utmost thanks to MDC Conqrete Inc. for letting me use their facilities for this research.

References

[1] Y. Agrawal, T. Gupta, R. Sharma, N. L. Panwar, and S. Siddique, "A Comprehensive Review on the Performance of Structural Lightweight Aggregate Concrete for Sustainable Construction," Jun. 2021, *Multidisciplinary Digital Publishing Institute (MDPI).* https://doi.org/10.3390/constrmater1010003.

[2] R. D. Rosario, A. D. La Cruz, and M. P. De Guzman, "A Review of Biomineralization as Solution for Roads and Infrastructures Concrete Sustainability," Aug. 2024, *Salehan Institute of Higher Education.* https://doi.org/10.28991/CEJ-2024-010-08-020.

[3] J. N. Bhaidas *et al.*, "Advancements in Lightweight Concrete: A Comprehensive Review," https://www.ijraset.com/best-journal/advancements-in-lightweight-concrete-a-comprehensive-review.

[4] M. Ibrahim, A. Ahmad, M. S. Barry, L. M. Alhems, and A. C. Mohamed Suhoothi, "Durability of Structural Lightweight Concrete Containing Expanded Perlite Aggregate," *Int. J. Concr. Struct. Mater.*, vol. 14, Dec. 2020. https://doi.org/10.1186/s40069-020-00425-w.

[5] M. Nepomuceno, L. Oliveira, and S. M. R. Lopes, "Methodology for mix design of the mortar phase of self-compacting concrete using different mineral additions in binary blends of powders," *Constr. Build. Mater.*, vol. 26, pp. 317–326, Jan. 2012. https://doi.org/10.1016/j.conbuildmat.2011.06.027.

[6] G. W. Leong, K. H. Mo, Z. P. Loh, and Z. Ibrahim, "Mechanical properties and drying shrinkage of lightweight cementitious composite incorporating perlite microspheres and polypropylene fibers," *Constr. Build. Mater.*, vol. 246, Jun. 2020. https://doi.org/10.1016/j.conbuildmat.2020.118410.

[7] M. C. Stratoura, G. E. D. Lazari, E. G. Badogiannis, and V. G. Papadakis, "Perlite and Rice Husk Ash Re-Use As Fine Aggregates in Lightweight Aggregate Structural Concrete—Durability Assessment," *Sustainability (Switzerland)*, vol. 15, Mar. 2023. https://doi.org/10.3390/su15054217.

[8] O. Sengul, S. Azizi, F. Karaosmanoglu, and M. A. Tasdemir, "Effect of expanded perlite on the mechanical properties and thermal conductivity of lightweight concrete," *Energy Build.*, vol. 43, pp. 671–676, Feb. 2011. https://doi.org/10.1016/j.enbuild.2010.11.008.

[9] D. Roberto Rosario and M. J. Viado, "Encapsulating immobilized ureolytic bacteria yields self-healing concrete apropos sustainable transportation materials: a review," in *E3S Web of Conferences*, EDP Sciences, Feb. 2024. https://doi.org/10.1051/e3sconf/202448803019.

[10] O. Benjeddou, G. Ravindran, and M. A. Abdelzaher, "Thermal and Acoustic Features of Lightweight Concrete Based on Marble Wastes and Expanded Perlite Aggregate," *Buildings*, vol. 13, Apr. 2023. https://doi.org/10.3390/buildings13040992.

[11] K. H. Yang, H. Y. Kim, and H. J. Lee, "Mechanical Properties of Lightweight Aggregate Concrete Reinforced with Various Steel Fibers," *Int. J. Concr. Struct. Mater.*, vol. 16, Dec. 2022. https://doi.org/10.1186/s40069-022-00538-4.

[12] M. Dwarampudi and B. Venkateshwari, "Performance of light weight concrete with different aggregates—a comprehensive review," *Discover Civil Engineering*, vol. 1, Aug. 2024. https://doi.org/10.1007/s44290-024-00015-9.

[13] A. J. Tenza-Abril, D. Benavente, C. Pla, F. Baeza-Brotons, J. Valdes-Abellan, and A. M. Solak, "Statistical and experimental study for determining the influence of the segregation phenomenon on physical and mechanical properties of lightweight concrete," *Constr. Build. Mater.*, vol. 238, Mar. 2020. https://doi.org/10.1016/j.conbuildmat.2019.117642.

[14] Y. A. Dolatabad, R. Kamgar, and M. A. Jamali Tazangi, "Effects of perlite, leca, and scoria as lightweight aggregates on properties of fresh and hard self-Compacting concretes," *Journal of Advanced Concrete Technology*, vol. 18, pp. 633–647, Oct. 2020. https://doi.org/10.3151/JACT.18.633.

[15] O. Szlachetka, J. Dzięcioł, and M. Dohojda, "Mechanical Properties of Perlite Concrete in Context to Its Use in Buildings' External Walls," *Materials*, vol. 17, Dec. 2024. https://doi.org/10.3390/ma17235790.

[16] A. El Mir, S. G. Nehme, and J. J. Assaad, "Durability of self-consolidating concrete containing natural waste perlite powders.," *Heliyon*, vol. 6, p. e03165, Jan. 2020. https://doi.org/10.1016/j.heliyon.2020.e03165.

[17] A. M. Solak, A. J. Tenza-Abril, and V. E. García-Vera, "Influence of the segregation phenomenon on structural efficiency of lightweight aggregate concretes," *Materials*, vol. 13, pp. 1–15, Dec. 2020. https://doi.org/10.3390/ma13245754.

[18] R. U. D. Nassar, S. S. Bhatti, M. Shahin, M. Hossain, and H. A. Al Slibi, "Lightweight aggregate concrete produced with crushed stone sand as fine aggregate," *Cogent Eng.*, vol. 7, Jan. 2020. https://doi.org/10.1080/23311916.2020.1792219.

[19] F. K. Alqahtani, "A Sustainable Alternative for Green Structural Lightweight Concrete: Performance Evaluation.," *Materials (Basel).*, vol. 15, Dec. 2022. https://doi.org/10.3390/ma15238621.

[20] O. Szlachetka, J. Dzięcioł, and M. Dohojda, "Mechanical Properties of Perlite Concrete in Context to Its Use in Buildings' External Walls," *Materials*, vol. 17, Dec. 2024. https://doi.org/10.3390/ma17235790.

[21] U. Mohammed Umar and K. Muthusamy, "Potential of Waste Material as Coarse Aggregates for Lightweight Concrete Production: A Sustainable Approach," *CONSTRUCTION*, vol. 3, pp. 87–114, Jun. 2023. https://doi.org/10.15282/construction.v3i1.9217.

[22] M. Khoshvatan and M. Pouraminia, "The Effects of Additives to Lightweight Aggregate on the Mechanical Properties of Structural Lightweight Aggregate Concrete," *Civil and*

Environmental Engineering Reports, vol. 31, pp. 139–160, Mar. 2021. https://doi.org/10.2478/ceer-2021-0010.

[23] L. Dvorkin, V. Zhitkovsky, R. Makarenko, and Y. Ribakov, "The Influence of Polymer Superplasticizers on Properties of High-Strength Concrete Based on Low-Clinker Slag Portland Cement.," *Materials (Basel).*, vol. 16, Mar. 2023. https://doi.org/10.3390/ma16052075.

[24] L. Domagała and E. Bryła, "The Properties of Lightweight Aggregates Pre-Coated with Cement Pastes and Their Suitability for Concrete.," *Materials (Basel).*, vol. 14, Oct. 2021. https://doi.org/10.3390/ma14216417.

[25] F. S. Barbosa, A. L. Beaucour, M. C. R. Farage, and S. Ortola, "Image processing applied to the analysis of segregation in lightweight aggregate concretes," *Constr. Build. Mater.*, vol. 25, pp. 3375–3381, Aug. 2011. https://doi.org/10.1016/j.conbuildmat.2011.03.028.

[26] S. Grzeszczyk and G. Janus, "Lightweight reactive powder concrete containing expanded perlite," *Materials*, vol. 14, Jun. 2021. https://doi.org/10.3390/ma14123341.

[27] P. Chen, W. Tan, J. X. Lu, Y. Sun, and C. S. Poon, "Enhancing the performance of lightweight high-performance concrete through an alternative pre-wetting method for lightweight aggregates using simulated cement pore solution," *Constr. Build. Mater.*, vol. 409, Dec. 2023. https://doi.org/10.1016/j.conbuildmat.2023.134024.

[28] C. Li, L. Miao, Q. You, S. Hu, and H. Fang, "Effects of viscosity modifying admixture (VMA) on workability and compressive strength of structural EPS concrete," *Constr. Build. Mater.*, vol. 175, pp. 342–350, Jun. 2018. https://doi.org/10.1016/j.conbuildmat.2018.04.176.

[29] C. Li, L. Miao, Q. You, S. Hu, and H. Fang, "Effects of viscosity modifying admixture (VMA) on workability and compressive strength of structural EPS concrete," *Constr. Build. Mater.*, vol. 175, pp. 342–350, Jun. 2018. https://doi.org/10.1016/j.conbuildmat.2018.04.176.

[30] Y. Boutouam, M. Hayek, K. Bouarab, and A. Yahia, "A Comprehensive Review of Plant-Based Biopolymers as Viscosity-Modifying Admixtures in Cement-Based Materials," May 2024, *Multidisciplinary Digital Publishing Institute (MDPI).* https://doi.org/10.3390/app14104307.

[31] D. K. Panesar and B. Shindman, "The effect of segregation on transport and durability properties of self consolidating concrete," *Cem. Concr. Res.*, vol. 42, pp. 252–264, Feb. 2012. https://doi.org/10.1016/j.cemconres.2011.09.011.

[32] X. Bai, H. Zhou, X. Bian, X. Chen, and C. Ren, "Compressive Strength, Permeability, and Abrasion Resistance of Pervious Concrete Incorporating Recycled Aggregate," *Sustainability (Switzerland)* , vol. 16, May 2024. https://doi.org/10.3390/su16104063.

Customization of 3D Printed Silicone Foams

Mae Hwa TAI[1], Jia Huey SIM[1,4,5*], Jing Yuen TEY[2,5], Law Yong NG[1,4], Hui San THIAM[1,4], Danny Wee Kiat NG[3,6], Wei Hong YEO[2,5], Zhi Hua LEE[1,4]

[1]Department of Chemical Engineering, Lee Kong Chian Faculty of Engineering & Science, Universiti Tunku Abdul Rahman, 43000 Kajang, Selangor, Malaysia

[2]Department of Mechanical and Material Engineering, Lee Kong Chian Faculty of Engineering & Science, Universiti Tunku Abdul Rahman, 43000 Kajang, Selangor, Malaysia

[3]Department of Mechatronics and Bio Medical Engineering, Lee Kong Chian Faculty of Engineering & Science, Universiti Tunku Abdul Rahman, 43000 Kajang, Selangor, Malaysia

[4]Centre for Advanced and Sustainable Materials Research (CASMR), Universiti Tunku Abdul Rahman, 43000 Kajang, Selangor, Malaysia

[5]Centre for Sustainable Mobility Technologies (CSMT), Universiti Tunku Abdul Rahman, 43000 Kajang, Selangor, Malaysia

[6]Centre for Healthcare Science and Technology (CHST), Universiti Tunku Abdul Rahman, 43000 Kajang, Selangor, Malaysia

[*]simjh@utar.edu.my

Keywords: Silicone, 3D Printing, Additive Manufacturing, PDMS, Porous

Abstract. The rheological properties of the PDMS inks significantly impact the creation of foam using direct ink writing (DIW). This study investigates the influence of glucose and nano-silica on the rheological properties of ink and the characteristics of its pores. Nano-silica significantly enhances yield stress of the ink due to its substantial and immediate network formation with PDMS chains, while glucose primarily acts as a passive filler. A minimum of 5.00phr of nano-silica is essential to maintain silicone ink in viscoelasticity condition, ensure a consistent ink extrusion rate and shape fidelity during direct ink writing (DIW). Additionally, glucose exerted pronounced effect on the porosity of PDMS foam's, with 42% porosity achieved with 10phr glucose increment. Consequently, the PDMS foam's porosity can be tailored to the specific application by varying the glucose dosage. The deduction of printability and porosity dependencies on the nano silica and glucose respectively allows the printing of tunable silicone foams for advanced process such as soft robotics, wearable sensors via IoT, drug delivery and etc.

Introduction

Polydimethylsiloxane (PDMS) is one of the most extensively studied silicone-based polymer known for its exceptional physicochemical properties, making it a very promising material for diverse industrial applications [1]. The porosity in PDMS has been reported to improve the surface area-to-volume ratio and mechanical flexibility [2]. Therefore, these porous PDMS has been studied for a variety of separation performances, notably the oil–water filtration in connection to increasing problems associated with water contamination [3-5]. In addition to separation applications, porous PDMS has also been widely utilized in tissue engineering, where scaffold structures require key properties such as biocompatibility, mechanical resilience, and porosity [6, 7]. The porous nature of these scaffolds plays a crucial role in promoting cell attachment and proliferation, facilitating oxygen diffusion, and supporting efficient waste removal [8]. Furthermore, the use of porous PDMS has expanded into the field of electronics, particularly in wearable pressure sensors [9]. These sensors operate by detecting variations in electrical resistance or capacitance when subjected to mechanical deformation [9, 10]. The incorporation of porosity into PDMS enhances flexibility, reduces its elastic modulus, and improves its ability to respond more precisely to electrical property changes [11].

Porous polydimethylsiloxane (PDMS) structures can be synthesized via multiple fabrication techniques. Gas foaming and templating methods are being widely employed due to their ability to tailor porosity and mechanical properties [12]. The gas foaming approach relies on the incorporation of chemical blowing agents that decompose to release gaseous by-products, or the direct injection of inert gases into the PDMS matrix under controlled conditions [13]. In contrast, templating techniques are extensively utilized due to their cost efficiency, operational simplicity, and minimal requirement for specialized equipment [12]. This method involves the dispersion of sacrificial templates such as solid particulates or structured networks within the uncured PDMS matrix, which are then subsequently extracted post-vulcanization through dissolution or thermal decomposition and lastly leaving behind a controlled porous morphology [14]. Though sacrificial templating offers enhanced structural control, its integration into additive manufacturing processes introduces inherent challenges. Extrusion-based printing of particle-loaded PDMS suspensions is limited by nozzle geometry, whereby the size distribution of particles must be precisely controlled to avoid clogging. Consequently, optimizing the formulation, rheological properties, and printing parameters is critical to ensure uniform material deposition while achieving high-resolution porous architectures for advanced biomedical, microfluidic, and wearable electronic applications [15]. The successful deployment of direct ink writing (DIW) for silicone-based materials requires inks with well-defined rheological properties. More precisely, these inks must exhibit viscoelastic behavior with suitable yield stress to maintain structural integrity during layer-by-layer deposition while simultaneously demonstrating shear-thinning characteristics to ensure smooth and continuous extrusion [16]. Incorporating porogens into DIW formulations enables the introduction of porosity within extruded filaments, facilitating the fabrication of tailored porous structures [17].

Compared to conventional casting and moulding approaches, this study demonstrated direct ink writing (DIW) of 3D porous PDMS foam with tuneable pores properties, structural stability and fidelity where the ink formulation integrated with glucose templating and mediated with nano-silica as the rheological control for printability [18, 19]. A stable glucose/PDMS composite ink was formulated with glucose particles well dispersed in the silicone matrix. Following vulcanization, selective glucose removal via ethanol immersion yielded a porous network. By systematically varying glucose and nano-silica concentrations, the study investigates the effects of templates loading on ink's rheological properties and pore properties. With the aids of comprehensive rheological and morphological characterization studies, feasible ranges of nano-silica and glucose dosage for successful printing of silicone foams with tunable pores properties can be established.

Methodology

Material

In this study, α,ω-silanol-terminated polydimethylsiloxane (PDMS) with a viscosity of 20,000 cPs was selected as the primary polymer. Silicone oil and nano-silica were added to impart shear-thinning properties of the silicone ink. The crosslinking process was facilitated using methyl tris(methyl ethyl ketoxime) silane (MOS) and vinyl tris (methylethylketoxime) silane (VOS). All reagents and chemicals were utilized as received from their respective suppliers, SiSiB® Silicones. Glucose, sourced from 99 Speed Mart Sdn. Bhd, was employed as sacrificial material for the formation of porous PDMS structure. Additionally, absolute ethanol from Merck was used as a solvent to extract porogens.

PDMS Foam Ink Formulation

The preparation of PDMS-based ink involved mixing of several components. PDMS, a silicone polymer, was first mixed with silicone oil to promote uniform dispersion. Two types of crosslinkers, VOS and MOS and catalyst of DBTDL were added to facilitate the curing process. Lastly, the glucose as porogens for foam formation and as reinforced agent was introduced to the ink. The glucose particles were pre-screen and controlled at 75 µm to ensure a more uniform and controlled pore structure formed. Homogeneous dispersion of glucose and other components were achieved using an overhead stirrer (IKA RW20 digital). Two sets of experiments were designed and conducted systematically by varying the concentration of nano-silica and glucose respectively, with details shown in Table 1 and 2.

Table 1. Ink Formulation for Printable Silicone Foam With Varying Nano-silica Loading.

Ink Designation	PDMS (phr)	VOS (phr)	MOS (phr)	Nano-silica (phr)	Glucose (phr)
S40-NS8.33	100	10.8	1.2	8.33	40
S40-NS6.67	100	10.8	1.2	6.67	40
S40-NS5.00	100	10.8	1.2	5.00	40
S40-NS3.33	100	10.8	1.2	3.33	40

Table 2. Ink Formulation for Printable Silicone Foam With Varying Glucose Loading.

Ink Designation	PDMS (phr)	VOS (phr)	MOS (phr)	Nano-silica (phr)	Glucose (phr)
S40-NS5.00	100	10.8	1.2	5	40
S50-NS5.00	100	10.8	1.2	5	50
S60-NS5.00	100	10.8	1.2	5	60
S70-NS5.00	100	10.8	1.2	5	70
S80-NS5.00	100	10.8	1.2	5	80

Direct Ink Printing of PDMS Foam

Direct ink writing (DIW) of the samples was carried out using a lead screw-driven extrusion printer. Printing was performed at a constant speed of 7 $mm \cdot s^{-1}$, with a nozzle diameter of 0.84 mm and a layer height of 0.4 mm. To ensure that the resulting porosity was solely attributable to glucose leaching during solvent extraction, all samples were printed with 100% infill density. The final dimensions of each printed sample were 2.5 mm (L) × 2.5 mm (W) × 0.5 mm (H).

PDMS Foam Formation via Solvent Extraction

The printed foam samples were cured for 48 hours under controlled conditions, and surface tack-free tests were subsequently carried out to verify the completeness of curing Porous PDMS samples were then formed through solvent extraction, wherein the embedded glucose particles within the printed structures were removed via ethanol solvent. All samples were first allowed to cure under ambient conditions. The samples were then immersed in 50 wt% ethanol solution under continuous stirring for 48 hours, followed by oven drying at 60 °C. The washing and drying cycles were repeated until a constant mass was achieved, confirming the complete removal of glucose. Fig. 1 presents the samples (S70-NS5.00) before and after the washing process, highlighting the appearance changes following the removal of glucose particles.

Fig. 1. Sample of S70-NS5.00 before and after solvent extraction.

Rheology Properties Characterization

The rheological properties of the ink formulations were characterized using an Anton Paar Physica MCR 301 rheometer. A 25 mm diameter parallel plate geometry with a 1 mm working gap was employed for all measurements. Flow behavior and viscosity profiles were assessed by varying the shear rate between 0.1 s^{-1} and 100 s^{-1}. To explore the linear viscoelastic region (LVR) of the ink formulation, amplitude sweep tests were performed at an angular frequency of 10 rad·s^{-1}, with shear strain ranging from 0.01% to 100%. To further evaluate the viscoelastic properties of the silicone ink formulations, frequency sweep tests were conducted across a frequency range of 0.1 rad·s^{-1} to 100 rad·s^{-1} at a constant shear strain amplitude of 0.01%.

Pores Properties Characterization

The morphology of the porous PDMS samples was characterized using scanning electron microscopy (SEM) with a Hitachi S-3400N model. The SEM analysis was conducted at a probe current of 20 mA and an acceleration voltage of 15 kV to ensure high-resolution imaging. Prior to imaging, all samples were coated with a thin layer of palladium and gold to improve surface conductivity and image quality. The characteristic of the pores, including pore size, pore size distribution, and porosity, were quantitatively analyzed using ImageJ software, in accordance with methods employed in the studies of other researchers [20].

Results and Discussion

Rheology Properties

Successful DIW of PDMS requires inks with viscoelastic behavior—providing yield strength for shape retention and shear-thinning properties for smooth extrusion.

Flow Properties

Fig. 2a illustrates the viscosity profiles of the formulated silicone inks as a function of shear rate, highlighting the effects of nano-silica (NS) concentrations on viscosity behavior. From Fig. 2a, all the inks formulated with different NS loadings exhibited a pronounced shear-thinning behavior. The ink with the highest NS content of 8.33 phr (S40-NS8.33) demonstrated the highest initial viscosity, exceeding 4×10^6 Pa·s, which progressively decreased with increasing shear rate. This trend indicated that the incorporation of NS had enhanced the structural integrity of the ink by increasing particle-particle interactions, thereby restricting the flow. As the NS concentration decreased, the viscosity of the ink declined, suggesting a reduction in internal resistance and a more fluid-like behavior presence. The effects of varying glucose concentration (40phr-80phr) at a fixed NS loading (5phr) were analyzed as shown in Fig. 2b. The initial viscosity increased with increasing glucose concentration, Ink formulation of S80-NS5.00 exhibited the highest viscosity, followed by S70-NS5.00, S60-NS5.00, S50-NS5.00, and S40-NS5.00. This trend aligns with the variations observed in NS content, indicating that a higher glucose fraction similarly contributed to the increment in ink viscosity. In contrast, the impact of glucose particles on ink viscosity, was less pronounced compared to NS. Despite the differences in significance level, all formulations exhibited desired characteristic of shear-thinning behavior, it is an essential property for achieving controlled extrusion in direct ink writing (DIW). Notably, the emergence of a secondary viscosity plateau at intermediate shear rates suggests potential microstructural rearrangements within the ink, which may significantly influence print resolution and fidelity.

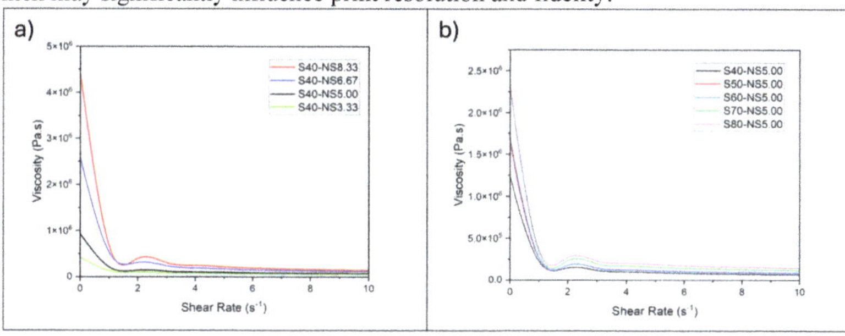

Fig. 2. Viscosity curve of the ink samples with a) varying nano-silica loading b) varying glucose loading.

Viscoelasticity

The frequency sweep test is a critical rheological assessment used to evaluate the viscoelastic behavior of ink formulations The test provides insights on the ink's properties of resisting deformation and the recovery of its original structure under the applied stress, essential for maintaining printing structure fidelity. The loss factor (tan δ), derived from the ratio of G" to G', serves as an indicator of the ink's viscoelastic balance. Fig. 3 illustrates the variation in storage modulus G' and loss modulus G" for silicone inks under different NS and glucose loadings. All inks displayed solid-like behavior, with the storage modulus (G') dominating over the loss modulus (G") (tan δ < 1) regardless of the NS and glucose loadings.

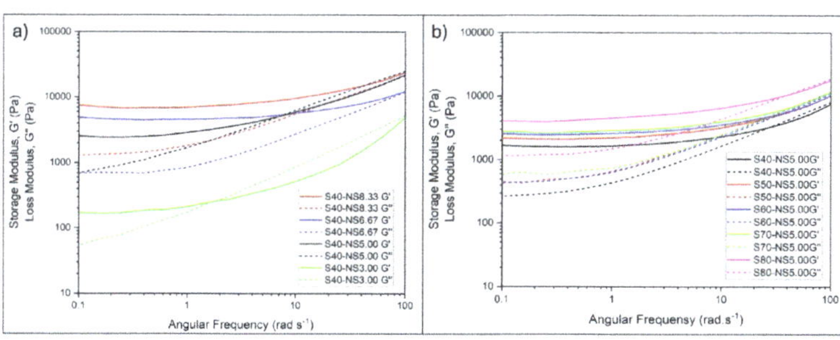

Fig. 3. Storage and loss modulus of the ink samples under a) varying nano-silica loading b) varying glucose loading.

Yield Stress

According to the *ASTM D7175* standard , the static yield stress was determined based on a 10 % drop in G′ from the plateau [21]. The shear stress at the linear viscoelastic region (LVR) limit corresponds to the yield stress of the sample. From Figure 4 (a), the static yield stress at varied NS dosages: S40-NS5.00, S40-NS6.67, S40-NS8.33 and varied glucose: S40-NS5.00, S50-NS5.00, S60-NS5.00, S70-NS5.00, S80-NS5.00 fall within 100 – 1000 Pa. In addition, those formulation displayed the static yield stress and storage modulus, G' were well above the dynamic yield stress and loss modulus, G" in the LVR which revealed the solid-like behaviour. The increased of NS and glucose particles loadings led to the increment in both G' and G" in LVE region. This indicated that ink added with both particles exhibited stronger mechanical strength than liquid phase, thus contributing to a higher storage modulus [22]. Referring to C. Perrinet et al., 90 Pa of yield stress was sufficient for 3D printing via DIW. Therefore, the ink samples with NS dosages from 5 phr to 8.33 phr, glucose dosages from 40 phr to 80 phr which achieved static yield stress of 100 Pa and above are printable [23]. The printability of the formulated ink was proven and discussed in the section of printed structure. It can be observed from Figure 4, the storage modulus (G') remained stable and forming a plateau within the LVR. However, once the shear stress exceeded this limit, a significant drop in G' was observed, indicating structural breakdown and permanent deformation. It is interesting to discover that for ink formulation S40-NS8.33 with 3.33 phr NS exhibited the dynamic yield stress and loss modulus, G" higher than the static yield stress and storage modulus, G' in the LVR and there was no crossover point between G' and G". The formulation ink with NS of 3.33 phr and below were still in the liquid-like behaviour and the dosage was insufficient to reinforce the formulation ink to be printable. This suggests that the threshold NS loading for maintaining solid-like behavior is 5.00 phr.

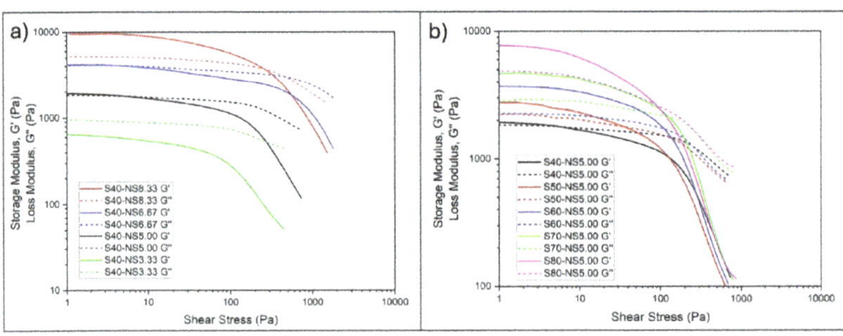

Fig. 4. Amplitude sweeps of the ink samples with a) varying nano-silica loading b) varying glucose loading.

Pores Properties

Among all the formulated PDMS inks, S40-NS8.33, S40-NS6.67, S70-NS5.00, and S80-NS5.00 demonstrated printability. To ensure consistency across formulations, all samples were printed using identical printing parameters, followed by a solvent extraction process to introduce porosity. Specifically, the printed structures were immersed in a 50 wt% ethanol solution under continuous stirring for 48 hours to selectively remove glucose particles from the bulk material. The removal of glucose effectively created an interconnected porous network within the silicone matrix, thereby influencing the material properties and permeability. This process highlights the importance of developing series of silicone ink for customization of printed silicone foams. To investigate the effects of glucose loading on porosity, SEM images of the samples had been displayed in Fig. 7. Scanning electron microscopy (SEM) images (Fig. 5) revealed distinct variations in pore morphology and size distribution, indicating that the porosity of the foams can be finely tuned by adjusting the composition of sacrificial template such as glucose. Glucose loading had severely affected the pores properties of the silicone structure. The 10 phr increment in glucose content, from 70 phr to 80 phr resulted in a nearly 15±5 % increase in porosity, from 26.5±5 %(S70-NS5.00) to 40.89±5 % (S80-NS5.00). These findings suggested that glucose acts as an effective sacrificial template for porosity modulation. The results align with previous studies which demonstrated that tunable pore structures can be achieved by varying the concentration and distribution of sacrificial components within templated silicone foams [24].

Besides, further pores characterization was carried out to evaluate the nature of the pores. Therefore, qualitative gas permeation was screened using a dead-end filtration cell setup based on standard bubble-point observations. For gas permeation, the wetted membrane was pressurized with nitrogen gas while the downstream outlet tube was submerged in water; the formation of a steady stream of bubbles was taken as evidence of gas passage through open pores. According to the IUPAC definition, an open pore is a cavity or channel that communicates with the external surface of a particle or material, thereby allowing fluid transport through the structure [25]. All printed silicone foam exhibited gas permeation behaviour, indicating that the presence of multiple pore networks interconnected across the printed silicone structures to allow gaseous flow.

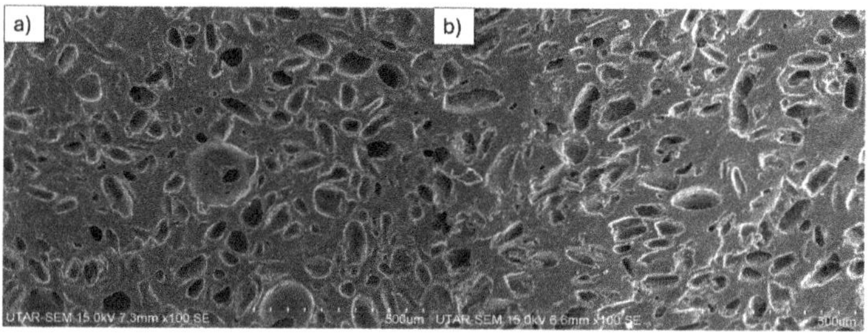

Fig. 5. SEM images of PDMS ink samples: a) S70-NS 5.00 and b) S80-NS5.00.

Mechanical Properties

The mechanical properties of the printed PDMS foams were evaluated through tensile testing to assess their structural integrity. As shown in Table 3, samples with 26.5 ± 5% porosity exhibited a Young's modulus of 0.201 MPa, ultimate tensile strength of 0.473 MPa, and elongation at break of 322.7%, while samples with 40.89 ± 5% porosity showed a Young's modulus of 0.194 MPa, ultimate tensile strength of 0.277 MPa, and elongation at break of 179.2%. The results indicate that increasing porosity leads to reduced tensile strength and ductility, consistent with typical porous elastomer behavior as reported by Q. Chen et al.[24]. These findings confirm that the printed foams maintain adequate mechanical integrity for wearable and low pressure systems while allowing tunable mechanical performance through porosity control [26].

Table 3. Mechanical properties of foams with various porosity %.

Samples	Porosity (±5%)	Young modulus (MPa)	Ultimate tensile strength (MPa)	Elongation at break (%)
S70-NS5.00	26.5	0.201	0.473	322.7
S80-NS5.00	40.89	0.194	0.277	179.2

Printed Silicone Foam Structures

The formulated inks satisfied the key rheological requirements for DIW, including sufficient yield stress, a dominant storage modulus (G′ > G″ at low shear), pronounced shear-thinning behavior, and viscosities compatible with lead-screw extrusion. This printability is demonstrated by the successful fabrication of silicone foam in hollow tube and scaffold structures. Using S70-NS5.00 as a representative of the printed silicone foam, dimensional deviation and shape fidelity were evaluated by comparing the printed structure with the original CAD design. The deviations in length, width, and height were 21.5 ± 5%, 23.5 ± 5%, and 12.5 ± 5%, respectively. Figure 6a shows the original CAD design, while Figure 6b presents the corresponding printed silicone foam.

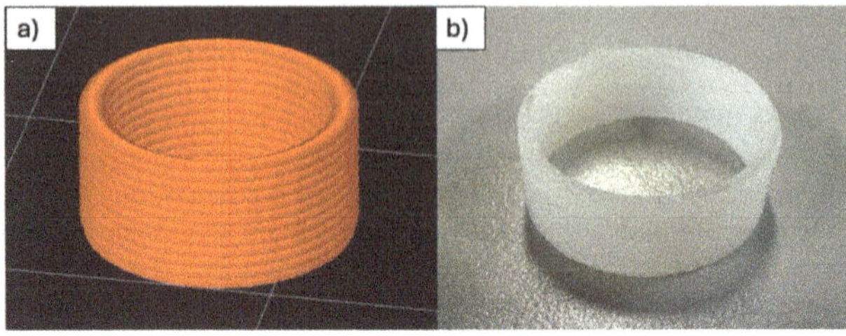

Fig. 6. Image of a) original CAD design of the hollow tube; b) printed hollow tube.

The same ink formulation was further employed for scaffold printing, where honeycomb toolpaths generated lattice-scale (interstrand) porosity and glucose porogens introduced intrastrand microporosity, resulting in hierarchically porous scaffolds (Fig. 7). Scaffold pore size and connectivity were governed by strand spacing, raster angle, layer height (mesostructural parameters), and glucose loading (microstructural control), enabling independent tuning of permeability, specific surface area and effective mechanical properties.

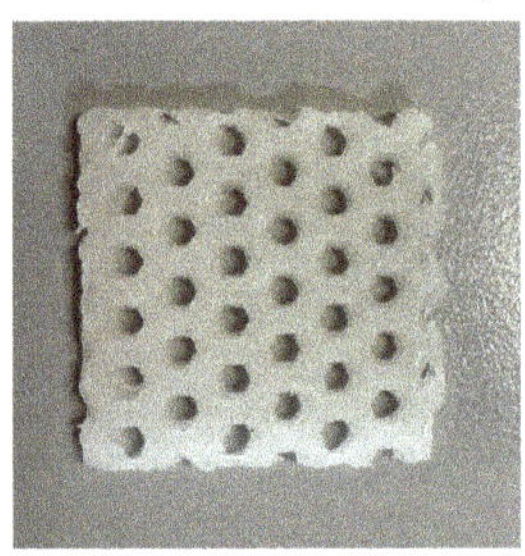

Fig. 7. Printed silicone foam in scaffold structure.

Conclusion

This study highlights the importance of optimizing nano-silica (NS) and glucose loadings to ensure stable extrusion, shape retention, and tunable microstructures of 3D printed silicone foam. Nano-silica, due to its high surface area and strong interactions with PDMS chains, significantly increased the yield stress, ensuring structural integrity during printing. A minimum threshold of 5.00 phr NS is required to maintain solid-like behavior ($G' > G''$), thereby preserving printability. In contrast, glucose primarily functions as a sacrificial filler, influencing the porosity of the silicone foam without substantially affecting yield stress. Notably, a 10phr increase in glucose dosage (from 70phr to 80phr) resulted in the achievement of 42% porosity for silicone foam. These findings provide a foundation for the development of customizable porous silicone elastomers with potential applications in biomedical scaffolds, flexible electronics, and soft robotics which were reported in several literature studies [27-29].

Acknowledgement

This research was supported by the Ministry of Higher Education (MoHE), through the Fundamental Research Grant Scheme (FRGS/1/2023/TK10/UTAR/02/1).

References

[1] S. Kramer, P. Krajnc, Hierarchically Porous Microspheres by Thiol-ene Photopolymerization of High Internal Phase Emulsions-in-Water Colloidal Systems, Journal. 13 (2021) http://10.3390/polym13193366

[2] R. Woo, G. Chen, J. Zhao, J. Bae, Structure–Mechanical Property Relationships of 3D-Printed Porous Polydimethylsiloxane, ACS Applied Polymer Materials. 3 (2021) 3496-3503. https://doi.org/10.1021/acsapm.1c00417
[3] J. Lv, Z. Gong, Z. He, J. Yang, Y. Chen, C. Tang, Y. Liu, M. Fan, W.-M. Lau, 3D printing of a mechanically durable superhydrophobic porous membrane for oil–water separation, Journal of Materials Chemistry A. 5 (2017) 12435-12444. http://dx.doi.org/10.1039/C7TA02202F
[4] K. Pandey, H. Bindra, D. Paul, R. Nayak, Smart multi-tasking PDMS Nanocomposite sponges for microbial and oil contamination removal from water, Journal of Polymer Research. 27 (2020) https://10.1007/s10965-020-02109-1
[5] A. Turco, E. Primiceri, M. Frigione, G. Maruccio, C. Malitesta, An innovative, fast and facile soft-template approach for the fabrication of porous PDMS for oil–water separation, Journal of Materials Chemistry A. 5 (2017) 23785-23793. http://dx.doi.org/10.1039/C7TA06840A
[6] A. Díaz Lantada, H. Alarcón Iniesta, B. Pareja Sánchez, J.P. García-Ruíz, Free-Form Rapid Prototyped Porous PDMS Scaffolds Incorporating Growth Factors Promote Chondrogenesis, Advances in Materials Science and Engineering. (2014) 612976. https://doi.org/10.1155/2014/612976
[7] J. Si, Z. Cui, P. Xie, L. Song, Q. Wang, Q. Liu, C. Liu, Characterization of 3D elastic porous polydimethylsiloxane (PDMS) cell scaffolds fabricated by VARTM and particle leaching, Journal of Applied Polymer Science. 133 (2016) https://10.1002/app.42909
[8] M. Razavi, R. Primavera, A. Vykunta, A.S. Thakor, Silicone-based bioscaffolds for cellular therapies, Materials Science and Engineering: C. 119 (2021) 111615. https://doi.org/10.1016/j.msec.2020.111615
[9] S. Masihi, M. Panahi, D. Maddipatla, A.J. Hanson, A.K. Bose, S. Hajian, V. Palaniappan, B.B. Narakathu, B.J. Bazuin, M.Z. Atashbar, Highly Sensitive Porous PDMS-Based Capacitive Pressure Sensors Fabricated on Fabric Platform for Wearable Applications, ACS Sensors. 6 (2021) 938-949. https://10.1021/acssensors.0c02122
[10] M. Abshirini, M.C. Saha, L. Cummings, T. Robison, Synthesis and characterization of porous polydimethylsiloxane structures with adjustable porosity and pore morphology using emulsion templating technique, Polymer Engineering & Science. 61 (2021) 1943-1955. https://doi.org/10.1002/pen.25710
[11] M. Abshirini, M. Saha, M. Altan, Y. Liu, 3D Printed Flexible Microscaled Porous Conductive Polymer Nanocomposites for Piezoresistive Sensing Applications, Advanced Materials Technologies. 7 (2022) 2101555. https://10.1002/admt.202101555
[12] T. Métivier, P. Cassagnau, New trends in cellular silicone: Innovations and applications, Journal of Cellular Plastics. 55 (2018) 151-200. https://doi.org/10.1177/0021955X18806845
[13] C. Yu, C. Yu, L. Cui, Z. Song, X. Zhao, Y. Ma, L. Jiang, Facile Preparation of the Porous PDMS Oil-Absorbent for Oil/Water Separation, Advanced Materials Interfaces. 4 (2017) 1600862. https://doi.org/10.1002/admi.201600862
[14] J. González-Rivera, R. Iglio, G. Barillaro, C. Duce, M.R. Tinè, Structural and Thermoanalytical Characterization of 3D Porous PDMS Foam Materials: The Effect of Impurities Derived from a Sugar Templating Process, Journal. 10 (2018) https://10.3390/polym10060616
[15] B.M. Rauzan, A.Z. Nelson, S.E. Lehman, R.H. Ewoldt, R.G. Nuzzo, Particle-Free Emulsions for 3D Printing Elastomers, Advanced Functional Materials. 28 (2018) 1707032. https://10.1002/adfm.201870141

[16] A. M'Barki, L. Bocquet, A. Stevenson, Linking Rheology and Printability for Dense and Strong Ceramics by Direct Ink Writing, Scientific Reports. 7 (2017) 6017. https://10.1038/s41598-017-06115-0
[17] Y.W. Ahmed, A. Loukanov, H.-C. Tsai, State-of-the-Art Synthesis of Porous Polymer Materials and Their Several Fantastic Biomedical Applications: a Review, Advanced Healthcare Materials. n/a (2024) 2403743. https://doi.org/10.1002/adhm.202403743
[18] J. González-Rivera, R. Iglio, G. Barillaro, C. Duce, M.R. Tinè, Structural and Thermoanalytical Characterization of 3D Porous PDMS Foam Materials: The Effect of Impurities Derived from a Sugar Templating Process, Polymers (Basel). 10 (2018) http://10.3390/polym10060616
[19] R. Iglio, S. Mariani, V. Robbiano, L. Strambini, G. Barillaro, Flexible Polydimethylsiloxane Foams Decorated with Multiwalled Carbon Nanotubes Enable Unprecedented Detection of Ultralow Strain and Pressure Coupled with a Large Working Range, ACS Applied Materials & Interfaces. 10 (2018) 13877-13885. https://doi.org/10.1021/acsami.8b02322
[20] R. Zargar, J. Nourmohammadi, G. Amoabediny, Preparation, characterization, and silanization of 3D microporous PDMS structure with properly sized pores for endothelial cell culture, Biotechnol Appl Biochem. 63 (2016) 190-9. https://10.1002/bab.1371
[21] R. Rial, Z. Liu, J.M. Ruso, Soft Actuated Hybrid Hydrogel with Bioinspired Complexity to Control Mechanical Flexure Behavior for Tissue Engineering, Journal. 10 (2020) 1302. http://10.3390/nano10071302
[22] L.Y. Zhou, Q. Gao, J.Z. Fu, Q.Y. Chen, J.P. Zhu, Y. Sun, Y. He, Multimaterial 3D Printing of Highly Stretchable Silicone Elastomers, ACS Appl Mater Interfaces. 11 (2019) 23573-23583. https://10.1021/acsami.9b04873
[23] C. Perrinet, E.-J. Courtial, A. Colly, C. Marquette, R. Fulchiron, An Emulsion Approach to Resolve the Paradox of 3D Printing of Very Soft Silicones, Advanced Materials Technologies. 5 (2020) 1901080. http://10.1002/admt.201901080
[24] Q. Chen, J. Zhao, J. Ren, L. Rong, P.-F. Cao, R.C. Advincula, 3D Printed Multifunctional, Hyperelastic Silicone Rubber Foam, Advanced Functional Materials. 29 (2019) 1900469. https://doi.org/10.1002/adfm.201900469
[25] K.S.W. Sing, Reporting physisorption data for gas/solid systems with special reference to the determination of surface area and porosity (Recommendations 1984), 57 (1985) 603-619. https://doi.org/10.1351/pac198557040603
[26] S. Zhang, C. Ge, R. Liu, Mechanical characterization of the stress-strain behavior of the polydimethylsiloxane (PDMS) substate of wearable strain sensors under uniaxial loading conditions, Sensors and Actuators A: Physical. 341 (2022) 113580. https://www.sciencedirect.com/science/article/pii/S0924424722002187
[27] I. Miranda, A. Souza, P. Sousa, J. Ribeiro, E.M.S. Castanheira, R. Lima, G. Minas, Properties and Applications of PDMS for Biomedical Engineering: A Review, J Funct Biomater. 13 (2021) http://10.3390/jfb13010002
[28] R.A. Lima, The Impact of Polydimethylsiloxane (PDMS) in Engineering: Recent Advances and Applications, Journal. 10 (2025) 41. http://10.3390/fluids10020041
[29] I. Reshma, G.N. Kasinathan, A. Tikoo, P. Meduri, S. Rath, S. Dutta-Gupta, Multifunctional flexible electrospun polydimethylsiloxane (ePDMS) membranes for soft robotics applications and photocatalytic conversion platforms, RSC Applied Polymers. (2025) http://10.1039/D5LP00354G

Examining the Internal Dynamics and Self-Healing Efficiency of Bio-Enhanced Rigid Pavement for Sustainable Infrastructure Solutions

Charmaine Kay GUÑABO[1,a*], Roberto D. ROSARIO[1,b], Meriam LEOPOLDO[2,c] and Razon DOMINGO[3,d]

[1]FEU Institute of Technology, Sampaloc, Manila, Philippines, 1008

[2]College of Engineering and Architecture, Mapua Malayan College, Mindanao, Philippines

[3]Polytechnic University of the Philippines, Sta. Mesa, Manila, Philippines

[a]cggunabo@feutech.edu.ph, [b]rdrosario@feutech.edu.ph, [c]mpleopoldo@mcm.edu.ph, [d]rcdomingo@pup.edu.ph

Keywords: Self-healing Concrete, Soptivity, Ultrasonic Pulse Velocity, Bio-Concrete, Sustainable Infrastructure

Abstract. Replacing and repairing cracked concrete can be a costly and time-consuming process that can also be disruptive, with adverse environmental impacts. Taking these challenges into consideration, the current study explores Bacillus subtilis as a bio-admixture to activate an intrinsic self-healing mechanism that allows concrete to heal its cracks by itself. More specifically, the use of Bacillus subtilis ATCC 6633, incorporated into the concrete at a dosage of 2.36 x 109 CFU to heal cracks in the size range of 1-2 mm were tested. Each of the bacteria was assessed at four concentrations—1%, 3%, 5%, and 10% were evaluated for their ability to enhance the self-healing property of the concrete. The self-healing performance and internal structure recovery were assessed using two different methods: the Sorptivity Test (ASTM C 1585) and the Ultrasonic Pulse Velocity (UPV) equipment PUNDIT PL200. The findings revealed that Bacillus subtilis significantly enhanced the self-healing process in 3-7 days, effectively repairing cracks within a 10-day period, as demonstrated by improvements in material recovery, absorption rates, and ultrasonic pulse velocity. The results suggest that 5% Bacillus subtilis concentration has considerable potential to improve the durability and sustainability of concrete, offering a more environmentally friendly and cost-effective solution for autonomous crack repair using the two distinct methods.

Introduction

Concrete keeps deteriorating naturally and fissures significantly lessen durability by oxidizing the reinforcing bars and weakening the integrity of the concrete [1]. Rebuilding and maintaining the cracked concrete element can be extremely expensive, labor-intensive, inconvenient, and sometimes detrimental to the environment. Repairs or strengthening techniques, as well as the design solution, are typically far more complicated relative to new construction [2], [3]. The development of novel materials that guarantees both ecological sustainability and structural integrity is essential to address the increasing environmental concerns associated with concrete-related structures [4], [5], [6]. In this regard, the adoption of self-healing concrete is a noteworthy development, providing an innovative approach to enhancing the resilience and sustainability of concrete infrastructure [7]. The introduction of self-healing concrete aligns with the United Nations Sustainable Development Goals (SDGs), particularly Goal 9, which focuses on building resilient infrastructure, and Goal 12, which promotes sustainable consumption through the reduction of cement production. Additionally, Goal 13 is addressed by advocating for immediate action to combat climate change, as self-healing concrete helps reduce the carbon footprint associated with cement manufacturing and repair processes, further supporting climate mitigation efforts [8], [9], [10].

Self-healing concrete is a novel innovation in construction materials that aims at addressing the primary problem concerning the longevity of structures and associated degradation [11]. By introducing bacterial spores to the concrete matrix, both permeability and porosity are reduced, decreasing the potential that materials could exposed with sulfate and accelerating the rate of chloride penetration [12]. It has been established that this system is an autogenous repair mechanism; without human interference, structures may be made considerably more secure and dependable, which can have a substantial advantageous ecological effect [11]. The autogenous process, however, only works on early cracks since materials may progressively heal in the presence of water. Furthermore, Crack capping is inadequate for the long-term healing of cracks because it was initially discovered during the early stage of recovery. [13].

Another approach has been known as "microbial-based self-healing," a self-sustaining technique which employs the biological characteristics of bacteria such as Bacillus pores to mend concrete cracks without requiring for an external sealing solution [11], [14] Bacillus subtilis an aerobic bacterium [15] improves the durability, fracture characteristics, and flexibility of concrete. Furthermore, the research demonstrated that integrating bacteria in reinforced concrete could effectively modify its fracture and deflection properties [16].

This study assesses the effectiveness of the self-healing mechanism in concrete by incorporating Bacillus subtilis into the concrete matrix. The evaluation is conducted using two procedures: The Ultrasonic Pulse Velocity (UPV) test using the PUNDIT PL200, which measures the velocities at which sound waves pass through concrete to assess its integrity and internal quality of concrete samples[17], and the sorptivity test (ASTM C 1585), which gauges the rate at which water is absorbed by the concrete samples and the increase of capillary rise [18]. These tests provide insights into how *Bacillus subtilis* enhances the self-healing properties of concrete, particularly in terms of reducing water absorption and improving the material's internal integrity over time.

The study aims to validate the efficiency of various concentrations of *Bacillus subtilis* (1%, 3%, 5%, and 10%) in promoting the self-healing of concrete using the modified ASTM C1585 as manual method and validation using the PUNDIT PL200 equipment. Specifically, to investigate the effects of varying concentrations on the healing process, particularly the duration at which concrete cracks completely mend. This research aims to bridge the gap in evaluating self-healing efficiency by analyzing results obtained from two different measurement methods.

Materials and Methods

The study employed a strain of bacteria from the genus *Bacillus Subtilis* (ATCC 6633), which has already been studied extensively for a range of application in both medical and engineering [19], [20]. The B. Subtilis is acquired from the University of the Philippines, Manila. The bacteria spore is rod-like, aerobic, gram-positive bacteria grows quickly. The cells typically range from 2–6 μm long and slightly smaller than 1 μm in diameter. Through a duration for doubling of as low as 20 minutes, the ideal growing temperature is between 30^0 and 35^0 C. Long chains of cells joined by deformed septal wall material are common under certain growth conditions [21]

Bacillus subtilis (ATCC 6633) is cultivated using an alkaline mineral medium enhanced with manganese to improve sporulation or in liquid media that contains 5-g peptone, 3-g meat extract, 0.42 g $NaHCO_3$, and 0.53 g Na_2CO_3 per liter of distilled water (pH 9.7). Erlenmeyer flasks on a shaker table should be used to incubate cultures aerobically for two hours at 150 rpm. After a day, the mixture is suspended in 0.10 L of sterile water after being centrifuged for 20 minutes at 2000 rpm [22], [23] The Shaeffer-Fulton staining method was adopted to determine and manually quantify the bacteria sporulation [24]. Considering the specified nutritional composition, the experiment showed an exponential growth having a cell concentration of 2.36 x10^9 cells/mL (see Fig 1).

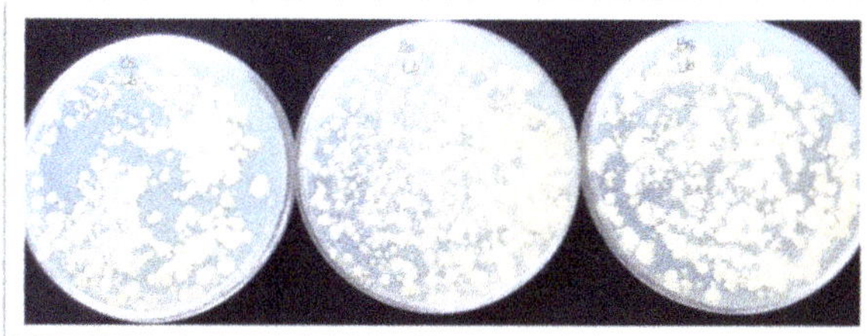

Figure 1. Bacillus Subtilis-grown Colonies

Water, Cement, Coarse and Fine Aggregate. In this study, Type I Portland Cement (Eagle Cement) from San Ildefonso, Bulacan, Philippines, was utilized. The criteria for severe exposure were used under the ACI 211-11 [20], [25] and required amount of water according to (ACI A1.53.3 ACI was employed [25]The cement content was 380 kg/m³, with a water-to-cement ratio (w/c) of 0.47. In accordance with the AASHTO criteria for minimum grading requirements for coarse aggregates, a blend of 60% 19mm (3/4") and 40% 12mm (3/8") aggregates was adhered with the 19 mm (3/4") largest nominal aggregate size (see Table 1). Additionally, the study used "vibro sand" for fine aggregates (Refer: with Table 2), which were collected in Porac, Pampanga, Philippines.

***Table 1.** Coarse of Aggregates (CA) Specifications*

C.A. Source	Max. Size	Sp. Gravity (SSD)	Absorption (%)	Unit Weight Rodded (kg/m3)
Angono Rizal, Philippines	19mm	2.8	1.0	1578

***Table 2.** Fine Aggregates (FA) Specifications*

Fine Aggregates	Fineness Modulus	Specific Gravity (SSD)	Absorption %	Unit Weight (Rodded)	Soundness Loss %	Mortar Strength %
Specs	-	-	-	-	12.0 max.	95 min.
Results	2.71	2.51	2.9	1676	2.74	0.28
Test Method					ASTM C-88	ASTM C-87

Concrete Design Mix. The study established the mass of fresh concrete using the ACI 211.1-91 Table A1.5.3.7.1 [25] accounted for the aggregate's maximum nominal size of 19 mm. The computation generated a concrete mass of 2345 kg/m³, the suggested weight of the concrete components per unit volume is calculated using the ACI 211.1-91 (see Table 3.0). Additionally, the study focused on different concentrations of Bacillus Subtilis applied directly to the concrete mix. The lower concentrations, which ranged from 0% to 1% (3.8 kg), 3% (11.4 kg), 5% (19 kg), and the higher concentration of 10% (38 kg) were applied based on the weight of the cement and water content. The portion of water content was replaced with a bacterial solution, with the aim of evaluating the impact of Bacillus Subtilis on the self-healing properties of the concrete [20].

Table 3. *Bacteria Concrete Design Mixtures in kg/m³*

Variant	Cement	Water	CA (19mm)	CA (12mm)	FA	Bacteria	Batch ID
0%	380.0	180.0*	798.4*	535.8*	599*	0	A
1%	380.0	176.2**	798.4*	535.8*	599*	3.8	B
3%	380.0	168.6**	798.4*	535.8*	599*	11.4	C
5%	380.0	161.0**	798.4*	535.8*	599*	19.0	D
10%	380.0	142.0**	798.4*	535.8*	599*	38.0	E

* *Adjusted weight based on material moisture condition during mixing period*
***Adjusted water content based on the added weight liquid Bacillus Subtilis*

Results and Discussion

Workability Test. The workability and consistency of fresh concrete were evaluated in accordance with ASTM C143 [26]. The study followed the ACI-recommended slump range of 25mm to 100mm (1 to 4 in.) for pavements and slabs [25] The reference samples (Batch A) resulted in a slump of 100mm to 110mm (4 to 4 1/3 in.), showing slight increased of the designed workability range (see Fig. 2). The samples containing 1%, 3%, and 5% Bacillus subtilis had slightly higher slumps, but consistently satisfied the design workability. The corresponding slumps were 85 mm (3 1/3 inches), 106 mm (4 in.), and 114 mm (4 1/3 in.) (see Fig. 2). However, the 10% Bacillus Subtilis samples, on the other hand, demonstrated a 135–145 mm (5 1/3–5 2/3 in.), which was behind the targeted range. This significant increase in slump suggests the possibility that the 10% Bacillus concentration will improve the fluidity of the concrete by acting as a plasticizer, however, results would affect the mechanical qualities of the concrete. The workability/slump beyond the design target could result in reduced strength and durability, potentially altering hydration processes and might affect the longevity of concrete.

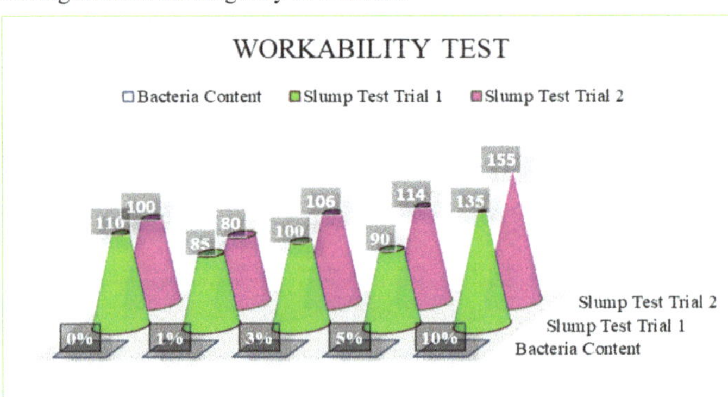

Figure 2. Workability (Slump) Test

Water Absorption Test (Sorptivity Rate). The sorptivity test procedure was employed in this investigation to measure capillary rise in concrete specimens and ascertain the sealing capability duration [24]. Each specimen's exposed broken section was partially submerged in water during testing to gauge the water intrusion or uptake. Capillary action is the primary mechanism that controls the velocity of water absorption in cracked concrete. Cracks have a major impact on this behavior because they increase the surface area that is exposed to water, which speeds up the capillary rise [24], [27], [28] Conversely, a reduction in capillary rise indicates improvement in the concrete internal structure. The enhancement can be attributed by bacteria concrete undergone biomineralization, a process by which microorganism deposits calcite within the pores and microcracks effectively reducing permeability and contributing to the self-healing mechanism of the material [7].

In this study, the sorptivity test is conducted using pre-cracked cube specimens (150 × 150 × 150 mm) from each variant to compare the differences in water uptake rates among bacterial concentrations. The procedure followed the standard method outlined in ASTM C1585 [29]. Before testing, the pre-cracked specimens were oven-dried at 55°C until the change in mass stabilized to less than 0.2%. The sides of each specimen were sealed with duct tape, and the top surface was covered with plastic to prevent water evaporation. The samples were then placed in a container with water, ensuring that the water level was maintained at approximately 4 mm above the base (see Fig. 3). Over a period of eight days, the mass change of each specimen was recorded at predetermined intervals of 0, 5, 15, 25, 45, 60, 120, and 180 minutes (see Table 4). After each measurement, the specimens were returned to the water to continue the test, following the procedure established by Rajczakowska et al. (2023) [30]. Capillary rise *(l)* is measured (Refer with: Equation 1) at specific time intervals to determine the rate of water absorption in concrete. Capillary *(l)* action-driven absorption primarily controls the velocity at which water or other liquids penetrate cracked concrete. Cracks have a significant effect on this phenomenon because they increase the surface area susceptible to water, which promotes capillary rise [20], [27]

$$I(t) = \frac{\Delta m}{A \bullet \rho} [\text{mm}] \qquad (1)\ [29]$$

Figure 3: Sorptivity Test of Bacteria Concrete

Table 4*. Water Absorption Test (Sorptivity Rate)*

Test time		√Time	Change of Mass (*g*) Bacteria Concrete %					I = ΔMass/area*/density of water** (mm)				
Day	Seconds	$sec^{1/2}$	0%	1%	3%	5%	10%	0%	1%	3%	5%	10%
	0	0	0	0	0	0	0	0.06667	0.08889	0.11111	0.15556	0.08889
	60	8	1.5	2	2.5	3.5	2	0.13333	0.08889	0.22222	0.15556	0.13333
	300	17	3	2	5	3.5	3	0.15556	0.13333	0.33333	0.28889	0.17778
	600	24	3.5	3	7.5	6.5	4	0.24444	0.15556	0.40000	0.37778	0.24444
	1200	35	5.5	3.5	9	8.5	5.5	0.40000	0.17778	0.44444	0.46667	0.28889
	2400	49	9	4	10	10.5	6.5	0.48889	0.24444	0.46667	0.57778	0.35556
	3600	60	11	5.5	10.5	13	8	0.57778	0.26667	0.48889	0.62222	0.35556
	7200	85	13	6	11	14	8	0.62222	0.26667	0.60000	0.71111	0.35556
	10800	104	14	6	13.5	16	8	0.73333	0.71111	1.02222	1.17778	0.71111
	14400	120	16.5	16	23	26.5	16	0.77778	0.77778	1.11111	1.24444	0.77778
3	86400	294	17.5	17.5	25	28	17.5	0.77778	0.84444	1.20000	1.33333	0.77778
4	172800	416	17.5	19	27	30	17.5	0.84444	0.91111	1.24444	1.44444	0.86667
5	259200	509	19	20.5	28	32.5	19.5	0.88889	0.91111	1.02222	1.42222	0.88889
6	518400	720	20	20.5	23	32	20	0.97778	0.95556	0.93333	1.15556	0.60000
7	604800	778	22	21.5	21	26	13.5	1.06667	0.91111	0.84444	1.06667	0.40000
8	691200	831	24	20.5	19	24	9	0.06667	0.08889	0.11111	0.15556	0.08889

**Area of specimen (wetted) = 22,500 mm^2*
***Density of water = 1×10^{-3} g/mm^3*

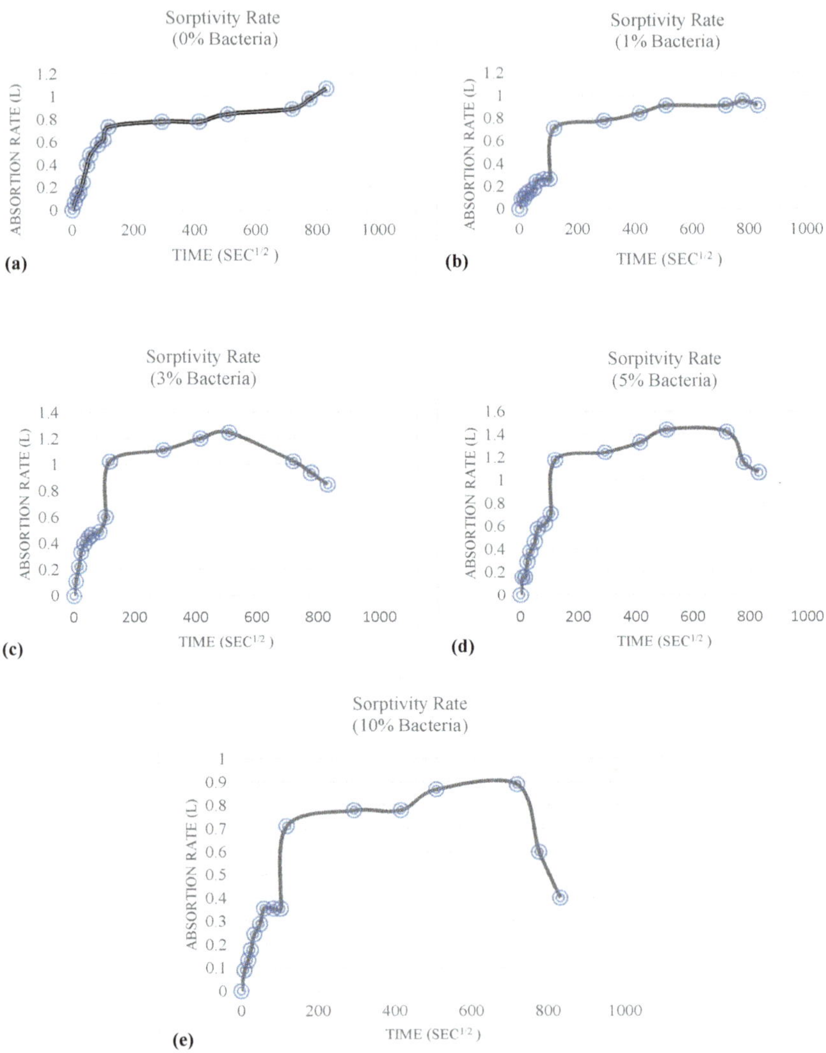

Figure 4: Capilliary Rise (l) of Concrete Samples

The reference sample showed a steady rise in capillary rise over time, consistent with the data (see Fig. 4a), which suggested that water was gradually infiltrating through the concrete matrix. In contrast, the specimen containing 1% Bacillus Subtilis (see Fig. 4b) demonstrated an apparent reduction in sorptivity (*I*) by the seventh day, suggesting the onset of self-healing activity. The samples with 3% and 5% bacterial concentrations (see Fig. 4c and 4d) showed a considerable decreased in capillary rise by the fourth day, followed by a substantial reduction by the seventh day, signifying an accelerated healing response. In contrast, the 10% bacterial specimen (Fig. 4e) showed a delayed decrease in capillary rise until the seventh day, but by the eighth day, water intrusion had significantly decreased. Therefore, higher bacterial concentrations could hamper the hydration and healing processes initially before having a substantial sealing effect over time.

Internal Healing Response using UPV Area Scanning. The internal properties of the samples were observed using two-dimensional visualization of Ultrasonic Pulse Velocity (UPV) area scanning on 3, 7 and 10 days following the onset of initiated cracks. The PUNDIT PL200 (see Fig. 5) is rendered to identify areas with different material densities and possible vulnerabilities by creating a spatial representation of the UPV measurements throughout the scanned area [10], [31], [32]. The technique made it possible to precisely monitor alterations in the internal structure of the concrete, providing a thorough evaluation of the tested samples' capacity for self-healing and recovery [31]. The study adopted Whitehurst P.A. (1971) velocity classifications to monitor the concrete internal quality.

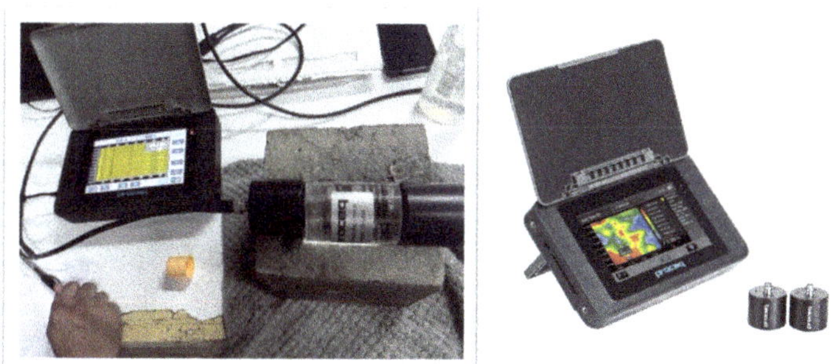

Figure 5: PUNDIT PL 200[32]

***Table 5.** Ultrasonic Pulse Velocity Qualitative Information by Whitehurst P.A. (1971)*

UPV (m/s)	Concrete Quality
< 2000	Very Poor
2000 - 3000	Poor
3000 - 3500	Questionable
3500 - 4000	Good
4000 - 4500	Excellent

The reference sample showed early signs of internal recovery 3 days after the initiated cracking (Refer with: Fig. 6c), likely due to the continued hydration of previously unhydrated cement particles, some part of samples velocities indicated from "Excellent" (before cracking) to "Poor" (after cracking) and resume to "Excellent" (after 3 days) (Refer with: Table 5). By the 7th day, the specimen shown noticeable recovery (Refer with: Fig. 6d), indicating a short-term sealing effect or internal structural adjustment within the concrete matrix. However, by the 10th day, a decline in internal uniformity was observed (Refer with: Fig. 6e) from "Excellent" to "Questionable' (Refer with: Table 5). This reduction may be attributed to the limited self-healing capacity of plain concrete, particularly its inability to close wider cracks or even prevent the propagation of microcracks over time.

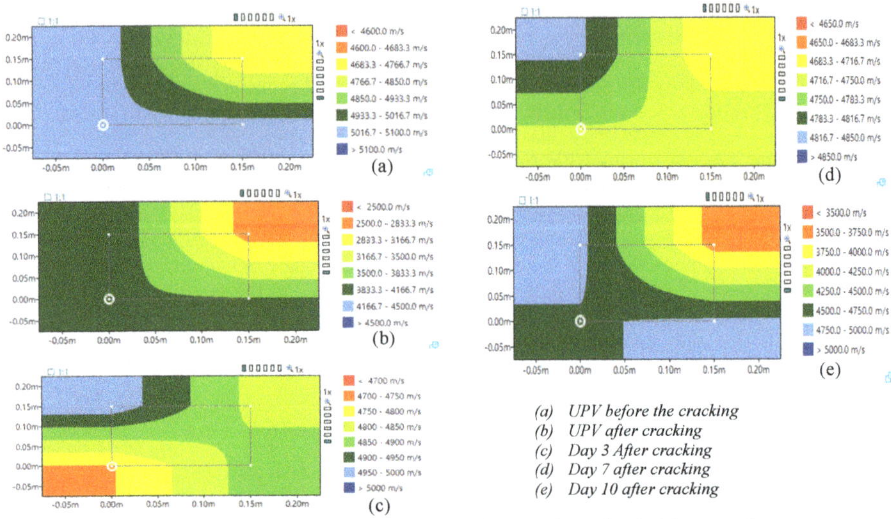

(a) UPV before the cracking
(b) UPV after cracking
(c) Day 3 After cracking
(d) Day 7 after cracking
(e) Day 10 after cracking

Figure 6: UPV- Area Scanning of Reference Sample (0% Bacteria)

The samples consisting of 1%, 3% and 5% Bacillus subtilis recovered shortly after 3 days of initial cracking, with notable development as concrete aged projected by UPV area scanning and velocities (Refer with: Figure 7). This demonstrates how effectively the bacterial growth functions for stimulating rapid self-healing mechanism and calcium carbonate precipitation. Further, the 5% sample demonstrated a notable healing on the 7th day projected by UPV intensities and crack closure of 2-mm width on the 10th day based on visual monitoring of samples. The consistent recovery over time demonstrates how to improve the concrete's structural performance and durability.

The sample containing 10% Bacillus subtilis, on the other hand, demonstrated ongoing degradation until the 3rd day, indicating an early loss of structural integrity. However, the specimen showed a noticeable improvement by the seventh day of curing, indicating that the concrete matrix's self-healing and strength restoration mechanisms had quickly been activated (Refer with: Figure 7). The recovery process advanced more slowly between days seven and ten, showing a steady and continuous improvement in the material's characteristics.

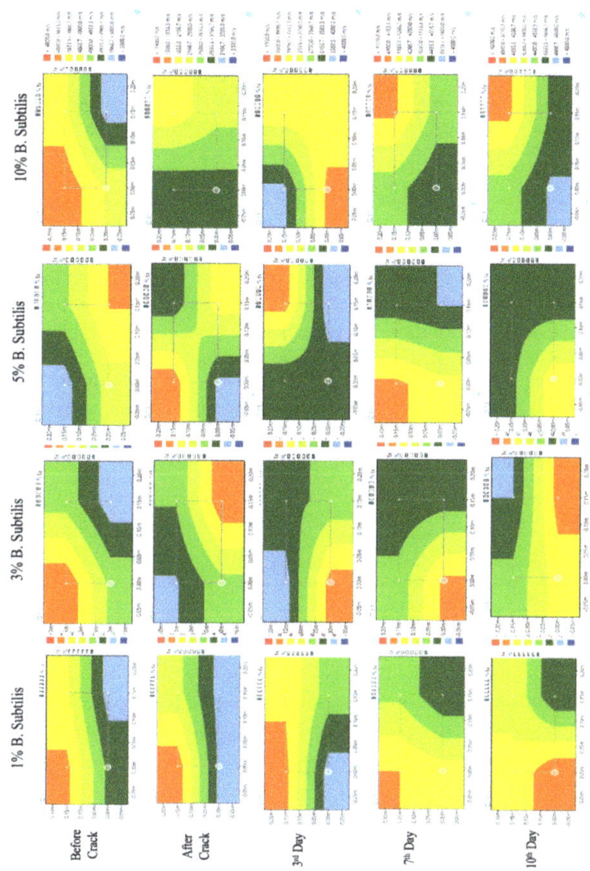

Figure 7: UPV- Area Scanning of Samples Containing Bacteria

Conclusion and Recommendation

The capillary rise and rate of fracture closure in bacterial concrete were assessed using the sorptivity test. Pre-cracked cube specimens (150 × 150 × 150 mm) from each mix were examined by partially immersing them in water and tracking changes in mass over time, in accordance with ASTM C1585. The results showed that adding Bacillus subtilis to concrete improves self-healing properties and successfully lowers capillary rise. Higher bacterial concentrations (10%) showed delayed but notable improvement at later curing phases, whereas lower concentrations (1–5%) encouraged earlier and more consistent reductions in sorptivity. This implies that while a high bacterial content may initially hamper healing and hydration, it eventually helps to seal microcracks over time.

Further, the Ultrasonic Pulse Velocity (UPV) area scanning provided valuable insight into the internal recovery and self-healing performance of the concrete samples over time. The reference sample showed partial recovery on early age of concrete due to continued cement hydration but later exhibited reduced internal integrity by the tenth day, confirming the limited self-healing capacity of plain concrete. In contrast, the samples containing 1%, 3%, and 5% Bacillus subtilis displayed significant improvement in internal integrity as early as the third day, with UPV velocities and visual observations confirming accelerated healing and calcium carbonate precipitation. Moreover, the 5% bacterial mix achieved the most notable recovery, including complete sealing of a 2-mm crack by the 10th day. Meanwhile, the 10% bacterial sample initially degraded up to the 3rd day but demonstrated rapid recovery by the 7th day and gradual improvement as concrete age.

The sorptivity test (ASTM C1585) and Ultrasonic Pulse Velocity (UPV) scanning results confirm that incorporating Bacillus subtilis enhances internal healing efficiency, promotes early strength restoration, and improves the durability of concrete especially at optimal bacterial concentrations around 3% to 5%. Future application might use the range 3% - 5% concentration per weight of cement to optimize the blending of direct mixture. Also, long-term investigation under varying environmental condition is recommended to gauge the effectiveness of bacteria concrete in extreme environment. Field application is also advocated to validate the results of laboratory experiments. Addressing the above challenges, the integration of Bacillus subtilis into concrete presents a promising advancement toward sustainable, durable, and self-healing construction materials, paving the way for longer-lasting infrastructure with reduced maintenance and environmental impact.

References

[1] X. Zhang, X. Fan, M. Li, A. Samia, and X. (Bill) Yu, "Study on the behaviors of fungi concrete surface interactions and theoretical assessment of its potentials for durable concrete with fungal-mediated self-healing," *J. Clean. Prod.*, vol. 292, (Apr. 2021). https://doi.org/10.1016/j.jclepro.2021.125870.

[2] K. Heiza, A. Nabil, N. N. Meleka, M. Tayel, K. Heiza, and N. Meleka, "State-of-the Art Review: Strengthening of Reinforced Concrete Structures-Different Strengthening Techniques," (2014). [Online]. Available: https://www.researchgate.net/publication/260991475

[3] R. Rosario D and M. J. Viado, "Encapsulating immobilized ureolytic bacteria yields self-healing concrete apropos sustainable transportation materials: A review," *E3S Web of Conferences*, vol. 488, p. 03019, (Feb. 2024). https://doi.org/10.1051/e3sconf/202448803019.

[4] V. C. Li and E. Herbert, "Robust self-healing concrete for sustainable infrastructure," *Journal of Advanced Concrete Technology*, vol. 10, no. 6, pp. 207–218, (Jun. 2012). https://doi.org/10.3151/jact.10.207.

[5] "Sustainable Transportation and Fuels | Department of Energy." Accessed: (Feb. 21, 2024.) [Online]. Available: https://www.energy.gov/eere/sustainable-transportation-and-fuels

[6] C. R. Rusnak, "Sustainable Strategies for Concrete Infrastructure Preservation: A Comprehensive Review and Perspective," *Infrastructures 2025, Vol. 10, Page 99*, vol. 10, no. 4, p. 99, (Apr. 2025). https://doi.org/10.3390/INFRASTRUCTURES10040099.

[7] R. D. Rosario, A. D. La Cruz, and M. P. De Guzman, "A Review of Biomineralization as Solution for Roads and Infrastructures Concrete Sustainability," *Civil Engineering Journal*, vol. 10, no. 8, pp. 2745–2760, Aug. 2024. https://doi.org/10.28991/CEJ-2024-010-08-020.

[8] "— SDG Indicators." Accessed: (Oct. 03, 2024). [Online]. Available: https://unstats.un.org/sdgs/report/2024/

[9] "Goal 9 | Department of Economic and Social Affairs." Accessed: (Oct. 03, 2024). [Online]. Available: https://sdgs.un.org/goals/goal9

[10] R. D. Rosario, A. Alvarez, R. C. Quinto, and M. de Guzman, "Intelligent Transportation Systems: Enabling Sustainable Transportation and Efficient Traffic Management—A Review," *Communications in Computer and Information Science*, vol. 2378 CCIS, pp. 311–323, (2025). https://doi.org/10.1007/978-3-031-84148-4_24.

[11] M. Sarkar, M. Maiti, S. Xu, and S. Mandal, "Bio-concrete: Unveiling self-healing properties beyond crack-sealing," *Journal of Building Engineering*, vol. 74, (Sep. 2023). https://doi.org/10.1016/j.jobe.2023.106888.

[12] A. Raza *et al.*, "Sustainability assessment, structural performance and challenges of self-healing bio-mineralized concrete: A systematic review for built environment applications," (May 01, 2023), *Elsevier Ltd.* https://doi.org/10.1016/j.jobe.2023.105839.

[13] G. Hong, C. Song, and S. Choi, "Autogenous Healing of Early-Age Cracks in Cementitious Materials by Superabsorbent Polymers," *Materials*, vol. 13, no. 3, (Feb. 2020,) doi: 10.3390/MA13030690.

[14] B. S. Shashank and P. S. Nagaraj, "Self-healing Behavior of Microcapsule-Based Concrete," *Lecture Notes in Civil Engineering*, vol. 260, pp. 3–13, (2023). https://doi.org/10.1007/978-981-19-2145-2_1.

[15] H. M. Son, H. Y. Kim, S. M. Park, and H. K. Lee, "Ureolytic/Non-ureolytic bacteria co-cultured self-healing agent for cementitious materials crack repair," *Materials*, vol. 11, no. 5, (May 2018). https://doi.org/10.3390/ma11050782.

[16]B. S. Shashank, P. Kumar.K, and P. S. Nagaraja, "Fracture behavior study of self-healing bacterial concrete," *Mater. Today Proc.*, vol. 60, pp. 267–274, (Jan. 2022). https://doi.org/10.1016/j.matpr.2021.12.520.

[17]"Ultrasonic Pulse Velocity Testers - NDT Concrete - Gilson Co." Accessed: (Oct. 22, 2025.) [Online]. Available: https://www.globalgilson.com/proceq-pundit-upv-test-instruments

[18]"Test Method for Measurement of Rate of Absorption of Water by Hydraulic-Cement Concretes," (Sep. 2020). https://doi.org/10.1520/C1585-20.

[19]J. Errington and L. T. van der Aa, "Microbe profile: Bacillus subtilis: Model organism for cellular development, and industrial workhorse," *Microbiology (United Kingdom)*, vol. 166, no. 5, pp. 425–427, (2020). https://doi.org/10.1099/mic.0.000922.

[20]R. D. Rosario, F. Poso, A. Victoria, and M. de Guzman, "Bio-Engineered Fiber-Reinforced Rigid Pavement: A Durability, Strength Recovery and Self-Healing Evaluation," pp. 625–637, (2026). https://doi.org/10.1007/978-981-95-4534-6_58.

[21] J. Errington and L. T. van der Aa, "Microbe Profile: Bacillus subtilis: model organism for cellular development, and industrial workhorse," *Microbiology (N Y).*, vol. 166, no. 5, pp. 425–427, (2020). https://doi.org/10.1099/mic.0.000922.

[22] L. Pawar *et al.*, "Physiology of spore formation for bacillus probiotic production," *Bacillus Probiotics for Sustainable Aquaculture*, pp. 64–86, (Dec. 2024). https://doi.org/10.1201/9781003503811-4/PHYSIOLOGY-SPORE-FORMATION-BACILLUS-PROBIOTIC-PRODUCTION-LOKESH-PAWAR-ARYA-SINGH-NAYAN-CHOUHAN-HADEER-AMER-LIKITHA-NAGA-PRASANTHMADDULURI-SOIBAM-KHOGEN-SINGH-MEHDI-SOLTANI.

[23] H. M. Jonkers, A. Thijssen, G. Muyzer, O. Copuroglu, and E. Schlangen, "Application of bacteria as self-healing agent for the development of sustainable concrete," *Ecol. Eng.*, vol. 36, no. 2, pp. 230–235, (Feb. 2010). https://doi.org/10.1016/j.ecoleng.2008.12.036.

[24] "Endospore Staining- Principle, Reagents, Procedure and Result." Accessed: (Oct. 24, 2025). [Online]. Available: https://microbiologyinfo.com/endospore-staining-principle-reagents-procedure-and-result/

[25] "ACI Mix Design – Pavement Interactive." Accessed: (Feb. 04, 2025). [Online]. Available: https://pavementinteractive.org/reference-desk/design/mix-design/aci-mix-design/

[26] "Test Method for Slump of Hydraulic-Cement Concrete," (Jun. 01, 2020), *ASTM International, West Conshohocken, PA*. https://doi.org/10.1520/C0143_C0143M-20.

[27] N. M. Alderete, Y. A. Villagrán Zaccardi, and N. De Belie, "Mechanism of long-term capillary water uptake in cementitious materials," *Cem. Concr. Compos.*, vol. 106, (Feb. 2020). https://doi.org/10.1016/J.CEMCONCOMP.2019.103448.

[28] R. D. Rosario, J. A. Padilla, N. G. Bonto, R. C. De Mesa, and O. G. Dela Cruz, "Value engineering on car curbing ownership in metro Manila," in *TRANSPORT, ECOLOGY - SUSTAINABLE DEVELOPMENT: EKOVarna2022*, AIP Publishing, (Aug. 2023), p. 020017. https://doi.org/10.1063/5.0162414

[29] "C1585 Standard Test Method for Measurement of Rate of Absorption of Water by Hydraulic-Cement Concretes." Accessed: (Feb. 22, 2024"). [Online]. Available: https://www.astm.org/c1585-13.html

[30]M. Rajczakowska, I. Tole, H. Hedlund, K. Habermehl-Cwirzen, and A. Cwirzen, "Autogenous self-healing of low embodied energy cementitious materials: Effect of multi-component binder and crack geometry," *Constr. Build. Mater.*, vol. 376, p. 130994, (May 2023). https://doi.org/10.1016/J.CONBUILDMAT.2023.130994.

[31]"(PDF) Application of ultrasonic pulse velocity to detect concrete flaws." Accessed: (Oct. 24, 2025). [Online]. Available: https://www.researchgate.net/publication/283352274_Application_of_ultrasonic_pulse_velocity_to_detect_concrete_flaws

[32]"Pundit 200 | Ultrasonic pulse velocity test of concrete." Accessed: (Oct. 24, 2025.) [Online]. Available: https://www.screeningeagle.com/en/products/pundit-200

Advanced Materials and Sustainable Energy Technologies – AMSET2025 Materials Research Forum LLC
Materials Research Proceedings 66 (2026) 367-380 https://doi.org/10.21741/9781644904152-34

A Cognitive Digital Twin for Smart Facilities Management

Axle Rhon Q. PURUGGANAN[1,a], Mayah Mae A. CATORCE[1,b], Sherwin M. VELMONTE[1,c], Jazteen Dave B. COMETA[1,d], Alonica R. VILLANUEVA[1,e*]

[1]Computer Engineering Department, Technological Institute of the Philippines, Quezon City, Philippines

[a]qarpurugganan@tip.edu.ph, [b]bqmmacatorce@tip.edu.ph, [c]qsmvelmonte01@tip.edu.ph, [d]qjdcometa@tip.edu.ph, [e]avillanueva.cpe@tip.edu.ph

Keywords: Digital Twin, Energy Forecasting, Machine Learning, Facility Management, Internet of Things

Abstract. Modern buildings face challenges in energy efficiency and occupant comfort due to non-data-driven management systems. A Building Management System (BMS) is one of the technologies that use data to efficiently manage facilities, such as buildings, in terms of energy consumption. A Digital Twin (DT) is an emerging technology, often embedded in the BMS, that automatically simulates and visualizes the physical environment, such as a building, in the digital environment. However, the traditional DT has reactive intelligence and requires frequent manual calibration, thereby limiting its ability to effectively monitor and optimize the building's energy consumption. Thus, this study aims to develop a Cognitive Digital Twin (CDT) that is capable of simulating the physical environment in a digital environment for smart facility management by designing a device that collects data from IoT sensors in the physical environment, and an automated control system that controls the Air Conditioner (AC) and lighting schedules based on real-time data and develop an artificial intelligence forecasting model to predict AC energy consumption. Test results show that the CDT successfully replicates and simulates the physical environment in a digital environment, including the detection of external environment conditions such as temperature, carbon dioxide, and humidity, the detection of the number of occupants, and real-time monitoring of energy consumption. The study also successfully simulates how the CDT adapts by dynamically controlling the AC and lightning systems based on learned data from the physical environment. Moreover, the study obtained an acceptable forecasting model for a 24-hour-ahead forecast, with a Coefficient of Variation of the Root Mean Square Error (CV RMSE) of 20.26% for predicting AC energy consumption. Thus, this study shows that the CDT significantly helps the owner and/or facility manager effectively control and manage a facility, such as a building, by enabling it to self-learn from data, dynamically adapt the AC and lighting systems, and forecast AC energy consumption, thereby optimizing building energy consumption.

Introduction

A traditional Digital Twin (DT) simulates and visualizes the physical environment in the digital environment. The DT was also used to manage facilities and buildings in real time. It utilizes IoT sensors, building management systems, and simulates environmental and building parameters. According to the study of Elyasi et. al.[1], The DT automatically updates the digital environment from the physical environment or condition of the actual facility or building. The Digital twin helps facility managers simulate the physical environment using its digital replica, thus improving the efficiency of facility management. Despite the advantages of DT, there are also several drawbacks, including limited (reactive) intelligence and requiring frequent manual calibration[2][3].

The Cognitive Digital Twin (CDT), on the other hand, is an emerging technology that represents an advanced type of traditional Digital Twin. The CDT adapts advanced AI, cognitive computing, and self-learning capabilities based on the facility's internal and external parameters [4][5][6][7].

This study employs the concept of the CDT in facility management. The CDT utilizes a cognitive engine to adapt to actual building conditions, including occupant, internal environmental, and a decision and control layer to optimize AC scheduling. Thus, optimizing the building's energy consumption and effectively managing the facility.

Review of Related Literature

Digital Twin Technology

The concept of a Digital Twin is to replicate the real world in a virtual environment in real-time [16][17]. According to Vrabic's study[8], a digital twin is a digital representation of a physical item or assembly that uses integrated simulations and service data, with information continuously updated and visualized in various ways to predict current and future conditions in both design and operational environments, thereby enhancing decision-making. The Digital Twin is a representation of both an existing physical object and its digital counterpart. They are fully integrated in both directions, so that changes made to the physical object automatically update the digital object, and vice versa.

Digital Twin Technology in Facility Management

Digital Twin (DT) Technology contributes to facility management by enhancing asset performance, improving building operations efficiency, and reducing maintenance costs. Several studies have supported the benefits of DT in facility management. As described by Ünal et al. [9], this framework enables real-time monitoring and control via a pipeline spanning multiple stages from data acquisition to visualization. In the study of Piras et al.[10], The process begins with Data Collection, where environmental data such as temperature, humidity, CO_2 levels, and occupancy are gathered from sensors installed throughout the facility. The next stage, Data Storage and Preparation, ensures that this data is securely organized and readily available for analysis. At the heart of the framework lies the Digital Twin Cognitive Analytics Model, which leverages artificial intelligence and machine learning to interpret data, forecast future energy consumption, and deliver actionable insights. Finally, the Visualization and Control stage delivers these insights to users through an interactive dashboard, enabling facility managers to monitor conditions and make informed decisions, such as adjusting lighting or air-conditioning systems. The study by Piras et al. demonstrates the DT's potential for virtually repressing physical building through real-time monitoring, predictive simulation, and optimization of building operations [18]. The DT can also monitor and optimize energy consumption in the building, thereby reducing energy use and achieving cost savings [19]. DT also enables managers to enhance occupant comfort by controlling environmental conditions, such as temperature, air quality, and lighting, thereby improving thermal comfort by up to 20% [20]. Despite the advantages of traditional Digital Twins for monitoring by replicating and simulating the physical environment in a digital environment, there are still disadvantages to the implementation in building management. Traditional digital twins have lacked intelligence, adaptability, and autonomous decision-making. The DT only includes limited (reactive) intelligence and requires frequent manual calibration[2][3]. It only replicates physical systems, but it cannot learn or make decisions. Thus, an advanced DT, such as a Cognitive Digital Twin, can enhance thinking and information processing capabilities in a manner similar to human beings. CDT can remember information, interpret data to make it useful, and identify unusual events or patterns.

Cognitive Digital Twin

To enhance traditional digital twin systems, the concept of the Cognitive Digital Twin (CDT) introduces advanced cognitive capabilities. Mokhtari et al. [12] define several cognitive functions within CDT, including perception, which refers to the system's ability to detect real-time changes in environmental and occupancy conditions. Attention allows the CDT to prioritize relevant data, such as a sudden rise in CO_2 levels, over less critical information. Memory stores historical data used for learning patterns and generating forecasts. Reasoning is used to infer potential outcomes from current trends, while problem-solving enables the system to suggest corrective actions when conditions deviate from optimal parameters. Lastly, learning empowers the CDT to adapt over time by improving its forecasts and recommendations based on previous experiences. In addition, according to Naveed et al. [11], a CDT can simulate not only the physical state of a building but also intelligently process data to mimic human-like decision-making. The study by Ghansan et al. [15] proposes a cognitive building system architecture (CBSA) that integrates machine learning, artificial intelligence, and big data analytics to replicate human-like reasoning and learning within the building management system. The study recommends that the future direction is the use of cognitive systems for energy efficiency and environmental sustainability. On the other hand, the study by Meshini et al. [21] explores the concept of CDT for large, spatially distributed infrastructures through the integration of BIM and GIS.

Methodology

Cognitive Digital Framework

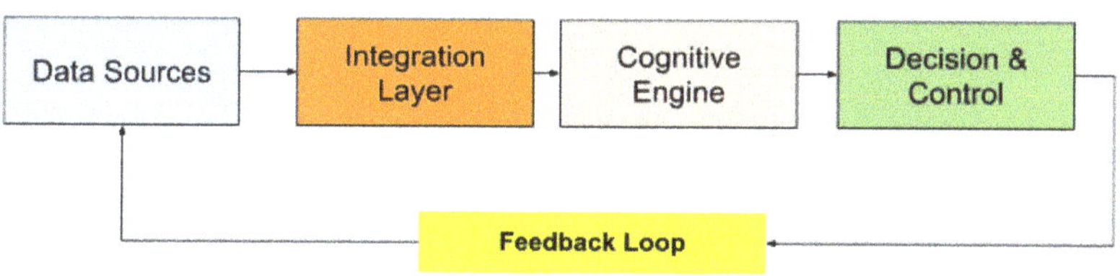

Fig. 1 Cognitive Digital Twin Framework

As shown in Fig. 1, the study gathers data from various data sources, including IoT sensors that collect data on temperature, humidity, and carbon dioxide. It also uses external APIs to gather weather data. A camera was also connected to the centralized module to detect the number of occupants. The integration layer consists of a centralized hardware module that integrates various data sources, APIs, and MQTT for communication. The cognitive engine of the CDT learns from the actual building and adapts by automatically optimizing the AC schedules based on the physical environment. It also utilizes an AI algorithm to predict the energy consumption associated with AC usage. Three AI algorithms were tested to find the most reliable and accurate for forecasting the AC energy consumption using the Coefficient of Variation of the Root Mean Square Error (CV RMSE). The Decision and Control employs control mechanisms to automatically send feedback to the AC based on the learn data of the CDT. The feedback loop is a mechanism that allows feedback from the system's output to its input.

Dataset

The dataset was acquired from the paper entitled "CU-BEMS, smart building electricity consumption and indoor environmental sensor datasets" [13], which includes electricity consumption and indoor environmental measurements of the seven-story 11,700-*m2* office

Materials Research Proceedings 66 (2026) 367-380 https://doi.org/10.21741/9781644904152-34

building located in Bangkok, Thailand, collected from July 1, 2018, to December 31, 2019, at one-minute intervals. There are two(2) categories of data collected, as shown in Table 1. The electricity consumption of the individual air conditioning units, lighting, and plug loads for 33 zones of the building, while the indoor environmental sensor data are collected from temperature, humidity, and ambient light (kux) in the same zones.

Table 1.

Category	Description
Electricity consumption data (kW)	Electricity consumption of individual air conditioning units, lighting, and plug loads in each of the 33 zones of the building.
Indoor environmental sensor data	Temperature (°C), relative humidity (%), and ambient light (lux) measurements of the same zones.

Data Pre-Processing

The first step is to choose an AC Unit that is similar to a typical room AC (1.0-1.5 ton) or an Average Power Consumption of 1500 watts. Then analyze the 2018 and 2019 CU-BEMS data. The CU-BEMS 2nd floor, z2_AC2, shows power consumption values that align with expected behavior for a typical 1.0-1.5 ton room AC. Thus, the z2-AC2 data will be used as training data for the energy consumption forecasting model.

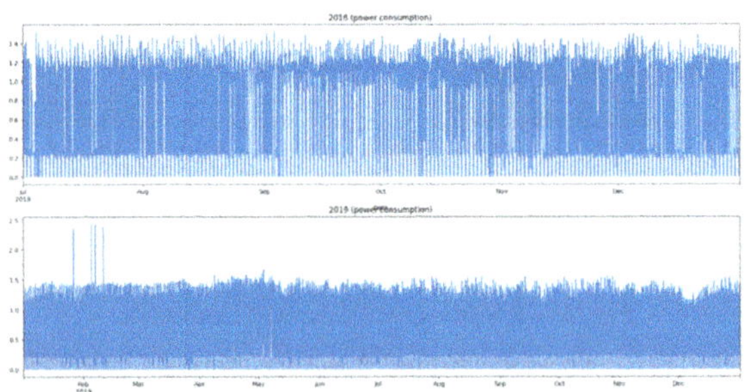

Fig. 2 z2_AC2 Power Consumption from 2018 to 2019

Fig. 2 shows the data visualization of the 2018 and 2019 power consumption of the z2_AC2 column. Based on data from October 2018 to October 2019, a consistent pattern in energy consumption is evident, indicating seasonal temporal behavior, as shown in Fig. 3.

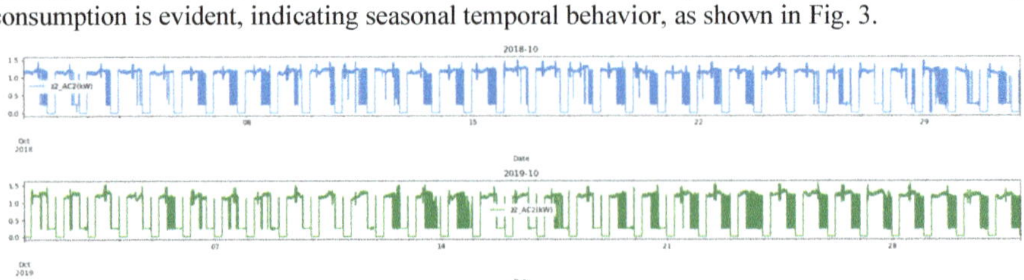

Fig. 3 Data Seasonal Temporal Pattern

Fig. 3 shows that the data exhibits strong daily seasonality. Thus, a seasonal imputation will be the approach used to replace the missing values. Seasonal Imputation is a technique that leverages dataset seasonality to replace missing values.

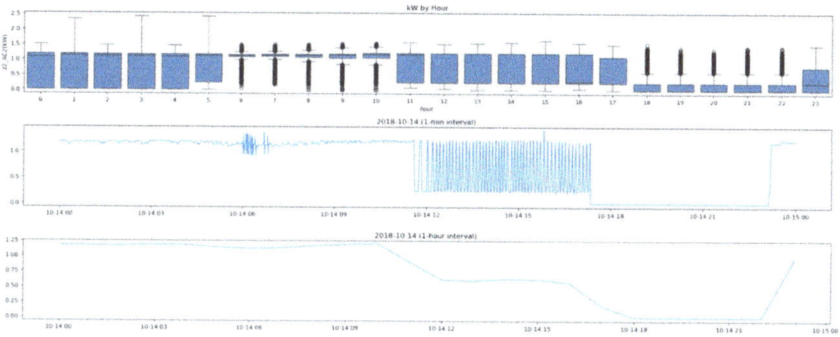

Fig. 4 Pattern in a Day

Fig. 4 shows the pattern in 1-minute, 1-hour, and 1-day intervals. The hourly aggregation smooths the short-term fluctuations observed at 1-minute intervals. It shows a stable high consumption from midnight until late morning. Near-zero consumption between 6:00 PM and 10:00 PM indicates that the AC is off, and a sharp increase at 11:00 PM indicates the AC restarting. Thus, it shows that converting from 1-minute to 1-hour intervals can help the forecasting process analyze historical data.

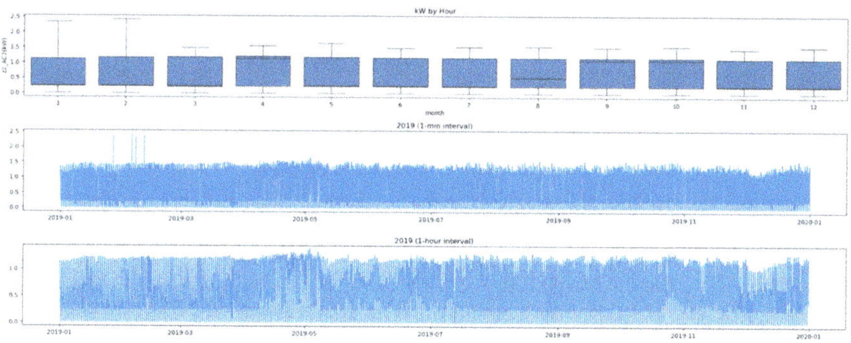

Fig. 5 Pattern in Months

Fig. 5 shows energy consumption patterns over 12 months, including 1-minute and 1-hour intervals for 2019. Months 1 and 2 show occasional peak AC operations, while Months 3 to 7 show moderate and consistent AC operations. Month 8 shows reduced AC operations, while months 9-12 show steady AC energy consumption. Overall, energy consumption is relatively stable month to month. On the other hand, short-term variability is high at 1-minute intervals, and there is a clear repeating pattern at hourly intervals.

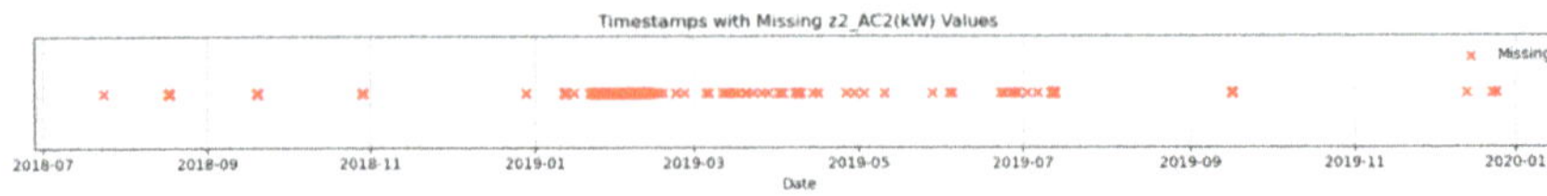

Fig. 6 Missing Data

Figure 6 shows that only about 0.3% of the data is missing, with the longest consecutive missing value being 319 minutes (approximately 5 hours). The missing values will be replaced by values that exist on other days, weeks, or even months at the same time.

Feature Extraction and Engineering
The first step is to convert the current data's power (kW) into energy (kWh). The data collection interval is 1 minute. To reduce the amount of data fed to the forecasting model, this study uses the equation below to resample from 1-minute to 1-hour intervals.

where:

Energy (kWh) = Power (kW) * Time (h)

Timeseries Frequency = 1-min interval = 1/60 Hour

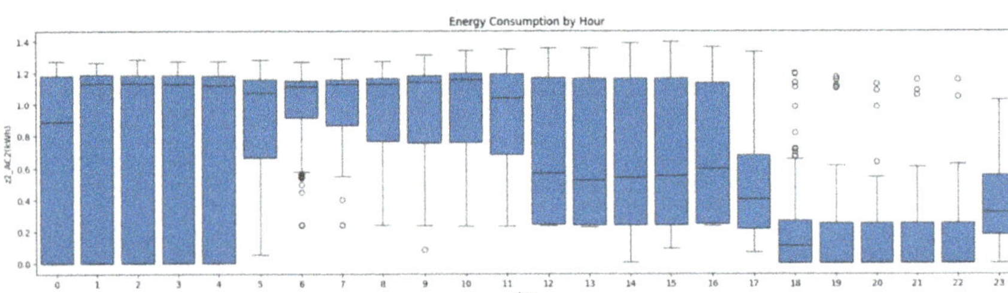

Fig. 7 Energy Consumption by Hour

Fig. 7 shows the energy consumption of z2_AC2 over one-hour intervals (24 hours in total) in a day. Consumption remains high from 12:00 AM to 4:00 AM, indicating the AC is running steadily overnight. On the other hand, consumption drops slightly from 5:00 AM to 7:00 AM, and the outliers indicate that the AC is either off or operating at reduced load. Consumption increases again from 8:00 AM to 11:00 AM, while it remains consistent from 12:00 PM to 4:00 PM. The consumption drops from 5:00 PM to 10:00 PM. The consumption starts increasing again at 11:00 PM. The evening usage shows occasional high outliers due to sporadic AC operation.

Forecasting Model Development and Evaluation
Three forecasting models, namely XGBoost, LSTM, and BI-LSTM, were trained. The dataset was split into 80% for training and 20% for testing. There are two (2) forecast horizons, 24-hour and 36-hour, used in developing the three models.

The following hyperparameters were used in training the model, as shown in Table 2.

Table 2: Hyperparameters of the Three Models

Forecasting Model	Parameters
XGBOOST. XGBRFRegressor	default
LSTM	Two LSTM layers with 50 neurons Dense Layer with 25 Neurons Optimizer = Adam Loss Function = MSE Batch_Size = 1
Bi-LSTM	Two LSTM layers with 50 neurons Bidirectional = Yes Dense Layer with 25 Neurons Optimizer = Adam Loss Function = MSE Batch_Size = 1

Their performance was evaluated using the Coefficient of Variation of the Root Mean Square Error (CV RMSE) metric.

According to ASHRAE Guideline 14-2023, a model prediction with a reliable and acceptable baseline for hourly data is one with a CVRMSE of less than 30%.

Software Design and Development

This section describes the design and development of the software integrated into the study. The software consists of embedded software and a web application that features a 3D simulation of a room in a building. The embedded software is deployed on the microcontroller in accordance with IEEE P2510, which defines quality measures, controls, parameters, and definitions for sensor data in Internet of Things (IoT) implementations.

The 3D simulation in the web application displays the energy computation, including the Coefficient of Performance (COP), defined as the ratio of useful heating or cooling provided to the work required (COP = Q/W). The computed value is used to monitor the AC's efficiency based on environmental parameters and conditions. This was also used to optimize AC energy use.

On the other hand, Unity is used to create a platform that monitors and controls the system for the user. React was used to develop the web application design. It displays the sensor's data, such as air quality, humidity, and temperature, in real-time. It also includes monitoring of the energy consumption and allows users to control the air conditioning and lighting systems.

The Digital Twin Dashboard, as shown in Fig. 8, displays a 3D simulation of the room within a building. The dashboard consists of (1) Facility-wide and (2) Element-wise. The Facility-wide dashboard displays data across the facility's rooms. The Element-wise dashboard displays data for individual elements, such as rooms, sensors, cameras, air conditioners, and lights.

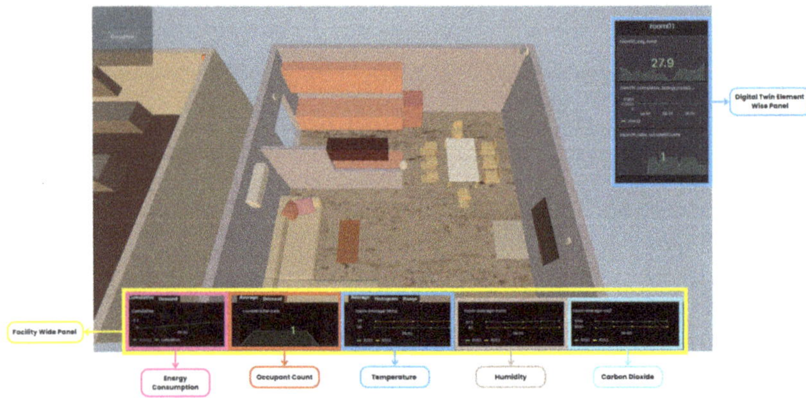

Fig. 8 Digital Twin Dashboard

Table 3: Digital Twin Dashboard

Software Parts	**Description**
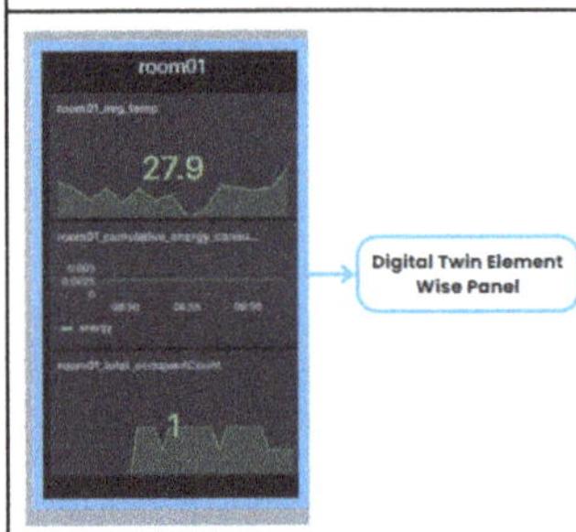	This part shows the panel of the specified element within the digital twin such as rooms, sensors, air conditioners, lights, and cameras.
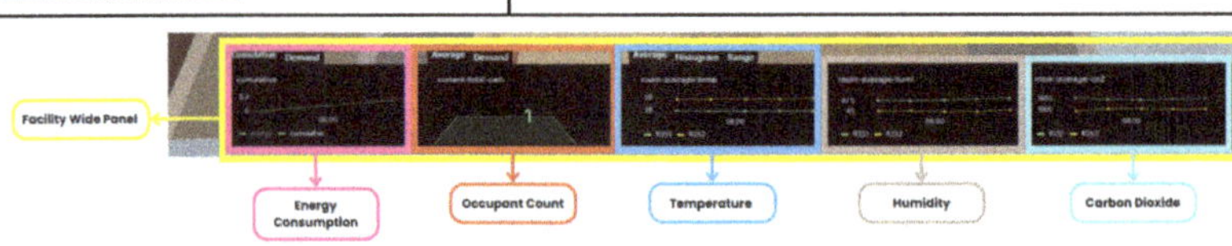 **Facility Wide Panel**	
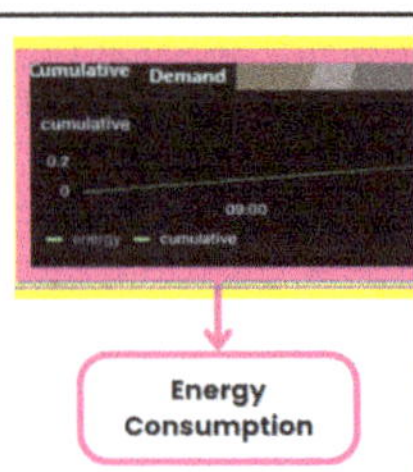	This shows the energy data used for the whole facility. tabs in this panel, namely Cumulative and Demand. The cumulative adds up the energy being consumed by the air conditioners in the facility, while the Demand shows the peak hours.

Software Parts	Description
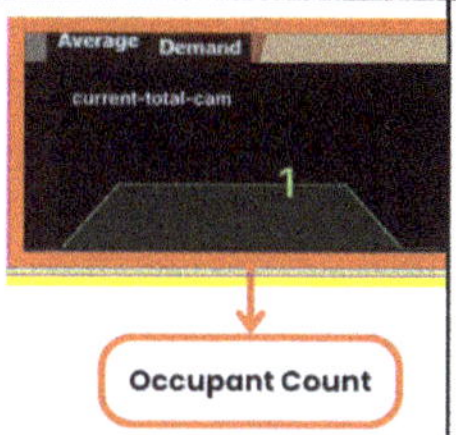	This panel shows the total number of people currently inside the rooms of the digital twin. It helps monitor how crowded or empty a room is.
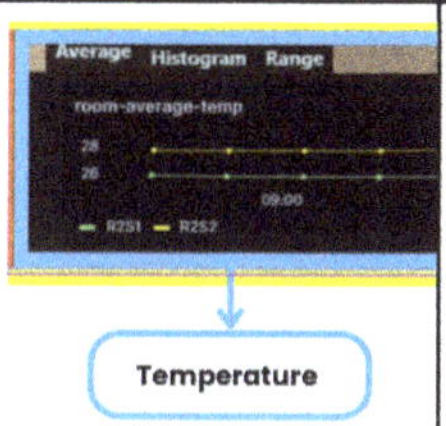	This part displays the average room temperature across the entire building. It collects data from all the temperature sensors in different rooms and shows a combined result.
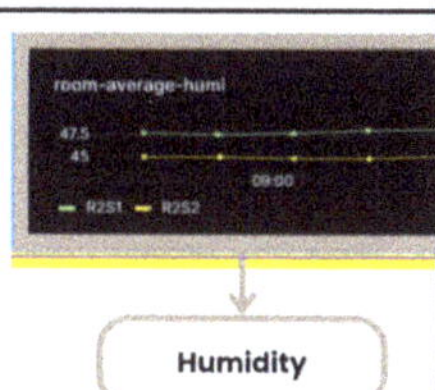	This shows the average humidity level inside the building. Humidity is the amount of moisture in the air, and this panel helps track how dry or humid the rooms are.win.
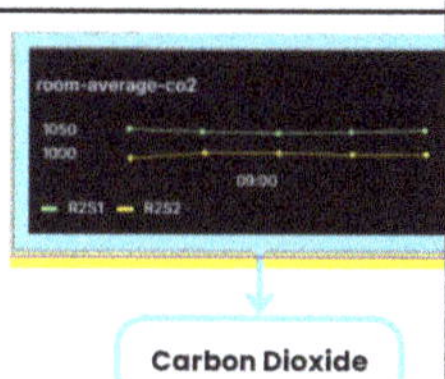	This panel shows the average carbon dioxide level in the facility's digital twin.

Hardware Prototype Design and Development
The hardware section of this system focuses on the physical components used in the physical environment of the cognitive digital twins. This section illustrates how the different physical components were interconnected and how sensor data was collected into the microcontroller for processing and display in the 3D dashboard of the web application.

Hardware Components
The illustrative diagram of the hardware components' interconnection, shown in Fig. 9, was designed to gather data, control the AC and lighting system, and process data sources.

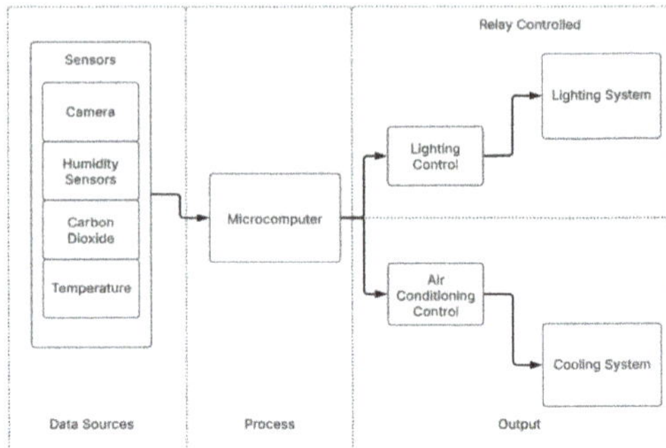

Fig. 9 Illustrative Diagram of the Hardware Components

In Fig. 9, the illustrative diagram shows different data sources gathered from the facility's sensors, such as humidity, air quality, and temperature sensors, which were collected as input to the microcontroller. Next, after the data are collected, it will be processed by the microcomputer. This study utilized Raspberry Pi (microcomputer) as the central processing unit, where it processed the raw data from sensors, user input to control the AC and lighting system, as well as triggered from the IoT sensors and no. of occupants, log data in the database, and interacted with the web application to visualize the facility information. The processed data from the Raspberry Pi is used to monitor air quality and alert facility managers and building occupants when temperature, humidity, and CO2 levels exceed predefined normal thresholds. The data is sent through a messaging service, such as MQTT, and then sent to Kafka, which organizes and manages it for forwarding to Telegraf. The Telegraf prepares the data for storage in InfluxDB. The data from the Influx database was displayed on the Digital Twin Dashboard. The Digital Twin Dashboard displays live information to the 3D simulation software via Grafana.

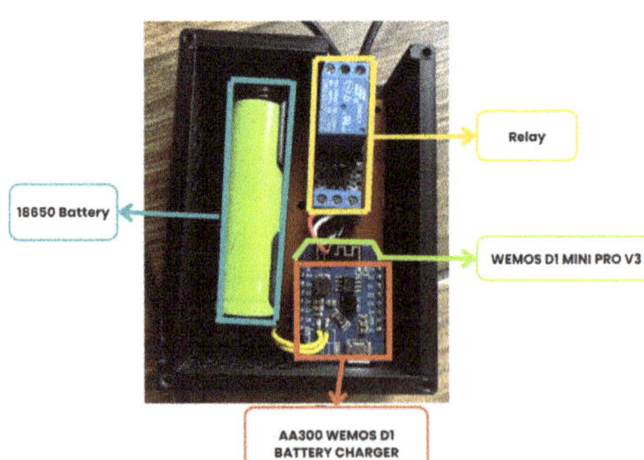

Fig. 10 Relay Module

The Relay Module, as shown in Fig. 10, is a battery-powered module composed of the WEMOS D1 Mini Pro V3 and an ESP8266 Wi-Fi microcontroller. Commands were transmitted using the MQTT protocol and then decoded to switch the light on or off. For power management, the AA300 WEMOS D1 Battery Charger was used as the charging port for the module, while the 18650 battery served as the primary power source.

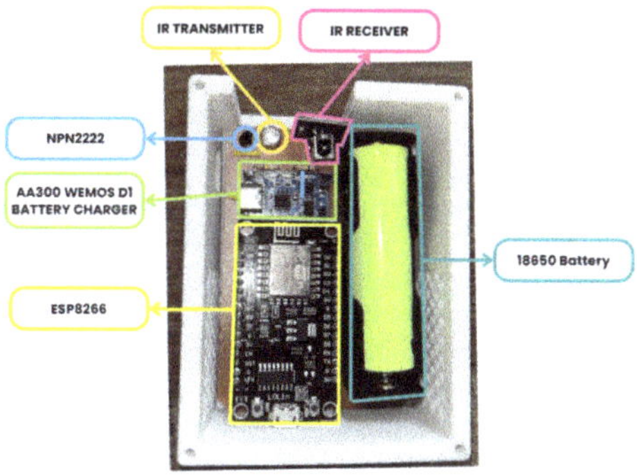

Fig. 11 IR Blaster Module

The IR Blaster Module, as shown in Fig. 11, is a battery-powered module composed of an ESP8266 microcontroller. The module transmits the IR sequence to the Digital Twin via MQTT. The air conditioning can be manually controlled using the web application. Its usage data is monitored and used by a forecasting model to predict future energy consumption, assisting in planning and efficiency. The command will be fed to the IR module to transmit and receive IR signals to the air conditioning system. The IR receiver encodes and decodes the IR signals.

The number of occupants in the room was estimated using an object detection algorithm. This information was fed to the Digital Twin Dashboard to adapt the AC system's temperature and monitor the room's energy consumption.

Results and Discussion

Test Results for Cognitive Digital Twin Functionality

Table 3 summarizes the functionality test of CDT to learn, recognize, and adapt to building conditions. There are seven (7) test cases conducted to test how the CDT operates based on the objectives of the study. Five (5) test cases were conducted to determine whether the CDT can learn from data collected by a device in a physical environment over 24 hours, including temperature, humidity, carbon dioxide, occupant presence, and the number of occupants. On the other hand, two (2) test cases were also conducted to determine whether the CDT adapts to the learned data, using the control mechanism employed in the device to dynamically control the air-conditioning and lighting systems based on real-time IoT data and the number of occupants detected by the camera. There are twenty (20) trials conducted for each test case, both in physical and digital environments.

Table 3. Summary of Functionality Test Results for Cognitive Digital Twin

Test Cases	Physical Environment	Digital Environment
Controlling the Air-conditioned System	Successful	Successful
Controlling Lighting System	Successful	Successful
Collecting Energy Consumption	Successful	Successful
Temperature Detection	Successful	Successful
Humidity Detection	Successful	Successful
Carbon Dioxide Detection	Successful	Successful
Occupant Detection	Successful	Successful

According to the test conducted, all test cases were 100% functional and successfully performed the assigned function of the CDT in both physical and digital environments. It shows that the CDT learned the internal environment by successfully detecting temperature, humidity, and carbon dioxide levels, and by counting occupants out of twenty (20) trials. In addition, the results show that the CDT adapts dynamically by controlling the air-conditioning and lighting systems based on user input and learned input data (temperature, humidity, carbon dioxide, and no. of occupants) across twenty (20) trials. The CDT alerts the user to the varying internal environmental conditions learned and automatically controls the AC and lighting systems, thus helping to efficiently reduce the building's energy consumption through real-time data simulation.

Overall, the CDT system successfully simulates, learns, and adapts in both physical and digital environments.

AC Energy Forecasting Model Performance

Table 4 summarizes the model's performance in forecasting AC energy consumption. Three AI algorithms, namely XGBoost, LSTM, and BI-LSTM, were trained and tested in forecasting to identify the most effective forecasting model. Overall, the LSTM outperformed the two models both in 24-hour and 36-hour forecasting. Thus, the LSTM model was identified as the best for forecasting AC Energy consumption.

Table 4. Summary of AC Energy Forecasting Model Performance in CV RMSE

Forecast Horizon	XGBOOST(%)	LSTM (%)	BILSTM (%)	Remarks
24-hour ahead	**24.24**	**20.26**	**21.01**	**Acceptable**
36-hour ahead	24.38	21.19	23.67	Acceptable

Moreover, according to ASHRAE Guideline 14-2023, a model prediction with a reliable and acceptable baseline for hourly data is one with a CVRMSE of less than 30%. Both 24-hour and 36-hour forecasting for the three (3) algorithms are acceptable and reliable. In addition, the 24-hour head of the LSTM is the most acceptable and reliable compared to the two algorithms.

Overall, the system can forecast 24 hours ahead using LSTM with an accuracy of 20.26%, which is acceptable and reliable. Since the owner and manager had prior knowledge of future energy consumption, they will optimize the building's air-conditioning energy consumption.

Conclusion

The study successfully designed and developed a smart facility management system by developing a Cognitive Digital Twin (CDT). The CDT successfully replicates the physical environment in a digital environment by collecting internal environment data, such as temperature, humidity, carbon monoxide levels, and the number of occupants, thereby dynamically adapting by controlling air conditioning and lighting systems. The CDT also learns from accumulated energy consumption and can forecast 24 hours ahead, achieving 20.26 CV RMSE, which is acceptable and reliable, thereby alerting the owner and/or facility manager to efficiently manage AC energy consumption within the building. The CDT in this study is to improve comfort, safety, and energy efficiency by automating decisions based on actual conditions inside the building. The cognitive digital twin transforms facility management from a reactive to a proactive approach, improving efficiency, reducing costs, and making facilities more sustainable and adaptable.

The study only utilizes indoor environmental data and the number of occupants. Thus, it limits the variables available for the CDT's decision-making. The fusion of other input parameters (data sources), such as external weather data from national agencies like PAGASA and occupant behavior and usage patterns, with advanced Artificial Intelligence can also be integrated into the study to further improve the reliability and functionality of CDT for facility management. In addition, agentic AI can automate workflows for facility management owners and managers.

Overall, the Cognitive Digital Twin for Smart Facility Management successfully built a working smart facility management system that can improve energy management and building comfort. It shows the potential of a cognitive digital twin to make facilities smarter, more sustainable, and more responsive to the needs of the people who use them.

References

[1] Elyasi, N., Bellini, A., & Klungseth, N. J. (2023). Digital transformation in facility management: An analysis of the challenges and benefits of implementing digital twins in the use phase of a building. IOP Conference Series Earth and Environmental Science, 1176(1), 012001. https://doi.org/10.1088/1755-1315/1176/1/012001

[2] Boje, C., Guerriero, A., Kubicki, S., & Rezgui, Y. (2020). Towards a semantic Construction Digital Twin: Directions for future research. Automation in Construction, 114, 103179. https://doi.org/10.1016/j.autcon.2020.103179

[3] Fuller, A., Fan, Z., Day, C., & Barlow, C. (2020). Digital Twin: enabling technologies, challenges and open research. *IEEE Access*, *8*, 108952–108971. https://doi.org/10.1109/access.2020.2998358

[4] D'Amico, R. D., Erkoyuncu, J. A., Addepalli, S., & Penver, S. (2022). Cognitive digital twin: An approach to improve the maintenance management. CIRP Journal of Manufacturing Science and Technology, 38, 613–630. https://doi.org/10.1016/j.cirpj.2022.06.004

[5] Mokhtari, K. E., Panushev, I., & McArthur, J. J. (2022). Development of a cognitive digital twin for building management and operations. Frontiers in Built Environment, 8. https://doi.org/10.3389/fbuil.2022.856873

[6] Jinzhi, L., Zhaorui, Y., Xiaochen, Z. et al. Exploring the concept of Cognitive Digital Twin from model-based systems engineering perspective. Int J Adv Manuf Technol 121, 5835–5854 (2022). https://doi.org/10.1007/s00170-022-09610-5

[7] Ghansah, F. A. (2024). Digital twins for smart building at the facility management stage: a systematic review of enablers, applications and challenges. *Smart and Sustainable Built Environment*. https://doi.org/10.1108/sasbe-10-2023-0298

[8] A. Fuller, Z. Fan, C. Day and C. Barlow, "Digital Twin: Enabling Technologies, Challenges and Open Research," in IEEE Access, vol. 8, pp. 108952-108971, 2020, doi: 10.1109/ACCESS.2020.2998358.

[9] Unal, P., Albayrak, Ö., Jomâa, M., Berre, A.J. (2022). Data-Driven Artificial Intelligence and Predictive Analytics for the Maintenance of Industrial Machinery with Hybrid and Cognitive Digital Twins. In: Curry, E., Auer, S., Berre, A.J., Metzger, A., Perez, M.S., Zillner, S. (eds) Technologies and Applications for Big Data Value . Springer, Cham. https://doi.org/10.1007/978-3-030-78307-5_14

[10] Piras, G., Agostinelli, S., & Muzi, F. (2025). Smart Buildings and Digital Twin to monitoring the efficiency and wellness of working environments: A Case study on IoT Integration and Data-Driven Management. *Applied Sciences*, *15*(9), 4939. https://doi.org/10.3390/app15094939

[11] Shahzad, Naveed and de Paula Ferreira, William and Deschamps, Fernando, Cognitive Digital Twins: A State-of-The-Art Review. Available at SSRN: https://ssrn.com/abstract=5085883 or http://dx.doi.org/10.2139/ssrn.5085883

[12] Mokhtari, K. E., Panushev, I., & McArthur, J. J. (2022). Development of a cognitive digital twin for building management and operations. *Frontiers in Built Environment, 8*. https://doi.org/10.3389/fbuil.2022.856873

[13] Pipattanasomporn, M., Chitalia, G., Songsiri, J., Aswakul, C., Pora, W., Suwankawin, S., Audomvongseree, K., & Hoonchareon, N. (2020). CU-BEMS, smart building electricity consumption and indoor environmental sensor datasets. *Scientific Data*, *7*(1), 241. https://doi.org/10.1038/s41597-020-00582-3

[14] Hou, H., Lai, J. H., Wu, H., & Wang, T. (2023). Digital twin application in heritage facilities management: systematic literature review and future development directions. *Engineering Construction & Architectural Management*, *31*(8), 3193–3221. https://doi.org/10.1108/ecam-06-2022-0596

[15] Ghansah FA, Lu W (2025), "Cyber-physical systems and digital twins for "cognitive building" in the construction industry". *Construction Innovation: Information Process Management*, Vol. 25 No. 3 pp. 787–818, doi: https://doi.org/10.1108/CI-07-2022-0164

[16] Qian, C., Liu, X., Ripley, C., Qian, M., Liang, F., & Yu, W. (2022). Digital Twin—Cyber replica of Physical Things: architecture, applications and future research directions. Future Internet, 14(2), 64. https://doi.org/10.3390/fi14020064

[17]Ding, Y., Zhang, Y., & Huang, X. (2023). Intelligent emergency digital twin system for monitoring building fire evacuation. Journal of Building Engineering, 77, 107416. https://doi.org/10.1016/j.jobe.2023.107416

[18] Piras, G., Agostinelli, S., & Muzi, F. (2024). Digital Twin Framework for Built Environment: A Review of Key Enablers. Energies, 17(2), 436. https://doi.org/10.3390/en17020436

[19] Jamil, S., Rahman, M., & Fawad. (2022). A comprehensive survey of digital twins and federated learning for industrial internet of Things (IIoT), Internet of Vehicles (IOV) and Internet of Drones (IOD). Applied System Innovation, 5(3), 56. https://doi.org/10.3390/asi5030056

[20] Khallaf, R., Khallaf, L., Anumba, C. J., & Madubuike, O. C. (2022). Review of digital twins for constructed facilities. *Buildings*, *12*(11), 2029. https://doi.org/10.3390/buildings12112029

[21]Meschini, S., Pellegrini, L., Locatelli, M., Accardo, D., Tagliabue, L. C., Di Giuda, G. M., & Avena, M. (2022). Toward cognitive digital twins using a BIM-GIS asset management system for a diffused university. *Frontiers in Built Environment, 8*. https://doi.org/10.3389/fbuil.2022.959475

Advanced Materials and Sustainable Energy Technologies – AMSET2025 Materials Research Forum LLC
Materials Research Proceedings 66 (2026) 381-389 https://doi.org/10.21741/9781644904152-35

Resiklo: An IoT Based Waste Management System Implementing Neural Networks and QR Codes for Recycling Incentives

Maria Allyza A. TOQUIRE[1,a], Adrian Cristopher NEBASA[1,b], Einid Hope H. TANDOG[1,c], Loise Poleen L. CULANNAY[1,d], Jenelyn M. ARANAS[1,e*]

[1]College of Computer Studies, Technological Institute of the Philippines, Manila, Philippines

[a]mmatoquire@tip.edu.ph, [b]macnebasa@tip.edu.ph, [c] mehhtandog@tip.edu.ph, [d] mlpculannay@tip.edu.ph, [e] jaranas@tip.edu.ph

Keywords: Waste Management, IoT, Neural Networks, QR Codes, Recycling Incentive

Abstract. Waste management remains a major challenge in many communities due to increasing waste generation, inefficient segregation practices, and limited participation in recycling programs. This study proposes RESIKLO, a smart waste management system developed for Guinayangan, Quezon. The system integrates Internet of Things (IoT)–enabled smart bins, machine learning–based waste classification, and QR code technology to support an incentive-based recycling program. Ultrasonic sensors monitor real-time waste bin fill levels, while a mobile application allows users to track waste disposal activities and redeem digital recycling rewards. The system was evaluated based on software quality characteristics, including functional suitability, performance efficiency, compatibility, usability, reliability, security, and maintainability. Results showed high user satisfaction with mean scores ranging from 4.3 to 4.6, interpreted as Very Satisfactory to Excellent, with usability receiving the highest rating. These findings indicate that RESIKLO effectively enhances waste monitoring, improves waste segregation practices, and encourages community participation in recycling initiatives.

Introduction

Solid waste management remains a significant challenge in many developing communities because of its impact on public health, environmental sustainability, and urban development. Rapid urbanization and population growth continue to increase the volume of waste generated worldwide, placing pressure on local governments to improve waste collection and disposal systems [1]. Unsustainable consumption and production patterns also contribute to environmental pollution, highlighting the need for more effective waste management strategies [2].

In many municipalities, inefficiencies in waste collection, poor segregation practices, and low community participation reduce the effectiveness of solid waste programs and increase operational costs [3], [4]. In the Philippines, the Department of Environment and Natural Resources emphasizes the need for stronger enforcement of solid waste management policies among local government units [5], [6]. Previous studies have also shown that financial incentives and pro-environmental strategies can significantly improve household participation in recycling initiatives [7], [8].

To address these challenges, this study proposes RESIKLO, a smart waste management system that integrates Internet of Things (IoT) technology, machine learning, and a mobile application to enhance waste monitoring and recycling participation. Smart waste management systems have been shown to improve segregation accuracy, optimize waste collection efficiency, and increase community engagement through incentive-based mechanisms [9], [10], [11]. By combining intelligent monitoring with a digital reward system, RESIKLO aims to support more efficient waste management and encourage sustainable environmental practices within the community.

Related Work

Municipal solid waste management remains a complex issue in developing regions due to its significant effects on public health, environmental sustainability, and urban development. Increasing waste generation and unsustainable consumption patterns have intensified the need for more effective waste management strategies worldwide [1], [2].

In the Philippines, waste management systems continue to face challenges such as inadequate infrastructure, poor waste segregation practices, and limited enforcement of environmental policies among local government units [3], [4]. Low public participation in recycling programs further reduces operational efficiency and increases environmental impacts [5], [6]. Studies have shown that community awareness, household participation, and incentive mechanisms play a crucial role in improving recycling behavior and promoting compliance with environmental regulations [7], [8].

Recent technological advancements provide promising solutions to these challenges. Smart waste management systems that integrate IoT technologies, artificial intelligence, and automated waste classification have demonstrated improvements in monitoring, waste segregation, and operational efficiency [9], [10], [11]. In addition, incentive-based recycling programs and collaborative digital platforms have been shown to increase community participation in sustainable waste management initiatives [12], [13]. Predictive modeling and artificial neural networks have also been applied to optimize waste collection routes and support data-driven decision-making for municipal waste management operations [14]. Building upon these developments, this study introduces RESIKLO, a smart waste management system designed to improve waste monitoring, automate waste sorting processes, and encourage community participation in recycling through digital incentives.

Methodology

To guarantee usableness and act, RESIKLO was made by an iterative design-test-validate cycle. The smart bin mechanically scans, classifies, and sorts waste by integrating four chambers, sensors, a camera, and Arduino-based controllers. Classification is powered by a MobileNetV2 convolutional neural network proficient on more than 4,000 images and optimized with TensorFlow Lite for real-time identification of plastic, paper, metal, and organic waste. A cohort mobile application provides QR code scanning, points tracking, bin locations, and reward redemption, with connections logged through Firestore and handled using the Giftaway API. These mechanisms are maintained by a cloud-based backend architecture that allows safe logging, monitoring, analytics, and incessant user engagement

A. System Architecture

Fig.1 follows the multi-tier design with a bidirectional flow of information between users, smart bins, the mobile application, and the cloud server. Users deposit recyclable materials into the smart bin, which is equipped with sensors and a camera for real-time detection and classification. The bin directs the item into the appropriate compartment using a stepper-motor mechanism guided by limit switches for precision. Classification data and usage records are automatically transmitted to the cloud, which synchronizes information with both the mobile app and the administrator dashboard. Administrators are able to monitor recycling activity, track performance, and oversee system operation through this platform, ensuring both transparency and accountability.

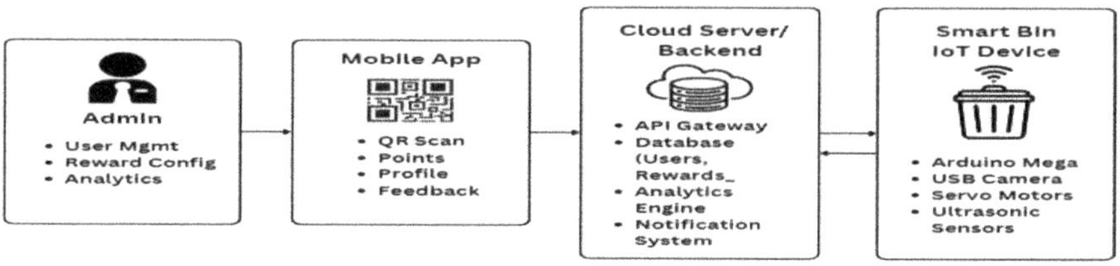

Fig. 1 RESIKLO IoT Architecture

B. Sensor Architecture

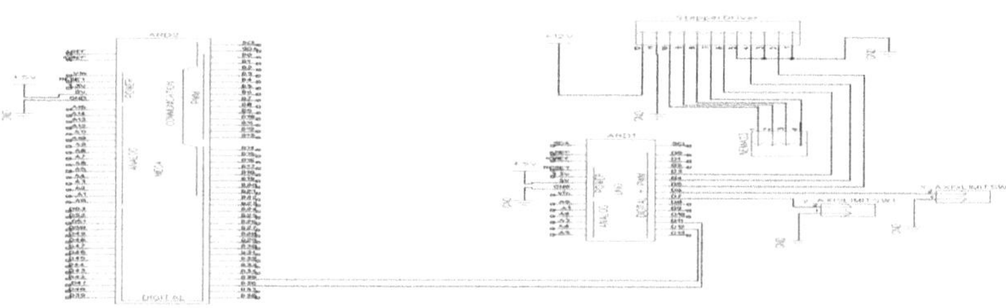

Fig. 2 Sensor Architecture

The sensor and hardware design (Fig.2) uses two Arduino boards (ARD1 and ARD2) as the core controllers. The boards manage motion control through a stepper motor driver, input from X and Y-axis limit switches, and communication with multiple sensors via digital and analog pins. The system operates on dual power rails (+5V and +12V) and employs I2C protocols for efficient data exchange. This setup allows the bin to accurately handle multiple sensor inputs while maintaining precise control over the segregation mechanism.

C. Hardware Implementation

Table 1 presents the hardware components unified into the RESIKLO system, characterized conferring to their types and efficient resolutions. These comprise structural materials, electronic controllers, sensors, mechanical subsystems, and user interface devices. Each constituent was cautiously selected to confirm resilience, cost-effectiveness, and technical reliability, supporting both the operational and structural necessities of the system. The inventory varieties from the smart bin's exterior construction materials to advanced electronic control systems, reflecting the methodical integration of diverse technologies vital for building automated waste management solutions.

Materials Research Proceedings 66 (2026) 381-389 https://doi.org/10.21741/9781644904152-35

Table 1. Hardware Component Specifications

Component	Hardware Component Specifications	
	Commercial Name	***Purpose in Project***
Smart Bin Housing	eSUN PETG+HS 3D Printer Filament 1.75mm 1KG	Durable exterior casing for waste receptacle
Sorting Mechanism	Wantai Stepper Motor Nema 23	Primary motor for the waste sorting system
Motor Control Unit	Wantai Stepper Motor Driver DQ542MA 4A	Controls stepper motor movement and speed
Conveyor System	DC Worm Gear Motor SGM-370	Powers waste transport mechanism
Secondary Motor Control	L298N DC Dual H-Bridge Motor Driver	Controls DC motors for auxiliary functions
Sorting Mechanism	Wantai Stepper Motor Nema 23	Primary motor for the waste sorting system
Motor Control Unit	Wantai Stepper Motor Driver DQ542MA 4A	Controls stepper motor movement and speed
Conveyor System	DC Worm Gear Motor SGM-370	Powers waste transport mechanism
Secondary Motor Control	L298N DC Dual H-Bridge Motor Driver	Controls DC motors for auxiliary functions
Secondary Controller	Uno R3 Board ATmega328 DIP Arduino	Auxiliary processing and sensor management
Sensor Interface	Sensor Shield V5.0 Arduino Compatible	Connects and manages multiple sensors
Support Structure	8mm stainless steel rod 4 meters	Framework for mechanical components
Drive System	GT2 Belt - 6mm 10mm width 1 meter	Motion transfer for sorting mechanism
Belt Drive Component	GT2 Aluminum Timing Pulley	Ensures precise belt movement control
Display Unit	7-inch LCD	User interface and system status display
Motion Detection	Infrared Sensors (6 pcs)	Waste detection and positioning
Display Connection	HDMI Cable	LCD screen connectivity
Main Controller	Arduino Mega	Primary system control and processing
Wireless Controller	ESP32	WiFi and Bluetooth connectivity
Processing Unit	Mini Computer	Advanced computations and data processing
Power Management	Buck Converter	Voltage regulation and conversion
Power Distribution	Extension Cord	Power supply management
Electrical System	Wirings	Internal electrical connections
Illumination	LED Strip White	Internal lighting system
Image Capture	Camera	Waste identification and monitoring

D. Mobile App Implementation

Table 2 outlines the key software components and technologies used in the development of the RESIKLO mobile application. The frontend was built using the Flutter SDK, enabling a cross-platform environment for efficient development and consistent performance across devices. Firebase authentication provides a secure login and verification system, while the Firebase Realtime Database ensures continuous synchronization and reliable data storage. The application also integrates a QR Code SCanner Library for processing recycling transactions directly from the bins and a Flutter Map SDK to assist users in locating and navigating to bin sites. To support its incentivization feature, the app utilizes the GiftAway API for managing voucher-based rewards, complemented by a Push Notification Service that delivers reminders and real-time alerts to users. Together, these technologies ensure a robust, secure, and user-friendly mobile platform that enhances community engagement in recycling.

Table 2. Mobile Application Components and Integration

Component	Mobile App Specifications	
	Commercial name	Purpose in Project
Frontend Framework	Flutter SDK & Development Tools	Cross-platform app development environment
User Authentication	Firebase Authentication SDK	Secure login and user verification system
Backend Service	Firebase Realtime Database	Real-time data synchronization and storage
Code Reader	QR Code Scanner Library	Scanning and processing QR codes on bins
Location Services	Flutter Map SDK	Bin location tracking and navigation
Voucher Service	GiftAway API	Management and distribution of recycling rewards and incentives
Alert System	Push Notification Service	User notifications and reminders

E. Neural Network Implementation

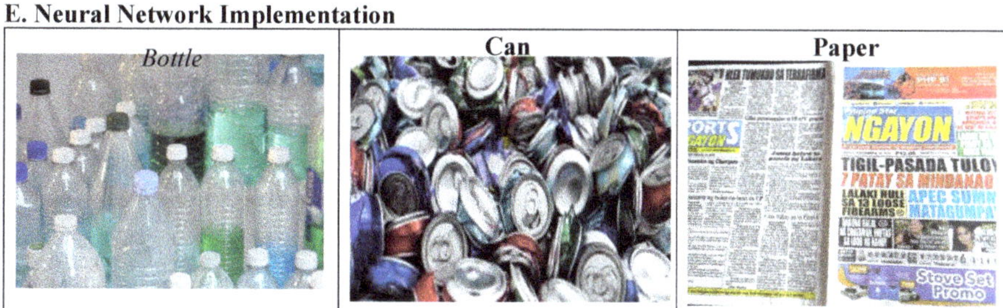

Fig. 3. Waste Classification Training Dataset with Multiple Categories and Angles

The above figure displays RESIKLO's waste classification ability utilizing a convolutional neural network (CNN) based on the MobileNetV2 architecture, selected for its effectiveness in embedded and resource-limited environments. As shown in Fig. 3, the training dataset comprised 4,314 images of recyclable items taken under controlled conditions with uniform lighting and various viewing angles. The dataset was evenly distributed across four categories: bottles (1,057 photos), cans(1,103 photos), paper (1,078 photos), and empty containers (1,076 photos). Among these, 3,452 images were allocated for training and validation, whereas the remaining 20% (approximately 800 images) precision. The model utilized transfer learning, beginning with a pre-trained MobileNetV2 that has 155 layers. The base model was augmented with six tailored layers: a GlobalAveragePooling2D layer, two fully connected Dense layers containing 1,024 and 512 units, two Dropout layers set at a rate of 0.2 to mitigate overfitting, and a concluding Dense layer for classification. This modification increased the depth of the architecture to 161 layers in total. Training was conducted with a batch size of 32 for 35 epochs, using early stopping mechanisms to prevent overtraining. The process involved about 25 test batches, with 6-7 batches dedicated to each category, ensuring balanced validation across all material types. The model was implemented on Python using the TensorFlow framework and trained with Google Colab with the support of a Google A100 High-RAMGPU, significantly accelerating computation, and it categorizes waste items in real time. When a user drops an item into the smart bin, the camera takes its picture and transmits it to the trained CNN model for identification. The system subsequently determines if the object is a bottle, can, paper, or container, and routes it to the correct compartment via the stepper-motor-operated sorting mechanism. After a successful classification, the bin produces a distinct QR code that the user scans using the RESIKLO mobile app. The cloud server logs the transaction, refreshes the user's point balance, and synchronizes the information with the mobile application and the admin dashboard. Administrators are able to track recycling activities and the overall performance of the system in real time. This process demonstrates how the integration of machine learning, IoT-based hardware, and cloud computing enables RESIKLO to deliver an automated, intelligent, and incentivized waste segregation solution.

F. Points System and Rewards System

The points system is implemented through the mobile application platform, which enables systematic tracking of recycling rewards. The system utilizes QR code scanning functionality for point collection, implements real-time tracking with merchant vouchers powered by gift-away integration for transaction history, and allows points to be converted into vouchers from various merchants.

This reward system has proven effective, with data showing that it has contributed to increased community participation in recycling programs. To earn points, users must scan the QR code displayed on the RESIKLO Bin after each successful waste recycling.

Table 3. Resiklo Points System and Awards

Activity	Points Awarded	Items	Peso Value(1point=₱1)
Waste Recycling	3	Paper	₱3
Waste Recycling	5	Cans	₱5
Waste Recycling	10	Plastic Bottles	₱10

The points system is implemented through the mobile application platform, which enables systematic tracking of recycling rewards. The system utilizes QR code scanning functionality for point collection, implements real-time tracking with merchant vouchers powered by gift-away integration for transaction history, and allows points to be converted into vouchers from various merchants.

This reward system has proven effective, with data showing that it has contributed to increased community participation in recycling programs. To earn points, users must scan the QR Code displayed on the RESIKLO Bin after each successful waste recycling.

Results & Discussion

Implementation of smart IoT-Based Trash Bin.

The research used smart trash bins that are fitted with IoT sensors to record the levels of waste and enhance better utilization of waste sorting. Real-time bin fill levels were measured using ultrasonic sensors whereas machine learning algorithms aided in the classification of wastes. System score (functional suitability) equated to 4.3, meaning that the smart bin system is useful in facilitating the process of waste monitoring and sorting. These findings indicate that the monitoring through an IoT can enhance the effectiveness of waste management activities.

Waste Tracking Mobile Application.

The mobile application was designed to enable users to monitor the waste recycling activities and their rewards in recycling. The application permits the scan of QR code to verify the waste disposal and ensures easy interface to follow the position of the recycling process. The results of the evaluation revealed that usability (4.6), compatibility (4.5), and system integration (4.1) are highly performed, which means that the application can be accessed, reliable, and easy to use with various devices.

Digital Rewards for Community Participation.

The system also established a digital reward system to help motivate correct waste management and motivate more people to engage in the system. The findings indicated that 78% of users who were registered engaged in the recycling activities on a weekly basis with the reward system registering a successful transaction rate of 99.97 using the integrated API. There was also 85 percent awareness amongst users that they were more aware of recycling compared to 73 percent who said their disposal behavior of waste had improved.

Web-Based System Evaluation and Monitoring.

An online application was created to assist in collecting, monitoring and analyzing waste management operations. The software quality model of ISO/IEC 25010 was used to evaluate the system. The results show good performance in the areas of key qualities usability (4.6), compatibility (4.5), security (4.5), performance efficiency (4.4), reliability (4.4), and maintainability (4.3) according to ISO/IEC 25010 standards (Table 4). Stable performance and high community engagement was evident in the deployment in Guinayangan, Quezon. The waste classification was rated at 4.3/5 on functional appropriateness and 78 percent of the users reported to be active every week, especially younger users who were driven by rewards. The mobile app had 4.6-usability and compatibility, 4.1-integration, and QR-based reward system had a success rate of 99.97 percent. In general, RESIKLO was a convenient, safe, and effective combination of IoT-based technology, machine learning, and mobile-based solutions, which can be successfully applied to waste monitoring, recycling operations, and engaging the community. These findings point to the possibility of smart waste management systems that will encourage the local communities to apply sustainable practices and the workload has the potential to be extended with more features and long-term changes.

Table 4. ISO/IEC 25010 Evaluation Summary

Quality Characteristics	Mean Score	Interpretation
Functional Suitability	4.3	Very Satisfactory
Performance Efficiency	4.4	Very Satisfactory
Compatibility	4.5	Excellent
Usability	4.6	Excellent
Reliability	4.4	Very Satisfactory
Security	4.5	Excellent
Maintainability	4.3	Very Satisfactory

Conclusion

This paper has assessed RESIKLO which is an IoT-based waste management system that is meant to enhance monitoring of waste, recycling activity and community involvement in waste management within Guinayangan, Quezon. The system combines smart bins with IoT-based solutions, waste classification via machine learning, and a mobile app and digital reward system based on QR codes. Its use was able to meet the study goals by offering the possibility to monitor the level of waste in real-time, offering a platform to monitor the activities related to recycling, and motivating the community to participate in it by using incentive-related systems. The results of the evaluation revealed that the system was well performing in regard to the features of the ISO/IEC 25010 with the rating of 4.3 to 4.6, which can be understood as the Very Satisfactory to Excellent level. The highest score (4.6) was in the usability, compatibility and security were rated 4.5. The reward system also proved to be very reliable and the success rate was 99.97 with a steady user engagement (78% weekly participation) which proves the effectiveness of the system. In general, RESIKLO is an effective and technological approach to waste management in the community. The system allows optimization of operational effectiveness and allows maintaining responsible environmental conduct by incorporating IoT tracking and smart classification of waste and digital incentives. It is also demonstrating some possible expansion to other municipalities in the quest to enhance sustainable waste management practices. The next step can be to improve on supported waste types, predictive analytics to support waste collection planning, and reward mechanisms to further empower community involvement.

References

[1] Chen, D. M., Bodirsky, B. L., Krueger, T., Mishra, A., & Popp, A. (2020). The world's growing municipal solid waste: Trends and impacts. Environmental Research Letters, 15(7), 074021. https://doi.org/10.1088/1748-9326/ab8659

[2] Glavič, P. (2021). Evolution and current challenges of sustainable consumption and production. Sustainability, 13(16), 9379. https://doi.org/10.3390/su13169379

[3] Coracero, E. E., Gallego, R. J., Frago, K. J. M., & Gonzales, R. J. R. (2021). A long-standing problem: A review of the solid waste management in the Philippines. Indonesian Journal of Social and Environmental Issues (IJSEI), 2(3), 213–220. https://doi.org/10.47540/ijsei.v2i3.144

[4] Dwivedi, A., Thakur, A. K., & Kesari, J. P. (2021). Feasibility study on sustainable solid waste management system in developing countries. International Journal of Advances in Engineering and Management, 3(7), 3038–3043.

[5] Environmental Management Bureau. (2020). Solid wastes. https://emb.gov.ph/wp-content/uploads/2018/09/3-Solid-Waste-1.8.pdf

[6] Dalugdog, W. D. (2021). Level of compliance of the local government units (LGUs) in the implementation and enforcement of R.A. 9003 (Ecological Solid Waste Management Act of 2000) in CALABARZON. Asian Journals of Management Studies.

[7] Vorobeva, D., et al. (2022). Adoption of new household waste management technologies: The role of financial incentives and pro-environmental behavior. Journal of Cleaner Production, 362, 132328. https://doi.org/10.1016/j.jclepro.2022.132328

[8] Shan, L., Xu, J., & Miao, Y. (2022). Investigating the influencing factors of incentive-based household waste recycling using structural equation modelling. Waste Management, 142, 120–131. https://doi.org/10.1016/j.wasman.2022.02.014

[9] Cheema, S. M., Hannan, A., & Pires, I. M. (2022). Smart waste management and classification systems using a cutting-edge approach. Sustainability, 14(16), 10226. https://doi.org/10.3390/su141610226

[10] Liang, S., & Gu, Y. (2021). A deep convolutional neural network to simultaneously localize and recognize waste types in images. Waste Management, 126, 247–257. https://doi.org/10.1016/j.wasman.2021.03.017

[11] Trushna, T., Krishnan, K., Soni, R., et al. (2024). Interventions to promote household waste segregation: A systematic review. Heliyon, 10(2), e24332. https://doi.org/10.1016/j.heliyon.2024.e24332

[12] Castro, R. C. C., et al. (2020). Development of a waste management system using the concept of "Basura Advantage Points" through an artificial neural network. IEEE Xplore.

[13] Sidhu, N., Pons-Buttazzo, A., Muñoz, A., & Terroso-Saenz, F. (2021). A collaborative application for assisting the management of household plastic waste through smart bins: A case study in the Philippines. Sensors, 21(13), 4534. https://doi.org/10.3390/s21134534

[14] Kridakorn, T., et al. (2024). Application of artificial neural networks for predictive modeling of municipal solid waste collection in tourist cities. Global Journal of Environmental Science and Management, 10(4), 1859–1876. https://doi.org/10.22034/gjesm.2024.04.22

A Review of Urban Flooding Management Approaches: Predictive Modeling for Stormwater Catch Basins

Roberto ROSARIO[1,a*], Sean Padolina PADOLINA[1,b], Meriam LEOPOLDO[2,c] and Razon DOMINGO[3,d]

[1]FEU Institute of Technology, Sampaloc Manila, Philippines

[2]College of Engineering and Architecture, Mapua Malayan College, Mindanao, Philippines

[3]Polytechnic University of the Philippines, Sta. Mesa, Manila, Philippines

[a]rdrosario@feutech.edu.ph, [b]smpadolina@feutech.edu.ph, [c]mpleopoldo@mcm.edu.ph, [d]rcdomingo@pup.edu.ph

Keywords: Stormwater Management, Urban Flooding, Machine Learning, Predictive Modeling, Flood Control System, Sustainable Water Management

Abstract. Urban flooding has become a major global problem that is caused by the fast growth of cities, climate change, and the lack of proper infrastructure to cope with the increased stormwater runoff. Therefore, an effective stormwater management system is necessary to alleviate flooding by gathering and rerouting rainwater to the drainage systems. Predictive modeling has turned into an important tool for stormwater management simulation, as it can demonstrate runoff behavior and the impact of catch basins. Different hydrologic and hydraulic modeling software each have their own features for studies related to stormwater, watershed, and floodplains. The EPA SWMM integrates green solutions, such as rainwater harvesting, bio-retention systems, permeable pavements, floodplains, and infiltration trenches. The PCSWMM makes it possible to predict floods in real time and design detention systems making it appropriate for large and multiple uses watersheds at the same time. HEC-HMS provides a solid base for event-based and watershed simulations that require a lot of data sources and different runoff diversion methods. Meanwhile, InfraWorks and XPSWMM are dedicated to integrated hydrologic-hydraulic modeling. Moreover, MIKE+ and OpenFlows FLOOD provide high resolution detailed flood simulations and sophisticated unified platforms. Nevertheless, there is not a single modeling platform that completely solves the problem of hydrological, hydraulic and urban variability. The review applied Latent Dirichlet Allocation (LDA) to select relevant papers from various databases and identify the research issues in the simulation software for urban flooding management, which in turn led to the future research and policy making in flood control systems. The research shows that existing modeling tools can only partially connect real-time monitoring information with climate changes and machine learning methods to create complete prediction systems. The research results give new understanding about creating adaptive data-driven flood modeling methods which will help improve stormwater management techniques and upcoming flood management policies for urban areas at risk of flooding.

Introduction

Urban flooding is a major problem in cities across the globe, particularly in regions experiencing rapid urbanization due to inadequate stormwater drainage systems, an increase in impermeable land areas, and climate change [1,2,3]. The current drainage systems might be unable to cater the significant runoff from urbanization. Extreme rainfall events have the potential to exceed conventional stormwater management systems, resulting in flooded streets, surface runoff, and substantial property damage [2,4,5]. Scholars have found that as the impervious surface ratio raised by 90%, the flooding volume increased by 108% [3,6]. Moreover, the surface run-off increased by 9326.9 $x10^3$ m^3 as the pervious surface was transformed to 60% impermeable [3]. Due to the growing frequency of urban flooding in recent years, flood-risk management has been recognized as a global concern which emphasizes the need for sustainable flood-mitigation measures [1].

Stormwater management includes integrated system that concentrate on infiltration, run-off reduction, retention, and improving water quality [3,7,8]. The conventional strategy is to accumulate surface runoff thru drainage systems and convey it to subsurface infiltration trenches or ponds [9]. This technique efficiently lowers surface runoff volumes while allowing groundwater recharge. In contrast, enlarging open pervious surfaces and improving infiltration rates of the surface using permeable pavement or porous materials, however, the implementation is often constrained in highly developed areas [10,11,12]. Moreover, there are large-volume catch basins at the receptor points that serve to control the directed and mixed water from the entire drainage system which often need pumping due to the flat area. One of the latest design innovations is setting aside the low places that are to be used as both temporary storage during the rains and public parks during the hot months, thus enhancing drainage and urban life [13,14].

The stormwater regulation systems are necessary to avoid water-logging in cities [8,15]. The changes in climatic patterns and the increase of mean sea level have made it progressively difficult to deal with stormwater problems, and thus the use of standard flood management methods has become useless [8,16]. Researchers have focused on innovative stream techniques that can delay runoff discharge and subsequently cause a reduction in peak flow during the heavy rainfall period thereby mitigating urban flooding. The influence of artificial intelligence (AI) and machine learning (ML) on stormwater management by way of developing more accurate and advanced predictive models has recently become a topic of interest [8,17]. Stormwater predictive modeling is the method of forecasting future runoff and flooding events using the help of artificial intelligence based on land use, rainfall intensity, and stormwater infrastructure capacity [17,18]. By using predictive models which incorporate real-time data, geographical stations, and satellites, the urban planners and engineers can foresee possible flooding and accordingly take preventive measures so as to be able to control urban runoff more easily and effectively [17,19]. The continuous development of comprehensive modeling tools presents an opportunity of a long-term flood risk management approach that can assist communities in becoming more resilient to both temporal weather variations and permanent environmental changes [2,8,17]. This study has investigated sustainable drainage systems together with machine learning methods. The research about predictive modeling tools which assess large stormwater catch basins and detention systems remains insufficient according to existing studies. This study overcomes that restriction through its systematic review which assesses urban flooding management predictive modeling tools by using Latent Dirichlet Allocation (LDA) to examine current research and discover vital study areas and technical deficiencies. The results offer guidance for creating adaptive data-driven predictive system modeling which enhance stormwater catch basin efficiency and urban flood control in areas with low elevation like the Philippines.

Methodology

The research has utilized the Latent Dirichlet Allocation (LDA) for topic modeling and MATLAB to analyze large datasets of published articles from 2020 – 2025 to identify research gaps in urban flooding management. The keywords used are "Stormwater Management", "Urban Flooding", and "Predictive Modeling". The methodology extracted relevant topics from a dataset of 206 Elsevier articles which provide valuable insights into the development of stormwater systems and the predictive modeling which help in the design of more resilient urban flood control systems. LDA revealed (refer with: Fig 1) top topics such as "Stormwater Management", "Flood Control Systems", "Rainwater Run-off" and "Predictive Modeling".

Figure 1. Topic Modelling using Matlab

A total of 206 articles were initially identified through the Scopus database. After removing duplicate and irrelevant records, 54 articles remained for screening. Following a comprehensive screening process, 29 articles were determined to be relevant and were included in this review. (Refer with: Fig 2)

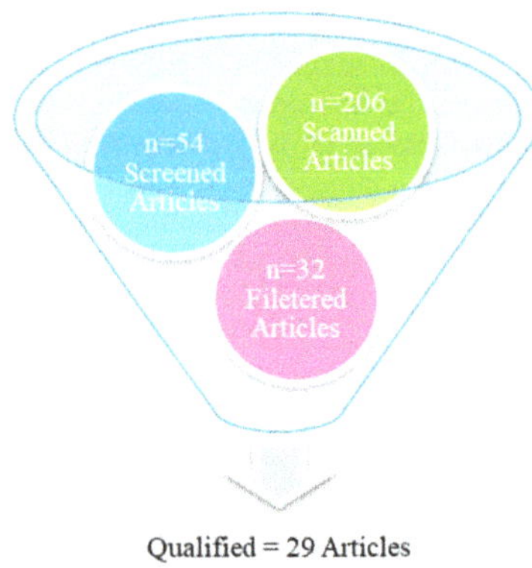

Figure 2: Literature Identification, Screening, and Inclusion

Urban Stormwater Management Systems and Contemporary Flooding Challenges

Urbanization drastically impacts the hydrologic behavior of catchments by expanding impermeable areas such as streets, roofs, and highways [20,21]. The growing frequency of intense rainfall events, rising urbanization, and climate change have all led to changes in urban stormwater management. Modern approaches generally use a systems-oriented framework that takes consideration for the generation, distribution, storage, and release of run-off throughout the whole urban catchment. The three (3) major approaches are: restoration, modification and engineering solutions, each offering distinct hydraulic and environmental benefits. However, several issues must be resolved to make sustainable stormwater management [21,22].

Restore: Reforestation and Subsurface Infiltration. Flood mitigation has come to rely on reforestation as the major technique. The latter is the method adopted in environmental conservation by the application of vegetation layering, mainly in the form of trees and plants [23]. The method is supported by safeguarding coasts with mangroves and increasing the water holding capacity of flood-prone areas by using vegetation blocks. This technique is indeed a very effective means to protect humans from floods while at the same time protecting nature [23,24].

The aim of the technology is to capture rainwater through the surface drainage and direct it to underground infiltration systems where it can be absorbed by the soil. This process helps to keep the surface runoff at very low rates and at the same time increases the groundwater level. Nevertheless, the application is limited by geological and hydrogeological factors in the area. Thus, poor soil permeability, high groundwater levels, or the risk of contamination of soil or groundwater could all cause a significant drop in infiltration rate. Moreover, underground systems must be designed in such a way that clogging and loss of capacity over time are prevented, which in turn requires regular maintenance and monitoring [25].

Modification. Embedding changes in ecological features strengthens the control of flooding. The enlarging of the area, in particular, the making of riverbanks more solid, and the creation of trenches and swales, as well as the provision of these features, will all lead to a significant increase in the capability of the river to slow down and to hold excess water [24].

The use of permeable pavements for paving, doing soil vegetatively, and using porous materials have all been mentioned as methods for increasing infiltration rates. The system offers a very simple but effective method which is infiltration at or very close around the point of rainfall. This method is known to minimize surface run-off. The challenges it poses to the structural design project, the requirement for durable permeable materials, and the difficulty of maintenance due to sediment blockage are some of the issues this method encounters. Consequently, although it is efficient in areas of low to medium density, sometimes the method is not feasible in densely populated areas due to the restriction on its scalability [26,27].

The scheme of modifying manages to reduce runoff while at the same time providing temporary water storage and treating the water by natural biological processes [28]. Nonetheless, the mentioned challenges are related to the operation of the project and include among others: guaranteeing sufficient surface area, integrating with the existing infrastructure, and winning public acceptance of open reservoirs. In addition to that, seasonal variations, silt build-up, and maintenance of the plants are factors that affect the functioning of the scheme and need to be managed through adaptive design and management measures [22].

Engineering Solutions. Engineered interventions such as floodwalls, embankments, and storage reservoirs are applied when natural restoration or modification approaches are inefficient. Structures are carefully built to prevent extreme flooding in metropolitan areas [29].

Retaining stormwater addresses the flooding in low-lying areas downstream where significant amounts of runoff converge [29]. Retention facilities are often large-scale installations designed temporarily hold rainfall and discharge controllably. The system frequently requiring pumping mechanism due to geographical limits [24,30] Although engineering scheme is very effective at lowering the risk of downstream flooding, they have presented significant challenges and could be vulnerable to power outages or catastrophic incidents that exceeded design capacity [14,31]. To address these issues, recent strategies include sacrificial low-ground areas that act as both public areas during dry conditions and flood storage during storms. However, safety, accessibility, and social acceptability must all be carefully considered when designing the water retention facility [13,14,29,31].

Meanwhile, the so-called "flood control tunnel" in Burgos Circle, Bonifacio Global City (BGC), Taguig, Metro Manila, Philippines (Refer with: Fig. 3 and Fig. 4), can manage pluvial rain. Indeed, the tunnel is an underground retarding basin built beneath Burgos Circle, which functions as temporary storing excess stormwater area during periods of intense rainfall [32]. The facility serves as a subterranean reservoir that collects and controls run-off to prevent the overwhelming

Materials Research Proceedings 66 (2026) 390-405 https://doi.org/10.21741/9781644904152-36

surface run-off. The holding tank can accommodate approximately 22,000 m^3 of water [33]. The water cannot flow out of the basin during the heavy storm but is released step by step into the nearby rivers after the rain has stopped, thus lowering the peak discharge and flood risk [24,34]. This method has proved the efficiency of retention facility in the metro area by marking a significant drop in flooding both within and outside the city [8,34]. The detention basin is an exemplary model of a modern metropolis flood-resistant design that has opened the door to the installation of similar systems in Metro Manila as a long-term solution to flooding [24,34].

Figure 3. Burgos Circle, BGC, Philippines, "Flood Control Tunnel" (Aerial View) [33]

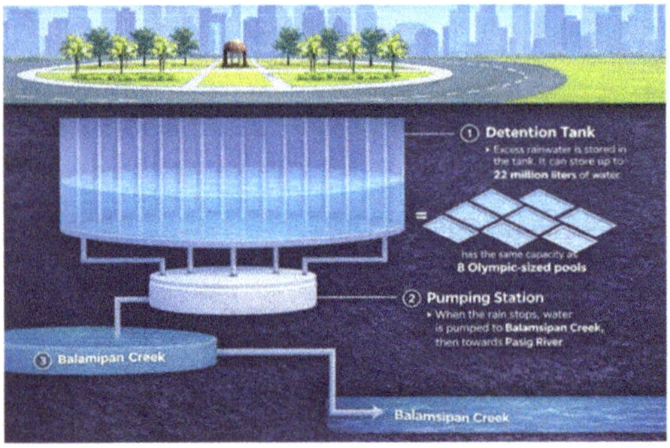

Figure 4. Underground Retention Basin Beneath Burgos Circle, BGC, Philippines (Cross-sectional detail) [33].

However, the climate variability significantly disrupt natural hydrologic processes, putting stormwater systems under unprecedented demand [17,35]. High-intensity rainfall events have become more frequent and intense, surpassing the design capacity of traditional drainage systems. Urban growth also shortens catchment reaction times, decreases infiltration, and speeds up runoff generation by substituting impermeable materials for permeable surfaces [36]. Indeed, Nature-Based Solutions (NBS) have shown that the development of predictive urban planning tools that incorporate GIS-based spatial planning, complex assessment frameworks, and geospatial simulation could help with the implementation and optimal positioning of NBS in stormwater management [37]. Furthermore, a transition from conventional hydrological modeling to intelligent and dynamic systems has also been advocated [17]. (Refer with: Table 1) shows some applications and challenges of NBS in stormwater management.

Table 1. The following projects utilized the NBS in stormwater management.

Location / Project	**Types**	**Scale / Capacity**	**Performance / Outcomes**	**Applicability of predictive modeling**	**Ref. No.**
Water Sensitive Urban Design - WSUD (Barcelona)	Application of swales and park vegetations for water retention.	Catchment Area (as shown)	Improved interception efficiency and reduced surface flooding and island heat effect	Most predictive models rely on static or simplified inlet efficiency parameters and have limited capability to dynamically simulate water accumulation leading to potential overestimation of interception and percolation efficiency in real storm events.	[38]
Chicago, USA	Deep tunnel and underground stormwater storage system (TARP)	Regional mega-infrastructure; deep tunnels and reservoirs (multi-million m^3 storage)	Long-term reduction in urban flooding and combined sewer overflow frequency	Large-scale predictive models require extensive calibration and long-term datasets; accurate simulation of pressurized flow, transient surcharge, and operational controls increases numerical complexity and uncertainty in extreme-event forecasting	[39]
Rotterdam, Netherlands (Benthemplein Water Square)	Water plaza with integrated stormwater detention via surface inlets	Local site-scale detention ($\approx 10^3$ m^3 order of magnitude)	Short-term detention and peak attenuation during cloudburst events	Model performance is highly sensitive to high-resolution surface topography, rainfall spatial variability, and outlet control assumptions; simplified boundary conditions may underestimate localized ponding or overflow behavior	[40]
South Korea (urban flood-prone cities)	Deep stormwater drainage tunnels and underground conveyance-storage facilities	City-scale infrastructure	Reduction in flood depth and duration shown in design and optimization studies	Predictive modelling accuracy depends on 1D–2D coupling strategies and assumed tunnel roughness; simplified representations may not fully capture inlet–tunnel interactions and transient storage behavior under extreme rainfall	[41]

Indeed, NBSs are crucial for managing surface runoff and mitigating flooding in urban areas [42]. The systems typically include catch basin inlets, subsurface pipe conveyance, and outfalls to manage runoff during rainfall. Contemporary predictive modeling studies using Storm Water Management Model (SWMM) and related software consistently demonstrated that a gradual rise in surface imperviousness ratio results a significant increase in surface rainfall volumes and flooded junctions that exceeded the surface drainage systems' capacity [43], [44].

Studies using predictive modeling revealed that the geographical extent and duration of surface ponding are underestimated the inlet capacity, clogged mechanisms, and surface–substructure interactions, even though such an approach has the benefit and capacity to evaluate systems from a wide perspective [45]. Therefore, when considering flood risk management simulations of urbanized catchments, explicit acknowledgment of inlet capacities and maintenance parameters is crucial [46].

Technical Challenges in Flood Management Approaches

The existing methods which have been created to control urban flooding face numerous execution difficulties which remain unsolved. Restoration and infiltration-based approaches face limitations because of soil permeability and groundwater conditions and the gradual clogging of their infiltration systems [56]. The hydraulic performance of permeable pavements and vegetated swales and infiltration trenches will decrease because their maintenance requires ongoing efforts which will result in sediment buildup [57,59]. The engineered flood control measures which include detention tanks and underground reservoirs, and pump-assisted drainage systems enable better runoff management, but their operation depends on mechanical systems and power sources which face risks during extreme rainfall events [56,57,58]. The systems' performance prediction remains challenging because of three main factors which include rainfall variability and inlet capacity and surface-to-subsurface flow patterns.

Stormwater Predictive Modelling Software Key Components and Challenges

The integration of predictive modeling in designing stormwater catch basin and urban flooding management represents a significant advancement in the effort to address the unresolve challenges of urban flooding. Recent studies have made significant effort in utilizing the software for predicting stormwater runoff (Refer with Table 2).

Table 2. Predictive Modelling Software Key Components and Challenges

Software	Type / Primary Use	Key Components	Challenges	Ref.
EPA SWMM (Storm Water Management Model)	Drainage, Water Detention facility and Hydrologic-Hydraulic Model	√ Run-off Simulations √ Spatial Variability √ Drainage system modeling √ Sizing Detention Facility √ Free Software	√ Hydrometeorological estimates of inflows √ Complex tunnel geometry √ Mixed flows	[47,48]
PCSWMM	Drainage, Watershed, sewer over flow mitigation,	√ Stand-alone GIS for water management	√ Data modeling complexity √ Model Calibration	[49,50]

Software	Type / Primary Use	Key Components	Challenges	Ref.
	floodplain delineation and integrated catchment analysis	√ Time series management and analysis √ Hydraulic and hydrologic modeling √ Evaluates Green infrastructure and water quality modeling √ Ease of configuration of detention pond	√ Inherent Software constraints √ Limited accessibilty √ Cost	
HEC-HMS	Watershed Hydrological Model	√ Open channel Simulations √ Synthetic Precipitation √ Hydrologic simulations √ Model Optimization √ Simulated Watershed response √ Sediment and water quality √ Stream Flow Forecasting √ GIS Connection	√ Complexity in handling large data √ Software integration compatibility √ Display issue using high-resolution images √ Calibration Complexity	[51]
InfoWorks ICM	Integrated Catchment Model (Autodesk)	√ Catchment runoff modeling √ Advanced calibration √ Seamless software integration √ Water quality	√ Simulation stability √ Date integration between complex detailed models √ Modeling urban environments variability √ Cost	[52]

Materials Research Proceedings 66 (2026) 390-405 https://doi.org/10.21741/9781644904152-36

Software	Type / Primary Use	Key Components	Challenges	Ref.
XPSWMM	Stormwater, wastewater and flood modelling	√ Hydrodynamics Modeling √ Simulate rainfall run-off √ Analyze Water Quality	√ Simulation stability √ Date integration between complex detailed models √ Modeling urban environments variability √ Cost	[52]
MIKE URBAN / MIKE FLOOD	Urban Water & Flood Modelling	√ Dynamic water distribution modelling √ Stormwater and waste water analyses √ River modelling √ Predicts the impact of flood event √ Evaluate flood mitigation measures	√ Software transition and integration √ Complex model coupling √ Calibration and Validation	[53,54]
OpenFlows FLOOD	Flood & Hydraulics Modelling	√ Analyze and mitigate flood risks in urban, riverine, and coastal areas √ Simulate hydrological and hydraulic processes √ Flood Forecasting √ Water quality issues	√ Computational intensity √ Data Interoperability √ Integration Complexity √ Model Stability	[55]

The U.S. Environmental and Protection Agency model (EPA SWMM) can simulate general green infrastructure measures designed to reduce runoff volume, delay peak flows, and improve infiltration and evapotranspiration at the source. These practices include bio-retention cells or bioswales, infiltration, and evaporation; continuous permeable pavement systems, and green roofs, which incorporate soil and vegetation above a drainage layer to retain and evapotranspiration roof runoff. The SWMM also includes infiltration trenches, which are gravel-filled narrow excavations that intercept runoff and promote infiltration [47]. However, the unpredictability of hydrometeorological data estimation, the difficulty of hydraulic calculations for tunnel geometry and related structures, and the existence of rapidly varying flow conditions during storms are some of the primary challenges. Furthermore, although realistic simulation of mixed-flow and two-phase flow regimes are currently unrealistic for use at large, system-scale modeling levels due to methodological and technical restrictions [48].

Further, the PCSWMM offers a thorough modeling framework that incorporates catchment and watershed-scale analysis, water quality modeling, low impact development (LID), and green infrastructure evaluation. The software includes sanitary sewer systems with sophisticated control and calibration tools, dual drainage systems that combine surface and subsurface flows. In addition, the platform provides real-time flood forecasts for both immediate and long-term planning, as well as stormwater management and sewer rehabilitation to reduce floods, overflows, and infiltration problems. Planners can assess future hydrologic impacts and create adaptive strategies for resilient, regulation-compliant urban water systems to advanced features that enable floodplain mapping and risk analysis, dynamic storm simulation using high-resolution radar rainfall, and climate change resilience assessments [50]. The software was able to accurately simulate larger multi-purpose watersheds [49]. However, data restrictions, model complexity, and innate software limitations are primarily constraints to adopting PCSWMM for floodplain modeling. Model validation and calibration might prove challenging due to these problems, especially in large and complicated research subject areas [49].

The Hydrologic Engineering Center – Hydrologic Modeling System (HEC-HMS) integrates the modern computational tools, meteorologic processes, and actual basin representation. The software provides an all-inclusive method to the simulation of watershed hydrology. The entire process for discharge modeling is carried out with the help of different techniques for infiltration, ground water flow, and channels routing which can be applied to live data and ongoing events at the same time. Moreover, the HEC-HMS accurate modeling a climatic variability which made possible by the comprehensive meteorologic inputs, such as historical and synthetic precipitation, and evapotranspiration. Flexible temporal specifications are used to govern hydrologic simulations to produce comprehensive tabular and graphical results. Moreover, integrated optimization tools of HEC-HMS can easily be calibrated for modeling parameters against the actual datasets [51].

InfraWorks creates accurate simulation of rivers, sewer networks, surface runoff, and floodplains by combining hydrologic and hydraulic components. The software facilitates the simulation of important aspects of water quality, including temperature fluctuation, sediment behavior, and pollution transport. Additionally, InfraWorks easily integrates with other Autodesk platforms, such as Civil 3D and InfoDrainage, allowing for the ease of integration of other designs. Further, XPSWMM software also provide thorough stream and floodplain simulation model and combines hydrologic and hydraulic assessments. To assess stormwater, sanitary, and sewage systems together. The XPSWMM allows for a thorough modeling of runoff, drainage patterns, and pollutant transfer. The software also opens up the possibility of doing scenario-based analysis just easier, since it compares the performance of green infrastructure, and storm occurrences and also allows for the use of different software [52]. However, the software was met with criticism over its ability to represent temporal and spatial variability, data integration coming from complex and difficult models, and simulation permanence [52].

MIKE+ software which is the model manager acts as the main user interface for the stormwater model setup and the water distribution simulations which are carried out with great efficiency. The software is capable of dynamic hydrological and water quality modeling. Under this circumstance, the system allows for the management of pumps and valves, leak minimization, network segmentation organization, flow volume assessment, and water source tracking. Further, the software simulate river current and enable detailed analysis of flows, hydrologic changes, and pollutant transport quality to guide in disaster mitigation and water quality improvement, while the 2D overland module offers sophisticated flood simulation features to predict risks of failure, and assess mitigation strategies for climate adaptation and disaster preparation [53]

OpenFlows FLOOD uses spatially distributed numerical analysis in modeling hydrologic and hydraulic processes within river basins. The software can be integrated with OpenFlows SewerGEMS to link surface flow with runoff and sewage network behavior, as well as to simulate coastal flooding produced by storm surges. OpenFlows models realistic predictions of flood occurrences, the platform uses real-world digital context and enhanced visualization, as well as ContextCapture 3D reality meshes. The software presents built-in real-time management to compare various mitigation techniques, delineates flood prone area and enhances emergency preparation [55].

The review of predictive modeling software for managing stormwater catch basins and urban floods has brought some important challenges that need to be solved in order to improve urban flood control systems and promote sustainable stormwater management. The key challenges are as follows.

1. Most existing stormwater models fail to capture the complexities of rapidly evolving urban environments, indicating a need for more advanced predictive tools that can integrate the increasing land used and terrain [2,19].
2. Lack of integration of climate predictability in stormwater models, which is crucial for simulating extreme weather and preparing cities for future challenges. Studies have shown that modeling the spatial and temporal climate variability remains a challenge [4,15].
3. The application of machine learning and AI to flood predictions has shown potential, but it is still not extensively developed especially in stormwater catch basin management. Furthermore, there are difficulties in working with large datasets and performing high-number analysis, together with limitations in the use of high-resolution images [8].
4. One of the challenges in software transition and integration is the software compatibility issue, the data interoperability constraint, and the complex model coupling. In addition to that, the high complexity in calibrating and validating models across different scales and datasets still poses a challenge.
5. One of the most significant challenges facing the water forecasting system is the high costs associated with software licensing and model development and maintenance.
6. The last point is that the integration of green infrastructure into predictive models and the scalability of smart technologies in large urban areas still needs to undergo research [17].

Conclusion.

The use of predictive modeling in the management of stormwater catch basin and urban flooding simulations is indeed a great progress to solve the unresolved problems associated with the old methods. There are several hydrologic and hydraulic simulation tools available, and each has its own strengths. The EPA SWMM is particularly adept at modeling the green infrastructure methods such as bioretention systems, permeable pavements, rooftop gardens, infiltration trenches, and stormwater harvesting, which are all aimed at reducing runoff, delaying peak flows, and enhancing infiltration. The PCSWMM extends the capabilities of SWMM by adding watershed-scale estimation, water quality simulation, LID evaluation, real-time flood forecasting, climate change assessment, and detention and sewer system design; hence, it becomes suitable for large, multi-purpose watersheds.

In addition, HEC-HMS provides a powerful and adaptable scenario for watershed hydrology based on meteorologic inputs, with complex runoff and routing algorithms, and time-integrated optimization tools in real-time simulations. On the other hand, InfraWorks and XPSWMM focus on the integrated hydrologic-hydraulic modeling, floodplain visualization, and pollutant transport analysis. In the same way, MIKE+ does offer advanced simulations of water distribution, collection systems, river modeling, and 2D overland flooding but OpenFlows FLOOD is more into fully distributed flood simulation, urban-coastal integration, and providing high-quality visualization in real-world digital environments.

The research shows that existing modeling platforms enable hydrologic and hydraulic simulations, yet no single tool can solve the complexities of urban variability and hydraulic and hydrological modeling requirements. The study demonstrated its importance through existing technological deficiencies in predictive flood modeling tools and demonstrated the urgent requirement for synchronized modeling systems that utilize current monitoring systems and weather changes and research-based insights.

References

[1] L. Cea and P. Costabile, "Flood Risk in Urban Areas: Modelling, Management and Adaptation to Climate Change: A Review," (Mar. 01, 2022), MDPI. https://doi.org/10.3390/hydrology9030050.

[2] K. Guo, M. Guan, and D. Yu, "Urban surface water flood modelling-a comprehensive review of current models and future challenges," Hydrol Earth Syst Sci, vol. 25, no. 5, pp. 2843–2860, (May 2021). https://doi.org/10.5194/HESS-25-2843-2021.

[3] T. S. Bibi, D. Reddythta, and A. S. Kebebew, "Assessment of the drainage systems performance in response to future scenarios and flood mitigation measures using stormwater management model," City and Environment Interactions, vol. 19, p. 100111, (Aug. 2023). https://doi.org/10.1016/J.CACINT.2023.100111.

[4] O. Tamm, T. Kokkonen, L. Warsta, M. Dubovik, and H. Koivusalo, "Modelling urban stormwater management changes using SWMM and convection-permitting climate simulations in cold areas," J Hydrol (Amst), vol. 622, (Jul. 2023). https://doi.org/10.1016/j.jhydrol.2023.129656.

[5] R. D. Rosario, J. A. Padilla, N. G. Bonto, R. C. De Mesa, and O. G. Dela Cruz, "Value engineering on car curbing ownership in metro Manila," in TRANSPORT, ECOLOGY - SUSTAINABLE DEVELOPMENT: EKOVarna2022, AIP Publishing, (Aug. 2023), p. 020017. https://doi.org/10.1063/5.0162414.

[6] D. Habete and C. M. Ferreira, "Impact of Forecasted Land Use Change on Design Peak Discharge at Watershed and Catchment Scales: Simple Equation to Predict Changes," J Hydrol Eng, vol. 21, no. 7, Jul. 2016. https://doi.org/10.1061/(asce)he.1943-5584.0001384.

[7] R. D. Rosario, A. D. La Cruz, and M. P. De Guzman, "A Review of Biomineralization as Solution for Roads and Infrastructures Concrete Sustainability," Civil Engineering Journal, vol. 10, no. 8, pp. 2745–2760, (Aug. 2024). https://doi.org/10.28991/CEJ-2024-010-08-020.

[8] Y. Qin and Y. Qin, "Urban Flooding Mitigation Techniques: A Systematic Review and Future Studies," Water 2020, Vol. 12, vol. 12, no. 12, (Dec. 2020). https://doi.org/10.3390/W12123579.

[9] T. D. Fletcher et al., "SUDS, LID, BMPs, WSUD and more – The evolution and application of terminology surrounding urban drainage," Urban Water J, vol. 12, no. 7, pp. 525–542, (Oct. 2015). https://doi.org/10.1080/1573062X.2014.916314.

[10] Q. Zhou and Q. Zhou, "A Review of Sustainable Urban Drainage Systems Considering the Climate Change and Urbanization Impacts," Water 2014, Vol. 6, Pages 976-992, vol. 6, no. 4, pp. 976–992, (Apr. 2014). https://doi.org/10.3390/W6040976.

[11] R. Rosario D and M. J. Viado, "Encapsulating immobilized ureolytic bacteria yields self-healing concrete apropos sustainable transportation materials: A review," E3S Web of Conferences, vol. 488, p. 03019, (Feb. 2024). https://doi.org/10.1051/e3sconf/202448803019.

[12] X. S. Wang, L. Wan, and B. Hu, "New approximate solutions of horizontal confined–unconfined flow," J Hydrol (Amst), vol. 376, no. 3–4, pp. 417–427, (Oct. 2009). https://doi.org/10.1016/J.JHYDROL.2009.07.050.

[13] C. L. C. García, Y. A. Domínguez, V. M. C. González, J. Jimenez, and L. González-Rodríguez, "Design and assessment of sustainable retention ponds for flood reduction in an urban drainage system in Cuba," Journal of Hydrology and Hydromechanics, vol. 73, no. 3, pp. 273–283, (Sep. 2025). https://doi.org/10.2478/JOHH-2025-0021.

[14] M. Starzec, J. Dziopak, M. Starzec, and J. Dziopak, "A Case Study of the Retention Efficiency of a Traditional and Innovative Drainage System," Resources 2020, Vol. 9, vol. 9, no. 9, pp. 1–19, (Sep. 2020). https://doi.org/10.3390/RESOURCES9090108.

[15] A. I. A. García, J. C. Santamarta, A. I. A. García, and J. C. Santamarta, "Scientific Evidence behind the Ecosystem Services Provided by Sustainable Urban Drainage Systems," Land 2022, Vol. 11, vol. 11, no. 7, Jul. 2022. https://doi.org/10.3390/LAND11071040.

[16] A. N. Angelakis et al., "Evolution of Floods: From Ancient Times to the Present Times (ca 7600 BC to the Present) and the Future," Land 2023, Vol. 12, vol. 12, no. 6, (Jun. 2023). https://doi.org/10.3390/LAND12061211.

[17] J. G. Gacu et al., "Artificial Intelligence (AI) in Surface Water Management: A Comprehensive Review of Methods, Applications, and Challenges," Water 2025, Vol. 17, vol. 17, no. 11, (Jun. 2025). https://doi.org/10.3390/W17111707.

[18] S. Shishegar, S. Duchesne, G. Pelletier, and R. Ghorbani, "A smart predictive framework for system-level stormwater management optimization," J Environ Manage, vol. 278, (Jan. 2021). https://doi.org/10.1016/J.JENVMAN.2020.111505.

[19] D. Chitwatkulsiri, H. Miyamoto, D. Chitwatkulsiri, and H. Miyamoto, "Real-Time Urban Flood Forecasting Systems for Southeast Asia—A Review of Present Modelling and Its Future Prospects," Water 2023, Vol. 15, vol. 15, no. 1, (Jan. 2023). https://doi.org/10.3390/W15010178.

[20] L. Locatelli et al., "Hydrologic impact of urbanization with extensive stormwater infiltration," J Hydrol (Amst), vol. 544, pp. 524–537, (Jan. 2017). https://doi.org/10.1016/J.JHYDROL.2016.11.030.

[21] C. O. Ongaga, M. Makokha, K. Obiero, I. Kipkemoi, and J. Diang'a, "Urbanization and hydrological dynamics: a 22-year assessment of impervious surface changes and runoff in an urban watershed," Frontiers in Water, vol. 6, p. 1455763, (Dec. 2024). https://doi.org/10.3389/FRWA.2024.1455763/BIBTEX.

[22] A. Sabut, A. M. Bhat, and P. K. Mohapatra, "Stormwater control measures, design aspects, and impact of climate change – A review," Water-Energy Nexus, (Dec. 2025). https://doi.org/10.1016/J.WEN.2025.12.001.

[23] M. S. B. Kusuma, F. Isma, E. O. Nugroho, M. B. Adityawan, and A. Azmeri, "An assessment of mangrove restoration contribution for compound flood risk reduction in built estuary area based on a hydrodynamic model," Nature-Based Solutions, vol. 9, p. 100295, (Jun. 2026). https://doi.org/10.1016/J.NBSJ.2025.100295.

[24] "Floods: Mitigation and Adaptation – Green Convergence For safe food, healthy environment and sustainable economy." Accessed: (Dec. 24, 2025). [Online]. Available: https://greenconvergencephil.com/floods-mitigation-and-adaptation/

[25] L. M. Ahiablame, B. A. Engel, and I. Chaubey, "Effectiveness of Low Impact Development Practices: Literature Review and Suggestions for Future Research," Water, Air, & Soil Pollution 2012 223:7, vol. 223, no. 7, pp. 4253–4273, (Jun. 2012). https://doi.org/10.1007/S11270-012-1189-2.

[26]: : "Pervious Pavement :: Pervious Concrete for Green, Sustainable Porous and Permeable Stormwater Drainage ::" Accessed: (Nov. 08, 2023). [Online]. Available: http://www.perviouspavement.org/materials.html

[27] Z. G. Niu, Z. W. Lv, Y. Zhang, and Z. Z. Cui, "Stormwater infiltration and surface runoff pollution reduction performance of permeable pavement layers," Environmental Science and Pollution Research, vol. 23, no. 3, pp. 2576–2587, (Feb. 2016). https://doi.org/10.1007/S11356-015-5466-7.

[28] M. Cotton, R. Rynk, L. Loder-Rossiter, and A. Carpenter, "Site planning, development, and environmental protection," The Composting Handbook: a how-to and why manual for farm, municipal, institutional and commercial composters, pp. 409–500, (Jan. 2021). https://doi.org/10.1016/B978-0-323-85602-7.00018-2.

[29] B. Chen and T. F. M. Chui, "A review on implementing sustainable drainage systems in sloping environments: understanding, approaches, and opportunities," J Hydrol (Amst), vol. 661, p. 133577, (Nov. 2025). https://doi.org/10.1016/J.JHYDROL.2025.133577.

[30] "What are Retention/Detention Systems & Why They are Important." Accessed: Dec. 24, 2025. [Online]. Available: https://www.bushmantanks.com.au/blog/what-are-retention-detention-systems-why-they-are-important?srsltid=AfmBOooWx-k4qmYh0wp7-4dtzMiFGlv-ALk4VOAxv6p-4ounlZFM9v1a

[31] G. McDowell, M. Koppes, G. McDowell, and M. Koppes, "Robust Adaptation Research in High Mountains: Integrating the Scientific, Social, and Ecological Dimensions of Glacio-Hydrological Change," Water 2017, Vol. 9, vol. 9, no. 10, Sep. 2017. https://doi.org/10.3390/W9100739.

[32] "Bonifacio Global City - DCCD Engineering Corporation." Accessed: Dec. 24, 2025. [Online]. Available: https://www.dccd.com/project/bonifacio-global-city/

[33] "Asian Gallery | The 'flood control tunnel' in Bonifacio Global City (BGC) is actually an underground retarding basin or detention tank located beneath Burgos Circ... | Facebook." Accessed: Dec. 24, 2025. [Online]. Available: https://www.facebook.com/groups/318122976168561/posts/1497477284899785/

[34] "StormChamber® Stormwater Detention Basin Guide | NDS." Accessed: Dec. 24, 2025. [Online]. Available: https://www.ndspro.com/us/en/resources/articles/stormchamber--guide

[35] K. Heiza, A. Nabil, N. N. Meleka, M. Tayel, K. Heiza, and N. Meleka, "State-of-the Art Review: Strengthening of Reinforced Concrete Structures-Different Strengthening Techniques," 2014. [Online]. Available: https://www.researchgate.net/publication/260991475

[36] G. Dharmarathne, A. O. Waduge, M. Bogahawaththa, U. Rathnayake, and D. P. P. Meddage, "Adapting cities to the surge: A comprehensive review of climate-induced urban flooding," Results in Engineering, vol. 22, p. 102123, Jun. 2024. https://doi.org/10.1016/J.RINENG.2024.102123.

[37] M. Wang, J. Zhao, Z. Xiong, M. Zhang, L. Wang, and S. K. Tan, "Strategic deployment of nature-based solutions for urban flood management in high-density urban landscapes," Ecol Indic, vol. 176, p. 113681, Jul. 2025. https://doi.org/10.1016/J.ECOLIND.2025.113681.

[38] M. del M. Pérez Cambra, M. D. Martínez Santafé, and J. Roca Cladera, "Exploring the mitigation of compound events in Barcelona: Urban water scarcity, flood risk and reduction of surface temperatures through water-sensitive urban design," Urban Clim, vol. 59, p. 102298, Feb. 2025. https://doi.org/10.1016/J.UCLIM.2025.102298.

[39] N. Oberg, C. E. Bondar, M. A. Hoy, and A. R. Schmidt, "Conveyance Analysis of Chicago's ""Deep Tunnel"" System," Examining the Confluence of Environmental and Water Concerns - Proceedings of the World Environmental and Water Resources Congress 2006, pp. 1–10, 2007. https://doi.org/10.1061/40856(200)10.

[40] S. Ilgen, F. Sengers, A. Wardekker, S. Ilgen, F. Sengers, and A. Wardekker, "City-To-City Learning for Urban Resilience: The Case of Water Squares in Rotterdam and Mexico City," Water 2019, Vol. 11, vol. 11, no. 5, May 2019. https://doi.org/10.3390/W11050983.

[41] Y. J. Cho, "Optimal design of deep stormwater drainage tunnels to address urban flooding in Korea," Scientific Reports 2024 14:1, vol. 14, no. 1, pp. 24896-, Oct. 2024. https://doi.org/10.1038/s41598-024-76471-1.

[42] M. Kõiv-Vainik et al., "Urban stormwater retention capacity of nature-based solutions at different climatic conditions," Nature-Based Solutions, vol. 2, p. 100038, Dec. 2022. https://doi.org/10.1016/J.NBSJ.2022.100038.

[43] J. Li et al., "An Event-Based Resilience Index to Assess the Impacts of Land Imperviousness and Climate Changes on Flooding Risks in Urban Drainage Systems," Water 2023, Vol. 15, vol. 15, no. 14, Jul. 2023. https://doi.org/10.3390/W15142663.

[44] S. Amerinia, "Assessment of low-impact development practices on stormwater management using SWMM 5.2: a case study of Shiraz, Iran," 2023. https://doi.org/10.48336/ZZ9D-D738.

[45] E. Sañudo, L. Cea, J. Puertas, E. Sañudo, L. Cea, and J. Puertas, "Modelling Pluvial Flooding in Urban Areas Coupling the Models Iber and SWMM," Water 2020, Vol. 12, vol. 12, no. 9, Sep. 2020. https://doi.org/10.3390/W12092647.

[46] R. D. Rosario, A. Alvarez, R. C. Quinto, and M. de Guzman, "Intelligent Transportation Systems: Enabling Sustainable Transportation and Efficient Traffic Management—A Review," Communications in Computer and Information Science, vol. 2378 CCIS, pp. 311–323, 2025. https://doi.org/10.1007/978-3-031-84148-4_24.

[47] "Storm Water Management Model (SWMM) | US EPA." Accessed: Dec. 25, 2025. [Online]. Available:https://www.epa.gov/water-research/storm-water-management-model-swmm

[48] V. G. Geller, Y. Tao, A. Lokhandwala, J. G. Vasconcelos, D. B. Wright, and B. R. Hodges, "Evaluating SWMM Modeling Performance for Rapid Flows on Tunnels with Geometric Discontinuities," Journal of Hydraulic Engineering, vol. 151, no. 1, p. 04024052, Jan. 2025. https://doi.org/10.1061/JHEND8.HYENG-14107;WGROUP:STRING:PUBLICATION.

[49] M. S. Ansari et al., "Exploring PCSWMM for Large Mixed Land Use Watershed by Establishing Monitoring Sites to Evaluate Stream Water Quality," Hydrology 2024, Vol. 11, vol. 11, no. 7, Jul. 2024. https://doi.org/10.3390/HYDROLOGY11070104.

[50] "SWMM5 modeling with PCSWMM." Accessed: Dec. 25, 2025. [Online]. Available: https://www.pcswmm.com/

[51] "HEC-HMS." Accessed: Dec. 25, 2025. [Online]. Available: https://www.hec.usace.army.mil/software/hec-hms/

[52] "Autodesk InfoWorks ICM 2026 | Get Prices & Buy Official InfoWorks ICM." Accessed: Dec. 25, 2025. [Online]. Available: https://www.autodesk.com/sg/products/infoworks-icm/overview

[53] "MIKE+ | Integrated Water Modelling Platform." Accessed: Dec. 25, 2025. [Online]. Available: https://www.dhigroup.com/technologies/mikepoweredbydhi/mikeplus

[54] M. Q. Mahmood, X. Wang, F. Aziz, and N. Dogulu, "Urban flood modelling: Challenges and opportunities - A stakeholder-informed analysis," Environmental Modelling & Software, vol. 191, p. 106507, Jun. 2025. https://doi.org/10.1016/J.ENVSOFT.2025.106507.

[55] "OpenFlows FLOOD | Hydraulics Modeling Software | Bentley Systems." Accessed: Dec. 25, 2025. [Online]. Available: https://www.bentley.com/software/openflows-flood/

[56] A. P. Monachese, M. T. Gómez-Villarino, J. López-Santiago, E. Sanz, A. F. Almeida-Ñauñay, and S. Zubelzu, "Challenges and Innovations in Urban Drainage Systems: Sustainable Drainage Systems Focus," *Water 2025, Vol. 17, Page 76*, vol. 17, no. 1, p. 76, Dec. 2024. https://doi.org/10.3390/w17010076.

[57] A. Sabut, A. M. Bhat, and P. K. Mohapatra, "Stormwater control measures, design aspects, and impact of climate change – a review," *Water-Energy Nexus*, vol. 9, no. 1, pp. 46–74, Dec. 2026. https://doi.org/10.1016/j.wen.2025.12.001.

[58] J. Xie *et al.*, "Enhancing Urban Drainage Resilience Through Holistic Stormwater Regulation: A Review," *Water 2025, Vol. 17, Page 1536*, vol. 17, no. 10, p. 1536, May 2025. https://doi.org/10.3390/w17101536.

[59] R. D. Rosario, F. Poso, A. Victoria, and M. de Guzman, "Bio-Engineered Fiber-Reinforced Rigid Pavement: A Durability, Strength Recovery and Self-Healing Evaluation," pp. 625–637, 2026. https://doi.org/10.1007/978-981-95-4534-6_58.

Materials Research Proceedings 66 (2026) 406-422 https://doi.org/10.21741/9781644904152-37

Feature Selection-Driven Supervised Machine Learning for Dissolved Oxygen Prediction Using Multidomain Environmental Parameters in a Major Fishing Lake

Al R. ROMANO[1,a], Eden May B. DELA PENA[2,b], Karl Ezra S. PILARIO[3,c]

[1]Technological Institute of the Philippines, Cubao, Quezon City 1109, Philippines

[2]Department of Chemical Engineering, University of the Philippines, Diliman, Quezon City 1101, Philippines

[3]Department of Mining, Metallurgical and Materials Engineering, University of the Philippines, Diliman, Quezon City 1101, Philippines

[a]aromano.ense@tip.edu.ph, [b]ebdelapena@up.edu.ph, [c]kspilario@up.edu.ph

Keywords: Dissolved Oxygen, Machine Learning, Water Quality Prediction, Aquaculture, Feature Selection

Abstract. Dissolved oxygen (DO) holds an important role in water quality monitoring due to its impact on the aquatic organisms inhabiting the water body. Sampaloc lake, located in San Pablo City in Laguna Province, Philippines, one of the major fishing lakes, has been affected by the decrease of DO concentration leading to continuous fish kill incidents. This study used supervised machine learning-based model utilizing the filter and wrapper method in feature selection to predict the DO levels in Sampaloc Lake. The goal of the study is to provide a baseline for building an early warning system for fish kills events in the lake. Water quality, physical and hydroclimatic parameters are the range of features selected as predictors to create a reliable DO predictive model. Based on the feature selection process, water temperature, atmospheric temperature, biochemical oxygen demand, surface pressure and pH are the consistent important feature in predicting DO for both filter and wrapper method. Overall, the result of the eXtreme Gradient Boosting Regression (XGBR) Model of the filter method with $R^2 = 0.95$, RMSE = 0.14, and MAE = 0.10 showed a reliable performance among the predicted models (Multiple Linear Regression, Random Forest, Kernel Ridge Regression, and Bootstrap Aggregating) tested in the study.

Introduction

Dissolved Oxygen (DO) refers to the presence of oxygen gas dissolved in water, which holds significance in evaluating water quality because of its impact on the aquatic organisms occupying the water body. Although various environmental factors and phenomena increase the concentration of dissolved oxygen [1], the unstructured, unpredictable, and complex nature of the aquaculture environment makes it readily susceptible to abnormal fish hypoxia when the concentration of water environmental variables exceeds a fish's tolerance threshold. For this reason, it is very crucial to monitor the abnormal fish hypoxia behavior [2].

Sampaloc Lake, located in the heart of San Pablo City, Laguna, is one of the seven crater lakes in the city and is primarily used for aquaculture [3] and ecotourism, which have a negative impact on the water quality [4]. Moreover, since 1988, fish cage operators have encountered fish incidents, which result in significant financial losses of millions of pesos. To this day, fish kill incidents due to dissolved oxygen depletion remain evident. In fact, on January 5, 2024, approximately two tons of tilapia raised through aquaculture were lost due to a fish kill that affected Sampaloc Lake in Laguna caused by a depletion of dissolved oxygen [5].

The prediction of DO is challenging in the Machine Learning (ML) process because DO is nonlinear and typically dependent on other water quality parameters and weather patterns [6]. A review of related literature reveals three key gaps in current ML applications for DO prediction. First, numerous studies investigated the DO content in lakes and other bodies of water without considering the site's fish farming activities and hydroclimatic characteristics. Second, utilization of different feature selection methods to determine most influential parameters has not been explored to achieve a higher model's accuracy. Lastly, there is a lack of studies that utilize a comprehensive set of predictor variables spanning water quality, physical, and hydroclimatic parameters. Most prior works have focused on a limited subset of these factors, potentially overlooking the complex and nonlinear interactions influencing DO levels.

The main goal of this study is to develop a supervised machine learning-based model for predicting DO in Sampaloc Lake using the filter and wrapper methods of feature selection. The main innovations of this study are as follows: (1) The study takes in a broader approach by combining physical and hydroclimatic parameters with water quality parameters as predictors of DO levels, highlighting land use and fish cage presence in the research locale. (2) The use of filter and wrapper methods in the feature selection process, and (3) the study utilized nonlinear machine learning models to effectively capture the complex relationships between predictors and DO.

Materials and Methods

Research Locale. The study was conducted at Sampaloc Lake, located at the center of San Pablo City in Laguna Province, Philippines (see Fig. 1). The coordinates of the Sampaloc Lake are 14.079524 degrees North Latitude and 121.328782 degrees East longitude, and a volcanic crater lake that spans a total surface area of 96.57 hectares with a perimeter of 3.748 kilometers. It is approximately 120 meters above sea level. The water is sourced from rainfall, surface runoff, and numerous natural springs and exits through seepage, evaporation, and outflow into Sabang Creek, the lake's sole water outlet [7].

Research Design. Historical observations, such as physical, water quality, and hydroclimatic parameters, as shown in Table 3, were collected from credible sources. The water quality parameters were collected monthly from 2010 to 2022 from designated sampling stations of Laguna Lake Development Authority (LLDA), resulting in 78 records. However, there are missing values because of the COVID-19 period. On the other hand, physical parameters were collected from the San Pablo Local Government Unit (LGU), while the hydroclimatic parameters were gathered from the National Aeronautics and Space Administration (NASA) Earth Science Research Program.

The descriptive analysis was conducted to determine the characteristics of the data, while the correlational analysis was used to examine the relationship between the identified parameters of the study. The filter and wrapper methods were used in the filter selection process to identify the most important features or parameters for predicting the DO value. In developing the model, the first step is to apply data preprocessing such as data normalization or scaling. After data preprocessing, the dataset will be divided into training and testing sets. Using Optuna, both the training set and testing set were used to train and optimize the different ML model algorithms. The performance of the predictive model was evaluated using the Coefficient of Determination (R^2), Root Mean Squared Error (RMSE), and Mean Absolute Error (MAE) (see Fig. 2). In this study, the seasonal variability was not considered in predicting the DO value.

Feature Selection. The Feature selection techniques were applied to ensure that the most relevant variables were retained while minimizing the inclusion of irrelevant features in the dataset. The filter and wrapper method was used as the basis for this selection process. In the filter method, Spearman correlation was computed for all features against the target variable (DO). The most correlated features were selected for further analysis. In the wrapper method, a feature selection method was applied using Sequential Feature Selector (SFS) with Linear Regression as the estimator. The forward selection approach iteratively added features to the model, selecting those that improved predictive performance. The process continued until seven (7) optimal features were identified.

Machine Learning Models. This section describes the ML models used in this paper for predicting DO concentrations (target, y) from the input multidomain environmental parameters (predictors, x), measured in the lake. A total of six (6) supervised machine learning algorithms were applied for analysis: Multiple Linear Regression (MLR), Kernel Ridge Regression (KRR), Support Vector Regression (SVR), Random Forest (RF), XGBoost Regression (XGBR), and Bagging Regressor. These models were employed using Python libraries for the regression tasks.

MLR was implemented to set a baseline model, using its simplicity and interpretability to understand linear relationships between DO and the predictors. KRR, SVR, RF, XGBR and Bagging regressor on the other hand, were chosen for its ability to model complex, nonlinear interactions using kernel functions and other hyperparameters through optimization, which MLR does not provide.

Baseline Model: Multiple Linear Regression (MLR). MLR, in this study serves as a reliable baseline model because it is simple enough to have interpretable results. The goal is to fit a linear model for predicting the DO content (y), using predictors (x) and linear coefficients, β [8]:

$$y = \beta_0 + \beta_1 x_1 + \beta_2 x_2 \ldots + \beta_n x_n + \epsilon \tag{1}$$

where *y* is the dependent variable, *x* is the explanatory or dependent variables *n* is the sample size and ϵ is the residual or error [8].

Kernel Ridge Regression (KRR) Model. To model the nonlinear relationships between the target variable (DO) and its predictors, KRR combines the principles of ridge regression with the kernel trick. In this study, a radial basis function (RBF) kernel was employed due to its effectiveness in capturing complex, nonlinear patterns within the data. The associated predictiom equation is given by [9]:

$$y = k\,(x, X)^T k\,(K + \alpha I)^{-1} y \tag{2}$$

where y is the dependent variable, $k\,(x, X) \in \mathbb{R}^n$, is a kernel function operating on any input x, $K \in \mathbb{R}^{nxn}$ is the kernel matrix, α is the regularization parameter and *I* is the identity matrix [15]:

Support Vector Regression (SVR) Model. The goal of SVR is to identify the smallest possible tube around the function's surface, minimizing the prediction error, the ε-insensitive tube, which is the gap between the predicted and actual outputs [10]. The hyperplane serves as the decision boundary that defines the optimal values. Adjusting the maximum ε attains the desired accuracy for the model. The slack variables (ξ) permit some training errors, ensuring the model's robustness against noise and outliers.

In this model, the predictive equation is [23]:

$$y = \sum_{i=1}^{n} \beta_i^* k(x, x') + b^* \tag{3}$$

where $k(x, x')$ is the kernel function that measures the similarity between two data points. β^* and b^* represent the optimal dual variables and the bias parameter, respectively. The support vectors correspond to the training data points for which $\beta^* \neq 0$ [23].

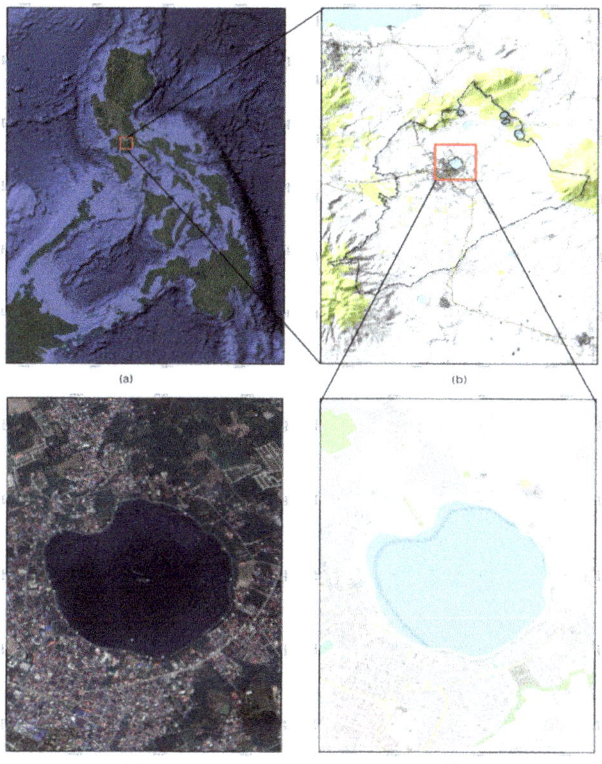

Fig. 1. Location of the study area: (a) Philippines, (b) San Pablo City vicinity, (c) Sampaloc Lake, and (d) Sampaloc Lake satellite view

Random Forest (RF) Model. RF is an ensemble learning technique that combines multiple decision trees to improve predictive accuracy and reduce overfitting. The base learners in the RF model are called decision trees. Each individual tree in the forest is built using a subset of the data, created through bootstrap sampling (sampling with replacement). The final prediction is obtained by averaging the outputs of the trees for regression tasks, or by taking the majority vote for classification tasks [11]. Mathematically, RF can be represented as follows [24]:

$$y = \frac{1}{B}\sum_{b=1}^{B} T_b(x) \quad (4)$$

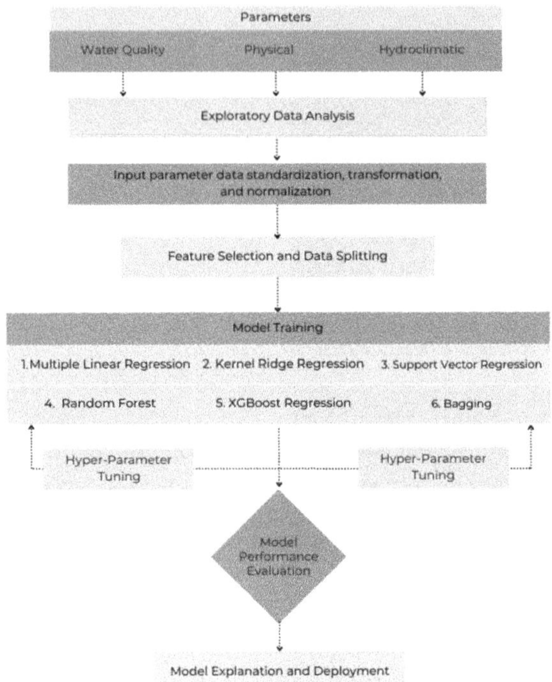

Fig. 2, Research Design

where $T_b(x)$ is the bth decision tree's output value and B is the total number of decision trees. The hyper-parameters that were tuned in RF in this work are the number of decision trees B and the splitting criterion which decides where to split the data space in each node.

eXtreme Gradient Boosting Regression (XGBR) Model. XGBR is a gradient boosting framework that has become popular for its high efficiency and performance [12]. In XGBoost, trees are constructed sequentially, with each new tree attempting to correct the residual errors from the previous one. The goal of the model is to minimize a loss function, L(), with respect to a set of parameters, θ. The model uses a regularization term to prevent overfitting and improve generalization [13].

The XGBR prediction equation can be represented as [25]:

$$y = \sum_{k=1}^{K} f_k(x_i) \tag{5}$$

where y is the predicted value for observation *i*, $f_k(x)$ represents the k-th regression tree, and K is the total number of trees. The objective function to minimize during training is:

$$L(\theta) = \sum_{i=1}^{n} L_i(f(x_i) + \sum_{k=1}^{K} \omega(f_k) \tag{6}$$

where $L_i(f(x_i)$ is the loss function for the i-th data point, and (f_k) is the regularization term where $\omega(f_k)$ helps control overfitting by penalizing the complexity of the tree f_k
Bootstrap Aggregating (Bagging) Model. XGBR is a straightforward ensemble technique. Using the original training set, several bootstrap samples (random samples with replacement) are created, and a different model is trained on each sample to accomplish bagging. A majority vote (for classification) or an average of the predictions (for regression) from each model in the ensemble yields the final prediction. One of the primary benefits of Bagging is its ability to reduce variance, especially when using high-variance models like decision trees [14]. Mathematically, the Bagging model predictive equation can be written as [24]:

$$y = \frac{1}{M}\sum_{m=1}^{M} f_m(x) \quad (7)$$

where y is the predicted value, $f_m(x)$ is the prediction from the m-th model, and M is the total number of models in the ensemble. Each model is trained on a different bootstrap sample, drawn randomly from the original training set.

Hyperparameter Tuning using Optuna. Bayesian Optimization (BayesOpt) is a cutting-edge method for tuning machine learning model hyperparameters using cross-validation [15]. This study utilizes Bayesian Optimization through the Optuna library available in Python, with an objective to find the hyper-parameter values that minimize the 5-fold cross-validation mean squared error on the training data [16,17].

Optuna is an open-source hyperparameter optimization framework designed to accelerate the process of finding optimal model configurations. It is selected over other optimization tools due to its flexibility, efficiency, and seamless integration with a wide range of machine learning frameworks. Unlike traditional methods like grid or random search, it focuses on the most promising regions, enabling faster convergence and better performance. Each model in this study except MLR, undergoes hyperparameter tuning with an equal number of optimization iterations, and only the best-performing configuration from these trials is used for comparison during the testing phase, ensuring an unbiased selection of the optimal model.
Model Performance Metrics. To quantify the reliability and accuracy of the models, three (3) performance metrics are considered: Coefficient of Determination (R^2), Root Mean Squared Error (RMSE), and Mean Absolute Error (MAE).

R^2 indicates the proportion of the variation in the dependent variable that can be explained by the independent variables. Thus, R^2 measures the goodness of fit [18]. RMSE measures the average magnitude of the errors between predicted values and observed values [19]. If R^2 measures the correlation between the predicted and observed values, RMSE provides a measurement of the difference between these values. On the other hand, MAE calculates the average of the absolute differences between predicted ($\hat{y}_i$) and actual (y_i) values. It assesses the accuracy of a model's predictions on a continuous scale [20].

Results and Discussion
Descriptive Statistical Analysis. Descriptive statistical analysis (see Table 3) revealed that in water quality parameters, Chloropyll-a (Chla) and Chloride (Chl) showed a high standard deviation, which potentially reduced model accuracy. On the other hand, due to limited data of the physical parameters, these may possess little information for the model to learn meaningful patterns. Precipitation (Pr) under hydroclimatic parameters has a significant variability making it a potential variable for DO predictive modeling. Other hydroclimatic parameters have a low standard deviation that contributed minimally to the model's predictive ability.

Table 1. Performance Metrics used for Model Evaluation

Performance Metrics	Range	Goal	Equation	Source	
R^2	0 to 1	High	$R^2 = 1 - \sum_{i=1}^{n} (\frac{y_i - \widetilde{y_i}^2}{y_i - \underline{y}^2})$	[18]	(8)
MAE	0 to ∞	Low	$RMSE = \sqrt{\frac{1}{n}\sum_{i=1}^{n}(y_i - \hat{y}_i)^2}$	[19]	(9)
RMSE	0 to ∞	Low	$MAE = \frac{1}{n}\sum_{i=1}^{n} \lvert\hat{y}_i - y_i\rvert$	[20]	(10)

Correlational Analysis. Correlational analysis of the predictors to the target variable, DO, was performed to identify the most influential predictors for DO, a Pearson correlation heatmap was generated (Figure 3) and can be interpreted as follows:

Water Quality Parameters. Parameters related to temperature, including water temperature (WT)(R=0.55) and air temperature (AT)(R=0.52), exhibit a positive correlation with DO levels. While Henry's law states that as WT increases, oxygen solubility decreases, supported by [20,21], the positive correlation observed suggests that thermal energy and solar radiation are likely driving primary productivity that enhances photosynthetic activity of algae, aquatic plants and phytoplankton during warmer periods thereby releasing oxygen into the environment [26, 27].

On the other hand, water quality parameters linked to nutrient enrichment and organic pollution, such as biochemical oxygen demand (BOD)(R=-0.52), phosphate (P)(R=-0.05), and nitrate (N)(R=-0.08), show a negative correlation with DO. These have repeatedly demonstrated consistently with the previous studies that nutrient over-enrichment and organic pollution are key drivers of hypoxia in aquatic ecosystems. Linking BOD and DO, microbial respiration is the main mechanism that drives DO depletion [28,29] Organic matter present in the water body are consumed by the microorganism requiring oxygen to survive [27]. Excessive nutrients (P and N) input in the water body stimulates rapid algal growth-a process called euthrophication. Although algae produce oxygen through photosynthesis, their death and decomposition by aerobic bacteria consumes DO leading to oxygen depletion [27].

Existing studies frequently associate anthropogenic impacts leading to DO depletion [30,31]. The result of the analysis support the idea by [32] that DO levels are not only influenced by pollution but a dynamic equilibrium between supply and demand.

Physical Parameters. Fish cage (FC) and land use land cover (LULC) was correlated with DO to show the anthropogenic impacts to DO levels. Nutrient and organic matter loading is the main mechanism linking DO, water body areas (WB)(R=-0.16) and agricultural areas (Ag) (R=-0.05). The negative correaltion indicates that agricultural and aquaculture activities are a primary source of nutrient and organic pollution, which leads to euthrophication and DO depletion [33]. The positive correlation of built-up areas (BU) (R=0.20) could be because of effective wastewater treatment operation that reduces organic loading as mentioned by [33]. Moreover, the positive correlation of vegetative areas (Vg) (R=0.05) indicates the effectives of natural filters in trapping organic matter, sediments and other pollutants before reaching the water body as mentioned by [34]. This correlation highlights that in certain context, LULC does not necessarily mean detrimental to DO levels as suggested by [35]. The results suggest that with effective wastewater treatment system, DO concentration can be maintained.

Overall, LULC showed a weak correlation with DO. These results might be attributed to the characteristics of the physical parameter data, wherein measurements were collected only every three years, resulting in low correlational interpretability. However, it is important to note that realizing these localized correlations is necessary for developing water qualtity interventions [31,36].

Hydroclimatic Parameters. Moisture-related parameters, specifically precipitation (Pr) (R=0.20), all-sky surface shortwave downward irradiance (SDI) (R=0.25) and relative humidity (RH) (R=0.15), positively correlate with DO concentrations. This means that higher precipitation leads to greater surface runoff and enhances water mixing, which dilutes pollutants and supports the oxygenation of surface waters. This is consitent with the principle that oxygen sollubility decreases as temperature rises [37]. RH is positively correlated with DO indicating that while increasing humidity indirectly lead to higher water temperature due to less evaporative cooling from water surface, this could also associated with increase cloud cover that could lower temperature by reducing solar radiation as discussed by [38]. The weak correlation also implies that other factors mitigate the direct positive effects of these variables to DO levels.

Parameters related to wind, namely wind speed (WS) and wind direction (WD), demonstrate a negative correlation with DO. WS (R=-0.24) correlation findings appears counterintuitive at first glance, as higher wind speed improves aeration process and increasing oxygen transfer from the atmospher into the water body [39]. However, this negative correlation may be attributed to the fact that strong wind can cause turbidity and sediment resuspension as mentioned by [40]. Based on the land cover around the Sampaloc lake, built up areas occupy the western part of the Sampaloc lake while agricultural areas are dominant on its southern part. Since these sections of the lake are relatively shallow. This could be the potential reasons why wind direction, positive towards east (WDx) and north (WDy) are negatively correlated with DO. Increase wind magnitude ans specific directions can lead to increased wave action and turbulence, causing the resuspension of bottom sediments.

Surface Pressure (SP) (R=-0.43) shows moderate negative correlation with DO. This mechanism can also be linked to Henry's Law, which states that DO is directly proportional to the partial pressure of oxygen in the atmosphere above it. However other factors can be associated might dominate such as weather conditions and wind resulting to a negative correlation. Reduced wind-driven mixing hinders oxygen transfer, clear skies can increased solar radiation, and consequently decreasing oxygen solubility.

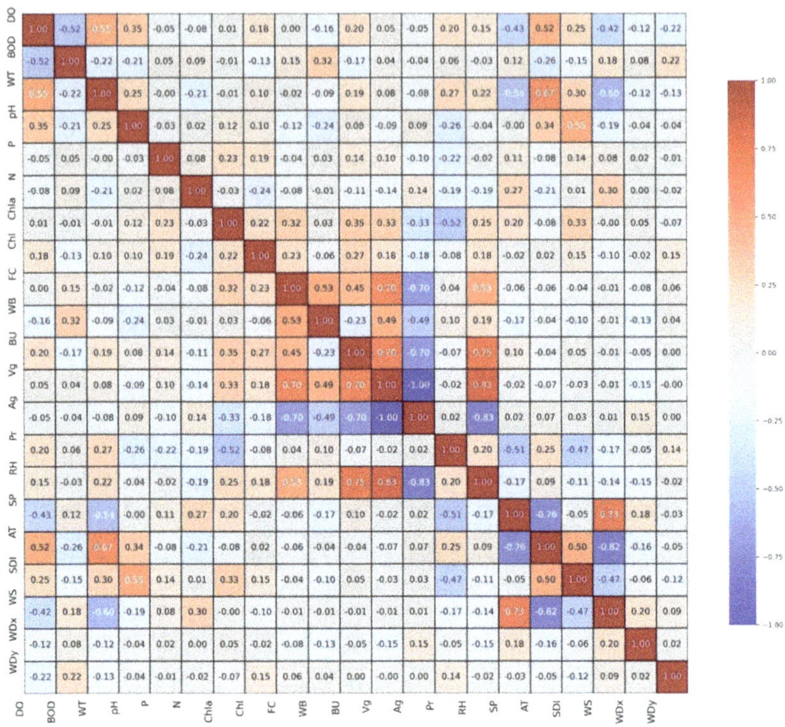

Fig. 3. Correlational Matrix

Feature Selection. Feature selection methods were performed to determine the best features to predict DO levels (see Table 2). Most of the features selected are from water quality and hydroclimatic parameters. The land use cover limits its contribution to the model because the unavailability of data reduces the statistical significance and variability of these parameters.

Table 2. Selected Features using Filter and Wrapper Method

Feature Selection Method	Selected Features
Filter	WT, AT, BOD, WS, SP, pH and RH
Wrapper	BOD, WT, pH, FC, Pr, SP, and AT

Table 3. Dataset Composition and Descriptive Statistical Analysis

Parameter	Role	Count	Completeness (%)	Std	Min	25%	50%	75%	Max
Dissolved Oxygen	Target	101	97.1	1.43	2.2	4.18	5.3	6.4	9.2
Biological Oxygen Demand	Input	90	86.5	2.65	2	4.31	7	8.77	13
Water Temperature	Input	95	98	2.37	15	26.5	28	29	35
pH	Input	97	93.2	0.51	6.5	7.4	7.7	8	9
Phosphate	Input	86	88.5	0.75	0.01	0.65	0.91	1.26	4.8
Nitrate	Input	82	79.8	0.23	0	0.05	0.11	0.31	1
Chlorophyll-a	Input	77	74	61.24	4.43	22.76	36.66	59.77	525.07
Chloride	Input	73	88.5	24.86	4	11	14	19	167
Agricultural Area	Input	104	100	1423.3	990.02	990.02	1327.2	3965.2	3965.2
Fish Cage Area	Input	104	100	2.3	9.73	10.02	10.34	10.98	18.63
Water Body Area	Input	104	100	29.81	115.56	126.72	126.72	186.57	186.57
Built-Up Area	Input	104	100	444.04	1046.1	1046.3	1471.28	1471.3	2171.3
Vegetative Area	Input	104	100	783.08	3103.5	3103.5	4626.8	4695.3	4695.3
Agricultural Area	Input	104	100	1423.3	990.02	990.02	1327.21	3965.2	3965.2
Precipitation	Input	104	100	175.54	6.17	147.41	221.9	366.36	931.21
Relative Humidity	Input	104	100	30.81	23.43	25.87	78.92	87.76	92.03
Surface Pressure	Input	104	100	0.2	98.9	99.08	99.22	99.38	99.68
Atmospheric Temperature	Input	104	100	1.3	22.32	24.53	25.5	26.36	28.61
Short Downward Irradiance	Input	104	100	2.86	9.7	13.25	15.4	18.16	21.81
Wind Speed	Input	104	100	0.25	0.86	1.19	1.42	1.59	1.88
Wind Direction (x-component)	Input	104	100	0.69	-1	-0.79	-0.04	0.58	1
Wind Direction (y-component)	Input	104	100	0.72	-1	-0.86	-0.11	0.62	1

Multiple Linear Regression. The MLR serves as a baseline model to determine a linear connection between DO levels and selected parameters based on filter and wrapper feature selection methods. The resulting predictive equation in the normalized variable space was determined as follows:

For filter method:

$$\mathrm{DO} = -2.75 \times 10^{-15} - 0.41\mathrm{BOD} + -0.15\mathrm{WT} + 0.19\mathrm{pH} + 0.08\mathrm{RH} - 0.20\mathrm{SP} - 0.14\mathrm{AT} + 0.10\mathrm{WS} \quad (11)$$

For wrapper method:

$$\mathrm{DO} = 6.11 \times 10^{-15} - 0.46\mathrm{BOD} + -0.13\mathrm{WT} + 0.22\mathrm{pH} + 0.10\mathrm{FC} + 0.14\mathrm{Pr} - 0.09\mathrm{SP} + 0.12\mathrm{AT} \quad (12)$$

As expected, linear regression is insufficient for modeling DO, since the accuracies are very low, as shown in Figure 4 for both feature selection methods. With R^2 values of 53.2% and 56.3% on the test data for the filter and wrapper methods, respectively, the model demonstrates weak explanatory power.

Nonetheless, Eq. 13 and 14 also reveal how influential each predictors by looking at the magnitude of their coefficients. The most critical and consistent predictor in both feature selection methods is BOD, followed by WT, then pH and SP. The model suggests that any increase in BOD concentration has the greatest impact on the increase in DO concentration.

Comparison of ML Models. The summary of the rest ML models trained and tested on the data are shown in Fig. 5. It is important to note that these models were already finetuned by Bayesian optimization with 150 iterations on Optuna.

With respect to performance, each model attained an accuracy exceeding 87% for the training set and above 89% for the testing set in the filter method of feature selection In contrast, the wrapper method resulted in over 87% accuracy for the training set and above 82% for the testing set. This indicates that all nonlinear models are reliable estimate DO concentration using the parameters listed in Table 3 for both methods. Fig. 6 shows the scattered plot of predicted and actual values for all models. The findings show that the sources of errors in the model predictions are primarily concentrated in the lower range of DO values, where the model tends to underestimate the observed DO concentrations.

On the training data, most models performed exceptionally well under the Filter Method, with XGBR achieving the highest accuracy based on the evaluation metrics, indicating an almost perfect fit. In contrast, RF showed better test performance under the Wrapper Method despite weaker training performance, indicating improved generalization. Based on the evaluation of the models using both the Filter and Wrapper feature selection methods, XGBoost Regression of filter method emerges as the best-performing model overall, with R^2=0.95, RMSE= 0.14 and MAE=0.10.

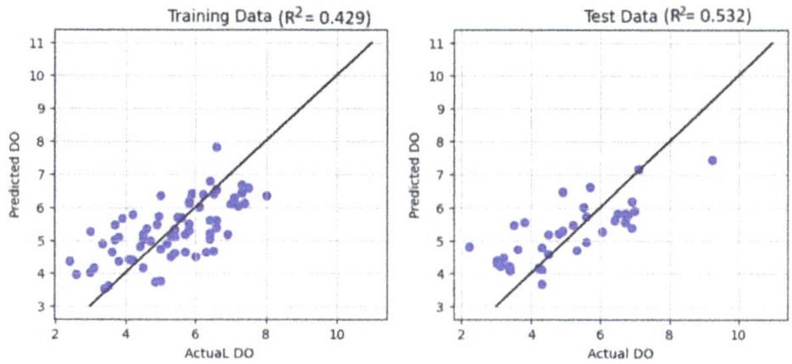

(a)

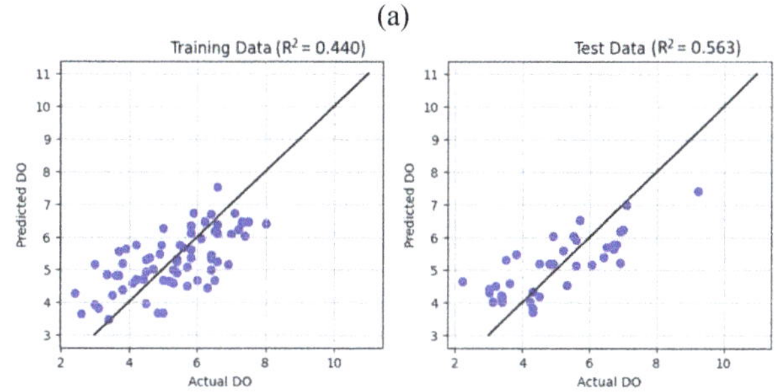

(b)

Fig. 4, Baseline results for DO prediction via Multiple Linear Regression using (a)Filter Method and (b) Wrapper Method: Predicted values vs. Actual values (in mg/L) scatter comparison plot for Training data and Test data

Conclusion and Recommendation

A total of 6 machine learning models were used in this paper to predict DO concentration from multidomain environmental parameters, including physical, water quality, and hydroclimatic parameters in Sampaloc Lake, Laguna, Philippines. The results show that (i) feature selection method is important to increase model accuracy; (ii) hyperparameter optimization is necessary to ensure a fair comparison across the models; (iii) XGBoost regression of filter method of feature selection offers good DO predictive power among the 6 ML models based on the performance metrics used. These procedures are recommended to improve the reliability of predictions from data-driven models, ensuring they can be more useful to guide policy decisions and improve interventions in the management of major fishing lakes.

To make sure that the developed DO predictive model becomes usable for policy-makers, it is recommended by the authors to operationalize the model through the development of a dashboard or mobile/web application. This interface would integrate the model's algorithm, allowing predictions across targeted major fishing lakes.

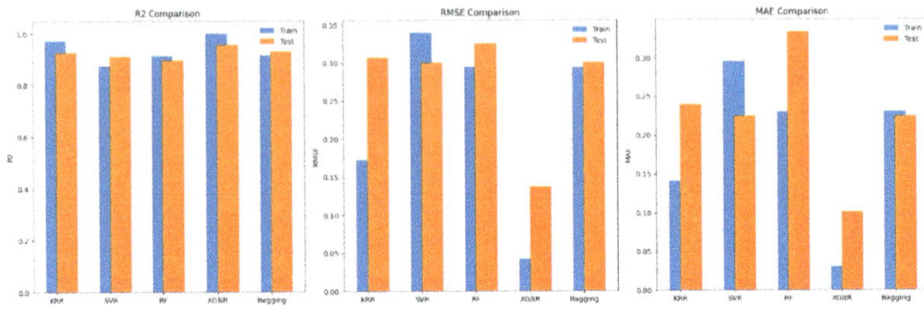

(a)

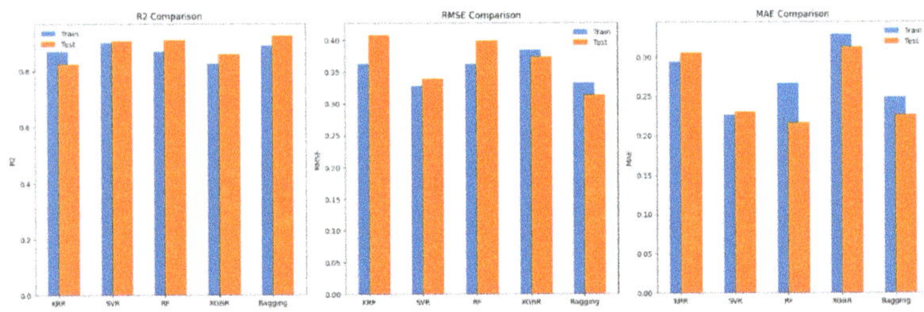

(b)

Fig. 5, Comparison of Evaluation Metrics (R^2, RMSE, and MAE) for (a) Filter Method and (b) Wrapper Method of Feature Selection

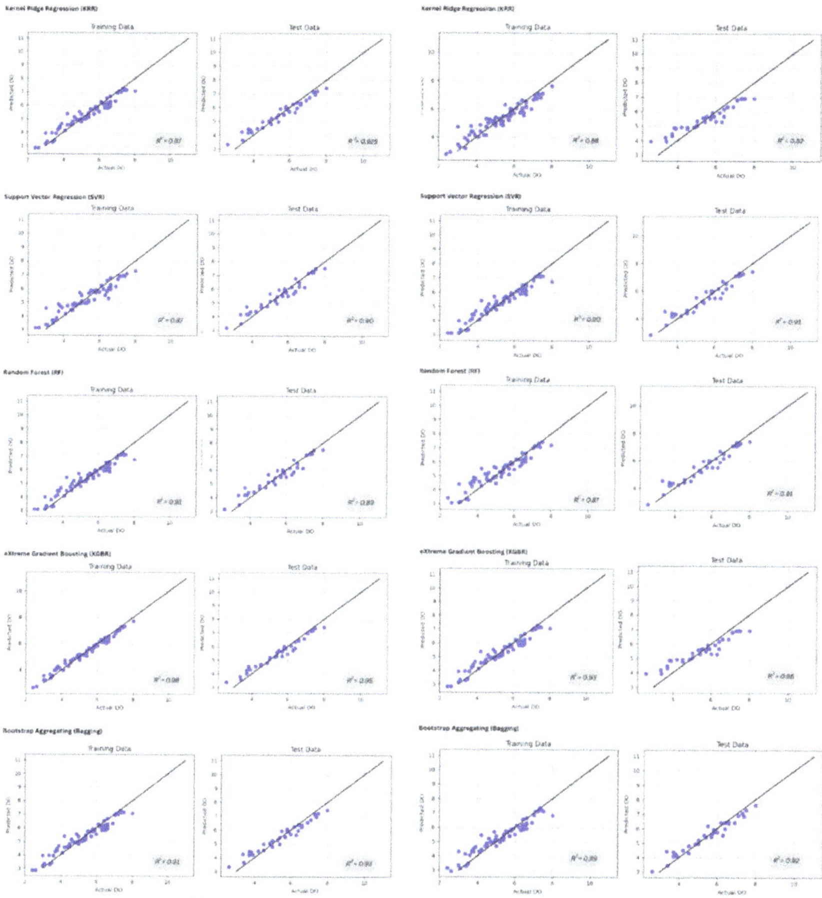

Fig. 6, Results on the prediction of DO concentrations in Sampaloc Lake using various ML models, presented as Predicted vs. Actual values (mg/L) plot for training and test data for (a) Filter Method and (b) Wrapper Method of Feature Selection

Acknowledgement

The authors would like to thank the Laguna Lake Development Authority(LLDA) and the San Pablo Laguna Local Government Unit (LGU) for the support in providing the data used in this study.

References

[1] Phan-Van, M., Rousseau, D., & De Pauw, N. (2008). Effects of fish bioturbation on the vertical distribution of water temperature and dissolved oxygen in a fish culture-integrated waste

stabilization pond system in Vietnam. Aquaculture, 281(1-4), 28–33. https://doi.org/10.1016/j.aquaculture.2008.04.033

[2] Wu, Y., Sun, L., Sun, X. et al. A hybrid XGBoost-ISSA-LSTM model for accurate short-term and long-term dissolved oxygen prediction in ponds. Environ Sci Pollut Res 29, 18142–18159 (2022). https://doi.org/10.1007/s11356-021-17020-5

[3] Brillo, B. B. B. (2017). The Politics of Lake Governance: Sampaloc Lake, Pandin Lake, and Tadlac Lake of the Laguna de Bay Region, Philippines. *Asia-Pacific Social Science Review*, *17*(1). https://doi.org/10.59588/2350-8329.1123

[4] Magcale-Macandog, D. B., Predo, C. D., Campang, J. G., Pleto, J. V. R., Perez, M. G. L. D., Larida, N. J. A., Natuel, F. A., Quiñones, S. G. L., & Cabillon, Y. C. L. (2021). Socio-economic and environmental impacts of COVID-19 pandemic: Building resilience of the seven lakes of San Pablo city, Philippines. In Elsevier eBooks (pp. 255–270). https://doi.org/10.1016/b978-0-323-85512-9.00012-7

[5] Cinco, M., January 5, 2021, Philippine Daily Inquirer. 2 tons of tilapia lost as fish kill hits Laguna's Sampaloc Lake.https://newsinfo.inquirer.net/1379395/2-tons-of-tilapia-lost-as-fish-kill-hits-lagunas-sampaloc-lake

[6] Zhu, M., Wang, J., Yang, X., Zhang, Y., Zhang, L., Ren, H., Wu, B., & Ye, L. (2022). A review of the application of machine learning in water quality evaluation. *Eco-Environment & Health*, *1*(2), 107–116. https://doi.org/10.1016/j.eehl.2022.06.001

[7] Laguna Lake Development Authority. (2005). Sampaloc Lake water quality report: 1996–2005. Environmental Quality Management Division. Retrieved from https://llda.gov.ph/wp-content/uploads/dox/7lakes/lakesampaloc.pdf

[8] Selim, A., Shuvo, S., Islam, M., Moniruzzaman, M., Shah, S., & Ohiduzzaman, M. (2023). Predictive models for dissolved oxygen in an urban lake by regression analysis and artificial neural network. Total Environment Research Themes, 7, 100066. https://doi.org/10.1016/j.totert.2023.100066

[9] Hofmann, T., Scholkopf, ¨ B., Smola, A.J., 2008. Kernel methods in machine learning. Ann. Stat. 36 (3), 1171–1220. https://doi.org/10.1214/009053607000000677

[10] Guo, H., Jinhui Jeanne Huang, Zhu, X., Wang, B., Tian, S., Wang, X., & Mai, Y. (2021). A generalized machine learning approach for dissolved oxygen estimation at multiple spatiotemporal scales using remote sensing. Environmental Pollution, 288, 117734–117734. https://doi.org/10.1016/j.envpol.2021.117734

[11] Breiman, L. (2001). Random Forests. Machine Learning, 45, 5-32. https://doi.org/10.1023/A:1010933404324

[12] Chen, K., Chen, H., Zhou, C., Huang, Y., Qi, X., Shen, R., Liu, F., Zuo, M., Zou, X., Wang, J., Zhang, Y., Chen, D., Chen, X., Deng, Y., & Ren, H. (2019). Comparative analysis of surface water quality prediction performance and identification of key water parameters using different machine learning models based on big data. Water Research, 171, 115454. https://doi.org/10.1016/j.watres.2019.115454

[13] an, Y., Jie, J., & Yin, M. (2025). Detection of event-related potentials as a biomarker in major depressive disorder using an XGBoost model. Biomedical Signal Processing and Control, 108, 107879. https://doi.org/10.1016/j.bspc.2025.107879

[14] J, A. O., Ogwueleka, F., & Odion, P. O. (2020). Effective and accurate Bootstrap Aggregating (BAgging) ensemble algorithm model for prediction and classification of hypothyroid disease. International Journal of Computer Applications, 176(39), 40–48. https://doi.org/10.5120/ijca2020920542

[15] Pilario, K. E., Escober, E. J., De Los Reyes, A., V., & Espino, M. P. (2024). Robust Prediction of Chlorophyll-a from Nitrogen and Phosphorus Content in Philippine and Global Lakes Using Fine-Tuned, Explainable Machine Learning. Environmental Challenges, 101056. https://doi.org/10.1016/j.envc.2024.101056

[16] Shahriari, B., Swersky, K., Wang, Z., Adams, R.P., De Freitas, N., 2016. Taking the human out of the loop: a review of Bayesian optimization. Proc. IEEE 104 (1), 148–175. https://doi.org/10.1109/JPROC.2015.2494218.

[17] Akiba, T., Sano, S., Yanase, T., Ohta, T., Koyama, M., 2019. Optuna: a next-generation hyperparameter optimization framework. In: Proceedings of the ACM SIGKDD International Conference on Knowledge Discovery and Data Mining, pp. 2623–2631. https://doi.org/10.1145/3292500.3330701.

[18] Chicco, D., Warrens, M. J., & Jurman, G. (2021). The coefficient of determination R-squared is more informative than SMAPE, MAE, MAPE, MSE and RMSE in regression analysis evaluation. PeerJ Computer Science, 7, e623. https://doi.org/10.7717/peerj-cs.623

[19] Hodson, T. O. (2022). Root-mean-square error (RMSE) or mean absolute error (MAE): when to use them or not. Geoscientific Model Development, 15(14), 5481–5487. https://gmd.copernicus.org/articles/15/5481/2022/

[20] Lai, S., Hu, X., Xu, H., Ren, Z., & Liu, Z. (2023). Multimodal sentiment analysis: A survey. Displays, 80, 102563. https://doi.org/10.1016/j.displa.2023.102563

[21] Jenny J-P, Normandeau A, Francus P, Taranu ZE, Gregory-Eaves I, Lapointe F, Jautzy J, Ojala AEK, Dorioz J-M, Schimmelmann A, et al. 2016b. Urban point sources of nutrients were the leading cause for the historical spread of hypoxia across European lakes. Proc Natl Acad Sci USA. 113(45):12655–12660.

[22] Hutchings, A. M., De Vries, C. S., Hayes, N. R., & Orr, H. G. (2024b). Temperature and dissolved oxygen trends in English estuaries over the past 30 years. Estuarine Coastal and Shelf Science, 306, 108892. https://doi.org/10.1016/j.ecss.2024.108892

[23] Cristianini, N., & Shawe-Taylor, J. (2000). An Introduction to Support Vector Machines and Other Kernel-based Learning Methods. Cambridge University Press.

[24] Hastie, T., Tibshirani, R., & Friedman, J. (2001). The Elements of Statistical Learning: Data Mining, Inference, and Prediction. Springer-Verlag. https://doi.org/10.1007/978-0-387-21606-5

[25] Grekousis, G. (2025). Geographical-XGBoost: A new ensemble model for spatially local regression based on gradient-boosted trees. Journal of Geographical Systems, 27(2), 169–195. https://doi.org/10.1007/s10109-025-00465-4

[26] Fusi, M., Barausse, A., Booth, J. M., Chapman, E., Daffonchio, D., Sanderson, W., Diele, K., & Giomi, F. (2024). The predictability of fluctuating environments shapes the thermal tolerance of marine ectotherms and compensates narrow safety margins. Scientific Reports, 14, 26174. https://doi.org/10.1038/s41598-024-77621-1

[27] Xiao, Q., Zhou, Y., Luo, J., Hu, C., Duan, H., Qiu, Y., Zhang, M., Hu, Z., & Xiao, W. (2024). Low carbon dioxide emissions from aquaculture farm of lake revealed by long-term

measurements. Agriculture, Ecosystems & Environment, 363, 108851. https://doi.org/10.1016/j.agee.2023.108851

[28] Dębska, K., Rutkowska, B., Szulc, W., & Gozdowski, D. (2021). Changes in selected water quality parameters in the Utrata River as a function of catchment area land use. Water, 13(21), 2989. https://doi.org/10.3390/w13212989

[29] Akinnawo, S. O. (2023). Eutrophication: Causes, consequences, physical, chemical and biological techniques for mitigation strategies. Environmental Challenges, 12, 100733. https://doi.org/10.1016/j.envc.2023.100733

[30] Dębska, K., Rutkowska, B., Szulc, W., & Gozdowski, D. (2021). Changes in selected water quality parameters in the Utrata River as a function of catchment area land use. Water, 13(21), 2989. https://doi.org/10.3390/w13212989

[31] Rather, I. A., & Dar, A. Q. (2020). Assessing the impact of land use and land cover dynamics on water quality of Dal Lake, NW Himalaya, India. Applied Water Science, 10, 219. https://doi.org/10.1007/s13201-020-01300-5

[32] Fajardo, C., Costa, G., Nande, M., Botelho, C. M., & Boaventura, R. A. R. (2023). Removal of pharmaceuticals from water by adsorption onto biochar produced from waste biomass. Environmental Challenges, 12, 100733. https://doi.org/10.1016/j.envc.2023.100733

[33] Akinnawo, S. O. (2023). Eutrophication: Causes, consequences, physical, chemical and biological techniques for mitigation strategies. Environmental Challenges, 12, 100733. https://doi.org/10.1016/j.envc.2023.100733

[34] Zhang, Y., Shen, Z., Wang, M., Guo, Y., & Xu, H. (2019). Adsorption of heavy metals from aqueous solution by biochar derived from agricultural waste: A review. Scientific Reports, 9, 1–12. https://doi.org/10.1038/s41598-019-56046-1

[35] Núñez-Delgado, A. (2024). Research on environmental aspects of retention/release of pollutants in soils and sorbents: What should be next? Environmental Research, 251(Part 1), 118593. https://doi.org/10.1016/j.envres.2024.118593

[36] Dey, S., Bera, T., & Bhattacharya, S. S. (2024). Biochar-based remediation of heavy metal–contaminated water and soil: Mechanisms and environmental applications. Environmental Monitoring and Assessment, 196, 12732. https://doi.org/10.1007/s10661-024-12732-w

[37] Kaur, R., & Singh, J. (2024). Biochar for wastewater treatment: A review of production, modification, and applications. Water, 16(24), 3604. https://doi.org/10.3390/w16243604

[38] Verma, A., Kumar, S., & Singh, R. (2024). Machine learning approaches for air quality prediction using regression techniques. In 2024 International Conference on Power Electronics, Intelligent Control and Energy Systems (ICPEICES) (pp. 1–6). IEEE. https://doi.org/10.1109/ICPEICES62430.2024.10719064

[39] Shen, Z., Zhang, Y., Jin, F., McMillan, O., & Al-Tabbaa, A. (2022). Characteristics and mechanisms of biochar for contaminant removal from water: A review. *Journal of Hydrology, 612*, 127711. https://doi.org/10.1016/j.jhydrol.2022.127711

[40] Xu, C., Luo, P., Wu, P., Song, C., & Chen, X. (2022). Detection of periodicity, aperiodicity, and corresponding driving factors of river dissolved oxygen based on high-frequency measurements. Journal of Hydrology, 609, 127711. https://doi.org/10.1016/j.jhydrol.2022.127711

Advanced Materials and Sustainable Energy Technologies – AMSET2025 Materials Research Forum LLC
Materials Research Proceedings 66 (2026) 423-429 https://doi.org/10.21741/9781644904152-38

Viability Assessment of the Philippines in Utilizing Solar Power Towers

Jon Andrei P. BATOHANON[1,a], Clarenz Angelo F. LANGIT[1,b], Jack Vincent R. MENDOZA[1,c] and Vhander T. RAMOS[1,d*]

[1]Electrical Engineering Department, Technological Institute of the Philippines, 938 Aurora Blvd., Cubao, Quezon City, 1109, Philippines

[a]qjapbatohanon@tip.edu.ph, [b]qcaflangit@tip.edu.ph, [c]qjvrmendoza@tip.edu.ph, [d]vramos.ee@tip.edu.ph

Keywords: Concentrated Solar Power, Philippines, System Advisor Model, Economic Viability

Abstract. In this study, to evaluate the feasibility of concentrated solar power (CSP) tower deployment in the Philippines, an economic and technical viability assessment was conducted. It is also the aim of this study to fill in the gap in local studies concerning CSP technologies. The Noor III solar tower was the base model for the techno-economic analysis of the thirteen selected sites across the Philippines via the System Advisor Model (SAM) software. Annual Energy Production, Capacity Factor (CF), and Levelized Cost of Electricity (LCOE) were evaluated using secondary data for site suitability and international financial metrics. Only two of the thirteen selected sites (Bolinao and Ilocos) show high potential for energy output due to the favorable Direct Normal Irradiation or DNI.

Introduction

Embracing sustainable, renewable energy sources remains challenging for every nation [1]. Reducing the effect of climate change and obtaining energy independence inherently requires integrating intermittent renewable energy sources and *dispatchable* power. One way to address this need is by deploying Concentrated Solar Power (CSP), particularly the Solar Power Tower (SPT) configuration, along with a *Thermal Energy Storage (TES)* that can store solar energy as heat and release electricity on demand, effectively operating as a reliable baseload source [2].

This study was conducted in response to the gap in local research and empirical data on CSP/SPT feasibility [3]. The available studies focused on site identification, investigating optimal sites for PV plants by employing geographical information systems and multiple-criteria decision-making approaches [4], or have been limited to PV technology. The absence of localized, quantitative assessment prevents the technology from moving from potential concept to practical deployment consideration [5].

The study utilized the established parameters of the *Noor III CSP plant* as the reference model and employed the *System Advisor Model (SAM)* software for simulation [6]. The primary objective is to evaluate the technical performance in terms of Annual Energy Production (AEP), Capacity Factor (CF), and economic competitiveness in terms of Levelized Cost of Electricity (LCOE) of this reference SPT model across 13 high-potential sites in the Philippines.

Review of Related Literature

The primary objective of this study is to identify the most effective approach and methodologies for determining the viability of a Solar Power Tower (SPT) in specific locations. Fortunately, international studies offer established methods for analyzing data and measuring SPT viability. This literature review will explain these diverse approaches to determine the best fit for the current study.

Chile is a promising South American country committed fully to renewable energy, which led to a study on the cost-performance of the Solar Power Tower (SPT). Quantifying the *Levelized Cost of Electricity (LCOE)* is the main focus of the study. They also conducted geographical and economic analysis utilizing critical parameters, such as *Direct Normal Irradiation (DNI)*, land topography, water and grid availability, wind speed, and city proximity. A *Solar Explorer Tool* and the Typical Meteorological Year (TMY) files were used in their analysis to develop a high-potential CSP map. To analyze the SPT's performance at different locations, they utilized SAM software and used an existing SPT model for the site selection as a reference. Three (3) locations have been identified with maximum, average, and minimum LCOE, and it is concluded that Chile's southern region was the most suitable for SPT installation [7].

A feasibility study for SPT was conducted in the Zahle Region of Lebanon. Unlike the broader Chilean study, it was focused on a single site: the *Electricité De Zahle*. The assessment included local weather, insolation conditions, and the financial costs of various SPT components. Their study included data restrictions such as a 1 million square meters land area, a minimum Internal Rate of Return (IRR) of 15%, and a Power Purchase Agreement (PPA) of $0.30 per kWh to minimize the suitable SPT types. The SAM software was used to determine the best output where the dependent and independent variables are the different SPT output energies and the site, respectively. The results concluded that it is best to install SPT with lower energy output but with greater hours of storage capability at the selected location [8].

Some international studies, such as the one conducted in Cameroon, employ advanced geospatial and multi-criteria analysis methods to select suitable sites for CSP plants. The Central African study involves eliminating unsuitable areas using restrictive criteria, followed by an evaluation of ten (10) sub-criteria grouped into climatic (solar irradiation, DNI, and temperature), orographic (slope and elevation), location (proximity to electricity networks, roads, residential areas, and population density), and watercourses (proximity to watercourses and water lines) aspects. With the expert judgment and application of the Analytical Hierarchy Process (AHP) method, their study reveals that the climate is the most impactful aspect, with a weighted score of 42% and orographic is the least impactful with a score of 10%. With these results, their study generated suitability maps, concluding that 42.35% of their country's land is unsuitable, 44% suitable, 13.64% very suitable, and only 0.01% as most suited in the standard scenario [9].

Furthermore, recent literature highlights that the economic viability of Concentrated Solar Power (CSP) and multi-energy systems is heavily dependent on the integration of Thermal Energy Storage (TES). A comprehensive systematic review of multi-energy systems demonstrated that TES plays a critical role in enhancing load shifting, seasonal balancing, and waste heat utilization [14]. The comparative scenario analysis in the review revealed that integrating TES can contribute to LCOE reductions of up to 25% in district energy applications [14]. More significantly for developing regions, the deployment of hybrid storage systems has been shown to reduce LCOE by up to 60% when compared to traditional baseload or diesel-based systems, with joint thermal and electrical storage deployments yielding 20% to 35% greater cost savings than single-storage configurations [14].

In addition to reducing overall LCOE, TES presents a highly competitive Levelized Cost of Storage (LCOS) when compared to purely electrical energy storage (EES) technologies, such as lithium-ion batteries [15]. Studies evaluating on-site energy storage underscore that TES often achieves an equal or greater cost-effectiveness than state-of-the-art Li-ion batteries due to its significantly lower capital costs and longer operational lifespans [15]. Unlike battery storage, TES also possesses the unique thermodynamic ability to harness diurnal ambient temperature swings, which further improves system efficiency and lowers the required charging costs [15]. Together, these international findings provide a crucial comparative framework: while baseline solar tower models without storage may struggle to meet global LCOE benchmarks, the strategic integration of TES serves as a primary mechanism to bridge the gap toward economic feasibility.

Methodologies

A techno-economic analysis and site suitability assessment for a feasible CSP tower in the Philippines was conducted for this study in reference to a previous local study [3]. The approach involves setting up appropriate criteria for the tower, establishing assumptions and limitations, and utilizing the SAM software for system modeling. As mentioned earlier, this study relies on secondary data to evaluate how feasible it is to adapt the Noor III base case model for the thirteen selected sites throughout the Philippines.

Site Suitability and Analysis

Data Collection and Processing

This study is based on secondary data from previous local research that identified potential solar tower sites throughout the Philippines for a site suitability analysis. The referenced study established ranking factors by evaluating key site parameters, such as Direct Normal Irradiance (DNI), typhoon frequency, grid capacity, slope, and proximity to water, the grid, and roads. Based on the results, the previous local study developed a comprehensive suitability map of the Philippines, highlighting the most advantageous provinces where CSP plants could be installed. These provinces are the following: Ilocos, Pangasinan, Pampanga, Southern Mindoro, Southern Masbate, and Maguindanao [3].

In accordance with the local study, the typical meteorological year files (TMY) of each selected region were gathered in preparation for the simulation. Such data includes solar irradiance, relative humidity, solar zenith, and other relevant data that evaluate the weather pattern of a specific region [3]. This is done using the National Solar Radiation Database (NSRDB) of the National Renewable Energy Laboratory (NREL). However, most of the six (6) general locations provided by the reference study were required to be specified by the NSRDB Viewer [10]. Hence, the following are the narrowed-down locations of each region for an even more precise weather information: Bolinao, Bulalacao, Alaminos, Ilocos, Concepcion, Domorog, Esperanza, Lubao, Mansalay, Monreal, Near Kalangan & Rio Grande River, Polloc Freeport, and Tubigan.

Techno-economic Analysis

Modeling Existing CSP Plants – Noor III

The Noor III Solar Tower in Ouarzazate, Morocco, is the reference model [11]. All of its detailed configurations and technical parameters will be used as input in the SAM software, including the TMY files of different selected sites.

Designed as a desktop application, SAM software creates models and runs simulations on diverse energy variables to assess how economically efficient systems are in the renewable energy sector [12].

Feasibility Analysis

To provide a comprehensive analysis and support future decision-making regarding Concentrated Solar Power (CSP) implementation, the study conducted a techno-economic analysis of solar tower potential in selected sites, evaluating the following three (3) main feasibility parameters, consistent with established reference studies: Annual Energy Production in kWh/GWh (total number of kWh or GWh produced by a solar tower in a year), Capacity Factor (CF) which is calculated as the ratio of solar tower's total energy output to its potential energy output at its rated capacity, and Levelized Cost of Electricity (LCOE) which refers to the total lifecycle costs and benefits ratio.

Materials Research Proceedings 66 (2026) 423-429 https://doi.org/10.21741/9781644904152-38

Assumptions and Limitations

To account for a realistic and representative possible performance, assumptions and limitations were applied:

a. The dependent and independent variables to be used in the SAM software are the TMY files and the technical parameters of Noor III, respectively.
b. Financial feasibility analysis relies entirely on the International Renewable Energy Agency (IRENA), limiting all cost data and solar tower criteria to international reported costs instead of actual Philippine costs.
c. Simulations were restricted to a single-year performance in SAM software, reflecting the solar towers' long-term output.

Criteria

To increase the reliability and credibility of the software results, the study included criteria entirely based on the IRENA and Power Purchase Agreement (PPA) database. The global and weighted-average total installation costs, Capacity Factors, and LCOE for CSP plants commissioned in 2020 were 4,581 $/kW, 42%, and 0.11 $/kWh, respectively [13]. Global standards were used in this study in the absence of local standards.

Additionally, as discussed in Section 3.1 on Site Suitability and Analysis, the six (6) general locations mentioned by a reference study [3] were required to be specified by the NSRDB Viewer. The thirteen (13) selected locations were based on the DNI values.

Interpretation of Results

The results from the SAM software for the selected sites are presented here, including four graphs displaying the:

- Direct Normal Irradiance (DNI)
- Capacity Factor (CF)
- Annual Energy Output
- Levelized Cost of Energy (LCOE)

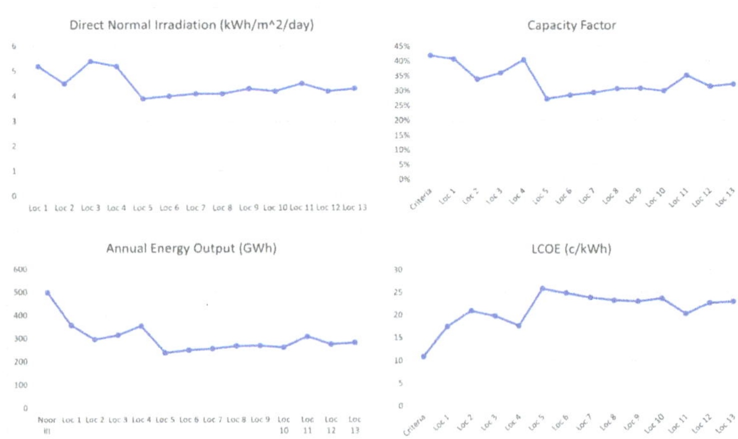

Figure 1. DNI, CF, Annual Energy, and LCOE of the 13 selected sites

Legends:
Loc 1-Bolinao, Pangasinan
Loc 2-Bulalacao, Mindoro
Loc 3-Alaminos, Pangasinan
Loc 4-Ilocos
Loc 5-Concepcion, Cataingan, Masbate
Loc 6-Domorog, Cataingan, Masbate
Loc 7-Esperanza, Masbate
Loc 8-Lubao Pampanga
Loc 9-Mansalay Mindoro
Loc 10-Monreal, Masbate
Loc 11-Near Kalangan & Rio Grande River
Loc 12-Polloc Freeport and Ecozone
Loc 13-Tubigan, Masbate

Thirteen (13) specific locations were simulated with Noor III as the base-case model. These 13 locations were particularly selected from the general locations referenced in the local study [3]: Pampanga, Ilocos, Southern Masbate, Mindoro, Maguidanao, and Pangasinan.

Figure 1 shows the annual energy output results for all selected sites. The lowest outputs were observed at Concepcion, Cataingan and Domorong, Cataingan, followed by Esperanza, Monreal, Lubao, Mansalay, Mindoro, Polloc Freeport, Tubigan, Masbate, Bulalacao, Mindoro, Near Kalangan, and Rio Grande River. Among the 13 locations, the top 3 with the highest energy output is Bolinao and Alaminos in Pangasinan, and Ilocos.

Figure 1 also shows the Capacity Factors (CF) for the 13 candidate locations. All the sites fall below the global average of 42% with Bolinao being the highest and Concepcion, Cataingan being the lowest.

Relatedly, Figure 1 also shows the LCOE values for the Noor III alongside the LCOE of the 13 potential locations. The graph shows that none of the 13 locations meet the desired economic criteria of 0.11 $/kWh, Concepcion, Cataingan being the least favorable, followed by Domorog, Cataingan. The findings collectively indicate that large-scale CSO deployment in the selected locations was not economically feasible.

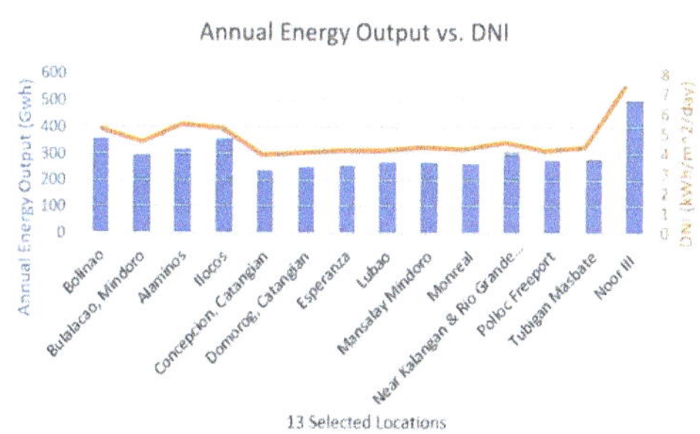

Figure 2. Annual Energy Output vs. Direct Normal Irradiation

Based on the results of techno-economic analysis, the provinces of Pangasinan and Ilocos exhibit the most significant potential for CSP deployment because of their DNI values. The absence of Thermal Energy Storage (TES) affected the capacity factor of the 13 selected sites to fall below the global criteria of 42%. Compared to the existing international benchmark, none of the 13 sites hit the economic feasibility mark of 0.11 $/kWh for LCOE. Despite these financial hurdles, the study confirms that deploying CSP in the Philippines is a crucial step forward for the country's renewable energy sector.

Conclusion

The Philippines has the technical capability to deploy CSP plants based on the DNI values exhibited by Bolinao and Ilocos, and a strong potential for annual energy production. The study concludes that installing SPT is not economically viable based on current global standards. The main factor is the lack of Thermal Energy Storage (TES). The result of the study shows that the capacity factor of Pangasinan and Ilocos is 40.8% and 40.5%, respectively, which is below the global standard.

Recommendations

Future local studies about CSP consider other types of CSP technologies and the inclusion of Thermal Sensitivity Analysis (TSA), which involves the impact assessment of Thermal Energy Storage (TES). The approach to site determination should use another methodology, such as a geographical information system in tandem with a multi-criteria decision-making method like the Analytical Hierarchy Process (AHP). To obtain an optimal LCOE result, consider increasing the financial data that serves as input in the SAM software. Lastly, further studies should consider all other variables that may be deemed impactful for the viability of installing SPT in the Philippines.

References

[1] Owusu, P. A., & Asumadu-Sarkodie, S. (2016). A review of renewable energy sources, sustainability issues and climate change mitigation. Cogent Engineering, 3(1), 1167990. https://doi.org/10.1080/23311916.2016.1167990

[2] Kennedy, K. M., Ruggles, T. H., Rinaldi, K., Dowling, J. A., Duan, L., Caldeira, K., & Lewis, N. S. (2022). The role of concentrated solar power with thermal energy storage in least-cost highly reliable electricity systems fully powered by variable renewable energy. Advances in Applied Energy, 6, 100091. https://doi.org/10.1016/j.adapen.2022.100091

[3] Levosada, A. T. A., Ogena, R. P. T., Santos, J. R. V., & Danao, L. a. M. (2022). Mapping of suitable sites for concentrated solar power plants in the Philippines using geographic information system and analytic hierarchy process. Sustainability, 14(19), 12260. https://doi.org/10.3390/su141912260

[4] Al Garni, H. Z., & Awasthi, A. (2018). Solar PV power plants site selection: A review. Advances in renewable energies and power technologies, 57-75.

[5] Olabi, A., Elsaid, K., Obaideen, K., Abdelkareem, M. A., Rezk, H., Wilberforce, T., Maghrabie, H. M., & Sayed, E. T. (2023). Renewable energy systems: Comparisons, challenges and barriers, sustainability indicators, and the contribution to UN sustainable development goals. International Journal of Thermofluids, 20, 100498. https://doi.org/10.1016/j.ijft.2023.100498

[6] System Advisor Model - SAM. (n.d.). Welcome. https://sam.nrel.gov/

[7] Hernández, C., Cerda, M. T., Felbol, C., & Dinter, F. (2023). Evaluation of the levelized cost of energy with new costs for concentrating solar power tower plants in northern Chile and impact of green taxes. SolarPACES Conference Proceedings, 1. https://doi.org/10.52825/solarpaces.v1i.615

[8] Zeaiter, J., Architecture, M. S. F. O. E. A., Azizi, F., & Ghaddar, N. (2025, April 25). Optimal Design and Feasibility Study of Utility-Scale Solar Power Plants in Lebanon. https://scholarworks.aub.edu.lb/handle/10938/34880

[9] Flora, F. M. I. (2024). Assessment of concentrated solar power (CSP) generation potential in Cameroon using a Multi-Criteria Analysis Method and Geographic Information System (GIS). Heliyon, 11(1), e41094. https://doi.org/10.1016/j.heliyon.2024.e41094

[10] NSRDB. (n.d.). https://nsrdb.nrel.gov/data-viewer

[11] Imam, A. A., Abusorrah, A. M., & Marzband, M. (2023). Potential of concentrated solar power in the Western region of Saudi Arabia: A GIS-Based land suitability Analysis and

Techno-Economic Feasibility Assessment. IEEE Access, 12, 1570–1598. https://doi.org/10.1109/access.2023.3344752

[12] Mohamed, L., Abdelkader, R., Laroussi, K., Singh, A. R., Bajaj, M., & Tuka, M. B. (2025). Comprehensive techno-economic optimization and performance analysis of molten salt concentrated solar power tower plants in Algeria. Scientific Reports, 15(1). https://doi.org/10.1038/s41598-025-97236-4

[13] IRENA (2021), Renewable Power Generation Costs in 2020, International Renewable Energy Agency, Abu Dhabi

[14] Jarwar, A. R. (2025). Advancing Multi-Energy Systems with Thermal and Electrical Storage: A Literature Review on Levelized Cost Reduction. International Journal of Sustainable Applied Sciences, 3(8), 541–552. https://doi.org/10.59890/ijsas.v3i8.141

[15] Odukomaiya, A., Woods, J., James, N., Kaur, S., Gluesenkamp, K. R., Kumar, N., Mumme, S., Jackson, R., & Prasher, R. (2021). Addressing energy storage needs at lower cost via on-site thermal energy storage in buildings. Energy & Environmental Science, 14(10), 5315–5329. https://doi.org/10.1039/d1ee01992a

Production of Radioisotope using DD Neutron Generators for Medical Applications

Muhammad Zubair*

Department of Mechanical & Nuclear Engineering, University of Sharjah, 27272, Sharjah, UAE

*mzubair@sharjah.ac.ae

Keywords: Radioisotopes, DD Neutron Generator, Technetium, Iodine

Abstract. Some isotopes including Iodine-131 and technetium-99m are radioisotopes but on other side these are important for medical treatment. In this research DD-109.4 neutron generator utilized to produce medically significant isotopes. Isotope yield and purity are evaluated under various operating conditions. The work offers a scalable and affordable substitute for nuclear reactors. Neutron generator available at university of Sharjah and MCNP software were used in this research.

Introduction

In this era nuclear isotopes were widely used as medicine. These isotopes are used for imaging and diagnosis purposes. Due to Tc- 99m half-life, gamma emission energy, and compatibility with imaging systems, this isotope is used at a large scale in the world. According to survey its utilization is more the 80 % around the globe. With six hours' half-life Tc-99m is an ideal isotope for medical imaging. It produces gamma rays of 140Kev [1]. The treatment of thyroid cancer is heavenly depending on I-131 due to its beta & gamma emission abilities. To use these isotopes in the medical field a very trusted equipment system is essential [2]. The isotopes production and transportation are an important issue. This includes safety issues, functioning of proper nuclear power plants (NPPs), and handling limitations. Highly Enriched Uranium (HEU) from old NPPs are a major component in traditional isotope production. On other hand expense, proliferation issues, and waste material put limitations on the production of radioisotopes. [3–4]. Large-scale nuclear reactors, which need substantial infrastructure and financial investments are a major component of traditional production processes. Tc-99m is usually made by fission process of Uranium-235 in reactors which produces a significant amount of radioactive waste and requires close regulatory monitoring [5]. Geographical and reactor maintenance are also potential causes in the production of radioisotopes. Another main issue is the planned and unplanned reactor shutdown. This process disturbs the production and supply of radioisotopes [6]. Therefor an alternate method or a source of radioisotopes production is essential.

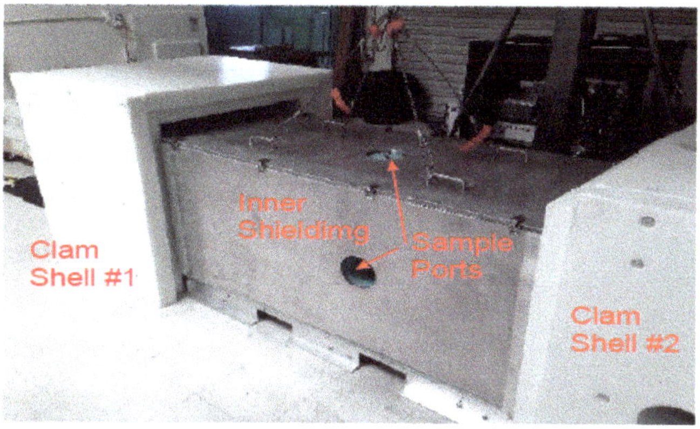

Fig. 1. Neutron generator with inner shielding and open clam shell

Methodology

DD neutron generator available at university of Sharjah is a viable and safe option available not only for the local community but also globally. It produces very large doses of gamma, fast and thermal neutrons. The moderator and two shielding layers protect the operator and public from the radiation. These options make them more demand-making. The generator uses a microwave-driven Electron-Cyclotron Resonance (ECR) plasma source to efficiently ionize the deuterium gas as shown in Fig. 1. This process provides high density for the output current. When the generator is operating, the titanium layer is initially being loaded with deuterium, and the neutron output is increasing with time. The warm-up period for the generator is approximately 20-30 minutes to reach a steady neutron output.

Specifications of DD Neutron Generator

The DD-109.4 requires a 30A 3-phase 380V circuit. Three 220 VAC 1 phase plugs are recommended for three other components: (1) the Koolance Heat Exchanger, (2) the Scroll Roughing Pump and (3) the Thermatron Heat Exchanger. The cooling system for the DD-109.4 is self-contained. The Thermatron Fluor inert heat exchanger fluid level needs to be maintained.

Flux Optimization

The neutron generator acceleration voltage (120–150 kV) and beam current (20–40 mA). The generator head consists of the Acceleration Chamber, ECR ion Source, Magnetron box, and Microwave Tuner. The neutrons are emitted from the interior of the target chamber. High purity ^{100}Mo and ^{130}Te samples are procured and prepared as thin foils or pellets to ensure uniform neutron activation. Samples are encased in aluminum capsules for enhanced heat dissipation and structural integrity during irradiation. Neutron flux is measured using a calibrated neutron detector (e.g., BF3 or He-3 detectors) placed adjacent to the target. Dose rate monitoring is conducted using a gamma-ray spectrometer to ensure compliance with safety thresholds

Results and Discussion

The simulation results are 5% lower than the experimental findings. The reason behind this is that the MC simulation did not consider the back scattering and leakage of neutrons. On the other side the results from MS simulation show that the correctness of the experimental arrangement. The experimental and simulation yield comparisons for TC-99m and I-131 are shown in Fig. 2 and 3, respectively.

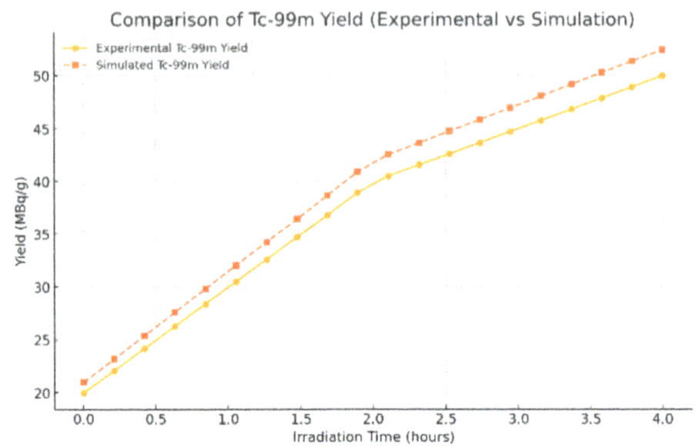

Fig. 2 TC-99m yield with MCNP

In the Initial Region (0–1 hour), the slope is steep because the production rate is high and radioactive decay is still small, so yield increases rapidly. In the Intermediate Region (1–3 hours), the curve still increases but more slowly because of continuous production of Tc-99m. In the last region (3–4 hours), the curve begins to flatten slightly. This indicates the system is approaching saturation activity, where: Production rate ≈ decay rate. So additional irradiation gives smaller increases in yield.

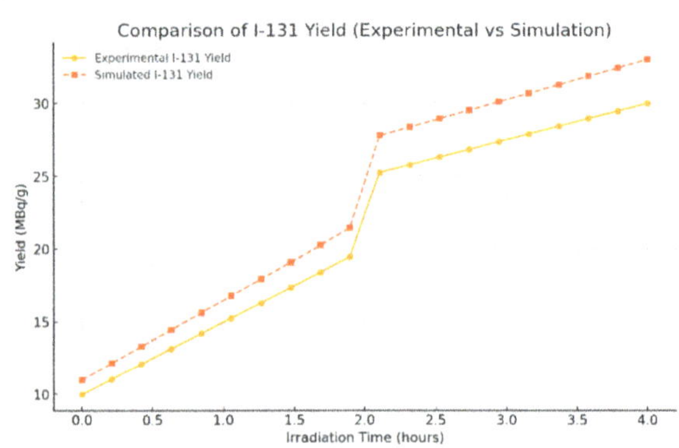

Fig. 3 I-131 yield with MCNP

The yield of I-131 increases steadily as irradiation time increases. This happens due to longer exposure of the target to the radiation field. As a result, more radioactive nuclei are produced. In the early irradiation period (0–2 hours), the increase in yield is almost linear because the production rate is dominant. Around the 2-hour mark, both curves show a noticeable increase in yield, which may indicate improved reaction efficiency. After this point, the yield continues to increase but at a slightly slower rate, suggesting that radioactive decay of the produced I-131 begins to counterbalance part of the production rate.

Conclusion

The present research assessed the viability of producing isotopes using the DD-109.4M neutron generator, with a focus on Tc-99m and I-131 via neutron activation of molybdenum-100 and

Materials Research Proceedings 66 (2026) 423-429 https://doi.org/10.21741/9781644904152-38

tellurium-130, respectively. Several important insights into the functionality, optimization, and suitability of tiny neutron generators for decentralized medical applications were revealed by a thorough comparison of experimental results with Monte Carlo simulations.

References

[1] International Atomic Energy Agency (IAEA). (2017). Cyclotrons and Medical Radioisotopes: Production and Applications. Vienna: IAEA Publishing.

[2] Boschi, A., Uccelli, L., & Martini, P. Technetium-99m radiopharmaceuticals: Challenges in production and potential solutions. Journal of Nuclear Medicine, 59 (2018) 329–337.

[3] Carroll, M. The Challenges of Medical Isotope Production without Nuclear Reactors. Applied Radiation and Isotopes, 85 (2014) 95–101.

[4] Moe, L., & Sopko, J. Decentralized Production of Medical Isotopes Using Neutron Generators: A Feasibility Study. Health Physics, 114 (2018) 230–239.

[5] Wilson, M. J., & Kaplan, A. Localized Isotope Production: A Review of Emerging Technologies. International Journal of Radiation Applications, 45 (2020) 124–133.

[6] Guertin, S. M., & Shultis, J. K. Neutron Generator Basics and Applications. Radiation Physics and Chemistry, 85 (2013) 56–65.

A New Bayesian Control Chart for the Monitoring of Parametric Uncertainty

Tahir ABBAS[1,a*], Muhammad ABID[2,b], Iffrah RASHEED[2,c]

[1]Department of Mathematics, College of Sciences, University of Sharjah, Sharjah 27272, United Arab Emirates

[2]Department of Statistics, Government College University, Faisalabad 38040, Pakistan

[a]Tabbas@sharjah.ac.ae, [b]mabid@gcu.edu.pk, [c]mhmmd_abid@yahoo.com

Keywords: Aerospace Manufacturing, Bayesian Charts, Posterior Distribution, Run Length Profiles, Statistical Process Control

Abstract. Bayesian control charts have proven to be powerful statistical process control tools for monitoring manufacturing processes and controlling variability. Control charts in a Bayesian setup are particularly advantageous in situations where parameter uncertainty is unavoidable in industrial engineering. Memory-type control charts, such as EWMA, HWMA, and DEWMA, are designed under a Bayesian framework to handle parametric uncertainty. This study introduces a new Double HWMA (DHWMA_B) chart within the Bayesian framework to further enhance the detection potential of the DHWMA chart. The Bayesian DHWMA chart is specifically designed to monitor the process mean and has effectively addressed the challenge of parametric uncertainty. The posterior estimates are obtained by carefully integrating prior knowledge with current information. The designed charts are evaluated using various performance measures of the run length (RL) profile, such as the average RL and the RL standard deviation. The RL characteristics of the proposed DHWMA_B and competing charts are computed and compared. The simulation study demonstrates that the DHWMA_B chart outperformed the competing charts in detecting the fault early. The proposal chart for DHWMA_B effectiveness shows effectiveness across all types of shifts in the process parameter. The real-world application in the aerospace manufacturing industry also demonstrates the superiority of the proposed DHWMA_B chart.

Introduction

Control charts (CCs) are considered vital statistical process control methods for monitoring the process parameters over time. Control charts are classified into two categories: memoryless and memory-based. The Shewhart chart (Shewhart, 1931) is memoryless, meaning it depends only on the most recent data point, and is effective at detecting large shifts. On the other hand, the CUSUM (Page, 1954) and EWMA (Roberts, 1959) are classified as the memory-based charts that integrate both past and current data. These charts are extensively used in manufacturing, finance, healthcare, and other related industries. The CUSUM and EWMA charts are often preferred over Shewhart for small to moderate shifts (Montgomery, 2009). The detailed studies, properties, and performance are examined under various conditions (Castagliola & Maravelakis, 2018; Maravelakis, 2012).

There are certain situations in the manufacturing industry where parameteric uncertainty is unavoidable. The Bayesian control charts are considered the best option for handling such parametric uncertainty, as they provide more reliable estimation through posterior distributions. Currently, the Bayesian control charts have been widely studied (Khan et al., 2023; Wang & Tsung, 2022; Iqbal et al., 2023; Jones et al., 2023; Wang et al., 2023; Noor et al., 2024). Building on this, we have presented a new Bayesian Double HWMA (DHWMA_B) chart to enhance fault-detection sensitivity for process parameters. The proposed Bayesian DHWMA_B control chart

Materials Research Proceedings 66 (2026) 434-439 https://doi.org/10.21741/9781644904152-40

introduces a fundamentally different monitoring structure from the existing EWMA, HWMA, and DEWMA charts. Unlike the HWMA chart, which relies on a single homogenous weighting between the current observation and the historical mean, the proposed method introduces an additional homogenous smoothing stage, thereby creating a double-layer memory structure. Furthermore, while the DEWMA chart utilizes two exponential smoothing stages, it does not incorporate Bayesian estimation. The DHWMA_B chart integrates Bayesian posterior estimation with the double homogenous weighting scheme, resulting in a distinct variance formulation of the monitoring statistic and corresponding control limits that explicitly account for posterior uncertainty and the homogenous smoothing parameters. This structural modification enhances the theoretical framework of the chart and improves its sensitivity in detecting small to moderate process shifts. The proposal chart is compared with EWMA, HWMA, and DEWMA charts under both Classical and Bayesian settings. The objective of this study is to design a Bayesian DHWMA chart that uses the posterior mean and posterior variance in its design, thereby enhancing its ability to handle parametric uncertainty.

Materials and Methods

(i) the statistical framework and assumptions, (ii) the design of the proposed Bayesian Double HWMA (DHWMA_B) chart, (iii) simulation setup for comparative evaluation, and (iv) the real-world dataset used.

Bayesian Methods

Bayesian analysis is considered the most vital branch of statistics due to its simplicity, utility, and the use of both prior information and current information in decision-making. Mathematically, it is a combination of two random information, the current *"x"* and the prior θ" such as;

$$f(\theta|x) = \frac{[\prod_{i=1}^{n} f(x|\theta)]f(\theta)}{\int [\prod_{i=1}^{n} f(x|\theta)]f(\theta)\, d_\theta} \tag{1}$$

Where $f(x|\theta)$ is called the likelihood function with prior distribution $g(\theta)$. It produces the posterior distribution as $f(\theta|x)$ (Hamada, 2002). In Bayesian methodology, we used the MLE (maximum likelihood estimation), conditional and marginal probabilities (Menzefricke, 2002). Let $X_1, X_2, \ldots, X_n$ be the random samples from the normal distribution, i.e., $X \sim N(\mu, \sigma^2)$. The prior for the unknown mean μ is assumed as normal, i.e., $\mu \sim N(a_0, b_0^2)$, then the posterior distribution of μ is also normal, with posterior mean and variance defined as:

$$\mu_n = \frac{nb_0^2\bar{x} + a_0\sigma^2}{nb_0^2 + \sigma^2} \tag{2}$$

$$\sigma_n^2 = \frac{\sigma^2 b_0^2}{nb_0^2 + \sigma^2} \tag{3}$$

Proposed Bayesian Structure

The designed structure of the DHWMA_B chart by using the idea of double HWMA statistics, such as:

$$HW_i = \phi\, \mu_{ni} + (1-\phi)\, \bar{\bar{\mu}}_{n(i-1)}$$
$$DH_i = \phi\, HW_i + (1-\phi)\, \bar{\bar{\bar{U}}}_{n(i-1)} \tag{4}$$

The corresponding control limits for the DHWMA-B chart are:

$$\text{LCL}_i = \begin{cases} \mu_o - k\sqrt{\phi^4\left(\frac{\sigma^2 b_0^2}{nb_0^2+\sigma^2}\right)} & \text{if } i = 1 \\ \mu_0 - k\sqrt{\phi^4\left(\frac{\sigma^2 b_0^2}{nb_0^2+\sigma^2}\right) + \frac{(1-\phi^2)^2(1+\phi^2)}{i-1}\left(\frac{\sigma^2 b_0^2}{nb_0^2+\sigma^2}\right)} & \text{if } i > 1 \end{cases} \tag{5}$$

$$CL_i = \mu_0$$

$$\mathrm{UCL}_i = \begin{cases} \mu_o + k\sqrt{\phi^4 (\frac{\sigma^2 b_0^2}{nb_0^2+\sigma^2})} & \text{if} i = 1 \\ \mu_0 + k\sqrt{\phi^4 \left(\frac{\sigma^2 b_0^2}{nb_0^2+\sigma^2}\right) + \frac{(1-\phi^2)^2(1+\phi^2)}{i-1}\left(\frac{\sigma^2 b_0^2}{nb_0^2+\sigma^2}\right)} & \text{if} i > 1 \end{cases} \quad (6)$$

Results and Discussion

Product quality is vital for organizations to meet customer requirements. To achieve that level of customer satisfaction, the quality manager takes various measures to ensure a satisfactory product is offered in the market. Average run length (ARL) is a measure of the performance of the statistical quality control charts. Before the first out-of-control signal, the average number of points on control limits is called the average run length. It can be written as $\mathrm{ARL} = \frac{1}{p}$. Where p is the probability of points that exceed the control limits of the chart. Further, there are two types of ARL: the first is in control $\mathrm{ARL}_0 = \frac{1}{\alpha}$, and the second is $\mathrm{ARL}_1 = \frac{1}{1-\beta}$ out of control. The performance of the control chart is evaluated using the Average Run Length (ARL). Another important measure is the Standard Deviation of Run Length (SDRL), which typically follows a normal distribution and tends to have smaller values when the process is out of control, assuming the process itself is normal. This analysis outlines the performance of the EWMA and synthetic charts. As the value of the shift parameter δ increases, both the ARL and SDRL decrease. $SDRL = \sqrt{E(RL)^2 - (E(RL))^2}$. ARL is not acceptable when the distribution is skewed, as it is influenced by the high values; therefore, MDRL = median (RL).

We use Monte Carlo simulations to generate random numbers, rather than collecting real data. A simulation study is conducted for the DHWMA control chart under the Bayesian approach using the posterior mean (a_0) and variance (b_0). Setting $ARL_0 = 500$, we determine the optimal values of a_0 and b_0 to evaluate the chart's performance for small shifts. The control limit coefficient is adjusted to minimize the ARL at small shifts. The study also reveals that each Bayesian-based chart has its own a_0 values, which vary with the smoothing constant.

A sensitivity analysis is conducted to determine the optimal hyperparameter values that achieve a minimum ARL for small shifts. A conjugate prior is used to derive the posterior mean and variance. The impact of hyperparameters on ARL, SDRL, and MDRL is also evaluated. It is observed that the value of a_0 varies across charts for small values of the smoothing constant. As the smoothing constant ϕ increases up to 0.25, the value of a_0 also increases and then stabilizes for DHWMA $_B$ and DEWMA $_B$ chart. With ARL $_0 = 500$, $n = 1$, and $\phi = 0.25$, the b_0 increases from 3.38 to 7, ARL $_1$ decreases until it plateaus, indicating $b_0 = 7$ as optimal. To identify a_0, values are varied from -0.5 to -2.5, and it is found that $a_0 = -2.5$ yields the minimum ARL_1, ensuring the mathematical stability of the posterior distribution and maintaining desirable statistical properties. However, this optimal a_0 does not generalize across all smoothing constants or chart types, so it must be evaluated separately for each case while keeping b_0 fixed.

TABLE 1 The RL profile for different charts under Classical and Bayesian setups when n=5, ø=0.05, ARL=500

CHARTS	L	δ	0.00	0.25	0.50	0.75	1.00	1.50	2.00	2.50	3.00	5.00
EWMA_C	2.645	ARL	501.71	19.75	6.09	3.20	2.05	1.27	1.04	1.00	1.00	1.00
		SDRL	524.79	14.63	3.88	1.77	1.01	0.49	0.19	0.03	0.00	0.00
		MDRL	332.50	16.00	5.00	3.00	2.00	1.00	1.00	1.00	1.00	1.00
HWMA_C	2.602	ARL	498.13	72.63	25.14	12.77	8.10	4.42	2.98	2.19	1.68	1.01
		SDRL	373.99	53.27	17.05	8.22	4.90	2.34	1.52	1.20	0.98	0.14
		MDRL	440.00	62.00	22.00	11.00	7.00	4.00	3.00	2.00	1.00	1.00
DEWMA_C	2.0791	ARL	507.13	17.20	5.06	2.59	1.68	1.12	1.01	1.00	1.00	1.00
		SDRL	541.82	13.82	3.99	1.75	1.00	0.35	0.10	0.00	0.00	0.00
		MDRL	354.00	14.00	4.00	2.00	1.00	1.00	1.00	1.00	1.00	1.00
DHWMA_C	1.3979	ARL	509.44	8.12	3.16	1.98	1.44	1.05	1.00	1.00	1.00	1.00
		SDRL	1897.74	7.09	2.33	1.37	0.89	0.31	0.06	0.00	0.00	0.00
		MDRL	12.00	5.00	3.00	1.00	1.00	1.00	1.00	1.00	1.00	1.00
EWMA _B	2.645	ARL	494.35	19.18	5.93	3.09	2.06	1.25	1.03	1.00	1.00	1.00
a_0	-2.5	SDRL	512.52	14.16	3.79	1.74	1.02	0.47	0.18	0.04	0.00	0.00
b_0	7	MDRL	340.00	16.00	5.00	3.00	2.00	1.00	1.00	1.00	1.00	1.00
HWMA_B	2.634	ARL	501.02	20.02	6.64	3.77	2.59	1.42	1.06	1.00	1.00	1.00
a_0	-2.5	SDRL	377.62	13.19	3.77	1.96	1.33	0.80	0.32	0.05	0.00	0.00
b_0	7	MDRL	432.00	17.00	6.00	4.00	3.00	1.00	1.00	1.00	1.00	1.00
DEWMA _B	2.121	ARL	500.45	15.55	5.07	2.59	1.69	1.12	1.01	1.00	1.00	1.00
a_0	-2.5	SDRL	536.25	12.58	3.93	1.75	0.99	0.35	0.08	0.00	0.00	0.00
b_0	7	MDRL	335.50	14.00	4.00	2.00	1.00	1.00	1.00	1.00	1.00	1.00
DHWMA_B	1.3966	ARL	507.18	6.72	3.08	1.99	1.42	1.05	1.01	1.00	1.00	1.00
a_0	-2.5	SDRL	1842.77	6.66	2.26	1.41	0.91	0.31	0.11	0.00	0.00	0.00
b_0	7	MDRL	12.00	5.00	3.00	1.00	1.00	1.00	1.00	1.00	1.00	1.00

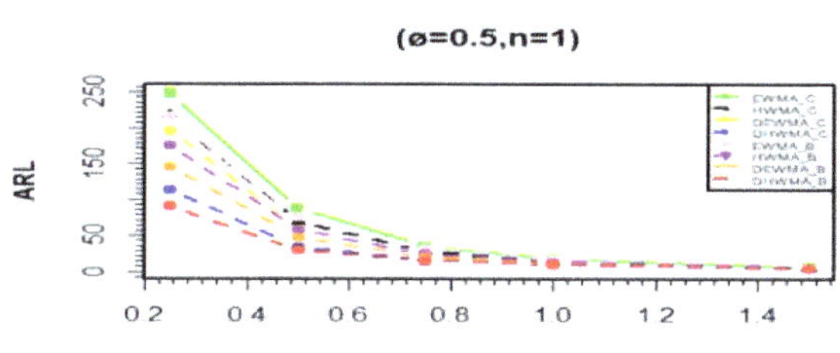

Figure 1. ARL curves of different control charts under Classical and Bayesian setups at different sample sizes with ϕ=0.5 and ARL_0=500.

Table 1 and Figure 1, based on the simulation study, indicate that the proposed DHWMA_B chart outperforms competing charts. For example, at δ=0.25, the ARL_1 values for EWMA_C,

HWMA_C, DEWMA_C, DHWMA_C, EWMA_B, HWMA_B, DEWMA_B, and DHWMA_B are 19.75, 72.63, 17.20, 8.12, 19.18, 20.02, 15.55, and 6.72, respectively. The results indicate that the proposed DHWMA_B chart yields the smallest ARL_1 value, and showed 66%, 91%, 61%, 17%, 65%, 66.5%, 57% improvement over the EWMA_C, HWMA_C, DEWMA_C, DHWMA_C, EWMA_B, HWMA_B, DEWMA_B charts. The HWMA_C chart maintains consistent performance regardless of sample size, whereas the DHWMA_B chart is efficient across different sample sizes.

In practice, these improvements indicate that the proposed chart detects process shifts faster while maintaining the desired in-control ARL. Smaller smoothing parameters improve sensitivity to small shifts, whereas moderate values provide more stable monitoring for larger shifts. Parameter selection can therefore be guided by the expected shift size, with recommended values supported by the sensitivity analysis.

Summary

The study proposes a new Bayesian control chart, specifically the DHWMA_B chart, for monitoring the process mean. The study evaluated the proposed and existing charts and found that the proposal chart performed well under a Bayesian setup for small to moderate shifts. Additionally, the values of the hyperparameter had a significant impact on ARL_1 across different charts, as they varied from chart to chart. This trend was also supported by the tables and graphs, which showed a decrease in average run length across all charts; however, HWMA_C exhibited unusual behavior, as it was not affected by sample size. It is concluded that DHWMA_B outperformed all the other charts discussed in the study.

Acknowledgement

The work is supported by the Office of the Vice Chancellor for Research and Graduate Studies (VCRG) at the University of Sharjah, United Arab Emirates, under Project 24021440167.

References

[1] Rakitzis, A. C., Castagliola, P., & Maravelakis, P. E. (2018). Cumulative sum control charts for monitoring geometrically inflated Poisson processes: An application to infectious disease counts data. Statistical Methods in Medical Research, 27(2), 622-641. https://doi.org/10.1177/0962280216641985

[2] Jones, C. L., Abdel-Salam, A. S. G., & Mays, D. A. (2023). Novel Bayesian CUSUM and EWMA control charts via various loss functions for monitoring processes. Quality and Reliability Engineering International, 39(1), 164-189. https://doi.org/10.1002/qre.3229

[3] Khan, I., Noor-ul-Amin, M., Khan, D. M., Ismail, E. A., & Sumelka, W. (2023). Monitoring of manufacturing process using Bayesian EWMA control chart under ranked-based sampling designs. Scientific Reports, 13(1), 18240. https://doi.org/10.1038/s41598-023-45553-x

[4] Maravelakis, P. E. (2012). Measurement error effect on the CUSUM control chart. Journal of Applied Statistics, 39(2), 323-336. https://doi.org/10.1080/02664763.2011.590188

[5] Montgomery, D. C. (2009). Introduction to Statistical Quality Control. 6 ed., John Wiley & Sons.

[6] Noor-ul-Amin, M., Khan, I., Iqbal, J., Rasheed, Z., Ismail, E. A., & Ahmad, B. (2024). Memory type Max-EWMA control chart for the Weibull process under the Bayesian theory. Scientific Reports, 14(1), 3111. https://doi.org/10.1038/s41598-024-52109-0

[7] Page, E.S. (1954). Continuous inspection schemes. Biometrika 41, 100-115. https://doi.org/10.1093/biomet/41.1-2.100

[8] Roberts, S.W. (1959). Control chart tests based on geometric moving averages. Technometrics 1, 239-250. https://doi.org/10.1080/00401706.1959.10489860

[9] Shewhart, W.A. (1931). Economic control of the quality of the manufactured product.

[10] Wang, K., & Tsung, F. (2022). Bayesian cross-product quality control via transfer learning. International Journal of Production Research, 60(3), 847-865. https://doi.org/10.1080/00207543.2020.1845413

Materials Research Proceedings 66 (2026) 440-445 https://doi.org/10.21741/9781644904152-41

Optimization of WS_2 Synthesis Parameters for Enhanced Structural and Optical Properties Toward Photovoltaic Application

Basra SULTANA[1,a*], R. SAIDUR[1,2,b*], Norulsamani ABDULLAH[1,c], M.A. ZAED[1,d]

[1]Research Centre for Nanomaterials and Energy Technology (RCNMET), Faculty of Engineering and Technology, Sunway University, Bandar Sunway, Petaling Jaya, 47500, Selangor, Darul Ehsan, Malaysia

[2]School of Engineering, Lancaster University, Lancaster, LA1 4YW, UK

[a]24104499@imail.sunway.edu.my, [b]saidur@sunway.edu.my, [c]nabdullah@sunway.edu, [d]zaed.sunway@gmail.com

Keywords: Photovoltaic, Solar Energy, Synthesis, WS_2

Abstract. Tungsten disulfide (WS_2) is a two-dimensional material that offers significant potential for photovoltaic (PV) application. The improvement of structural and optical properties remains a challenge. The aim of this study is to enhance the ability of WS_2 to utilize the solar energy through controlled sulfur supply. This study used a hydrothermal method to synthesize the material and applies a single-factor precursor variation by systematically changing the WCl_6:TAA molar ratio while keeping the other synthesis parameters constant and it evaluates the impact on optical properties, which represents the novelty of this work. The finding indicates that enriching the sulfur ratio leads to clustered WS_2 and improves its ability to absorb light. This notable improvement highlights the suitability of WS_2 as an excellent candidate for PV technology.

Introduction

Transition metal dichalcogenides (TMD) have recently used for PV due to their attractive chemical and physical characteristics. This study synthesized tungsten disulfide (WS_2) using the hydrothermal method due to its simplicity and low cost. Researchers have utilized various precursors to synthesize WS_2. Sodium tungstate and thiourea are used as the precursors by [1]. This process produced a nano-tumbleweed WS_2 morphology. In contrast, 2H-phase WS_2 nanorods were synthesized using tungsten hexachloride (WCl_6) and thioacetamide (TAA) [2]. Likewise, using the ultrasonic spray pyrolysis method, [3] synthesized 2H–WS_2 nanosheets utilizing ammonium tetrathiotungstate. Furthermore, WCl_6 and TAA were used as precursors, where ethanol served as the solvent by [4], and post-synthesis drying optimization (60–80^0 C, 3–24 h) enabled the conversion of metallic 1T-WS_2 into semiconducting 2H-WS_2. Another study performed by [5], used identical precursors (WCl_6 and TAA) but varied the solvent, achieving metallic 1T-WS_2 with DMF and semiconducting 2H-WS_2 with water.

Although WS_2 synthesis uses a variety of precursors, the effects of precursor concentration on optical features have not been investigated. Previous studies concentrated on the solvent, synthesis temperature [5], post-synthesis temperature and time [4]. Therefore, the specific role of sulfur precursor availability (TAA) remains unclear because earlier strategies mainly altered solvent, temperature, or post-treatment conditions, which can affect phase and morphology at the same time. To address this gap, this work uses single-factor precursor variation and systematically changes the WCl_6:TAA molar ratio while keeping the hydrothermal protocol constant, which strengthens the novelty claim compared with earlier multi-parameter variation approaches. The chosen ratio range covers a practical shift from sulfur-lean to sulfur-rich conditions at fixed WCl_6, which should influence sulfidation completeness and growth kinetics and produce measurable changes in morphology and optical response. This study aims to (i) synthesize WS_2 nanostructures across the WCl_6:TAA ratio range, (ii) evaluate the resulting structural and optical changes using SEM, UV–Vis, and FTIR, and (iii) determine an optimal precursor ratio that improves light absorption behavior for PV integration.

Materials and Methods

Materials and Consumables. Tungsten chloride (WCl_6, ≥99.9 %, Sigma-Aldrich) and thioacetamide (TAA, 98 %, Sigma-Aldrich).

Synthesis Method. Fig. 1 illustrates the process where WCl_6 and TAA are added with 50 ml of deionized water and stirred at 450 rpm for 50 minutes. The resulting solution was hydrothermally treated at 180^0 C for 20 hours, then repeatedly washed through centrifugation, and finally dried overnight at 60^0 C in a vacuum oven. Four samples are prepared with molar ratios of WCl_6 to TAA as 1:8.5, 1:11, 1:13.5, and 1:16, labeled as s1, s2, s3, and s4, respectively.

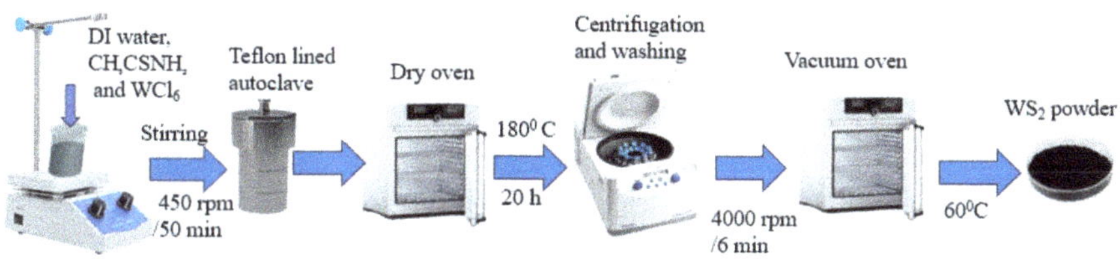

Fig. 1. Research methodology flow diagram

Characterizations. Surface morphology analysis was conducted using scanning electron microscopy (SEM) (Tescan Vega 3, Czech Republic). Light absorption and materials bonding were examined by UV-Visible (Perkin Elmer, USA) and FTIR (Perkin Elmer, USA), respectively.

Results and Discussion

The SEM image (Fig. 2a) shows aggregated WS_2 nanoparticles forming irregular nanoclusters, indicating incomplete sulfurization, which matches the nanoparticle WS_2 reported by [6]. Fig. 2 (b)-(d) shows that as the molar ratio of WCl_6:TAA changes the structure changes to cluster nanoparticles [7].

Fig. 2. SEM image of WS_2

The absorption spectrum (Fig. 3a) consists of three peaks, with interband absorption occuring at 200 nm as observed by [8] while band-to-band transitions occur around 290 nm [9] and 320 nm due to oxidized WS_2 forming WO_3 quantum dots [10]. In FTIR spectroscopy (Fig. 3b), the broad band at 636 cm^{-1} indicates the W-S bond [5] while the sharp band at 1045 cm^{-1}, 1086 cm^{-1} are ascribed to the S-S bond and at 879 cm^{-1} relates to the W-O-W bond [11].

This study strengthens the optical analysis by establishing a quantitative relationship between the WCl_6:TAA precursor ratio and UV-Vis-derived optical parameters. Tauc plots provide the optical bandgap energy (Eg) of samples S1-S4, with values of 3.91 eV for S1, 3.92 eV for S2, 4.05 eV for S3, and 3.99 eV for S4. These values show that the precursor ratio affects the optical transition behavior of WS_2 in a non-linear manner, with the highest bandgap in S3 (WCl_6:TAA = 1:13.5). The absorption spectra show the highest absorbance intensity for S3, followed by S4, while S1 and S2 show lower absorbance intensity. This also confirms that sulfur enrichment has a significant effect on the light absorption behavior of WS_2. Together with the Tauc-derived bandgap values, these results indicate a clear precursor-dependent structure-property relationship, where the TAA ratio affects both the optical transition edge and the absorption strength. Therefore, the optical results offer useful guidance for selecting the precursor ratio in WS_2-based materials for PV integration.

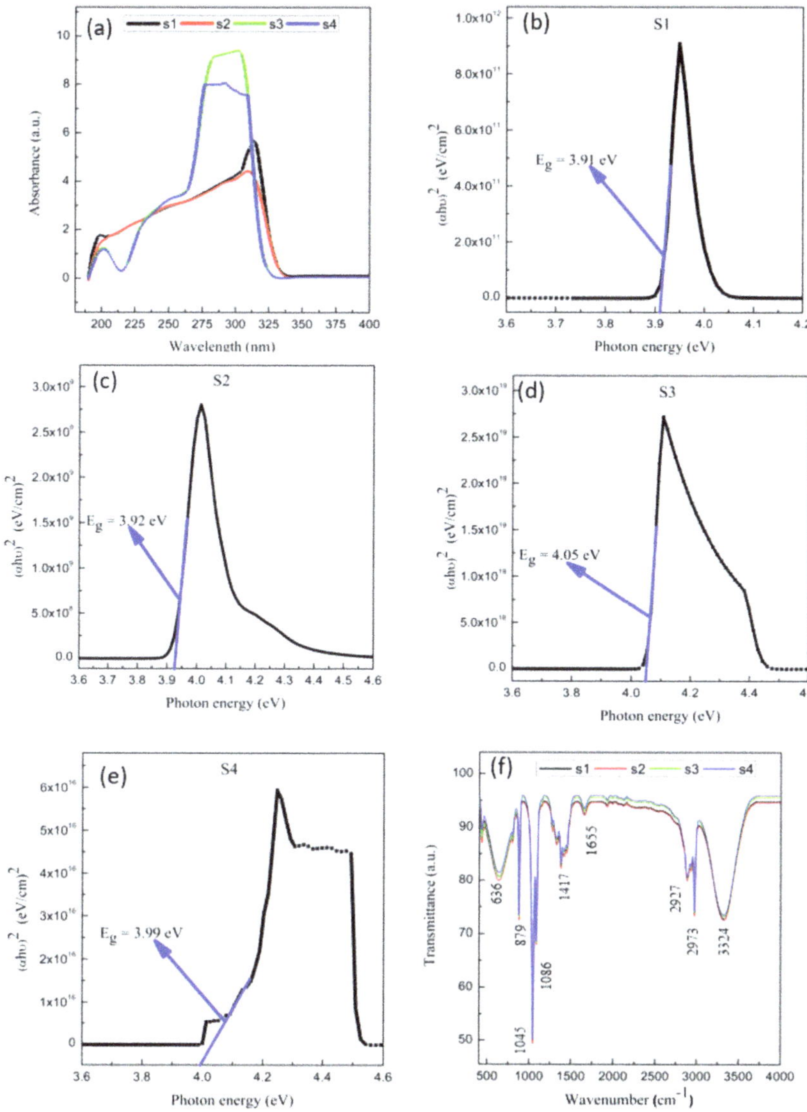

Fig. 3. (a) UV–visible absorption spectra, (b-e) corresponding Tauc-plot for determining band-gap and (f) FTIR spectra of WS_2.

Conclusion

This study successfully synthesized WS_2 using the hydrothermal technique with WCl_6 and TAA as precursors. A single-factor precursor variation is applied by systematically changing the WCl_6:TAA molar ratio (1:8.5, 1:11, 1:13.5, and 1:16) while keeping the synthesis conditions constant. The absorption spectra are strongly impacted when the molar ratio of WCl_6:TAA is changed. Quantitative optical analysis with Tauc plots shows bandgap energies of 3.91 eV, 3.92 eV, 4.05 eV, and 3.99 eV for S1-S4, respectively, which indicates a non-linear precursor-dependent trend with the highest bandgap in S3 (WCl_6:TAA = 1:13.5). The s3 sample displayed an almost perfect W:S ratio, significant vibrations in the W-S bond, and enhanced absorption of visible light with the highest absorption intensity among all samples. These improved properties demonstrate a clear precursor-dependent structure-property relationship and support the PV-relevant optical potential and integration feasibility of optimized WS_2 at the materials level, rather than indicating direct photovoltaic device performance. Device fabrication and photovoltaic performance evaluation are planned for future work.

Acknowledgment

The authors gratefully acknowledge the support of Sunway University's studentship scheme.

References

[1] M. Siddiq, A. Gunasekaran, A. Al Souwaileh, A. Sorrentino, A. Sambandam, Nano tumbleweed-like tungsten disulfide as a viable surrogate for platinum counter electrodes in dye-sensitized solar cells, Optical Materials 152 (2024) 115446. https://doi.org/10.1016/j.optmat.2024.115446

[2] D.S. Ahlawat, D. Singh, J. Gangwar, Enhanced structural, morphological and optical properties of WS2 nanorods synthesized via hydrothermal methodology, Materials Science and Engineering: B 313 (2025) 117935. https://doi.org/10.1016/j.mseb.2024.117935

[3] N.A.A. Malek, N. Alias, A.A. Umar, X. Zhang, X. Li, S.K.M. Saad, N.A. Abdullah, H. Zhang, Z. Weng, Z. Shi, Enhanced charge transfer in atom-thick 2H-WS2 nanosheets' electron transport layers of perovskite solar cells, Solar Rrl 4(10) (2020) 2000260. https://doi.org/10.1002/solr.202000260

[4] P. Bhatt, N. Halder, Systematic improvement of excitonic features in hydrothermally synthesized WS2 due to post-synthesis drying, Journal of Luminescence 266 (2024) 120293. https://doi.org/10.1016/j.jlumin.2023.120293

[5] D.S. Ahlawat, D. Singh, B. Singh, J. Gangwar, Insights into phase engineering and morphological tailoring of WS2 nanostructures via temperature controlled hydrothermal synthesis, Nano Express 6(2) (2025) 025010. https://doi.org/10.1088/2632-959X/addada

[6] G. Shen, Y. Yan, J. Tan, K. Hong, Oxygen mediated growth of flower-like tungsten disulfide nanosheets as highly active photocatalysts, Journal of Materials Science 57(40) (2022) 18819-18826. https://doi.org/10.1007/s10853-022-07831-3

[7] S.J. Hazarika, D. Mohanta, Inorganic fullerene-type WS2 nanoparticles: processing, characterization and its photocatalytic performance on malachite green, Applied Physics A 123(5) (2017) 381. https://doi.org/10.1007/s00339-017-0965-7

[8] L. Hristova, R. Martínez-Moro, E. Fernández-García, L. Vázquez, P. Atienzar, M.D. Petit-Domínguez, E. Casero, C. Quintana, M. Del Pozo, Optical Sensor Based on WS2 Quantum Dots for Lamotrigine Determination, ACS omega 10(17) (2025) 17257-17268. https://doi.org/10.1021/acsomega.4c09476

[9] D.-R. Hang, D.-Y. Sun, C.-H. Chen, H.-F. Wu, M.M. Chou, S.E. Islam, K.H. Sharma, Facile bottom-up preparation of WS2-based water-soluble quantum dots as luminescent probes for hydrogen peroxide and glucose, Nanoscale Research Letters 14(1) (2019) 271. https://doi.org/10.1186/s11671-019-3109-5

[10] K.H. Park, N.C. Kim, S.H. Song, Fabrication of WO3 Quantum Dots with Different Emitting Colors and Their Utilization in Luminescent Woods, Nanomaterials 14(11) (2024) 936. https://doi.org/10.3390/nano14110936

[11] P. Bhatt, A.K. Mondal, N. Halder, Observation of anomalous excitonic emission in V-doped WS2 QDs synthesized by hydrothermal method and subsequent mechanical exfoliation, Journal of Luminescence 277 (2025) 120904. https://doi.org/10.1016/j.jlumin.2024.120904

Keyword Index

About the Editors

About the Editors

Prof. Adarsh Kumar Pandey is a Professor at the United Arab Emirates University, Dubai, UAE. He is listed in the top 2 % of scientists in the world by Stanford University and Elsevier BV since 2021. He received his PhD in the field of Energy from Shri Mata Vaishno Devi University, India. His research involves, but is not limited to, Sustainable Energy Technologies, Advanced Energy Materials, Nano Enhanced Phase change materials (NePCMs), Solar PV and PVT, Dye sensitized solar cells. He is among the Top Researcher since 2020 with over 139 research and review articles under the keyword "Phase Change Materials" as per Researcher Discovery of Scopus database. He has also published over 360+ research/review article in reputed international journals with 10+ patents and few book chapters. He has also secured more than 3 million Malaysian Ringgits worth of research grants as principal/co-principal investigator.

Prof. Saidur Rahman is a Distinguished Research Professor and Head of the Research Centre for Nano-Materials and Energy Technology (RCNMET) at Sunway University, Malaysia. He obtained his PhD from the University of Malaya, Malaysia. Prof. Saidur Rahman is an internationally recognized researcher in the areas of renewable energy, nanofluids, heat transfer, energy efficiency, and energy policy. He has published extensively in high-impact journals and is a highly cited researcher, ranking among the leading scholars in his field. Prof. Saidur Rahman was recognized by Thomson Reuters as a Highly Cited Researcher for being among the top 1% of researchers with the most cited publications in his research area from 2014 to 2021. He is also listed among the top 2% of scientists worldwide by Stanford University and Elsevier BV. To date, Prof. Saidur Rahman has published more than 800 journal articles, the majority of which appear in top-ranking international journals. His work has received over 88,000 citations, with an h-index of 148 according to Google Scholar. In addition to his research contributions, he has supervised numerous postgraduate students, secured substantial research funding, and served actively in editorial and reviewer roles for reputable international journals.

Dr. Kalidasan Balasubramanian is a Lecturer at Sunway University in Malaysia. He holds a PhD in Sustainability in Science and Technology in specific to "thermal energy storage" from Sunway University, Malaysia. He is basically a Mechanical Engineer graduate, with Master of Technology in Energy Engineering from VIT University-Vellore, India. He holds expertise on thermal energy storage, nanomaterials, solar thermal system and more; as well he has published research/review papers in high impact journals. He has published over 100+ research/review article in reputed international journals with 5 patents and few book chapters. He has also secured more than 750K Malaysian Ringgits worth of research grants as principal investigator. Dr. Kalidasan actively contributes to international research collaborations and scholarly peer review and is committed to advancing high-quality, impact-driven engineering research aligned with global sustainability goals.

Dr. Norulsamani Abdullah is a Lecturer at the Research Centre for Nano-Materials and Energy Technology (RCNMET), Sunway University, Malaysia. She obtained her Bachelor with Honours (Chemical Engineering) and PhD (Fuel Cell Engineering) from Universiti Kebangsaan

Malaysia (UKM), Malaysia. Dr. Norulsamani is an experienced academic and researcher specializing in nanomaterials, catalysis, energy storage, fuel cells, and hydrogen-related technologies. She has published extensively in peer-reviewed, high-impact international journals, and her research work has received strong recognition within the scientific community. She has published more than 31 peer-reviewed journal articles, many of which appear in high-impact international journals and her research work has received strong recognition within the scientific community. She has successfully secured over RM 610,000 in competitive research grants as a principal and co-investigator. Dr. Norulsamani Abdullah is actively involved in research, manuscript evaluation, and academic publishing, and she contributes to maintaining high standards of scholarly quality and research integrity.

www.ingramcontent.com/pod-product-compliance
Ingram Content Group UK Ltd.
Pitfield, Milton Keynes, MK11 3LW, UK
UKHW020659041026
15590UKWH00001B/361

9 781644 904145